7/08

Extinction Is Forever

Threatened and Endangered Species of Plants in

the Americas and Their Significance in

Ecosystems Today and in the Future

Proceedings of a symposium held at the

New York Botanical Garden, May 11-13, 1976, in commemoration

of the Bicentennial of the United States of America

Edited by

Ghillean T. Prance

The New York Botanical Garden

Bronx, New York 10458, U.S.A.

and

Thomas S. Elias

The Cary Arboretum of

The New York Botanical Garden

Millbrook, New York 12545, U.S.A.

772348

Dedicated to the Memory of
Carlos Muñoz Pizarro 1913 - 1976

Published by

The New York Botanical Garden

Bronx, New York 10458, U.S.A.

Library of Congress Catalog Card Number: 77-302

International Standard Book Number: 0-89327-196-9

Printed in the United States of America

Design, Composition and Printing by
Maar Printing Service
Poughkeepsie, New York, U.S.A.

Publication of this book was made possible by the support
of the U. S. Department of Interior through the National
Parks Service and the U. S. Fish and Wildlife Service.

Printed on 100% recycled paper.

772348

TABLE OF CONTENTS

PREFACE

In the decade preceding this bicentennial year of the United States, much of the Western world has been confronted by a series of ominous, seemingly insoluble crises. The first of these was the threat to environmental survival, which began to reach public consciousness in the late 1960's. The second was the apparent shortage of energy resources, dramatized by the Mid-East oil embargo and greatly increased petroleum costs. The third was the economic downturn from which there still is incomplete and uneven recovery, with continuing inflation and persistent unemployment.

In the minds of most people, these problems are usually regarded as separate, unrelated disorders, each to be solved on its own terms: environmental degradation by imposing pollution controls; the energy shortage by finding new energy sources and new methods to conserve energy; the economic headache by manipulating national budgets, taxes, and interest rates. But each effort to solve one crisis soon comes into conflict with the solutions to others: for example, pollution controls frequently require additional energy; energy conservation often means fewer jobs. Inevitably, the advocates of one solution become the opponents of others. The results are that government policy stagnates and remedial action is either slow or, at best, timid. This tangled mass of problems perplexes both legislators and administrators, as well as academically-trained specialists like ourselves.

Pundits advance a variety of remedies for our national ills: Barry Commoner tells us we must lower our standard of living; the Ford Foundation believes that government may have to regulate the economy even more than it does at present. The Club of Rome at first clamored for a world economy with no growth, but now that august body opts for slow growth. Robert Heilbroner asserts that 20th-century Promethian man must in the future become Atlas, bearing on his shoulders the accumulated weight of all the deleterious impacts he has brought to this world, and learn from them the possibilities for survival.

This is the world setting in which we consider the subject before us — threatened and endangered species and ecosystems in the Americas. In the United States, the Smithsonian Report asserts that today 1/20th of our extant higher-plant diversity is variously endangered or threatened with imminent or ultimate extinction, unless steps are taken to arrest the causes leading to biological impoverishment. These causes — namely population growth, advancing industrial technology, and social affluence — are pressuring world botanical diversity in several significant ways: the world requirement for food production is steadily rising; forests are being felled at a rate far exceeding their regeneration; industrialization, mining, urbanization, and road building are reducing the areas formerly dominated by plants; human capacity for rapid and irrational destruction of plant communities continues unabated; chemical pollution of air, water, and land increases year by year; and monocultures promote the spread of pests and diseases.

As a result of these incessant and mounting pressures, a number of symptoms are apparent to the botanical observer: climax forest and grassland floras have been greatly reduced and locally extirpated throughout the Americas; island floras, with their high endemism and relatively low numbers of individuals per plant species, are severely threatened by soaring human impacts; weedy, ruderal, and marginal-land floras which have spawned so many of our food crops, are constantly pressured by land clearance, mechanized agriculture, and chemical herbicides; wetland floras, especially in temperate areas, are being decimated by drainage, landfill, and intensive land use; hydrophytic vegetation of fresh waters is increasingly subject to the ravages of eutrophication; coastal and estuarine floras are being degraded and reduced by amenity development and pollution; fragile desert floras are being severely damaged and diminished by land settlement, alteration in hydrology, and off-road vehicles; and everywhere rare and endangered species are subject to depredation by commercial collectors and individual devotees.

Well, "so what?" is the question that the lay public asks us. As professional botanists, it behooves us to be able to give a strong, clear answer in terms the public can easily understand and accept.

In this, as in all recent convocations focussing on the subject of the importance of biotic diversity, it is generally assumed by all of us as scientific professionals that preservation of such diversity matters. But the public wants to know if preserving botanical

1

diversity *really* matters, and if so, *how much* does it matter? Why is it important?

If we accept the basic tenet that botanical diversity is a legitimate concern for the future well-being of mankind, we must say so in terms that the public can comprehend and with facts upon which politicians can act. Mere lists of endangered species are not enough. Assertions of concern by ecologists and systematic botanists are not enough. Legalistic resolutions by disciplinary brethren are not enough. Dire forecasts by professional Cassandras are not enough. Without sacrificing scientific objectivity or demeaning the facts at hand, we urgently need comprehensible arguments linking mankind's present and future well-being to plants. We must find ways of interpreting economics and engineering in ecological terms. We must point out the largely unperceived importance of botanical diversity to the aesthetics of day-to-day living. We must define effective roles for institutions of botanical competence. We must be receptive to new ideas and new approaches to support our case. We must be up-to-date on how we are faring. And we must engage spokesmen from outside the mold of systematics to advance the cause, lest we be looked upon as so many nervous academics wringing our hands over the loss of our livelihood.

This symposium, as it relates to the present state and future of botanical diversity in the Americas, is an important step toward these ends. With the Americas as its geographic context, the conclusions and resolutions reached at this convocation should carry considerable weight in the chancels of legislation and decision-making in the individual countries represented at this meeting. But the process has only begun. Until Man adopts a partnership *with* Nature, rather than continuing as Nature's historical and traditional adversary, botanists must take the lead in continuing to press for the proper recognition and responsible management of our most precious natural resource — the earth's green mantle.

Howard S. Irwin

INTRODUCTION

This volume records the proceedings of an international symposium on threatened and endangered species of plants and ecosystems in the Americas which took place May 11-13, 1976. The symposium, in commemoration of the bicentennial of the United States of America, was sponsored by the New York Botanical Garden and was attended by 159 participants, speakers and observers, mainly from the Americas, and with a few Europeans present. Fourteen countries were represented at the three-day convocation.

The program placed special emphasis on the American tropics, and many of the speakers and delegates came from that region. This was particularly encouraging, because it prevented the symposium from becoming a forum for preaching temperate-region ideas and solutions to the tropical countries. Instead, the temperate-region delegates were able to learn about the problems of endangered species directly from the various Latin American experts who were present.

The subject of the symposium is a vast one and many different opinions and ideas are expressed in the papers which follow. However, a few themes kept recurring in many of the papers which were presented. The serious threat to vegetation in tropical America was soon apparent as each speaker showed his own slides and revealed the destruction of natural vegetation in the area of his concern.

Many speakers, from North America to Argentina, emphasized the need for ecosystem conservation rather than for the preservation of individual species. One of the most urgent problems facing biologists today is the selection of areas for perpetual protection which include as wide a species representation as possible. Several speakers discussed the need to conserve more widespread or disjunct species with a large amount of genetic variability in different parts of their range. Many speakers drew attention to lists of threatened and endangered species in their countries and appealed for assistance with the compilation of such lists. Others pointed out some of the inherent dangers of such lists because they draw attention to the species in question. At least one example was cited of an area which was destroyed because it was mentioned on an endangered species list. It was ravaged by the owners before the legislation was enacted to protect the area. This enabled the owners to develop the land rather than to leave it earmarked as a reserve for an endangered species. One of the most challenging lists in North America is that of extinct species. This has stimulated many botanists to look for "extinct species" and has resulted in the rediscovery of several.

The subject of human population control was also a frequent topic of discussion. Many delegates expressed the view that such control is the only realistic, though long-range, solution to save endangered plant species.

The theme of development versus conservation was fully aired in a number of presentations and discussions, notably that of Dr. Paulo de T. Alvim of Brazil and the closing address by Dr. Gerardo Budowski of Costa Rica. Botanists eager to save species in the tropics should study the uses of the forest as well as methods to conserve it, to achieve their goals of preservation of the natural gene pool of the Americas. The topic of the use of *Eucalyptus* in the tropics led to several spirited discussions. In defense of the practice, it was pointed out that the large-scale planting of Pines and *Eucalyptus* in already impoverished secondary areas can be useful in preserving other undisturbed areas, because the production from the plantations reduces the necessity to cut down more virgin forest.

The symposium was made possible by financial support from The National Science Foundation, The Organization of American States, The World Wildlife Fund, and The Achelis Foundation, to whom our thanks are due.

Thanks are also due the many people who contributed to the organization and smooth running of the symposium. I would like to express especial gratitude to the following individuals: Mr. Jim Gugluzza (my symposium assistant), Mr. Herbert Bijur (business manager), Mrs. Frances Maroncelli (secretary), Dr. Kent P. Dumont, Miss Dawn Beazer, and the marvellous group of volunteer assistants: Miss Majorie Brenner, Miss Julia Slocum, Mrs. Pearl Willis, Mrs. Edith Alexander, Mr. Sal Cimilluca, Mr. Lee Erde, Mrs. Marie Fluri, Mrs. Karla Freeman, Mrs. Betty Sokol, Miss Marierd Ward; also to my coeditor and organizer of the North American part of the symposium, Dr. Thomas Elias, and to Dr. Robert A. Goodland for conducting the panel discussion and for many helpful suggestions.

We would like to express our special appreciation to Mrs. Janet Bookman for copy-editing and proofreading the entire book, and for much help and encouragement with the production of this volume.

I am most grateful to Mrs. Anne E. Prance for much editorial assistance and to Mrs. Frances Maroncelli, Mrs. Joan Biringer, Miss Maria Galli, and Mr. Edward Dooner for typing the manuscripts.

Also included in this volume is a Bibliography on Endangered Plant Species and Habitats which was compiled by the library staff of the New York Botanical Garden, under the supervision of Charles R. Long and M. Miasek. An earlier version was distributed at the symposium. I would like to thank Dr. Ronald R. Stuckey, of Ohio State University, for making available his bibliography on endangered plant species, which provided some of the references included in this book.

The otherwise successful symposium was overshadowed by the death of the well-known Chilean botanist Dr. Carlos Muñoz Pizarro, who collapsed suddenly while presenting his paper on endangered species of Chilean plants. The symposium carried on after his death, and his contribution to Chilean conservation was mentioned by many delegates. It was resolved to dedicate the present volume to the memory of this great botanist who died in action doing what he most liked to do, lecturing on Chilean plants.

Ghillean T. Prance

THE CHALLENGE OF ENDANGERED SPECIES

George M. Woodwell

Marine Biological Laboratory, Woods Hole, Massachusetts 02543, U.S.A.

The earth is in the throes of a series of biotic changes that are unprecedented in human history. In magnitude and potential effects on the biosphere, these changes are equivalent to the other great biotic revolutions of geological time — the evolution of aerobic respiration, the appearance of the Angiosperms, and the splitting and fusion of continents — except that today's dramatic changes are occurring much more rapidly and are caused by a single species, Man. The process that leads to the endangerment and loss of species is part of a general pattern of change that can best be described as "biotic impoverishment," which is the basic issue of this symposium.

One might dream that on the only green planet we know, life would have a special value of its own, just as books and works of art do in our culture. And if the interest in life *per se* were not sufficient to protect it, one might hope that simple, selfish interest in human comfort and sustenance might confer a special status on living systems and force their conservation. Unfortunately, neither occurs. The stacks are open in the world's great library of life and we advertise to the vandals. The task of conservation is hopeless, of course, unless we can recognize that the function of government is to provide protection for precisely this type of essential resource in the public interest. The issue is political, and to carry it through successfully we must accept the challenge of defining the resources in terms that are understandable and usable in politics. Failure to do so will allow short-term commercial interests of diverse types to destroy or diminish resources irreversibly in violation of both basic principles of science and common sense.

The crucial biotic changes attributable to human activities are more pervasive and drastic than ever before, and they are continually accelerating. The causes are various. In our segment of the industrialized world the cause is overindustrialization with a leaky, toxic and still-growing industry. Elsewhere, it is more clearly assignable as overpopulation and overuse born of intractable greed and ineffectual government. The results are similar: a loss in renewable resources in support of the very people and industries that are despoiling them.

The loss can be measured in several ways that have been developed recently. The studies help to show how the diversity of species and human welfare are inextricably related.

Over the past decade or so, ecologists have made substantial progress in measuring the function of entire units of landscape as diverse as the wet forests of the Amazon Basin, the high Andean tundra or Puna that has been used by man for several thousand years, and the plant and animal communities of the open sea. Long experience has shown that disturbance of such landscapes usually shifts the arrays of species according to predictable patterns. The shift has two phases. First, the disturbance-sensitive species, those that are highly-specialized as members of an earlier "undisturbed" community, are either totally eliminated or substantially reduced in abundance. Second, hardy invaders, ubiquitous weedy species, replace them. Diversity, or the number of species per unit area, may drop at first but subsequently rise as exotics accumulate. This pattern has been described previously along the gradient of disturbance caused by ionizing radiation, and has been discussed as a general series of changes in the structure of ecosystems (Woodwell, 1962, 1967, 1970). In general, the species that are lost in this exchange are large-bodied, long-lived species with low reproductive rates. Those that are conserved are small-bodied, rapid reproducers that are hardy under a wide range of conditions, require low nutrient availability and do not support complex food webs.

Another equally important aspect is the shift in organic production brought about by these disturbances. I dwell on this latter aspect in the following discussion because it is intrinsically important and shows one further consequence of the loss of species. The emphasis has one great strength: it allows us to talk to the technologists in terms that are familiar to them, those treating energy fluxes.

In the past, I have emphasized that the major function of technology is to enable the rapid transformation and transport of basic resources around the world (Woodwell, 1974). The most important basic resource transformed is solar energy fixed in the biota and available as what ecologists call "net primary production," the energy or dry organic matter remaining at the end of one annual cycle of growth after the needs of the plants for metabolism have been met. This is the energy that is available to support animal populations and includes harvest by man. In agriculture we harvest only a small fraction of the net primary production, commonly less than 30-40 percent of the total. The remainder appears in roots, stems and leaves of plants that are not a part of the harvestable crop but are a part of the net primary production.

Table I. Net primary productivity of the world's major ecosystems (Adapted from R. H. Whittaker and G. E. Likens, 1975).

TYPE OF VEGETATION	NET PRIMARY PRODUCTION (Dry organic matter)			
	AREA (10^6km^2)	RANGE (g/m^2/yr)	MEAN (g/m^2/yr)	TOTAL (10^9 T/yr)
CONTINENTAL				
Tropical Forest — Wet	17.0	1000-3500	2200	37.4
— Seasonal	7.5	1000-2500	1600	12.0
Temperate Forest — Evergreen	5.0	600-2500	1300	5.5
— Deciduous	7.0	600-2500	1200	8.4
Boreal Forest	12.0	400-2000	800	9.6
Woodland	8.5	250-1200	700	6.0
Savanna	15.0	200-2000	900	13.5
Grassland	9.0	200-1500	600	5.4
Tundra	8.0	10-400	140	1.1
Cultivated Land	14.0	100-4000	650	9.1
Swamp and Marsh	2.0	800-6000	3000	6.0
Lake and Stream	2.0	100-1500	400	0.8
Other: Desert	18.0	10-250	90	1.6
: Extreme Desert	24.0	0-10	3	0.07
Total:	149.0	—	782	116.5
MARINE				
Open Ocean	332.0	2-400	125	41.5
Upwelling Zones	0.4	400-1000	500	0.2
Continental Shelf	26.6	200-600	360	9.6
Algal Beds	0.6	500-4000	2500	1.6
Estuaries	1.4	200-4000	1500	2.1
Total:	361.0	—	155	55.0
WORLD TOTAL:	510.0	—	336	171.5

The net productivity of major terrestrial ecosystems is indicated in Table I. These data are from a summary of recent detailed studies of terrestrial ecosystems carried on by scientists around the world under the auspices of various institutions. The table shows the ranges and averages for these ecosystems. The most important conclusion to be drawn from the figures is that net primary productivity is generally higher in forests than in grasslands and higher in natural vegetations than in agriculture. The replacement of forest by shrubland or grassland may improve the utility of the vegetation for man but does not increase the net primary productivity in general. The natural communities maintain themselves without an expenditure of energy by man; the man-modified communities, while their yield to man may be high, are maintained only by an appreciable expenditure of energy from outside the system. It seems clear that one should avoid a casual or incidental transformation of forest to shrubland or savanna. One possibility is that the best interests of man might be served, not by transformation of forest to vegetations of lesser stature and lesser production, but by managing the forests themselves for the diversity of products they can yield.

Marine ecosystems also can be considered in the same context. Table I summarizes the net primary productivity of marine ecosystems and shows that coastal ecosystems have the highest production. The coastal ecosystems are, of course, the ones that are at present under greatest pressure from direct exploitation, mechanical disturbance and pollution.

The net primary production of the world is summarized in the last column of Table I. This shows the sums obtained when the rates of productivity are applied to the areas of the earth represented by the various plant communities. More than two-thirds of the total is terrestrial; most of the energy through marine systems is concentrated in the coastal zones. The world total is equivalent to about 900×10^{12} kwh/year. This sum is about fifteen times greater than the 60×10^{12} kwh currently used by industry worldwide.

These issues may seem remote from the problems of endangerment and loss of species until we recognize that on a worldwide basis most of the energy used by man flows to him through natural communities, not through either technology or through agriculture. This statement is true if we consider primary energy sources such as oil or coal and solar energy trapped in photosynthesis and made available as net primary production. The energy from primary sources is transformed in various ways before being used by man. This transformation may be either technological and applied through agricultural machinery to the growing of food, or it may be biotic, effected through various other plants or animals that are themselves used by man. The biotic transformations obviously dominate on a worldwide basis and we recognize immediately that the availability of the energy to man depends on the diversity of species that perform the transformations.

The importance of individual species becomes most clear where we rely on animals to transform otherwise inedible organic matter into food for man. Virtually the only harvest we make of the oceans is through the fisheries. The oceans cover approximately two-thirds of the earth's surface and fix about one-third of the total carbon fixed on earth annually. Reduction or elimination of fish populations reduces the potential for turning this net production to human use. A similar circumstance, still more vivid as an illustration, exists in the high Arctic tundra, where large herds of caribou provide the major mechanism for transforming tundra plants into food for man. Elimination of the caribou would eliminate this flux of energy; there is no other organism available to replace the caribou in the food chain. The cameloids of the Andean Puna have provided a similar service for man for several thousand years, enabling human survival in an otherwise virtually uninhabitable area. Their loss through extinction or reduction in numbers reduces the potential of the area for support of man and there is at present no innovation to restore it. A parallel harvest occurs in the savannas and grasslands of the world through terrestrial food chains, also tapped by man for protein. We recognize now that the flux of energy through net primary production to man as food is much greater than that shown by consideration of agriculture alone; it probably includes 30-50 percent of the total net primary production of the earth when the use of protein is included. While there are several possibilities for increasing the supply of food — by improving management of fisheries, by intensification of agriculture, and by consuming protein from lower levels in the trophic structure and thereby avoiding the losses to respiration that occur with each change of trophic level — the fact remains that at present a major segment of the net primary production of the earth is used directly or indirectly as food for man and that this flux is strongly dependent on natural communities including animal populations, whose diversity determines both the extent of utilization and the number of opportunities for harvests.

Human uses of the earth's net primary production are not limited to food, of course. Fiber and fuel constitute the second major category and we recognize that most of the earth's forests are being harvested for one or the other or both. The third major use that man makes of the biota is in a vague set of service functions whose measurement is a contemporary challenge. These are the functions that the biota performs in stabilizing water flows and maintaining the quality of the environment. We might call them "public service functions". We scarcely recognize that they are real until they are irrevocably lost. Nonetheless, we live in an environment that is ameliorated importantly by plants and animals and, while we tend to think of the only essential biotic resource as food, this conclusion is scarcely sustainable in the light of our needs for fiber, fuel, and for a stable and congenial environment.

What are the contemporary trends in the amounts and distribution of net primary production? If the foregoing discussion of the uses of net primary production around the world seems subjective and difficult to analyze with precision, then analyses of the trends in patterns of use of net primary production are even more difficult. Two trends are apparent: first, a shift from forest to agriculture, and second, an increasing degree of toxification.

The shift from forest to agriculture has followed the growth of civilization around the world. The trend has commonly been one of progressive impoverishment, conspicuous over centuries but difficult to appraise on a year-by-year basis. The entire Mediterranean Basin was at one time forested, as was much of Asia Minor. Virtually all of the European continent also was forested during the first millenium after the birth of Christ, and it has become deforested only during the most recent millenium (Darby, 1956).

The current tendencies are toward the harvest of tropical rain forests around the world and its replacement by agriculture. A study by Dr. Laurence Hamilton (1976) of Cornell University has shown a 0.6-1.5 percent per year reduction in the areas of forests in various sections of South America. With these transitions comes a concomitant loss in net primary production. In certain circumstances the vegetation that replaces the forest may have a net production of a hundredth or less of the original vegetation.

The second trend in contemporary times is a trend toward changes in the chemistry of environment over large areas. This trend is the product of the basic approach used in all societies toward the regulation of toxins. The assumption is made that the environment has an "assimilative capacity" for all wastes. Though the assumption is convenient because it allows the release of small quantities of toxins into the general environment, it is also troublesome, because the small quantities of toxins may accumulate and affect the biota. The assumption is appropriate only when two criteria are met: the world must be large in proportion to the toxins released, and secondly, the toxins released must not be persistent. As both the demands on environment and the intensity of industrialization increase, one would expect that the concept of assimilative capacity would come under increasing criticism. The number of toxins, the quantities released and their persistence are simply too great. As a result, extensive areas of the earth are being subjected to basic changes in chemistry such as those described as "acid rain". The acidity is due to the presence of oxides of sulfur and nitrogen in rain. The concentrations are great enough to lower the pH to 3-4 over extensive areas of northeastern North America and Scandinavia. The effects are difficult to predict but there is little question that the effects are deleterious. A few years ago I suggested that a reduction in the net primary productivity of the forests of New England would constitute a loss of energy equal to the annual output of fifteen 1,000-megawatt reactors. This loss is taken from a most fundamental resource that includes agricultural production, forest production and fisheries. The burden falls on all, not simply on the industrial segment. It falls most heavily on those who are directly tied to basic resources and who are least capable of avoiding it by moving to another section of the country or the world (Woodwell, 1974).

These trends suggest that both the standing crop of organic matter and the net production of the world's plant communities must be declining as harvests progress and toxic substances spread more widely around the globe. The challenge is to measure the change. Despite the importance of the question there has been no adequate appraisal of the extent of it. The amount of CO_2 exchanged annually between the atmosphere and the biota is great enough to cause an annual oscillation in the CO_2 content of air of 5-15 ppm in the middle latitudes of the northern hemisphere, but there are simply too many factors that affect the CO_2 content of air to ascribe small changes year-by-year to any single cause. If

8

there is an effect of changes in the biota, the effect cannot yet be isolated (Hall *et al.,* 1974; Woodwell, unpublished).

Solutions to the problems of management of essential resources are not yet a part of our culture. We have made substantial progress in improving our sensitivity to these questions in the last decade through the development of public interest law groups such as the Natural Resources Defense Council and the Environmental Defense Fund. These groups enhance the ability of government to obey its own laws but they do not generate new laws, nor can they be strong influences internationally. Increasingly, the problems of the destruction of biotic resources are international problems that must be addressed by new and effective international protocols.

The endangerment and destruction of species is clearly an integral part of this problem. We see how important it is to human welfare when we consider that the loss of a species of fish or the reduction of a fishery to a point where it is not exploitable means that the net primary production potentially available to man through the harvest of that fish species must either flow through another fishery to man or be lost to decay. The possibility that it may flow through another fishery is real but we do not always see a surge in another fishery as one fishery is depleted. If it is lost because of the extinction of a species, it may be lost irreversibly.

The problems I have enunciated are large and have proven intractable so far. Virtually every country in the western world, including towns and cities in the United States from colonial times, all had strict conservation laws designed to preserve their forests, streams and other essential resources. But despite these laws, some of which extend back into the Middle Ages in Europe, the forests were destroyed and the lands were impoverished. The existence of an economic gradient over time seems enough to assure the destruction of resources regardless of the laws or the common sense. Are there other solutions?

Several years ago, the well-known nuclear physicist and Director of Oak Ridge National Laboratory, Alvin Weinberg, suggested that the best approach to many public issues is to sort out the technological segments of solutions and attempt to resolve those segments first, thereby reducing the problem to one of more manageable size. I would suggest that we have now reached the point where the problems of ecology can be reduced through science and technology. If we can agree on an objective such as preservation of essential resources over centuries or decades, then the steps necessary to achieve that preservation become clear. We recognize our dependence on energy flows through both the technological segment of society and through the biota. We also recognize a basic requirement for air that contains oxygen and nitrogen and carbon dioxide in constant proportions, for nutrient elements, and for water. The requirements of structure and function of man-dominated systems such as cities, urban landscapes, and farms to produce these stable relationships with essential resources, including the biota, can be defined with considerable precision. The patterns in which the resources can be preserved are known and are predictable over time. It is possible now to design cities that reuse their water supplies and that do not contaminate the air and that maintain a stable relationship with their sources of net primary production, no matter whether the use be for food or fiber or public services. We are rapidly developing ways of interpreting the exchanges of resources, one for another, and recognizing the dependence of our technology on energy, minerals, and on the biota.

With this recognition it is now appropriate to move toward stabilizing the relationship between man and resources. The first challenge for the new applied science of the design of ecosystems has precisely this objective. When this objective also becomes a major objective of governments, we will have passed that time when a symposium on the subject of endangered species has any practical meaning in our world. We hope that this symposium will move us rapidly toward that day.

LITERATURE CITED

Darby, H. C., 1956. The clearing of the woodland in Europe. *In:* Man's Role in Changing the Face of the Earth. Vol. 1. William L. Thomas, Ed., p. 183.

Hall, C. A. S., C. A. Ekdahl and D. E. Wartenberg, 1975. A fifteen-year record of biotic metabolism in the Northern Hemisphere. Nature **255:** 136-138.

Hamilton, L. S., 1976. Tropical rain forest and preservation: a study of problems and practices in Venezuela. Office of International Affairs, Sierra Club. International Service No. 4. 72 pp.

Whittaker, R. H. and G. E. Likens, 1975. The Biosphere and Man. *In:* H. Lieth and R. H. Whittaker, Primary Productivity of the Biosphere, Springer-Verlag, New York, pp. 305-328.

Woodwell, G. M., 1962. Effects of ionizing radiation on terrestrial ecosystems. Science **138:** 572-577.

———. 1967. Radiation and the patterns of nature. Science **156:** 461-470.

———. 1970. Effects of pollution on the structure and physiology of ecosystems. Science **168:** 429-433.

———. 1974. Success, Succession and Adam Smith. BioScience **24:** 81-87.

Section 1

Threatened & Endangered Species
Problems in North America

AN OVERVIEW

Thomas S. Elias

Cary Arboretum, Millbrook, New York 12545, U.S.A.

The continental United States is endowed with a rich and diverse flora of approximately 20,000 species. Unfortunately, at the present time 1,200 of these are threatened, approximately 750 species are in danger of becoming extinct and, according to the Smithsonian Institution, perhaps as many as 100 species have already become extinct. The threatened and endangered species are not uniformly distributed throughout the country, the greatest numbers occurring in Florida, Texas, Arizona, California, Oregon and Utah.

The rapid westward movement of people during the first hundred years of this nation's history was the driving force responsible for the modification of habitats and the destruction of plant species which continues virtually unabated in spite of recent legislation designed to halt this trend. Until recently, few people were aware of the altered or destroyed habitats and the plant species that perished with them. If we study the last recorded dates that our country's extinct plants were known or collected, we find that during the half-century between 1800 and 1850 only four species of plants were known to have become extinct. However, between 1851 and 1900, 41 species of plants became extinct. Between 1901 and 1950, 45 species succumbed. Assuming that these figures provide an accurate gauge, this accelerating rate of extinction indicates that a proportionately greater number of species will become extinct between 1951 and 2000. But it has only been within the last five years that a nationwide effort has been mounted in the United States to deal with the problem of endangered species.

On October 12, 1940, the Convention on Nature Protection and Wildlife Preservation in the Western Hemisphere was organized at the Pan American Union, in Washington, D.C. This was a pioneering effort on the part of the American Republics to protect our hemisphere's habitats and their floral and faunal components. Early in 1973, another major conference was held in our nation's capital, in an effort to preserve species of plants and animals which were endangered due to international trade. The convention of March 3, 1973 on International Trade in Endangered Species of Wild Fauna and Flora was the result. This international agreement marked the first effort to regulate trade in endangered species through a system of import and export permits. Under terms of this international pact, trade in endangered species is almost prohibited, while trade in threatened species is subject to careful regulation. By July 1, 1973, ten member nations of the Pan American Union had ratified this treaty. During that same year, the Endangered Species Act was passed by the United States Congress. This legislative milestone provides for the conservation of all plant and animal species that are determined to be endangered or threatened. In accordance with the Act, the Smithsonian Institution was directed to prepare a list of the threatened and endangered species of plants of the United States. The initial report was transmitted to the House of Representatives on January 9, 1975. The report is undergoing revision at the present time. It is expected to be considered by Congress shortly, and upon ratification, the list will be incorporated into the Endangered Species Act, at which point, all species so listed will be protected by federal law. Unfortunately, during the three years since the Endangered Species Act of 1973 was passed by Congress, no plant species have been officially protected.

There are many reasons for the decline of a plant species in the continental United States. The most obvious reason, and the cause of perhaps 90 percent of all endangered species, is habitat destruction or modification. Extensive road-building, shopping centers, and the ubiquitous urban sprawl, all serve as prime examples of the alteration.

When a marsh or swamp is drained, the habitat is modified in such a way that the plants can no longer survive there. In other areas, the commercial exploitation of plants is the principal cause for their decline. For example, the cacti in the southwestern states are being gathered by collectors and by wholesalers, who then sell them to dealers and collectors in other parts of the country. Commercial trade in native wild flowers is a big business. The excitement of buying an insectivorous plant at the local supermarket promotes wholesale gathering of sundews and pitcher plants, for example. In the northwestern United States, the very attractive *Lewisia tweedyi,* native to rock-slide areas of the Wenatchee Mountains of central Washington, is also endangered from overcollecting.

Fig. 1. The bright yellow Trollius laxus *grows in bogs or swamps throughout the northern United States, a habitat that is fast disappearing. An endangered species.*

We all agree that wholesale and retail trade and interstate commerce involving rare and endangered species of plants in the United States must be controlled, and the existing legislation is expected to help greatly to bring this about. But there are other reasons for a decline in species which are neither obvious nor well-documented. Approximately 22 percent of the flora of the northeastern United States has been introduced and naturalized, and we know little about the competitiveness of these species with our native species. One can see throughout that area, for example, the spread of purple loosestrife *(Lythrum salicaria),* which was introduced from Europe. It is rapidly taking over our northeastern wetlands.

Diseases and insect pests also account for the decline of some of our plant species. We are all familiar with the effect of the Dutch elm disease and the American chestnut blight on the American elm and chestnut trees. These are the very obvious examples. Many other less familar species are also being affected by pathogens and threatened with possible extinction. Many of the most harmful insects and diseases have been introduced into this part of the world through human actions.

Another cause of species decline is the loss or displacement of pollinators. If an introduced species of bee displaces a native species and is less effective as a pollinator, the reproductive success of the dependent plant species will be lessened. As a result, the populations will slowly decline. If species are unable to stabilize at lower population levels, the affected populations may face extinction.

Other factors may also be contributing to the decline of species and these warrant further concerted study. For example, in the northeastern United States we know that the acidity of the rainfall has been gradually increasing during the past twenty or thirty years. What effect is this having on the pH of the soil? Is the soil slowly being modified as a result? If so, we know this would have a dramatic effect on some species of plants with limited pH tolerance.

The threatened and endangered species are one of the poorest-known assemblages of plants in the United States. Little is known about their natural history, their reproductive mechanisms or their life cycles. But as we learn more about these plants, we may be able to identify other causes for their decline.

Admittedly, extinction is a natural phenomenon. One has only to compare the fossil forms from past geological ages with modern plants and animals. By doing so, we can see that large numbers of plants and animals became extinct long before man was a factor. Many species of dinosaurs ruled the world during the Jurassic and Cretaceous periods, when relatives of our herbaceous club mosses and ground pines were the size of trees. But for some reason, they completely died out, and surely urbanization was not responsible for their demise. Perhaps it was due to the changing climate, the changing topography, competition from newly-evolved groups, or some combination of new environmental conditions. The ability of populations of organisms to respond to environmental modification is dependent upon the extent of their gene pool, the genetic variability that is inherent in all populations. Since these gene pools are finite and response to change often evolves over generations, populations cannot adapt to all changes in their environments, especially sudden changes.

When faced with alterations in its environment, a population can respond either by evolving in a new direction, by adapting to the alterations, or by dying out. But there is a significant difference between the natural extinction of an individual species and the extinction caused by man. As already pointed out, the rate of extinction has accelerated to such alarming levels in recent years that unless strong positive actions are taken to preserve the great diversity of species, we increase the chance of upsetting the biological stability it provides.

How can we take positive action to prevent the further decline of plant species? The only reliable method is to preserve habitats containing those that are rare or endangered. Efforts have been made by many groups to purchase and set aside lands. The most active and most effective of these groups is the Nature Conservancy, which has played a major role in private land acquisition in this country during the past twenty years. Since the Endangered Species Act was passed in 1973, that organization has acquired several hundred parcels of critical habitats which will now be preserved in perpetuity. Furthermore, some states are taking decisive action to set aside critical habitats within their borders. For example, Oregon has established a preserve solely for the protection of a single species, *Darlingtonia californica.* In the Southwest, the Saguaro National Monument is a preserve for the giant cacti. Although the Endangered Species Act of 1973 specifies that federal assistance would be available to the individual states for acquisition and preservation of lands, funding for this purpose has not yet been allocated. Since no lands have yet been set aside as a result of the Act, we obviously cannot rely upon the federal government to assume the primary responsibility for preserving critical habitats. It is up to us to continue to promote, encourage and support private conservation agencies in their efforts to conserve these habitats and species.

Botanical gardens and arboreta also can play a major role in the preservation of endangered species. These organizations should assume a leadership role in the study of the natural history and life cycles of such endangered plant groups as the cacti, orchids, cycads, some of the attractive wild flowers, and others which serve as rock garden plants. Plants in these groups are being exploited by collectors and commercial dealers and as a result, such plants are more vulnerable even if their habitats are preserved.

Botanists and horticulturists associated with botanical gardens and arboreta should be studying the life cycles of these plants, concentrating on increasing their seed production in the field and learning to vegetatively propagate them. If the species has to be grown under greenhouse or nursery conditions, a program of reintroduction should be instituted to place the propagules into habitats that once contained the species. It is important to note here that original habitats must be preserved or attempts at reintroduction will probably fail. An example of habitat modification is the introduction of goats on the Channel Islands off the coast of California, an action which made the islands unsuitable for the reintroduction of plants. Newly-transplanted herbaceous species cannot hold their own against goats!

In the Waimea Arboretum in Haleiwa, Oahu, Hawaii, efforts are under way to determine the best methods of propagating a number of threatened species of flowering plants. The nearly-extinct *Sesbania tomentosa,* a coastal species whose ranks have been decimated by introduced grazing animals, perhaps has been saved. Seeds were collected from one of the remaining trees and were successfully germinated. These plants now have matured, flowered and produced fruits with viable seed. Establishment in gardens and reintroduction into suitable native habitats may insure the continued survival of this species. At present, the horticulturists at the Waimea Arboretum are optimistically focusing their attention on other endangered Hawaiian species.

In summary, we must recognize that extinction will continue not only through the measured ways of nature but on a much greater scale through the reckless ways of man. It will be only by concerted and constant vigilance that we can maintain most of the great diversity of plants in natural systems.

CANADA

George W. Argus

National Museum of Natural Sciences, Ottawa, Canada K1A OM8

The present knowledge of rare and endangered plants in Canada is inadequate for the task of preserving this important facet of our natural heritage. Much of what is known is anecdotal, hearsay, or poorly documented. In cases where there are valid reasons for believing that a taxon is rare and potentially threatened we usually do not have sufficient information about the species' ecological requirements and biology to provide for its protection.

There are indications that about 440 species (10 percent of the flora) in Canada are rare (Kershaw and Morton, 1976) and that there are centers of endemism characterized by taxa of limited geographical range (Argus and McNeill, 1974). Detailed information on these taxa, however, either does not exist or has not been assembled; therefore we are unable to do more than speculate about the endangered status of these rare Canadian plants.

The purposes of this paper are: (a) to present a review of the measures taken to protect rare and endangered plants in Canada; (b) to describe some of the successes that have already been achieved; and (c) to outline some of the problems with which we are confronted. Before commencing, though, I would like to consider briefly what is being done to improve the state of our knowledge about rare and endangered plants.

In 1974, the Systematics and Phytogeography Section of the Canadian Botanical Association formed a Rare and Endangered Species Subcommittee in response to an increasing general interest in the subject. Two projects developed out of this subcommittee: first, a list of the rare plants of Canada; and second, lists of the rare plants of the provinces and territories. The Canadian list, initiated by Dr. J.K. Morton of the University of Waterloo, Waterloo, Ontario, was completed by an M.Sc. student, Mrs. Linda Kershaw, who has produced a computerized list, including 1,364 taxa with notes on their status and distribution (Kershaw, 1976).

The provincial lists (Argus, et al. 1975-1976, unpublished) are being compiled as a project of the National Museum of Natural Sciences, Ottawa. Preliminary lists have been completed for Alberta, British Columbia, Manitoba, Newfoundland (including Labrador), Nova Scotia, Ontario, Quebec, and Saskatchewan, and lists for the other provinces and territories are planned. These preliminary lists, input on Alphatext for ease of revision, have been circulated among Canadian botanists for comments, and six of the lists already have undergone one or more revisions. Copies of these unpublished lists are available from the author.

For the purposes of the preliminary lists, a rare taxon was defined as: (1) one known from only a few localities within the province; or as (2) one occurring in a small geographical area within the province. These criteria, as imprecise as they are, were not always applicable to the taxa under consideration and the lists were based largely on published information or on communications from individuals. Such information is difficult to standardize because of differing concepts of rare, the relating of rare to different geographical areas, the misdeterminations of specimens, or statements based on inadequate data. In addition, there are the taxonomic problems associated with synonymy and the recognition of minor variants and hybrids as species, each of which poses its own unique difficulties. The long-range plan of the National Museum's program is to map and document those taxa which are included on the preliminary lists, to gather and verify unpublished data, and to recognize centres of rare species in Canada. Although this work may be coordinated and encouraged at the federal level, much of the input must of necessity emanate from local field botanists and naturalists.

LEGISLATIVE ACTION

In order to understand Canadian legislation designed to protect rare and endangered plants, three factors must be emphasized: (1) species of plants are a provincial responsibility except in the Northwest and the Yukon Territories, where they are under federal control; (2) the undeveloped Crown Lands within the provinces are under provincial control and the only major holdings of federally-controlled Crown Lands are in the territories; and (3) the concept of a governmental body holding land in trust is virtually unknown in Canada (Franson, 1972).

Legislation can deal with the problem of rare and endangered plants by declaring certain species endangered and protecting them as individuals. As pointed out throughout this Symposium, however, the most effective way to protect rare and endangered plants is through the protection of their habitats or natural ecosystems. In an excellent paper describing the background of the British Columbia Ecological Reserves Act, R.T. Franson (1972) discussed model ecological reserves legislation and outlined its desirable characteristics. See also Franson (1975).

Such legislation should:
1. provide for the acquisition and designation of Crown Land for the program;
2. give the power to expropriate or to accept gifts of private land conditioned on use as an ecological reserve;
3. provide protection against arbitrary removal of reserved lands from the program;
4. exclude the operation of other statutory powers that would be inimical to the program;
5. create opportunities for coordination of the provincial program with programs run by other provinces, the federal government, and private organizations;
6. provide the means for obtaining continued scientific input for the resolution of questions concerning the management of the reserves;
7. provide for the management of the reserves.

One important concern expressed by Franson was the need for some guarantee of permanent protection for the reserves. This could be accomplished by requiring consultation with an advisory committee prior to removing land from the act. The advisory committee would be a key part of the legislation in that it would also advise on other facets of the program, such as suggesting what areas should be included in the program, what scientific research should be undertaken, and on various management and administrative problems. The committee would also be empowered to hold public hearings and would be assured of an adequate budget to carry out these duties. These characteristics should be borne in mind while considering the existing federal and provincial legislation.

Federal Legislation

The only federal agreement dealing specifically with endangered plants is the International Convention on Trade in Endangered Species of Wild Fauna and Flora which, through an amendment to the Export and Import Permits Act (Canada, 1975), protects *Panax quinquefolius* against further exploitation by international trade. Under the Territorial Lands Act (Canada, 1970b) the Minister of Indian and Northern Affairs may establish land management zones "where he deems it necessary for the protection of the ecological balance . . ." and may "make regulations respecting protection, control and use of the surface of territorial lands." In relation to the Territorial Land Use Regulations (Canada, 1971), a Working Group on Proposed International Biological Program Ecological Sites has been formed under the Department of Indian and Northern Affairs to characterize ecological sites, to establish site evaluation criteria and to propose sites for special protection.

The National Parks Act (Canada, 1970a), also administered by the Department of Indian and Northern Affairs, is one of the most important federal tools for the protection of natural areas. In addition to providing for the establishment of traditional National Parks, many of which contain rare or endangered plants, it also provides for the establishment of special purpose parks such as the National Landmark Parks or the zoning of areas within parks to give varying degrees of protection. Lists of rare and endangered plants in the National Parks have been compiled by the Department, but no policy decision has been made concerning exactly how these species are to be protected. At the present time, permits are required for all botanical collecting within the National Parks.

In addition, the federal government occasionally provides special funds, such as the

National Second Century Fund of British Columbia to be used "to establish and develop nature conservation areas in every region of this province" (Krajina, 1973).

Provincial Legislation

Two provinces, Ontario and New Brunswick, have endangered species acts. Under both of these acts the endangered status of species may be qualified or limited by restricting it to particular geographical areas within the province or to a particular time.

The provinces of Alberta, British Columbia, New Brunswick, Ontario, and Quebec have ecological reserves or wilderness areas legislation that recognizes the need to reserve areas requiring protection because of the presence of rare or endangered native plants. A comparison of these acts with Franson's (1972) criteria for ecological reserve legislation (Table I) reveals that only the Alberta Wilderness Area Act and the New Brunswick and Quebec Ecological Reserves Acts embody most of the criteria. They differ from the British Columbia and Ontario acts mainly in: (1) providing for the acquisition of private lands by expropriation or by gift; (2) providing for a public advisory committee which can propose reserves and advise on their management; and (3) in requiring consultation with the advisory committee and a public announcement of intent before lands are withdrawn from the program. The latter provision is particularly important because it ensures a greater degree of permanency for the reserves than does legislation that permits withdrawal by the Lieutenant Governor in Council. It should be pointed out that the British Columbia act was a forerunner of the Alberta, New Brunswick, and Quebec acts and undoubtedly much was learned from the British Columbia experience. It falls short of the latter acts primarily on the question of reserve withdrawal and the acquisition of private lands. Although there is no provision for a public advisory committee in the British Columbia act, there is a voluntary committee that serves this function and receives some financial support from the province (Krajina, pers. comm.). The Ontario act is the weakest of the four and its function largely has been superseded by a natural areas classification of Ontario provincial parks.

The other provinces have provincial parks acts that could be used to protect natural areas but they do not specifically address themselves to the problems of rare and endangered plants or threatened habitats. Some provinces, including Manitoba, Ontario, and Saskatchewan, provide for zoning within the parks, or classify their parks into various types, depending on the proposed use of each park. A comparison of these acts with Franson's criteria (Table I) shows that all the designated parks fall short of the best existing natural areas legislation. None of them provides for a public advisory committee (the committee provided for under the Ontario and Newfoundland acts are to advise on the administration of particular parks), none prevents the arbitrary removal of lands from the program, and none provides for scientific input on management questions. More importantly, these acts do not specifically recognize the need to protect rare or endangered species or natural areas and this recognition seems to have an important psychological effect on the department administering the act. In this regard, it is significant to note that the "nature conservancy" areas designated under the British Columbia Park Act were transferred to the jurisdiction of the Ecological Reserves Act. If the original Park Act could have equally well served the function of an ecological reserve system, through its designation of nature conservancy areas, there would have been no need for a special act. Evidently it was felt that a special purpose act would be more effective. Franson (1972) sheds some light on this in pointing out that the question was hotly debated by members of the legal research team. Some of them felt that ecological reserve areas administered by park departments eventually would be used for recreational development and that the staffs of park departments, trained to think in terms of natural resource exploitation, would not be likely to provide effective protection against exploitation.

All Canadian provinces have enacted legislation that touches on one or another aspect of the general question of the protection of rare and endangered plants and all could serve in specific cases to protect certain species or habitats. A general view of apparently relevant provincial legislation may be useful to persons interested in the problem in a particular province.

In this overall review, I have not included some acts which may seem to be

TABLE I

A COMPARISON OF PROVINCIAL LEGISLATION PERTINENT TO THE CONSERVATION OF NATURAL AREAS COMPARED WITH AN ECOLOGICAL RESERVES MODEL

Desirable Characteristics of Ecological Reserves Legislation (Franson, 1972)

	Provide for acquisition and designation of Crown Lands.	Can expropriate or accept gifts of private lands.	Prevents arbitrary removal of lands from program.	Excludes the operation of inimical statutory powers.	Creates opportunities for co-ordination with other programs.	Continuing scientific input on management questions.	Provides for the management of reserves.	Provides for a public advisory committee.
ALBERTA								
Wilderness Areas Act	+	+	+	+	−	+	+	+
Provincial Parks Act	+	+	−	+	−	−	+	± a
BRITISH COLUMBIA								
Ecological Reserves Act	+	−	−	+	−	−	±	± b
MANITOBA								
Provincial Park Lands Act	+	+	−	+	−	−	+	−
NEW BRUNSWICK								
Ecological Reserves Act	+	+	+	+	−	−	+	+
Provincial Parks Act	+	+	−	± c	−	−	+	−
NEWFOUNDLAND								
Provincial Parks Act	+	+	−	−	−	−	+	± a
NOVA SCOTIA								
Provincial Parks Act	+	+	−	+	+	−	+	−
ONTARIO								
Wilderness Areas Act	+	+	−	−	−	−	+	−
Provincial Parks Act	+	+	−	+	−	−	+	± a
PRINCE EDWARD ISLAND								
Recreational Development Act	+	+	−	−	−	−	+	−
QUEBEC								
Ecological Reserves Act	+	+	+	+	−	+	+	+
SASKATCHEWAN								
Provincial Parks . . . Act	+	+	−	± c	−	−	+	−

a Committee to advise on the administration of a particular park.
b Voluntary advisory committee.
c Some inimical powers excluded.

appropriate, but were not deemed relevant. For example, most environmental protection acts are designed to control pollution, wildlife acts relate to animals only, forestry acts are oriented toward resource utilization, and all provinces have Crown Lands acts which can be used to withdraw lands for specific purposes. Most of these acts are not included in the review. When they are included it is on the authority of provincial officials who specifically mentioned these acts as being potentially useful.

Alberta

The Wilderness Areas Act (Alberta, 1971, 1972b) is concerned directly with reserving lands for the management and preservation of animal and plant life and protecting the environment of wilderness areas. It is administered by the Department of Recreation, Parks, and Wildlife.

The Environment Conservation Act (Alberta, 1970b, 1972a) establishes an Environment Conservation Authority which is concerned with the conservation of natural resources, including the protection of plant life. A public advisory committee on the environment has been appointed under this act. Although the protection of rare and endangered plants is only a small part of this committee's overall responsibilities, it has addressed the problem in some of its reports and recommendations (Trost, 1974). The Environment Conservation Authority is empowered under the Wilderness Areas Act to hold public hearings respecting proposals for additional wilderness areas or the withdrawal of existing reserves. This act is administered by the Department of the Environment.

The Provincial Parks Act (Alberta, 1970c) permits the designation of lands as parks, natural areas, or wilderness areas. It also provides for the establishment of a board and advisory committees to aid in the formulation of policies for the administration and development of these areas. The act is administered by the Department of Energy and Natural Resources.

The Public Lands Act (Alberta, 1970a) provides for natural areas that may be designated by order-in-council. Since 1969, following recommendations of the Natural Areas Committee, 288 natural areas have been reserved and others are under temporary reservation (Steele, pers. comm.). These natural areas do not enjoy the same degree of protection as do wilderness areas and the classification of natural areas presently is under review by the Department of Energy and Natural Resources.

British Columbia

The Ecological Reserves Act (British Columbia, 1971a) is designed to reserve Crown Land for various ecological purposes including the preservation of "rare or endangered plants and animals in their natural habitat". The act is administered by the Ecological Reserves Unit of the Department of the Environment. The act does not provide for an advisory committee, but a voluntary committee does receive some government support (Krajina, pers. comm.). The Park Act (British Columbia, 1965, 1973) formerly made provision for nature conservancy areas which were to be retained in a natural condition. Such areas as were designated under this act have been placed under the jurisdiction of the Ecological Reserves Act.

The Environment and Land Use Act (British Columbia, 1971b) established a committee that was to address itself to increasing public concern for and awareness of the environment and to insure the preservation and maintenance of the natural environment.

Manitoba

The Crown Lands Act (Manitoba, 1970c) enables lands to be set aside by the Lieutenant Governor in Council for a variety of public purposes including provincial parks, forest reserves and game reserves. It is also possible for ecological reserves to be established under this act and an Ecological Reserves Advisory Committee is studying the question.

The Provincial Park Lands Act (Manitoba, 1972) provides for various types of provincial parks including "natural parks, wilderness parks and heritage parks," any of which would be suitable for protecting rare and endangered plants. There is also a zoning program within provincial parks intended to be used to protect unique natural areas.

The Resource Conservation Districts Act (Manitoba, 1970a) is oriented toward resource control within each district for "the purpose of conserving, controlling, developing, protecting, restoring or using," forest, wildlife, and recreational resources. This act, which could be used to set up districts that function as ecological reserves, is administered by the Department of Tourism and Cultural Affairs.

The Wildlife Act (Manitoba, 1970b) provides for the preservation, maintenance, and restoration of wildlife habitat and could be used to protect certain natural areas. It is administered by the Department of Mines, Resources, and Environmental Management.

New Brunswick

Through the Endangered Species Act (New Brunswick, 1975b) administered by the Department of Natural Resources, a species may be declared endangered in a part of or throughout the province or at a particular time. The habitat of the species so designated is also protected from destruction or interference.

The Ecological Reserves Act (New Brunswick, 1975a) permits the reservation of lands for various ecological purposes including the protection of rare and endangered native plants and animals. The Environmental Council (New Brunswick, 1971) is designated under the act to advise on areas to be reserved, on changes in the reserves and on the management of the reserves. The act is administered by the Department of Natural Resources.

The Park Act (New Brunswick, 1973) could be used to protect natural areas but it is framed in very general terms. It may be significant that, in Section 13, it refers to measures that may be taken to protect "fish, animals and birds," but plants or habitats are not mentioned. This act is administered by the Department of Tourism.

Newfoundland

There are two provincial acts that could be used to protect natural areas: the Provincial Parks Act (Newfoundland, 1970) and the Crown Lands Act (Newfoundland, 1952). Both acts are very general and do not specifically refer to preserving natural areas for other than normal public use. The Provincial Parks Act is administered by the Department of Tourism.

A proposal for an act respecting wilderness and ecological reserves is being brought forward by the Minister of Tourism (Hickey, pers. comm.). This act, if passed, would permit the establishment of ecological reserves for the purposes of: (a) safeguarding animal and plant species threatened with extinction, and (b) safeguarding areas that contain unique or rare examples of botanical, zoological, or geological phenomena.

Nova Scotia

Nova Scotia has two acts that provide for the preservation of plants and animals in their natural state. The Provincial Parks Act (Nova Scotia, 1967b) is directly applicable in that it enables the Minister of Lands and Forests to take steps to ensure the security of animal and plant life in a park and the preservation of the park in a natural state. The Lands and Forests Act (Nova Scotia, 1967a) allows for the setting apart of Crown Lands for the preservation of forests or game. However, this act apparently has an applied intent.

Ontario

The Endangered Species Act (Ontario, 1971a) provides for the protection of species and their habitats when they are threatened with extinction by various factors including the destruction of habitats and overexploitation. It is administered by the Ministry of Natural Resources, Division of Fish and Wildlife.

The Wilderness Areas Act (Ontario, 1970b) was enacted to set apart public lands for the preservation of areas in a natural state in which research and educational activities may be carried on, as well as for the protection of flora and fauna. This act has been supplanted by the nature reserves classification within the Provincial Parks Act (Ontario, 1970a). Both of these acts are administered by the Ministry of Natural Resources, Division of Parks.

The Environmental Protection Act (Ontario, 1971b, 1973), whose purpose is to provide for the protection and conservation of the natural environment, enables the Department of the Environment to conduct studies on the quality of the natural environment and to establish an Environmental Council to advise on the results of current research on the natural environment. Although this act is primarily oriented toward problems of pollution, it could provide the means for studying the problems of rare and endangered species and threatened habitats.

Prince Edward Island

There are four acts that may have some bearing on the setting aside of natural areas for the protection of rare and endangered species, but none of them deals specifically with the problem. The Recreational Development Act (Prince Edward Island, 1969) replaces the Provincial Parks Act and permits the designation of areas as "protected" for purposes of aesthetic, educational, or scientific interests. This act is administered by the Department of Environment and Tourism.

The Planning Act (Prince Edward Island, 1974b), in conjunction with the Land Use Commission Act (Prince Edward Island, 1974a), provides for the setting up of a commission to advise and implement guidelines for land use, including the reservation of land for public purposes. A sympathetic commission could be instrumental in implementing an ecological reserves program. This act is administered by the Department of Community Services.

The Fish and Game Protection Act (Prince Edward Island, 1959) may be applicable in that Wildlife Management Areas could be designated for the management and conservation of wildlife. This act applies primarily to animals and only insofar as animal habitats could be protected would it have any bearing on the protection of rare and endangered plants.

Quebec

The Ecological Reserves Act (Quebec, 1975) is specifically designed to establish ecological reserves for: (a) the preservation of areas in their natural state; (b) scientific research and education; and (c) the protection of threatened or endangered plants and animals. This act is administered by the Department of Lands and Forests.

The Provincial Parks Act (Quebec, 1964) established four provincial parks under the administration of the Department of Tourism, Fish and Game. Control is exercised over the areas and plants may be collected only by permit.

Saskatchewan

The Provincial Parks, Protected Areas, Recreational Sites and Antiquities Act (Saskatchewan, 1965, 1973-1974, 1974-1975) provides for the establishment of "protected areas for the purpose of protecting and preserving therein their scenic, historic or scientific interest or significance." This act is administered by the Department of Tourism and Renewable Resources.

Local Legislation

It is impossible here to review all the legislative initiatives taken at local levels to protect rare species or habitats. However, two examples will be given of the kinds of actions that have been taken.

A study of environmentally sensitive natural areas was made for the Regional Municipality of Waterloo, Ontario, by the Department of Man-Environment Studies of the University of Waterloo (Francis and Eagles, 1975). In the study, financed by the Ontario Ministry of the Environment, 70 environmentally sensitive areas which they proposed to maintain in a natural state, were surveyed. The existing provincial legislative framework through which provisions for the effective protection and management of environmentally sensitive areas might be developed was reviewed and provisions to be considered for a natural area by-law were outlined. The proposals made in the report were officially accepted by the regional municipality as part of its Official Policies Plan.

In Manitoba, one of the last remaining areas of the true grass prairie in Canada is being protected as the St. James-Assiniboia Living Prairie Museum by the City of Winnipeg as one of its municipal parks (Johnson, 1975). Although it may prove to be difficult to preserve all of the natural features of this 42-acre prairie in an urban setting, its use as a public education center may be instrumental in stimulating the conservation of other natural areas in the province.

The International Biological Program

The Canadian Committee for the International Biological Program, Conservation of Terrestrial Communities Sub-Committee (CCIBP-CT), played an important role in protecting rare and endangered species through its organization of a survey of existing and potential ecological reserves and in the promotion of the establishment of a system of ecological reserves for the conservation of natural ecosystems (Fuller, 1973). It focused on habitats rather than on species because its primary concern was with the preservation of natural areas. However, the occurrence of rare, endangered, or otherwise significant species in an area was one of the criteria for the recognition of the biological importance of the area.

The IBP-CT played a major role in the establishment of Ecological Reserves Acts in British Columbia (Krajina, 1973) and in Quebec (Lemieux, 1972). In other provinces and territories, notably Alberta, Manitoba, Ontario, and the Northwest Territories, there has been close cooperation between government agencies and the IBP-CT and some IBP areas have received some form of protection or are being considered for ecological reserve status (McLaren and Peterson, 1975).

With the termination of the IBP-CT in 1974, the National Research Council of Canada established an Associate Committee on Ecological Reserves (ACER) to carry on some of the activities of CT (Anon., 1976). The purposes of ACER are: (1) to provide information and advice to governments and other interested parties, on the protection and preservation of ecological reserves (specifically with information on areas surveyed by the IBP-CT); and (2) to encourage legislative and regulatory action on behalf of ecological reserves. ACER is an interim committee and will serve only until its role can be taken over by a permanent committee. Evidently it is hoped that each of the original IBP-CT provincial and regional panels will secure government sponsorship of its activities or that the federal government will take steps toward a national ecological reserves system. The Alberta Ecological Survey is continuing the survey started there by the IBP-CT, with funding from the provincial government.

The IBP-CT has been an important catalyst in the movement to preserve ecologically significant areas, including those containing rare and endangered plants. Its vital role in supporting surveys of potential ecological reserves and in coordinating a national effort on behalf of an ecological reserves system must be continued.

Canadian Institute of Forestry

In 1971 the Canadian Institute of Forestry (CIF), an organization of professional foresters, advocated a program concerning the selection, protection, and management of representative examples of all significant forest and forest-related vegetation in Canada (Weetman and Cayford, 1972; Weetman, 1975). They gave a high priority to natural areas containing commercially important forest types and rare and endangered species of plants and animals. According to the program, proposed natural areas would be reserved for observational and nondestructive research and would not be used for experiments in forest management. The plan, similar in many ways to that of the IBP-CT, is a good one but its success has been impaired by the lack of an effective implementing organization. The CIF is able to serve only in an advisory capacity through the affiliation of its members with public and private agencies controlling land. Furthermore, it seems that a significant element in the forestry community may not agree with the plan. In British Columbia, the Council of Forest Industries vigorously opposed the IBP requests for ecological reserves, including timber of commercial quality. The result was a severe reduction in the acreage of some ecological reserves (Franson, 1972). However, the CIF has established a Natural Areas Committee that is maintaining a national register of forested natural areas that have legal protection and is attempting to move toward a national network of reserves (Weetman, 1975). Also, through some of its members, it played an important role in passing the New Brunswick Ecological Reserves Act and it continues its noteworthy efforts to reserve areas under the act (Moller, pers. comm.).

Private Conservation Groups

There are a large number of private naturalist and conservation groups that are also contributing to the protection of rare and endangered species. The Canadian Nature Federation is an important organization which is actively engaged in public education through its magazine "Nature Canada." The Nature Conservancy of Canada is involved in the purchase of private lands which it then turns over to appropriate management groups. Each of the provinces has one or more such groups which support efforts to establish natural areas and to protect rare and endangered species.

In some instances, private individuals have dedicated lands to natural area preservation and some industries (Johnson, 1975; Lewis, 1974) have cooperated in the preservation of ᵛ important natural areas.

SUCCESSES AND PROBLEMS

The degree of success in efforts to protect rare and endangered plants can be gauged by the legislation passed by the provinces of Alberta, British Columbia, New Brunswick, Ontario, and Quebec in direct response to the need for establishing ecological reserve systems or protecting individual species and their habitats. However, our enthusiasm about these successes is somewhat tempered when the specific efforts that have been made under this legislation on behalf of rare and endangered plants are evaluated. For example, Ontario and New Brunswick have endangered species legislation, but no species of plants have been protected under these acts although there are species that could be considered for inclusion.

In terms of the actual protection of natural areas, one of the most conspicuous success stories was the establishment of 52 ecological reserves in British Columbia (Krajina, 1973; Krajina et al., 1974). Of these, at least four were regarded as exceptional because of the presence of rare, threatened or relict flora, and others were established as examples of natural vegetation. However, as Franson (1972) points out, it should be noted that the amount of land devoted to the ecological reserve system in British Columbia is very small in proportion to the size of the province, and that subsequent proposals for reserves have been meeting heavy opposition from the mining and lumber industries. Since the reserves that have been established can be removed from the system by order-in-council, the situation in British Columbia bears close watching.

Three wilderness areas have been established in Alberta but none of them includes rare or endangered plants. The Natural Areas Committee of the Alberta Department of Energy and Natural Resources has established 288 natural areas and has placed a temporary reservation on other sites proposed by the IBP-CT. These natural areas do not enjoy the same degree of protection afforded to the wilderness areas and the entire system is under review. Under the Ontario Wilderness Areas Act some 39 areas have been designated, and under the Nature Reserves classification of the provincial parks nine areas have been reserved. Of the latter nine at least two areas have been reserved because of rare or relict flora. In Quebec one ecological reserve, containing relict stands of white pine *(Pinus strobus),* has been designated and others are pending. Under the New Brunswick Ecological Reserves Act some eight areas have been reserved and others are under consideration. In Saskatchewan 21 areas have been established as "protected areas," and of these, six are protected for their biological value but none of these is known to contain rare or endangered plants.

The successes that have been achieved can be largely attributed to the survey of potential ecological reserves supported by the IBP-CT and to the focusing of energy on the problem by that organization. The political adroitness of some of the proponents of ecological reserves systems, the availability of expert legal advice, the receptivity of some governments and the support of local naturalists and conservation groups also contributed to the successes. However, many of the achievements have been uncoordinated, the protection afforded many areas is inadequate and few areas receive the long-term protection that is essential both for ecological study and for the protection of rare and endangered species.

McLaren and Peterson (1975) in their review of the activities of the IBP-CT paint a favourable picture of provincial action in the protection of natural areas proposed by the

IBP-CT. However, I am not convinced that their view is a realistic one.

The failure of governments to use the existing legislation in more than a nominal way can be attributed to many factors. In some instances, the responsible government department has been committed to a multiple land use philosophy and therefore it is reluctant to reserve lands for what it considers a single use. The stronger the legislation in terms of ensuring long-term protection of the reserves, the more difficult is the decision for the department concerned. This problem is compounded by the fact that some of the responsible departments are "mission oriented" and are biased in favour of conservation for more efficient resource utilization rather than conservation for resource protection. Furthermore, the frequent changes in administrators often require "reeducation" to the importance of natural area preservation. This requires the time and commitment of dedicated private individuals who are often frustrated by the slow pace of the government decision-making process.

The endangered species acts were conceived primarily for the protection of animal species and are administered by the wildlife divisions of departments of natural resources. These departments usually do not have a staff which is either competent or inclined to study the problem seriously in order to present a case for the protection of plant species. Interdepartmental rivalry sometimes arises when more than one department administers legislation that could be used to protect natural areas. The existence of legislation such as some Noxious Weeds Acts, which seeks to eradicate the very species that are threatened by extinction, is inimical to protection.

The exercise of political and economic power by groups interested in blocking the reservation of lands that have potential mining, petroleum, forestry, or agricultural uses has been seen many times in the efforts of Parks Canada to establish national parks. Sometimes these efforts have been successful and other times not. However, the proponents of ecological reserves or of the protection of rare or endangered species usually do not have the political or economic backing, even as much as is available to Parks Canada, to ensure a fair contest.

Those lands set aside by individuals or private organizations for conservation purposes do not have the legal protection necessary to ensure their long-term dedication. Such lands could be subjected to expropriation for a variety of purposes including resource exploitation, without the owners having effective legal recourse (Lucas, 1974).

Finally, there is also a problem related to the adequate documentation of ecological reserves or of endangered species. If proposals for species or ecological area protection are to receive serious consideration from governments they must be well-documented, and such documentation is expensive, particularly in the case of species or natural areas in remote parts of Canada. As important as the IBP-CT survey of potential natural areas was, its coverage was incomplete and in some areas it was little more than cursory. Further studies of potential ecological reserves are needed and a more complete documentation is necessary of some areas which already have been surveyed.

Equally serious is our lack of accurate knowledge about the rare and endangered plants of Canada. The existing lists are the result of elementary surveys. The species included on them require careful checking to determine their present status, and undoubtedly it will prove necessary to add other species to the lists.

In this paper I have focused on the protection of natural areas as the most effective way to protect rare and endangered plants, and on government legislation as the most effective way to protect natural areas. Undoubtedly, a more assiduous and systematic implementation of existing legislation would contribute much to the success of efforts to protect rare and endangered plants. Although it is recognized that the ultimate responsibility for the flora and fauna of Canada is provincial, nevertheless one of the most compelling solutions to the problems of their conservation is through a federal initiative to establish a national conservation organization with public advisory committees on natural areas and rare and endangered species. The coordination of existing federal, provincial, municipal, and private efforts, and the focusing of public awareness on the problem, would have inestimable value.

Short of a national initiative, the provincial governments could make a significant contribution by forming and supporting public advisory committees with responsibilities: (a) to study potential ecological reserves and rare and endangered species; (b) for proposing areas and species for protection; (c) to advise on the management of ecological reserves; and (d) to educate the public about the problems and what they can do to help solve them.

ACKNOWLEDGEMENTS

The preparation of this paper would have been impossible without the help of a large number of individuals to whom I am indebted for their comments, ideas, and information, including: Mr. J. Bain, Prince Edward Island Dept. of the Environment; Mr. T. Beechey, Ontario Ministry of Natural Resources; Mr. A.M. Cameron, Nova Scotia Ministry of Lands and Forests; Mr. J.A. Carruthers, Parks Canada; Mr. T.A. Drinkwater, Alberta Dept. of Recreation Parks and Wildlife; Mr. P. Eagles, Univ. of Waterloo; Ms. M.L. Goatcher, Manitoba Dept. of Renewable Resources and Transportation Services; Hon. T.V. Hickey, Newfoundland Dept. of Tourism; Dr. G.W. Hodgson, Univ. of Calgary; Dr. K. Johnson, Manitoba Museum of Man and Nature; Mr. D.R. Johnston, Ontario Ministry of Natural Resources; Mrs. L. Kershaw, Univ. of Waterloo; Dr. V.J. Krajina, Univ. of British Columbia; Mr. G. Lemieux, Univ. Laval; Mr. A.S. Matsella, Saskatchewan Ministry of Tourism and Renewable Resources; Mr. R. Moller, New Brunswick Dept. of Natural Resources; Dr. J.K. Morton, Univ. of Waterloo; Dr. D. Nettleship, Environment Canada; Dr. N. Novakowski, Environment Canada; Dr. E.B. Peterson, Western Ecological Services Ltd.; Dr. J.M. Shay, Univ. of Manitoba; Mr. R.G. Steele, Alberta Dept. of Energy and Natural Resources; Mr. R.D. Thomasson, Manitoba Dept. of Renewable Resources and Transportation Services; Dr. W. R. Trost, Alberta Environment Conservation Authority; and Mr. R.J. Vrancart, Ontario Ministry of Natural Resources.

LITERATURE CITED

Alberta. 1970a. The Public Lands Act. Statutes of Alberta, ch. 80.

_____. 1970b. The Environment Conservation Act. Revised Statutes of Alberta, ch. 125.

_____. 1970c. The Provincial Parks Act. Revised Statutes of Alberta, ch. 288.

_____. 1971. The Wilderness Areas Act. Statutes of Alberta, ch. 114.

_____. 1972a. The Environment Conservation Amendment Act. Statutes of Alberta, ch. 38.

_____. 1972b. The Wilderness Areas Amendment Act. Statutes of Alberta, ch. 122.

Anonymous. 1976. Associate Committee on Ecological Reserves (ACER). Bull. Can. Bot. Assoc. **9**(2): 18-19.

Argus, G. W. and J. McNeill. 1974. Conservation of evolutionary centres in Canada. pp. 131-141. _In_ J. S. Maini and A. Carlisle. Conservation in Canada. A conspectus. Dept. of Environment, Can. Forest. Serv. Publ. 1340.

British Columbia. 1965. The Park Act. Statutes of British Columbia, ch. 31.

_____. 1971a. The Ecological Reserves Act. Statutes of British Columbia, ch. 16.

_____. 1971b. The Environment and Land Use Act. Statutes of British Columbia, ch. 17.

_____. 1973. An Act to Amend the Park Act. Statutes of British Columbia, ch. 67.

Canada. 1970a. The National Parks Act. Revised Statutes of Canada, ch. N-13.

_____. 1970b. The Territorial Lands Act. Revised Statutes of Canada, ch. T-6 and ch. 48.

_____. 1971. The Territorial Land Use Regulations. Canada Gazette Pt. II, **105**: 1908-1927.

_____. 1975. Amendments to the Export and Import Act. Canada Gazette Pt. II, 109(14).

Francis, G. R. and P. F. J. Eagles. 1975. A study of the environmentally sensitive areas for the Environmental Policy Plan of the Regional Municipality of Waterloo. 99 pp., multilith. Dept. Man-Environment Studies, Univ. Waterloo, Ont.

Franson, R. T. 1972. Legislation to establish ecological reserves for the protection of natural areas. Osgood Hall Law Jour. **10** (3): 583-605.

_____. 1975. The legal aspects of ecological reserve creation and management in Canada. Inter. Union for Conserv. of Nature Environmental Policy and Law Paper No. 9. Morges, Switzerland.

Fuller, W. A. 1973. The Conservation of Terrestrial Communities Programme in Canada. Syesis **6**: 11-16.

Johnson, K. 1975. Preservation of native plants. Manitoba Nature, Winter 1975, pp. 20-23.

Kershaw, L. J. 1976. A phytogeographical survey of rare, endangered and extinct vascular plants in the Canadian flora. M.Sc. thesis, Dept. of Biology, Univ. of Waterloo, Waterloo, Ont. Unpublished.

_____. **and J. K. Morton.** 1976. Rare and potentially endangered species in the Canadian flora — A preliminary list of vascular plants. Can. Bot. Assoc. Bull. **9**(2): 26-30.

Krajina, V. J. 1973. The conservation of natural ecosystems in British Columbia. Syesis **6**: 17-31.

_____., **P. A. Larkin, J. B. Foster and D. F. Pearson.** 1974. Ecological Reserves in British Columbia. Can. Comm. Int. Biol. Prog., Cons. Terr. Comm. Subcomm. Reg. 1, B. C., Vancouver, Multith, 185 pp.

Lemieux, G. 1972. Une prospective de la Conservation des espèces végétales au Québec. De Toute Urgence **3**: 17-30.

Lewis, F. A. 1974. The Nature Conservancy of Canada. Annual Report 1974. Toronto, Ont.

Lucas, A. R. 1974. Legal problems in Canadian conservation. pp. 375-391. _In_ J. S. Maini and A. Carlisle. Conservation in Canada. A conspectus. Dept. of Environment, Can. Forest. Serv. Publ. 1340.

Manitoba. 1970a. The Resource Conservation Districts Act. Statutes of Manitoba, ch. R-135.

_____. 1970b. The Wildlife Act. Revised Statutes of Manitoba, ch. W-140.

_____. 1970c. The Crown Lands Act. Revised statutes of Manitoba, ch. C-340.

_____. 1972. The Provincial Park Lands Act. Statutes of Manitoba, ch. 67.

McLaren, I. A. and E. B. Peterson. 1975. Ecological reserves in Canada: The work of
    the IBP-CT. Nature Canada, April/June pp. 22-32.
New Brunswick. 1971. The Clean Environment Act. Statutes of New Brunswick, ch. 3.
_____. 1973. The Parks Act. Revised Statutes of New Brunswick, ch. P-2.
_____. 1975a. The Ecological Reserves Act. Statutes of New Brunswick, ch. E-1.1.
_____. 1975b. The Endangered Species Act. Statutes of New Brunswick, ch. E-9.1.
Newfoundland. 1952. The Crown Lands Act. Revised Statutes of Newfoundland,
    ch. 17.
_____. 1970. The Provincial Parks Act. Revised Statutes of Newfoundland, ch. 312.
Nova Scotia. 1967a. The Land and Forests Act. Revised Statutes of Nova Scotia,
    ch. 163.
_____. 1967b. The Provincial Parks Act. Revised Statutes of Nova Scotia, ch. 244.
Ontario. 1970a. The Provincial Parks Act. Revised Statutes of Ontario, ch. 371.
_____. 1970b. The Wilderness Areas Act. Revised Statutes of Ontario, ch. 498.
_____. 1971a. The Endangered Species Act. Statutes of Ontario, ch. 52.
_____. 1971b. The Environmental Protection Act. Statutes of Ontario, ch. 86.
_____. 1973. Amendments to the Environmental Protection Act. Statutes of Ontario,
    ch. 84.
Prince Edward Island. 1959. The Fish and Game Protection Act. Statutes of Prince
    Edward Island, ch. 13.
_____. 1966. Amendments to the Fish and Game Protection Act. Statutes of Prince
    Edward Island, ch. 15.
_____. 1969. The Recreational Development Act. Statutes of Prince Edward Island,
    ch. 45.
_____. 1974a. The Land Use Commission Act. Statutes of Prince Edward Island, ch. 22.
_____. 1974b. The Planning Act. Statutes of Prince Edward Island, ch. 81.
Québec. 1964. The Provincial Parks Act. Revised Statutes of Québec, ch. 20.
_____. 1975. The Ecological Reserves Act. Gazette Officielle du Québec 107(6): 517-520.
Saskatchewan. 1965. The Provincial Parks, Protected Areas, Recreational Sites and
    Antiquities Act. Revised Statutes of Saskatchewan, ch. 54.
_____. 1973-1974. An Act to Amend the Provincial Parks, Protected Areas, Recreational
    Sites and Antiquities Act. Revised Statutes of Saskatchewan, ch. 74.
_____. 1974-1975. An Act to Amend the Provincial Parks, Protected Areas, Recreational
    Sites and Antiquities Act. Revised Statutes of Saskatchewan, ch. 34.
Trost, W. R. 1974. Land use and resource development in the eastern slopes. Report
    and Recommendations. 224 pp. Alberta Environment Conservation Authority,
    Edmonton, Alta.
Weetman, G. F. 1975. Forested Natural Areas in Canada. Forest. Chron. 51:297-298.
_____. and J. H. Cayford. 1972. Canadian Institute of Forestry policy for selection,
    protection and management of natural areas. Forest. Chron. 48:41-43.

THE NORTHEASTERN UNITED STATES

William D. Countryman

Aquatec, Inc., Box 2223
South Burlington, Vermont 05401, U.S.A.

The most populous section of the United States, the Northeast, has considerable pressures on fragile plant populations, and in particular on their habitats. In a number of these states, legislative machinery has been created that recognizes this threat, yet the measures taken so far generally have been inadequate. A close look at the situation in this critical part of the country indicates that ultimate answers require a very broad approach to the problem.

The nine northeastern states (New York, Pennsylvania, New Jersey and the six New England states) comprise an area of approximately 440,000 km^2. The region has over 1,200 km of Atlantic Ocean coastline, or nearly 17,000 km when measured along the deep indentations of its rugged shores (Delury, 1975). In addition, extensive shorelines border Lakes Ontario, Erie, Champlain and the St. Lawrence River. The highest elevation in the area is the summit of Mt. Washington in New Hampshire, 1,917 m above mean sea level. All of the New England States, most of New York and the northern portions of Pennsylvania and New Jersey were glaciated by the Wisconsinan Ice Sheet (Embleton and King, 1968).

The region is a part of the Gray's Manual range which, in its entirety, contains slightly over 5,500 species of vascular plants (Fernald, 1950). Approximately 3,200 species occur in New England (Seymour, 1969). The area contains a human population in excess of 50 million and includes a major part of the burgeoning East Coast megalopolis (Boston-New York City-Newark-Philadelphia, etc.

In pre-Columbian times the area was largely forested. Extensive lumbering and the clearing of land for agricultural pursuits have destroyed all but a trace of the area's primeval forests. Wetlands, both coastal and inland, have been greatly modified by draining, dredging, and filling. Alpine areas in the region have been extensively developed for recreational purposes. Many such areas are now reached easily by toll roads, ski lifts, or by readily accessible hiking trails. Large tracts of land have succumbed to urban development. Most of the region's larger rivers are heavily polluted by industrial and domestic wastes. Dams and hydroelectric generating stations have drastically altered the natural flow patterns of many of these rivers.

All states in the northeastern region have laws against the taking of plants and trees. Most of these laws are essentially concerned with the criminal theft of crops and timber and are not relevant to the topic discussed here. Many state and federal agencies have regulations prohibiting the picking or removal of wild flowers on lands under their jurisdiction. Federal agencies such as the U. S. Forest Service, National Park Service, U. S. Fish and Wildlife Service, and state forest and parks departments own or manage many thousands of acres. Though regulations of such agencies could be helpful in protecting endangered plant species, most such regulations are not actively enforced.

Four of the nine northeastern states have statutes specifically relating to the protection of rare or endangered plant species. For the most part, these laws are unenforceable or otherwise inadequate.

Maine, New Hampshire, New Jersey and Pennsylvania have no laws which provide specifically for the protection of endangered plant species. In Massachusetts and Rhode Island there are long-standing statutes which provide for the protection of a few, usually showy, species. Both Vermont and New York have laws designed to protect a large number of species. Many of these species are truly rare or threatened, but in the judgement of many botanists, these laws also cover many common species which are not threatened at the present time and thus are not now in need of protection.

More than a century ago, Connecticut became the first state in the nation to enact a law for the protection of an endangered plant species (1869 Ct. Acts 291[1]). Because of

---

[1] State laws, acts and statutes are cited in parentheses utilizing abbreviations generally accepted by the legal profession. For an explanation of these abbreviations see "A Uniform System of Citation," Twelfth Edition, 1976. The Harvard Law Review Association, Cambridge, Massachusetts.

its historic significance it is reproduced as follows in its entirety.

Be it enacted by the Senate and
House of Representatives, in General Assembly convened:

That any person who shall willfully and maliciously
sever or take away from the lands of another, any of the
species of plant known as "Lygodium palmatum" or
"Creeping fern," growing and being thereon, shall be
punished by a fine not exceeding seven dollars, or by
imprisonment in the common jail, not exceeding thirty
days, or by such fine and imprisonment both, at the
discretion of the court having cognizance of the offense.
Approved, July 8th, 1869.

Protection of
creeping fern

In 1925, Massachusetts enacted a law entitled "An Act to Prevent the Extinction of
the Mayflower." The penalty for picking or injuring the mayflower, *Epigaea repens* L., is
a fine of not more than $50 unless a person ". . . does the aforesaid acts while in disguise
or secretly in the night time . . .," in which case the fine is to be doubled (2 M.G.L.A. §7).

A Massachusetts law passed in 1935 provides for the "Protection of Wild Azaleas, Wild
Orchids and Cardinal Flowers"; persons to be punished by a ". . . fine of not more than
five dollars" if they dig or injure such species (266 M.G.L.A. §116A).

A 1939 Rhode Island statute lists about a dozen, mostly showy, plants (mountain
laurel, mayflower, flowering dogwood, great laurel, etc.) and specifies a fine of from
$10 to $50 for the taking of these species (1939 R.I. Acts 453-4).

In 1921 the Vermont Legislature enacted a law ". . . designed to provide protection
to rare plants and to protect these plants from being sold for commercial purposes"
(37 P.L., §8578). This statute listed 42 species by common and scientific names. The
statute provided that a person ". . . shall not take in any one year, except on lands owned
or occupied by him, more than a single uprooted specimen or two cuttings . . ." of any
of the species listed. The penalty for violators was specified as a fine of ". . . not more
than ten dollars for each plant or additional cutting so taken." This law was carried
forward in the revision of the Vermont Statutes in 1947 (41 V.S. 1947 §8443). In 1957
this statute was repealed and a new statute was enacted by the Vermont Legislature
(13 V.S.A. §3614). The new law added several species to those listed in the 1921 statute
and brought the taxonomic designations up to date. Other than those changes, however,
the 1957 law was essentially identical to the 1921 law. The 1957 law was repealed in
1972 and replaced by a new statute (13 V.S.A. §3651-3). The new law listed no species
by name but instead provided that the Secretary of Environmental Conservation shall
". . . adopt rules . . . making any . . . plant, flower, tree or shrub an endangered species
when he finds that the species is in danger of becoming extinct or its existence in the
state or a part thereof is threatened." The new law provides that a person ". . . who takes
or possesses an endangered species shall be fined not more than $1,000. Each individual
specimen or an endangered species so taken or possessed shall constitute a separate
violation." After a duly advertised public hearing (which, incidentally, no one from the
public attended) the Secretary adopted an "Endangered Species List" on 1 July 1975.
The newly adopted regulation expanded to 88 the number of listed species. As viewed by
a professional botanist, the Vermont endangered species list is indeed a curious document.
There are plant species on the list which do not occur in Vermont, introduced species and
species which are abundant and not in need of protection. The list omits species which,
in the opinion of many experienced botanists, are truly threatened or endangered.

New York's current Endangered Plant Law was passed in 1974 (E.C.L. §9-1503). In
many ways this New York law is reminiscent of Vermont's statute, but in other ways it is
distinctly different. New York's law empowers the Department of Environmental
Conservation to adopt a list of protected plants and provides for a fine, not to exceed
$25, for those who ". . . knowingly pick, pluck, sever, remove or carry away without the
consent of the owner thereof, any protected plant."

A comparison of the Vermont and New York laws and regulations provides some
interesting insights. The Vermont list includes 88 species, New York's list 225 species;

Vermont law provides a penalty of $1,000 for each plant or plant part taken, New York's penalty is $25 per offense but *only* if the violator "knowingly committed the offense."

I find no record of arrests or convictions for violations of any of the laws pertaining to the protection of rare or endangered plant species in any of the nine northeastern states. From this fact it may be concluded that the threat of fines or imprisonment has indeed provided protection to the plants. Or one *might* conclude that the laws provide little, if any actual protection because they are unrealistic and unenforceable. Few, if any, law enforcement officers could accurately identify all of Vermont's 88 protected species or the 225 species in New York's list. Who in New York could prove that a violator "knowingly" plucked a protected species? Would a Vermont judge fine a violator $1,000 for each blossom of a listed species in a Spring bouquet which had been innocently picked from the locally common flora?

I conclude that our laws and regulations designed to protect rare and endangered plant species do not, in fact, provide such protection. I further conclude that there is little threat from the random collection of plants nor do I now find any serious threat to such plants from commercial exploitation although this may have been the case in the past.

Are, then, our rare plants indeed threatened and if so, by whom or what? By what means may they be protected?

The pressure of ever increasing land development is unquestionably the most serious threat to rare plant species. It is, therefore, the habitats of rare plants which are endangered and which must be protected. Comprehensive land use planning legislation is, I believe, the most effective means of providing the needed protection.

It is beyond the scope of this study to describe in detail all the laws of the nine northeastern states which pertain to land use zoning, protection of wetlands, conservation easements, etc. A comprehensive compendium of laws relating to the protection of natural areas has been prepared by the Conservation Law Foundation of New England (CLFNE [1972?]). The Natural Resources Defense Council has prepared a handbook of land use controls in New York State (Moss, 1976).

Two examples of laws which have proven effective in preserving the habitats of rare and endangered plant species are the Massachusetts Conservation Act and Vermont's Act 250.

In Massachusetts, the Coastal Wetlands Act (12.06 G.L. §8c) and the Inland Wetlands Act (12.07G.L. §40A) are largely administered by town and city conservation commissions. These commissions, established by an act of the legislature (12.01 G.L. §8c), are empowered ". . . to develop programs of resource protection through planning, land acquisition and restriction, regulations, and development for multiple-use activities" (Hansen and Goodno, 1975). At the present time, 330 of the 351 cities and towns in Massachusetts have established conservation commissions. Members of these commissions are appointed by town selectmen and serve without compensation. Each commission consists of seven members with staggered terms. The functions of these commissions have been outlined by Ellis and Dawson (1973).

In the eighteen years since the establishment of the first Massachusetts Conservation Commission, an enviable record of accomplishment has been attained. Prior to World War II there were 60,000 acres of salt marsh in Massachusetts; today there are only 45,000 acres. At present, 31,000 of these dwindling acres have had development restrictions placed upon them by conservation commissions. Such restrictions may be temporary but are more often permanent, thus providing for the perpetual protection of conservation features. Typical restrictions forbid or limit filling, dumping, excavation, dredging, or construction or placing of buildings, roads, utilities or other structures on or above the ground. Conservation commissions have acted on 2,500 applications to modify wetlands. Commission decisions have been tested in three court cases, all of which were settled in favor of the wetlands.

It is interesting to note that the Massachusetts law defines bogs, marshes, and other wetlands botanically, on the basis of an established, detailed list of aquatic plant species. Thus the question of whether or not a site comes under the jurisdiction of a conservation commission is determined by the presence or absence of wetland species growing there, rather than by some vague or ambiguous definition of the area.

In addition to protecting habitats by restrictions, conservation commissions are empowered to acquire land. At present, conservation commissions in Massachusetts own over 34,000 acres of land.

In 1970 the Vermont legislature passed a comprehensive bill entitled "An Act to Create an Environmental Board and District Environmental Commissions" (10 V.S.A. 6001-91). This law, better known as Act 250, provides for a detailed review by District Environmental Commissions of all major and many minor developments within the state. There are nine local district commissions serving Vermont's fourteen counties. A nine-member state-level Environmental Board makes general policy decisions and hears appeals from decisions of the district commissions. The members of this board are appointed by the Governor with the advice and consent of the Senate. District commissions review applications on a site-by-site basis for such diverse developments as the construction of housing subdivisions, condominium complexes, shopping centers, highways, mobile home parks, camping areas, transmission lines, sanitary landfills, mining projects, etc. To obtain a permit for a development in Vermont, it must be demonstrated that a proposed project will not have an undue adverse impact on the environment. There are ten specific criteria in the law which must be satisfied before a permit may be granted. Act 250 requires that any development above an elevation of 2,500 feet may not be made without a permit. Only about three percent of Vermont's land is situated above that elevation. As all of Vermont's alpine and subalpine habitats occur well above this elevation, they are effectively protected by this law.

One criterion in Act 250 requires that a development will not cause an adverse impact on ". . . rare and irreplaceable natural areas. . ." nor will a permit be granted ". . . if it is demonstrated by any party opposing the applicant that a development or subdivision will destroy or significantly imperil necessary wildlife habitat or any endangered species . . .". The burden of proof for this requirement lies with the applicant. Many years ago a hydroelectric dam was built across a gorge of the Winooski River near the City of Burlington. In that gorge grew *Astragalus Robbinsii* (Oakes) Gray, an endemic species known only from this single station. Construction of the dam destroyed the habitat of this interesting species, thus causing its extinction. Shortly before the enactment of Vermont's Act 250, the last known Vermont station for *Hudsonia tomentosa* Nutt. was destroyed by a housing development. Today, with Act 250 in effect, a permit application for a development in Vermont which would endanger a rare species would be denied.

Other states in the region have laws which are perhaps equally as effective as those of Massachusetts and Vermont. In Maine, a law known as the Great Ponds Act (38 M.R.S.A. §380-424) and several wetlands laws (12 M.R.S.A. §4701-9 and 4751-8) serve to protect the inland waters of that state. Coastal wetlands in Maine are protected by a separate statute (38 M.R.S.A. §471-748). Rhode Island has a Coastal Resources Management Council Act (46 G.L.R.I. §23-1-16).

The Federal Coastal Zone Management Act of 1972 requires that state coastal management programs develop inventories of natural areas and provides for federal grants to establish estuarine sanctuaries.

A senate bill in Pennsylvania (No. 115, Session of 1975) will be, if eventually passed, one of the most all-encompassing statutes yet conceived for the protection of rare and endangered plant species. It provides for the establishment of wild plant sanctuaries, prohibits the sale or transportation of listed species, establishes a licensing procedure for "Wild Plant Conservationists," establishes a "Public Rare and Endangered Wild Plant Conservation Fund," and provides for the management of state parks and forest lands for rare and endangered wild plant areas. The bill also establishes an annual appropriation to support the program ($50,000 was suggested for the year 1975-1976).

Inventories of natural areas and lists of rare and endangered plants have been prepared for many of the northeastern states. These include New Hampshire (Lyon and Reiners, 1971; Hodgdon, 1973), Vermont (Vogelmann, 1964 and 1969; Countryman, 1971), Rhode Island (Seavey, 1975), Connecticut (Siccama et al., [1973?]), New Jersey (Fairbrothers and Hough, 1973) and Pennsylvania (Erdman and Wiegman, 1974).

Perhaps the most comprehensive natural area inventory yet prepared is that of the New England Natural Areas Project (NENRC, 1972). This project was undertaken by the

New England Natural Resources Center in 1970-1971 at a cost of approximately $175,000. It has been estimated that approximately 5,000 individuals were involved in developing this inventory, which includes selected archeological, geological, and biological sites and a category termed "esthetic visual sites." This assessment identified over 4,000 sites in the New England States. Maine has 2,025 sites, New Hampshire has 476, Vermont 517, Massachusetts 461, Connecticut 361, and Rhode Island 219. On a regional basis, the number of natural areas identified in the New England inventory is approximately two per 100 km^2. About 67 percent of these sites are in private ownership and ninety-two of them were identified as habitats of rare plants. This huge inventory, a computer-printed document approximately two feet thick, is essentially a compilation of raw data. Almost all areas proposed by the many authors of the inventory were included in the compilation. No attempt was made to evaluate the areas. However, concurrently with the preparation of the inventory, a committee of the New England Botanical Club developed a method for evaluation of natural areas on the basis of their botanical features. The work of the committee culminated in a document entitled "Guidelines and Criteria for the Evaluation of Natural Areas" (Countryman, 1972).

Inventories of natural areas and lists of rare and endangered plants provide guidance to legislative bodies and to the various commissions which act on applications for development. A published account of threatened species often provides a compelling argument for the denial of a permit for a development which would destroy a listed rare plant or its habitat. Similarly, applications to modify sites recognized in published accounts as natural areas may also justifiably be denied. In cases where development threatens plants or habitats not listed in existing inventories, permit-granting authorities may find guidance for decision in the guidelines and criteria developed by the New England Botanical Club.

LITERATURE CITED

Aldrich, J. W. 1975. Our wild flowers and a program to protect them. The Conservationist **29**(5): 23-29.

CLFNE (Conservation Law Foundation of New England). [1972?]. Identification, explanation, and analysis of natural area protection techniques. Conservation Law Foundation of New England; New England Natural Resources Center, Boston.

Countryman, W. D. 1971. Rare and uncommon Vermont flora. Pages 53-58 in Vermont interim land capability plan, Appendix D. Vermont State Planning Office, Montpelier.

_____, chairman. 1972. Criteria and guidelines for the evaluation of natural areas. Natural Areas Criteria Committee, New England Botanical Club, Cambridge. 11 pp.

Delury, G. E., ed. 1975. The world almanac and book of facts. Newspaper Enterprise Association, Inc., New York. 984 pp.

Ellis, R. J. and A. D. Dawson. 1973. Massachusetts Conservation Commission handbook. Massachusetts Association of Conservation Commissions, Boston. 92 pp.

Embleton, C. and C. A. M. King. 1968. Glacial and periglacial geomorphology. Edward Arnold, Ltd., Alva, Scotland. 608 pp.

Erdman, K. S. and P. G. Wiegman. 1974. Preliminary list of natural areas in Pennsylvania. Western Pennsylvania Conservancy, Pittsburgh. 106 pp.

Fairbrothers, D. E. and Mary Y. Hough. 1973. Rare or endangered vascular plants of New Jersey. Science Notes No. **14,** New Jersey State Museum, Trenton. 53 pp.

Fernald, M. L. 1950. Gray's manual of botany. Eighth edition. American Book Company, New York. 1,632 pp.

Hansen, D. L. and R. H. Goodno. 1975. Conservation commissions in Essex County, Massachusetts. Publication No. 109, Cooperative Extension Service, University of Massachusetts, Amherst.

Hodgdon, A. R. 1973. Endangered plants of New Hampshire. Society for the Protection of New Hampshire Forests, Forest Notes **114**:2-6.

Lyons, C. J. and W. A. Reiners. 1971. Natural areas of New Hampshire suitable for ecological research. Revised edition. Department of Biological Sciences, Dartmouth College, Hanover. 75 pp.

Moss, E., ed. 1976. Land use controls in New York State. Natural Resources Defense Council, The Dial Press-James Wade Books, New York. 368 pp.

NENRC (New England Natural Resources Center). 1972. New England natural areas project, phase I. New England Natural Resources Center, Boston. 40 pp.

Seavey, G. L. 1975. Rhode Island's coastal natural areas: priorities for protection and management. Marine Technical Report No. 43, Coastal Resources Center, University of Rhode Island, Kingston. 60 pp.

Seymour, F. C. 1969. The flora of New England. Charles E. Tuttle Company, Rutland, Vermont. 596 pp.

Siccama, T. G., W. A. Beales and J. E. Hibbard. [1973?]. Connecticut natural areas. Connecticut Forest and Park Association, Inc., East Hartford. 49 pp.

Vogelmann, H. W. 1964. Natural areas in Vermont. Vermont Agricultural Experiment Station, University of Vermont, Burlington. 29 pp.

_____. 1969. Vermont natural areas. Central Planning Office and Interagency Committee on Natural Resources, Montpelier. 30 pp.

THE SOUTHEASTERN UNITED STATES

James W. Hardin

*Department of Botany, North Carolina State University,
Raleigh, North Carolina 27607, U.S.A.*

The southeastern part of the United States comprises fourteen states. It is bounded by and includes Delaware, Maryland, West Virginia, Kentucky, Arkansas, and Louisiana. The other eight states are Virginia, North Carolina, South Carolina, Georgia, Florida, Alabama, Mississippi, and Tennessee. This area forms a fairly natural floristic region — much more natural than the old "Small's Manual" range implied. The vascular flora is estimated to include approximately 5,500 species. Based upon the 1975 Smithsonian list, 413 species, or 7.5 percent of the southeastern flora, are either recently extinct, endangered, or threatened. These 413 species represent about 20 percent of the extinct, endangered, or threatened species in the continental United States, including Alaska, again based upon the Smithsonian list. Of particular concern are the centers of endemism in the Southeast, namely the Southern Appalachian Mountains, the southeastern Coastal Plain including Florida, and specific unique habitats with high endemism such as limestone glades, granite flat-rocks, and shale barrens.

Since the Endangered Species Act of 1973, the Southeastern states have progressed in varying degrees toward developing programs to identify and protect these rare species, as indicated by Table I. Further comments will be based on this Table.

TABLE I.

STATUS OF STATE PROGRAMS WITHIN THE SOUTHEASTERN U.S.A.

	AL	AR	DE	FL	GA	KY	LA	MD	MS	NC	SC	TN	VA	WV
Organized committee	+	+		+	+	+	+	+		+				
Species list	p	+		+	p	p	p		p	+	p	p		p
Species accounts	p	+		p		p				+	p	p		p
Legislation														
Introduced	+									+				+
Passed					+			+			+			
Habitat Preservation		+	+	+	+			+	+	+	+	+		+
Research on species										+				
Propagation methods				+						+	+	+		

+ = completed   p = in preparation

AL = Alabama	GA = Georgia	MD = Maryland	SC = South Carolina
AR = Arkansas	KY = Kentucky	MS = Mississippi	TN = Tennessee
DE = Delaware	LA = Louisiana	NC = North Carolina	VA = Virginia
FL = Florida			WV = West Virginia

Organized committees — Most states have appointed committees which are charged with the responsibility of establishing official lists of rare and endangered species. These committees frequently include the most knowledgeable plant taxonomists in each of the states. Although there is no officially designated committee in West Virginia, activities there are coordinated through the Heritage Trust Program of the Department of Natural Resources. In Tennessee, there are several persons or groups working, unfortunately, more or less independently. We hope that eventually these people will be brought together to work as a unified group. Cooperation and coordination at the state level is imperative if there is to be an effective proposal submitted to the state legislature.

The "lead" state agency has not yet been designated in a few states. Where such have been named, the type of agency with the responsibility for endangered plant species varies tremendously from state to state. The departments which are concerned with this work include such agencies as Wildlife Resources, Natural Resources, Wildlife and Marine Resources, Fish and Wildlife, Heritage Programs, Forestry Commission, Game and Fresh Water Fish, etc. In about half of the states with designated lead agencies, both animals and plants are handled by the same office while other states have split the responsibility. For example, the Arkansas Department of Natural and Cultural Heritage is concerned with both; the Alabama Department of Conservation handles only the animals while the Department of Agriculture of that state is responsible for the plants; in Louisiana the State Forestry Commission supervises the plants and the Wildlife and Fisheries Commission administers the programs concerning the animals. Not enough time has yet elapsed to determine whether one situation is any better than another. The chief concern is to have the plants under the jurisdiction of the agency with the greatest botanical expertise, as well as the legal authority to protect plant species. This latter point is critical; for example, the Florida Game and Fresh Water Fish Commission has been in court to see if their authority extends to fresh water plants. No other state agency has a protective function assigned to them by the legislature. Protection of terrestrial plants will be an additional problem. I feel certain that other states will face this same type of dilemma when they reach the stage of implementing protection of plant species.

Regardless of which state agency ultimately is given jurisdiction, nearly every state is seriously hampered by lack of funding. The designated lead agencies have taken on this additional responsibility without sufficient appropriations to cover the increased scope of their activities. In many cases, the agency already handling fish and wildlife (meaning animals) has suddenly been given the responsibility for plants also, though the personnel is totally untrained in botany.

Species list — For the past three or four years, most of the southeastern states have been compiling lists of rare species. Three states — Arkansas, Florida, and North Carolina — are past the point of adding, deleting, or shifting species from one category of the list to another, and they can now move on to different aspects of the program. The other states, however, are still in the preliminary stage of compiling such lists. Some states are bogged down by lack of organization, as previously indicated. The Smithsonian list has been an important impetus and departure point, but most states are assigning species to categories solely on the basis of the status of the populations in the state. Though this is a logical step, it has led to some local disagreements concerning the inclusion of rare long-range disjuncts and rare peripheral species which undoubtedly both represent unique genotypes as well as pinpoint unique habitats which need to be preserved. North Carolina, for example, has included rare, peripheral species. Botanists concerned with the problem argue that since such species represent an integral part of the flora, the state, therefore, is the "keeper" of this one important segment of the entire range, which is probably a significant element of the total genetic diversity of the species. However, we have given the peripheral species a low priority since it will be more difficult to justify protection for some 300 species that are more or less common in adjacent states. The long-range disjuncts found in North Carolina, on the other hand, are given a much higher priority.

One problem with all lists is the lack of real documentation for the assignment of a species as either endangered or threatened. Most such assignments are based upon a subjective judgement by one or two knowledgeable individuals. In relatively few cases do we know the real extent of populations, numbers of individuals, reproductive capacity, and the nature of the immediate threat to continued existence. The need for baseline data to document these subjective decisions is obvious.

Pressure from various groups of interested parties to include the popular, showy, or dramatic plants regardless of their relative rarity may damage the validity of the lists and thereby seriously jeopardize their official acceptance. The zeal to protect all trilliums, or all orchids, or all carnivorous species, for example, must be moderated by careful botanical evaluation of each species.

Although species lists have been completed in Arkansas, Florida, and North Carolina, only in Florida has the lead agency indicated that the prepared list will be accepted. In North Carolina, the list prepared by the committee established by the State Museum of Natural History still has to be formally adopted by the Wildlife Resources Commission. Thus, in the three years since enactment of the federal Endangered Species Act, none of the southeastern states yet has an "official" list of threatened and endangered plant species.

No attempt has been made to include the lower plants (algae, fungi, lichens, or bryophytes) in the lists. This should be done eventually since such plants are undoubtedly indicators of unique habitats which should be preserved.

Species accounts — Species accounts, including descriptions, habitats, general distributions, and illustrations, have been published for Arkansas and North Carolina, and are in various stages of preparation for six other states. This information is needed by numerous groups and agencies right now.

Legislation — Of the fourteen southeastern states, only three have recently enacted laws protecting threatened and endangered plants: The Georgia Wild Flower Preservation Act of 1973, the South Carolina Nongame Endangered Species Conservation Act of 1974, and the Maryland 1975 Nongame and Endangered Species Conservation Act. Legislation was introduced in the 1975 sessions of the Alabama and North Carolina legislatures, but did not progress beyond committee hearings. It is hoped that these states, plus West Virginia, will act on similar bills next year.

Many years ago, North Carolina enacted legislation which protected certain showy species and Venus' fly-trap, but as far as I am aware, these laws have never been enforced. The law is designed to protect Venus' fly-trap from commercial exploitation of its natural populations. However, the majority of the plants purchased throughout the United States on the retail market have been collected illegally from natural populations by local collectors in the state.

Habitat preservation — Obviously, the most successful means of protecting rare species is through habitat preservation and management. Ten of the fourteen southeastern states have natural area or natural heritage programs. Six of these states cooperate with The Nature Conservancy, although these programs are largely in their infancy. The most effective program thus far is the Florida Endangered Lands Program, funded by a success-ful $200 million bond issue which has been used for the purchase of several fine tracts of land. North Carolina started a Natural Areas Program several years ago, and has just recently designated its first natural area.

Communication within the entire Southeast will be helped to a large extent by the efforts of the Conservation Committee of the Association of Southeastern Biologists. This committee has been developing an environmental alert system containing a network of state correspondents in key areas to promote identification of critical habitats and rapid response from the scientific community regarding threats to specific habitats.

One serious question which has not been answered is what is to be done with specific tracts of land when they are finally acquired as natural areas. How are such tracts to be managed? It has become quite clear in North Carolina, for example, that more than 50 percent of the endangered and threatened species exist not in the mature forests but rather in earlier successional stages which will have to be maintained by periodic cutting or burning. But, what time of the year should these areas be burned, and how often? Management practices for various habitats and communities will have to be established.

Research on species — To date, very few species have been studied in depth in terms of detailed environmental characterization or ecological life history, including reproductive strategy and capacity, seed germination, seedling establishment, etc. Such studies would make ideal thesis problems for students in ecology and systematics. They would provide the type of information needed to determine, for example, exactly why a species is rare, what sites and populations should be preserved, the size of population necessary for such preservation, and the management practices required to maintain these populations.

The Highlands Biological Station in the mountains of North Carolina is at present coordinating and supporting research on threatened and endangered animal and plant species of the Southeast. These activities are being conducted with the cooperation of the Southeastern Forest Experiment Station of the U.S. Forest Service. The initial emphasis is on endemics, and qualified pre- and post-doctoral investigators are encouraged to submit proposals for support of such research.

Propagation methods — Limited work is being done on methods of propagation, and only a very few attempts have been made as yet to reintroduce any native plants back into natural habitats. Some propagation has been attempted at the Horticultural Gardens at Clemson University. The North Carolina Botanical Garden has an ongoing program to develop propagation methods for all the native flora, and particularly for that state's rare species. The most extensive program of this type at present is that sponsored by the Tennessee Valley Authority. This program is now in its third year and includes publication of the first "Rare Endangered Threatened Flora Propagation Newsletter" published in the fall of 1975. This program should be expanded with additional funding, and other botanical gardens and horticulture departments would do well to follow its example and begin similar research.

Problems — Though the problems of dealing with rare and endangered species of the Southeast are not unique, I will attempt to summarize the major ones which must be solved on the state level before the programs can become effective:

1. *Lack of communication, cooperation, and coordination of efforts among all interested groups and government agencies:* There still exists, in some states, a proliferation of species lists, and a "tug-of-war" and petty bickering between proponents of certain lists and over the status of particular species. Improved communication and cooperation between professional taxonomists, knowledgeable amateur botanists, local wild flower or conservation societies, and the personnel of various state agencies will have to be established, lest the above-described infighting either seriously delays or even destroys some state programs.

An example of this lack of cooperation and coordination of efforts is the situation in which one state department charged with the responsibility to protect endangered species has absolutely no legal power to deal with the highway department concerning construction of roads through areas where endangered plants and animals exist. In another case, a state department of agriculture publishes its farmer's market bulletin advertising the sale of a certain endangered species, while just down the hall is the office in which the staff is desperately working to save this particular species.

2. *General public apathy:* As one respondent to my questionnaire so aptly expressed this major problem, we are dealing with "an unenlightened public, an indifferent press, and a rather venal body politic." If there were more public awareness as to what plants are endangered, and if the general public were given cogent arguments for the plants' protection, they would be more apt to agree to and work for their survival. "Why save some little inconspicuous plant that only grows in one or two bogs, and which is known by only a few experts?" This is a valid question, and will have to be answered to the satisfaction of the general public if we are going to make much progress. I think it is rather difficult for most people to become excited over a plant or animal they do not know or have never heard about. We need a massive, well-illustrated, education program to help eliminate this general apathy and lack of awareness.

Another problem we are confronted with is that of increased population in certain critical areas. Unfortunately, most of the vacationers and retirees who have flocked to Florida and to the Southern Appalachians in recent years are most apathetic about protecting the very environmental features that probably brought them to these areas in the first place.

3. *Direct opposition:* We can already begin to sense a growing opposition from individuals, major industrial landowners, and various professionals in the land-use business. Some of this early opposition could be eliminated or diminished by effective educational programs, and a genuine attempt on our part to work with these people rather than against them. For instance, foresters at present are afraid that someday they might not be able to harvest timber due to the presence of a few endangered herbs on the forest floor. Actually, cutting the trees might be the single management plan necessary to save those herbaceous species. Again, the solution boils down to formulating an effective educational program for all facets of society. We must start cooperative work on management plans which would be acceptable to all parties concerned.

4. *Lack of funding:* This is the basic perennial problem. If states are going to have an effective program of species protection and management of natural areas, increased funding is an absolute necessity.

ACKNOWLEDGEMENTS

I am indebted to twenty-four individuals who took the time and effort to answer my questionnaire or to write letters detailing the status of the program in their states. Their remarks have been the basis for this status report, and I deeply appreciate their cooperation.

THE MIDWESTERN UNITED STATES

Robert H. Mohlenbrock

Southern Illinois University, Carbondale, Illinois 62901, U.S.A.

In this paper, I will discuss the status of legal protection for threatened and endangered species of plants which occur in the midwestern states. Those states which are included herein are Ohio, Michigan, Indiana, Wisconsin, Illinois, Minnesota, Iowa, Missouri, North Dakota, South Dakota, Nebraska and Kansas. Information was gathered from the Department of Conservation or equivalent agency of each state. Additional information was obtained from private organizations and individuals.

No state laws exist in Ohio that specifically protect plants, although the Department of Natural Resources, Division of Wildlife enforces state laws protecting fish and wildlife. The future of endangered plant legislation in Ohio is uncertain. Though it is generally considered desirable, legislation will probably not be proposed until completion of an annotated list of threatened and endangered plants by the Ohio Biological Survey. This list is scheduled for completion during the current calendar year.

Ohio's Natural Areas Act provides a mechanism for the preservation of unique habitats through the creation and administration of a statewide system of nature preserves. Public hearings are required prior to the adoption of endangered wildlife regulations in Ohio. The staffs of universities, conservation organizations, and museums also are consulted for input into any endangered species programs considered by the state.

The legislature in the State of Michigan passed an act in 1962 which related to the disposition of certain native wild flowers. This act made it illegal for the public to remove these plant materials from lands under control of the Department of Natural Resources.

There is no specific division of the Michigan Department of Natural Resources which is charged with the responsibility of coordinating programs related to unique botanical communities. However, for more than twenty years, the Department has worked closely with the Michigan Natural Areas Council in identifying and managing lands which support unique plant communities. The Natural Areas Council includes many highly-trained experts in fields of natural science.

In 1972, legislation was adopted which established a procedure whereby unique natural areas could be dedicated and managed as preserves. Previous to this, the Michigan Natural Resources Commission had dedicated over 70,000 acres of publicly owned land in various natural area categories. An organization known as the Eastern Michigan Nature Association also has acquired a considerable number of acres which are being managed as nature preserves.

The State of Indiana does not have a law which provides protection for rare and endangered plant species. If such legislation was enacted, it is assumed that the Indiana Department of Natural Resources would be the administering agency. The Indiana Nature Preserve Act, however, has made it possible to designate all or parts of tracts administered by the Division of State Parks, Division of Nature Preserves, State Forests, and State Fish and Wildlife Areas as nature preserves. The rules and regulations for State Nature Preserves prohibit the reintroduction of plants or animals, endangered or otherwise, into suitable natural areas.

A Flora Indiana Program was authorized in 1975 by the Indiana Academy of Science, and as a result, a list of rare and endangered plants in Indiana has been prepared. The Indiana Natural Resources Commission is authorized to dedicate state nature preserves. One hundred seventy-seven natural areas totaling 17,954 acres are now included in the nature preserve system.

Illinois has an Endangered Species Act of 1972 which is administered by the Department of Conservation. However, this act pertains to animals only. Several wild flowers are "protected" under a Wild Flowers Preservation Act which was enacted during the 1920's.

A Nature Preserves Commission, whose members are appointed by the governor, was authorized by the state legislature during the 1960's. This commission is charged with designating nature preserves within the state. There are about forty such preserves which already have been set aside as a result. The commission also is in the process of completing a preliminary list of rare and endangered species of vascular plants in the state, publication of which is expected during this year.

The Illinois Department of Conservation is funding a three-year inventory designed to identify and characterize all natural areas in the State.

Two projects of interest should be mentioned here. The St. Louis District, U.S. Army Corps of Engineers, has prepared a list of possible threatened and endangered species of plants known to occur in its district, along with the known localities of each of these. The Shawnee National Forest is preparing a study on the threatened and endangered species of the forest. This project includes a life history of each of the involved species and a detailed map showing precise locations where such species can be found.

In September, 1975, the State of Wisconsin approved an endangered species law which conforms to federal legislation, though it is inadequate insofar as it concerns fish and wildlife only. A new endangered species law has been introduced into the State Legislature, but it has not yet been enacted and preliminary hearings are still being held on it. If finally approved, this legislation will follow the federal law providing the same protective status to plants as now exists for animals and establishing a threatened species list. The law will be administered by the Bureau of Fish and Wildlife Management of the State's Department of Natural Resources, through an endangered species coordinator. The Wisconsin Department of Natural Resources also is responsible for preservation of natural areas.

The State of Minnesota enacted a law entitled Protection of Threatened and Endangered Species, which conforms to the Federal law. Programs under this legislation are administered by the Commissioner of Natural Resources. Very little direct action has been taken in developing a program, but thus far the Division of Fish and Wildlife has handled matters pertaining to both plants and animals. A beautiful and comprehensive publication has been issued which reviews the status of uncommon species of plants and animals. Private groups and organizations were asked to comment on the draft proposal of this document.

Late in 1975, the Iowa legislature passed an act "relating to the conservation management and protection of fish, plant life, and wildlife species endangered or threatened with extinction and prescribing penalties." A list of endangered plant species is currently being prepared in consultation with university communities and members of various state groups. The law has given the Conservation Commission authority to "perform those acts necessary for the conservation, protection, restoration and propagation of endangered and threatened species." Both plant and animal programs are administered by the same section. Status reports are anticipated this year on several endangered or threatened organisms, including the monkshood and the pink lady's-slipper orchid. The state's preserve system will play an important role in the endangered species program. These areas will be surveyed to determine what populations exist so that maximum protection can be provided. No propagation or reintroduction studies have been done or are being contemplated at this time.

Missouri's endangered species statute is limited to animals, but an article in the Missouri Constitution stipulates the authority of the Conservation Commission in broad terms which include "forestry and all wildlife resources of the state." The Department of Conservation has been designated to handle matters related to endangered plants. One of the efforts of the Department of Conservation has been the publication of a bulletin in 1974 on Rare and Endangered Species of Missouri. Animals and several plant groups, including Bryophytes and Pteridophytes, are listed. Many agencies, private organizations, and individuals aided in the compilation of material for this bulletin.

The Missouri Department of Conservation also operates a natural areas program which provides for preservation of unique habitats.

In South Dakota, legislation conforming to Federal regulations is currently being drafted to protect, fund, and establish programs for all threatened and endangered species. This legislation would enable the Department of Agriculture and the Department of Game, Fish and Parks to administer endangered species programs for plants and animals, respectively.

An Endangered Species Committee has been established in South Dakota. The initial objectives are (1) to instigate a coordinated effort to obtain fundamental, baseline information for all forms of endangered species; (2) to develop a practical and efficient approach for the formation of a multidiscipline field study team which will serve as a model in conducting future baseline endangered species work; and (3) to develop a listing of possible sources for funds to support the studies needed. A preliminary draft of the Endangered Plants of South Dakota has been prepared.

South Dakota also is active in promoting natural prairie vegetation along major highways in the state. The Department of Transportation and Highways has a continuing program of seeding at least five native grass species along all new highways.

In 1975, the State of Nebraska passed a Nongame and Endangered Species Conservation Act. Its key sections pertaining to plants are important enough to warrant their being quoted here:

> "Endangered species shall mean any species of wildlife or wild plants whose continued existence as a viable component of the wild fauna or flora of the state is determined to be in jeopardy, or any species of wildlife or wild plants which meets the criteria of the Endangered Species Act.

The Legislature finds and declares:

    (1)   That it is the policy of this state to conserve species of wildlife for human enjoyment, for scientific purposes, and to insure their perpetuation as viable components of their ecosystems;

    (2)   That species of wildlife and wild plants normally occurring within this state which may be found to be threatened or endangered within this state shall be accorded such protection as is necessary to maintain and enhance their numbers;

    (3)   That this state shall assist in the protection of species of wildlife and wild plants which are determined to be threatened or endangered elsewhere pursuant to the Endangered Species Act by prohibiting the taking, possession, transportation, exportation from this state, processing, sale or offer for sale, or shipment within this state of such endangered species and by carefully regulating such activities with regard to such threatened species.

In addition to the species determined to be endangered or threatened pursuant to the Endangered Species Act, the commission shall by regulation determine whether any species of wildlife or wild plants normally occurring within this state is an endangered or threatened species as a result of any of the following factors:

    (a)   The present or threatened destruction, modification, or curtailment of its habitat or range;

    (b)   Overutilization for commercial, sporting, scientific, educational, or other purposes;

    (c)   Disease or predation;

    (d)   The inadequacy of existing regulatory mechanisms; or

    (e)   Other natural or man-made factors affecting its continued existence within this state. Except with respect to species of wildlife or wild plants determined to be endangered or threatened species under the provisions of sub-section (1) of this section, the commission may not add a species to nor remove a species from any list published pursuant to subsection (5) of this section unless the commission has first:

    (a)   Published a public notice of such proposed action;

    (b)   Notified the Governor of any state sharing a common border with this state, in which the subject species is known to occur, that such action is being proposed; and

    (c)   Allowed at least thirty days following publication for comment from the public and other interested parties. The commission shall issue regulations containing a list of all species of wildlife and wild plants normally occurring within this state which it determines to be endangered or threatened species and a list of all such species. Each list shall refer to the species contained therein by scientific and common name or names, if any, and shall specify with respect to each such species over what portion of its range it is endangered or threatened.

Except with respect to species of wildlife or wild plants determined to be endangered or threatened pursuant to the Endangered Species Act, the commission shall, upon the petition of an interested person, conduct a review of any listed or unlisted species proposed to be removed from or added to the lists published pursuant to subsection (5) of this section, but only if the commission publishes a public notice that such person has presented substantial evidence which warrants such a review. With respect to any endangered species of wild plants, it shall be unlawful for any person subject to the jurisdiction of this state to:

(a)   Export any such species from this state;

(b)   Possess, process, sell or offer for sale, deliver, carry, transport, or ship, by any means whatsoever, any such species; or

(c)   Violate any regulation pertaining to such species or to any threatened species of wild plants listed pursuant to this section and promulgated by the commission pursuant to this act.

The commission shall establish such programs, including acquisition of land or aquatic habitat or interests therein, as are necessary for the conservation of nongame, threatened, or endangered species of wildlife or wild plants.".

The State of Kansas has a statute on nongame animals which designates the Forestry, Fish, and Game Commission to implement the act. This Commission is effectively establishing advisory boards and committees to assist in complying with the law. The director of the State Biological Survey of Kansas asserts that, as the federal regulations on plants become more clear, it will be easy to incorporate plants into the legislation.

According to this same source, Kansas is fundamentally interested in preserving various types of habitat as a means of preserving critical elements of fauna and flora. The Advisory Board of the Natural and Scientific Area Preserves is actively involved in an inventory of existing areas and will make recommendations on areas in need of preservation.

North Dakota has not yet enacted any legislation to protect its endangered plant species.

In summary, only three midwestern states — Iowa, Minnesota, and Nebraska — have enacted statutes which protect threatened and endangered species of plants in accordance with Federal legislation. Most of the midwestern states either have prepared lists of threatened and endangered species of plants or are in the process of doing so. Efforts have been made in several states to establish some form of a nature preserves system.

What is needed is a concentrated effort by botanists to assist each of the state governments in drawing up a list of threatened and endangered species. The states must then work for the passage of legislation to protect these species. Cooperation between private and public agencies is essential, as is cooperation between neighboring states.

THE NORTHWESTERN UNITED STATES

Kenton L. Chambers

*Department of Botany, Oregon State University,
Corvallis, Oregon 97331, U.S.A.*

The Pacific Northwest region of the United States has only recently developed widespread interest in the conservation of threatened and endangered plants. Until about three years ago, activities in this section of the country on behalf of rare plants were sparse and ineffective; nevertheless, those who were concerned with the situation did discern a slow but definite evolution of public attitudes in regard to environmental problems.

It was during this comparatively quiescent period that organizations such as The Nature Conservancy, and several Federal committees concerned with Research Natural Areas, took the decisive action of setting aside, as parts of their natural preserves, a few significant habitats for rare plants. In spite of such official steps, however, there still was public apathy regarding plants themselves. It was difficult to achieve even such minimum gains as publicity for the obviously endangered species, or for the habitat destruction which threatened others. Botanists, aware of the growing concern for endangered animal species, were so discouraged over the dismal prospect that the best they felt could be hoped for was the conservation of samples of significant terrestrial and aquatic habitats, some of which might, by design, contain adequate populations of endangered plant species.

But in 1973 the tide began to turn, and since that time there has been steady progress in the Pacific Northwest. The plant conservation activities of the Federal government, involving the Smithsonian Institution and the Fish and Wildlife Service, as well as other agencies, are all familiar to most readers of this review, and therefore they need not be detailed here.

The preparation and release of the familiar Smithsonian Institution report (1975), and the subsequent publishing of the species list in the Federal Register by the Fish and Wildlife Service, have both been, beyond doubt, the major stimuli toward the greatly increased public interest in endangered plants in the Pacific Northwest, as well as in other parts of the country. Professional plant taxonomists, including this writer, can testify to the concern expressed via voluminous correspondence, personal inquiries, agency contacts, requests for information, and offers of participation by organizations and individuals which the Smithsonian Institution report engendered. Botanical personnel in educational and research institutions, as well as governmental agencies, are suddenly under increased pressure to provide answers for this new public awareness. Yet, for much of the region under review, there still exist only very incomplete data on the occurrence and distribution of most threatened and endangered taxa. Such concerns are by no means unique to this region, of course, and on a larger scale there may ultimately develop improved attention to plant taxonomy itself, and increased support for taxonomic research of all types.

As in other parts of the country, current efforts in the Pacific Northwest on behalf of threatened and endangered plant taxa are conditioned by specific characteristics of the region's flora. The states of Washington, Oregon, Idaho, and Montana contain large expanses of territory that are inaccessible, and for that reason, are poorly explored by taxonomists. In even the best herbaria of the four states, collections of many interesting species are spotty and incomplete. There is an almost total lack of local and county floras and checklists throughout the region — in sharp contrast to California and many eastern states. The area contains environmental extremes ranging from desert to alpine conditions with much edaphic and climatic diversity. Hence endemism, often combined with disjunct occurrences, is favored. Because of inadequate herbarium records, however, many species may be more common than the data in published records would suggest. On the other hand, herbarium data may lead to an overestimation of occurrence in those large areas where severe habitat modifications — agriculture, urbanization, range improvement, forest management practices, brush control, grazing, etc. — have rendered obsolete the localities cited by collectors of many years ago. In all likelihood, field checking of the majority of presently-listed threatened and endangered species will be required, although significant populations in remote sites still may be overlooked.

There is one outstanding recent floristic work (Hitchcock et al., 1955-1969) and a companion one-volume manual (Hitchcock and Cronquist, 1973) that encompass most of the region. Though the southern portions of Oregon and Idaho, as well as the plains of Montana, are excluded, these two books provide a uniform taxonomic treatment of the Pacific Northwest. The advantage this offers botanists in communicating both among themselves and with the public, is important during the developmental phase of programs on threatened and endangered species. Wrangles over nomenclature have mostly been avoided up to now, although there probably will be future disagreements over taxa that some experts may wish to resurrect from synonymy, or recognized taxa whose validity may be questioned in more recent monographs.

GENERAL STATUS OF PROGRAMS IN THE PACIFIC NORTHWEST

This region, on the whole, has been rather slow in organizing botanical programs aimed at the Endangered Species Act of 1973, the Smithsonian Institution's preliminary lists of threatened and endangered species, or the Fish and Wildlife Service's call for state assistance in implementing the 1973 law. Activities that were under way prior to the recent burst of national concern were mostly state-oriented, and dealt primarily with either habitat preservation or with the preparation of initial lists of putatively rare taxa. Important regional efforts existed in Federal interagency committees such as the Pacific Northwest Natural Area Committee, comprising representatives of the Forest Service, Bureau of Land Management, Bureau of Sport Fisheries and Wildlife, and National Park Service. By 1975 there were sixty Federal Research Natural Areas in Oregon and Washington (Franklin et al., 1972; Dyrness et al., 1975). The Nature Conservancy, which is also affiliated with the above Committee, has about twenty preserves either already established or in the process of establishment in these two states. The concept that threatened or endangered plant species might gain protection through Research Natural Areas was brought out frequently in the Committee's meetings. However, this has remained an incidental consideration because natural area needs have mostly been defined as aggregations of "cells" — ecosystems, communities, habitats, or endangered animal species. Nonetheless, one cannot overlook the fact that botanical data tend to predominate in defining natural area needs, and that preservation of habitats was emphasized as the conservation measure of choice in the report by the Smithsonian Institution (1975).

On a region-wide basis, no steps have yet been taken to list rare, threatened, or endangered plant taxa. All such efforts have been carried on within individual states, by professional botanists or by groups of amateurs and professionals working together. The state lists are still tentative and undergoing revision, but work on them has been greatly stimulated by the Smithsonian Institution's publication. Disagreements over concepts, terms, and criteria for rare plants, which were, initially, so troublesome to state workers, now seem to have been resolved.

Several significant points have become apparent during the compiling of state lists. The task of assembling fully satisfactory data for all the taxa seems too great to be accomplished by other than a long-term effort. The preferred approach, therefore, has been to identify, within each state, as many as possible of the taxa whose status is defined as "rare," "unusual," or "of special interest." This includes disjunct populations and populations which are marginal to the range of a species. Further research on selected species is necessary to confirm which ones are threatened or endangered according to the universally accepted criteria. Emphasis on the states' geographical boundaries merely concedes the fact of political independence of laws and programs in adjacent states, and admits to the frequent intrastate focus of activities that are state-sponsored and supported by public funds. Moreover, provincial viewpoints sometimes appear regarding the very large Federal ownership of lands in the western states. In some cases there is a tendency to place the burden for species conservation on the Federal landowners, rather than allow state and private lands to be "locked up" for the benefit of endangered organisms.

Each of the four states of this region has assigned to a particular state agency the responsibility of coordinator for threatened and endangered plants. Only in Oregon, however, is there a law actually "on the books" (titled, "Wild Flowers") that relates to plant conservation, but the law does not conform fully to Federal legislation. Responsible officials in all four states have willingly cooperated with the committees or task forces working on rare plants, and the possibilities now seem good that proposals for conforming laws will be submitted to the next legislature in one or more states.

Washington

Activities aimed at identifying and protecting the threatened and endangered plants of Washington State have been scattered and slow to develop, but it now appears that there is a good chance for statewide coordination through a recently appointed committee. Up to now, conservation of native plants has been the concern of the federally-sponsored Pacific Northwest Natural Area Committee, and the Washington State Natural Preserves Advisory Committee, which was established by state law in 1972 (Dyrness et al., 1975). As mentioned earlier, the Federal Research Natural Areas program involves both Washington and Oregon. In the state government, responsibility for natural areas lies with the Department of Natural Resources, and this office was also recently made the liaison agency for plants under the Endangered Species Act of 1973.

An interagency Task Force has been appointed to advise the Department of Natural Resources in the preparation of an illustrated publication on threatened and endangered plants. Funding for this project has yet to be granted, however, and the Department also has not yet been given a mandate to prepare conforming legislation which might make the state eligible for Federal support. Washington State does have laws concerning endangered animal species, with the administrative responsibility residing in the Department of Wildlife.

An unpublished partial inventory of "rare, threatened, and unique plants of Washington," was compiled by Dr. A.R. Kruckeberg in 1974 for a draft report on Research Natural Area needs. In the final report (Dyrness et al., 1975) the plant species are arranged in separate lists for the six physiographic provinces in the state and are referred to as "vascular plants of special interest."

Further progress in Washington appears to depend on successful coordination through a committee at the state level, the development of a broadened base of citizen input and support, and sufficient funding to carry out the planned program. It is encouraging to note that a Washington Native Plant Society, formed in 1976, is working actively on behalf of plant conservation.

Oregon

Like other northwestern states, Oregon contains extensive Federal lands, the future management of which will have major effects on numerous species of rare plants. The recently completed survey of needs for Research Natural Areas projected a total of some 376 areas for Oregon and Washington combined. State and private lands are also significant, and where there are natural areas worth setting aside from those sources, data on threatened and endangered plants are taken as important criteria. Oregon's state law regarding natural areas is similar to Washington's, involving a State Natural Area Preserves Advisory Committee (Juday, 1975). Preserves to be recommended by this committee in the future may serve to protect endangered plant species, although there are no present plans to use the system explicitly for that purpose. An inventory of natural areas on private lands in the state is now being prepared by The Nature Conservancy through its Oregon Natural Heritage Program which is funded by the state's Land Conservation and Development Commission.

Late in 1975 an Oregon Rare and Endangered Plant Task Force was formed under the leadership of Mrs. A.C. Siddall, a member of the State Natural Area Preserves Advisory Committee. This group hopes to compile Oregon lists of rare, threatened, and endangered plants, to identify those taxa most in need of protection, and to draft and recommend state legislation conforming to the Federal statute, Public Law 93-205. Oregon's present law on the protection of wild flowers places administrative responsibility in the State Department of Agriculture, which also is the Governor's designated liaison agency for threatened and endangered plants. In addition to being poorly enforced, this statute is inadequate and should be amended or replaced. Officials of the State Department of Agriculture have offered the task force full cooperation in advising on the form of legislation to be submitted. It remains to be seen whether the proposed bill will gain the support of Oregon's legislators during the 1977 session.

In 1974 Mrs. Siddall, in collaboration with this writer, prepared a list of the "rare, threatened, and unique plants of Oregon." This report was published by Dyrness et al. (1975) in the same format as the Washington list by Dr. Kruckeberg. A major effort to improve and refine this list is now under way, and is being coordinated by the Oregon Rare and Endangered Plant Task Force. In a cooperative effort, amateur and professional botanists throughout the state are assembling data on a select list of candidate species. At a conference in March, 1976, attended by 90 persons, a list of about 600 plants was reviewed. Following field checking during the summer, a first attempt at an "official" Oregon list of threatened and endangered plants will be written. The state has allocated no funds for the task force, however, and its work is on a voluntary basis, except for some financial help from the U.S. Fish and Wildlife Service for secretarial and office expenses.

Idaho

A program to develop a statewide system of Research Natural Areas in Idaho was begun in 1974 at a workshop held under the auspices of the University of Idaho's College of Forestry, Wildlife, and Range Science. Patterned after the workshop of the previous year, which discussed area needs in Oregon and Washington, the Idaho session involved 50 people representing governmental agencies, academic institutions, industry, and other private groups. Although Idaho has no legal provision for an official advisory committee, a formal organization was established in 1975. It is known as the Idaho Natural Areas Council, and consists of a Coordinating Committee and several technical committees.

The report issued after the 1974 workshop (Wellner and Johnson, 1974) contains two lists of plant species compiled by participants in a working group for rare plants. One list, "Rare and Endangered Plants of Idaho," by Douglass Henderson, contains 107 species which the group felt were in need of some degree of protection. Two items on the list were proposed as endangered species. A second list, "Uncommon Plants of Idaho," was based on literature, herbarium records, and distribution maps prepared by F.D. Johnson and R.W. Steele. Disjunct populations and outliers (those at the edge of a species' range) were included, as well as endemics. At succeeding workshops in 1975 and 1976, the lists were modified and refined by new field observations, and a third species was added to the state's endangered list.

Recently, the Governor's office has shown an interest in the work of the Idaho Natural Areas Council, and the State Planning Department made a grant in support of the 1976 workshop. No steps have been taken yet to develop legislation for threatened and endangered plants, however. Because much of the land in Idaho is under Federal ownership, there may be a reduced impetus for the state to promote plant preservation, per se. Under some conditions, Research Natural Areas can be effective in bringing rare plant species under protection. This is shown by two newly-designated areas in the Clearwater National Forest which contain clusters of woody and herbaceous species that are disjunct from Coast Range forests far to the west.

Montana

The problem of conserving threatened and endangered plants in Montana appears to be small compared to other areas of the Pacific Northwest, since only ten species are cited in the Smithsonian Institution's list for the state (1975). The State Government shows little interest in endangered plants, the Governor having designated the State Weed Control Supervisor, in the Department of Agriculture, as liaison officer on this subject. No state legislation is in prospect. Nonetheless, there are several botanists in Montana with strong interests in the native flora and a desire to see some program develop that would call attention to the state's endangered species. Dr. John Rumley, of Montana State University, expects to complete a list of threatened and endangered species by the end of 1976. This list will then be made available to state agencies, land planning groups, the Montana Natural Areas Committee, Federal research biologists, and other interested persons and organizations. Botanists at the other state and Federal laboratories are doing field studies from which information on rare plants will be gained. However, even the alpine areas, which are of greatest interest, have been incompletely surveyed up to now.

A coordinated planning effort for natural areas in the state was begun in 1974. The main impetus for this movement came from three professional societies — the Society of American Foresters, the Society for Range Management, and the Soil Conservation Society of America. A Montana Natural Area Committee was formed during a planning workshop (Schmidt and Dufour, 1975). This Committee held further workshops to inventory the natural area needs for the state, and as was the case in Idaho, a number of working groups were formed as a result. No group was assigned to rare or endangered plants, however, presumably because these were deemed of too little importance to natural area considerations in Montana. Official state sanction for a natural areas inventory came in 1974, through the Montana Natural Areas Act. Although only a few natural areas have been formally designated to date, many excellent candidate areas exist from which future choices can be made.

LITERATURE CITED

Dyrness, C. T., J. F. Franklin, C. Maser, S. A. Cook, J. D. Hall and G. Faxon. 1975. Research natural area needs in the Pacific Northwest. A contribution to land-use planning. USDA Forest Service General Technical Report PNW-38. 231 pp.

Franklin, J. F., F. C. Hall, C. T. Dyrness and C. Maser. 1972. Federal research natural areas in Oregon and Washington. A guidebook for scientists and educators. USDA Forest Service, Pacific Northwest Forest and Range Experiment Station, Portland, Oregon. 498 pp.

Hitchcock, C. L., A. Cronquist, M. Ownbey and J. W. Thompson. 1955-1969. Vascular plants of the Pacific Northwest. 5 vols., University of Washington Press, Seattle.

Hitchcock, C. L. and A. Cronquist. 1973. Flora of the Pacific Northwest. An illustrated manual. University of Washington Press, Seattle. 730 pp.

Juday, G. P. 1975. Oregon's natural area preserves program. Oregon State Land Board, Natural Area Preserves Advisory Committee, Salem, Oregon. 64 pp.

Schmidt, W. C. and W. P. Dufour. 1975. Building a natural area system for Montana. Western Wildlands, winter issue, 10 pp. unnumbered.

Smithsonian Institution. 1975. Report on the endangered and threatened plant species of the United States. U. S. Congress, Committee on Merchant Marine and Fisheries, Serial No. 94-A, U. S. Government Printing Office, Washington. 200 pp.

Wellner, C. A. and F. D. Johnson, eds. 1974. Research natural area needs in Idaho. A first estimate. College of Forestry, Wildlife and Range Sciences, University of Idaho, Moscow. 179 pp.

THE WESTERN UNITED STATES

James L. Reveal

Department of Botany, University of Maryland,
College Park, Maryland 20742, U.S.A.

The concept of protecting and maintaining endangered and threatened species of vascular plants is simple, but the reality of attempting to bring it to fruition is so tortuous and so politically infected with bureaucratic "red tape" that frustration is often the only reality known.

During the past few years, I have been working within the framework of the Endangered Species Act of 1973 at the federal level, initially in playing a minor role regarding Section 12 of the Act, which called for the Secretary of the Smithsonian Institution to report to Congress within one year on the "species of plants which are now or *may become* endangered or threatened." This was followed by my helping to prepare the so-called "Smithsonian Report (1975)." Recently I have become involved in the problems of implementation of the 1973 Act at local and state levels. It is in this connection that I have been asked to comment on the existing laws and present policies regarding endangered plants for five western states of the United States. These five states are California, Nevada, Utah, Colorado and Wyoming. According to the Smithsonian Institution list (1975) of endangered and threatened vascular plants, these five states harbor nearly one-third of all such plants for the continental United States. California has the most species listed (648 taxa), followed by Utah (157), Nevada (127), Colorado (40), and Wyoming (21). A new list is expected to be released shortly though the percentages and rankings of these states will probably not be altered.

In order to determine the laws and policies of the above-mentioned states, I wrote to each of their agencies responsible for endangered and threatened plants. In all cases, I received an answer from the state. In addition, I also wrote letters to various members of botanical communities in each of the states, and taxonomists in all states except Utah responded. I am grateful to the many individuals who have provided information for this report.

STATE LAWS AND POLICIES

Only two of the five western states have either directly or indirectly provided statutory provisions for endangered or threatened plant species. The Attorney General of California, through a Memorandum of Law, has demonstrated that his state has the legal authority to comply with the federal Endangered Species Act of 1973. The Attorney General has stated that in his opinion plants are covered under the present state law, along with "other invertebrates." In Nevada, a 1969 Act provides for the protection of endangered species, and while this act (Nevada Revised Statutes 527.270) states that any "species declared to be threatened with extinction shall be placed on the list of fully protected species," no plant has ever been listed by the State Forest Fire Warden, who is the official responsible for such action.

The remaining three states have no existing statutory provisions to protect endangered or threatened plants. Utah and Wyoming do not anticipate enactment of such state laws in the near future. Colorado has not ruled out the possibility, but a jurisdictional impasse at present seems to be the stumbling block there. The Colorado Fish and Game Department would prefer that it not be given responsibilities for the state's plant problems, and as a result, plants have been given that agency's lowest priority — below insects!

Although these states often do not have statutory jurisdiction over plants, all are in some way involved with a problem of endangered and threatened plants. Fish and Game departments are involved to some extent with endangered plants in California, Utah and Colorado. However, only California's Department of Fish and Game is actively involved with a plant program and that is associated with its review of environmental impact statements. If given the responsibility and adequate funding, the Utah department would also assume a more active role in this connection. In Nevada, the Division of Forestry is the responsible protective agency which is actively involved in supporting programs to survey the state's flora. The remaining states have no ongoing programs at the present time. However, each state is aware of its plant problems, and even Utah, Wyoming and Colorado are taking some positive, albeit minimal, action.

The policies of each state have been difficult to determine, largely because the Secretary of the Interior has not yet listed any plants as endangered or threatened. As a result, the states have not yet been required to take any action.

California appears to be well-equipped and ready to assume an active role in the protection and preservation of endangered and threatened plant species. The state has provided minimal funding to the California Native Plant Society (CNPS), which in turn has prepared a listing of rare and endangered California plants (Powell, 1974). The state will spend about $1,000,000 on endangered species (including plants) during the 1976-77 fiscal year *if* a cooperative agreement with the federal government can be concluded. It already has spent over $1,500,000 for acquisition of critical habitats under its Ecological Reserve Program, although none of this land was acquired specifically for plants. No other state has spent that much money for this purpose.

One possible problem with the California program is with the responsible agency, the California Department of Fish and Game. Since most of its funding comes from sportsmen's license fees, the agency is basically oriented towards hunting and fishing. It is making an effort to handle plants, but this requires some staff members trained in botany and plant ecology, which the Department lacks at present.

Nevada, like the remaining states, is strapped for funds to carry out its legal mandates simply because the legislature has not provided funds to do what it wanted done in the first place. The state has made clear that it wants to protect its native flora. It has people who seem capable of managing any program, and it needs only pressure — perhaps the listing of endangered and threatened species by the Secretary of the Interior — to motivate the state into action.

The major problem with the existing Nevada programs seems to be the general philosophy of the administrative agency, the Division of Forestry. This agency is primarily concerned with control of forest production, work with erosion control, reforestation, fire control, and similar duties. It is interesting that the State Forest Fire Warden is the official who determines whether or not a plant in Nevada is endangered. Yet, it is a representative of the state's Department of Agriculture who has been meeting with federal agencies on endangered plant species in the state, and not a representative from the Division of Forestry. It is essential that Nevada adjust its program to provide for a single administrative unit to handle the endangered species program.

Colorado is at present formulating a policy regarding endangered and threatened plant species. The state seems to be reluctant to formally list species of plants as endangered, as it would prefer to have "sensitive" plants which must be protected, but which can be violated if economic demands call for it. Since the Fish and Game Department does not want to include plants in its responsibilities, the Agriculture Department will probably assume authority. The state currently has no plans to press for legislation to protect its unique plants, although this may change in the immediate future.

Utah and Wyoming are the most difficult states to summarize because the problem in the formulation of their policies lies in the eternal conflict over states' rights. Correspondents in Utah have told me that "the federal agencies have taken upon themselves to take over the Endangered Species Act and to administer its provisions," and "Congress, in its wisdom [has] usurped state' [sic] authorities." Wyoming officials have been less specific in their comments to me in writing.

As near as I have been able to determine, Utah seems to have no immediate plans for formulating a cooperative policy for the protection of endangered plant species. Their opinion can be summarized in one quote:

"We do not understand the implication of or action necessary under the Endangered Species Act when a plant is placed on the endangered or threatened list, nor have the federal agencies been able to give us any answer to our questions."

I do not know what questions Utah authorities have asked, but a reading of the

Endangered Species Act of 1973, and the subsequent statements in the Federal Register (Greenwalt, 1975; Greenwalt and Gehringer, 1975) should have provided answers to nearly all questions, especially if those readings are judged against the comments of Lachenmeier (1974). However, a spokesman from the Division of Wildlife Resources has stated "that as soon as a specific list of endangered plants is published [by the Secretary of the Interior], the Utah Legislature would probably take action to provide protection and designate a state agency to administer the program."

While Utah has been specific in stating that it feels the federal government is usurping its rights, and seemingly is not about to provide any protection for endangered and threatened plants until forced to do so by the federal government, Wyoming has been only vague in this regard. Perhaps my initial reply received from Wyoming's Department of Agriculture best summarizes its policy: "This Department is involved in predator control and is also involved in protection of rangeland in Wyoming."

Wyoming does not feel that any of the plants listed by the Smithsonian Report (1975) for the state are "endangered in any way" although "in some instances we could find no information whatsoever relative to that plant."

Comments from persons interested in the flora of Wyoming, however, have indicated that efforts by individuals to supply some professional assistance have not been accepted by the state, especially in attempts to publish a flora of the state. I am not in a position to evaluate the reasons why the state has relied upon a non-taxonomist for professional advice instead of a plant taxonomist.

SYSTEMATICS EFFORTS

The plant taxonomists of the five western states are making a concerted effort to understand, document, and otherwise determine what species of their states' flora are endangered and threatened. As with all efforts, not everyone has been involved, and the efforts have not been uniformly applied in each state.

In California, which contains a large number of endangered and threatened species, the systematics community has been actively concerned with that state's unique plants. The report prepared by the California Native Plant Society (Powell, 1974) acknowledges the assistance of eighty botanists in determining what species should be listed for the entire state. Probably no other state has had such widespread help from both its own and outside taxonomists as California.

In its report (Powell, 1974), the California Native Plant Society (CNPS) indicated the status and location of each listed species, and each species is mapped (but these maps are purposely not published as a protective measure). The work of CNPS was partially supported one year by a contract from the state, but the six-year program was sustained basically by the volunteer efforts of professional botanists and numerous paraprofessionals who gave generously of their time and money.

Nevada is now compiling a survey of its flora. Although a group was formed in 1969 to undertake a review of the state's native plants, little progress was made until the Northern Nevada Native Plant Society (NNNPS) was formed in 1975. Through that organization, a checklist of the vascular plants of the state has now been started using the volunteer help of several professional taxonomists, experienced amateurs, and various other interested and concerned citizens. Part of the effort is to determine what species are endangered or threatened in Nevada. No recommendations regarding possible endangered or threatened species of plants have yet been made to the State Forest Fire Warden for protection at the state level.

In Utah, a recent publication (Welsh et al., 1976) reviews the status of that state's vascular plant flora, and also includes a list of plants which are endangered, threatened, rare or endemic to the state. A list of those species which are considered to be endangered or threatened was sent to the Smithsonian Institution, and this will be included on the revised Smithsonian list, much as CNPS has done with California plants. In both cases, the present list (Smithsonian Report, 1975) required only slight modification to bring the state and federal lists into harmony.

Unlike California and Nevada, which have native plant societies to sponsor their work, the study in Utah was carried out solely by individual volunteers.

In Colorado, efforts by the systematics community are just beginning. A recent meeting explored the possibility of establishing a native plant society, and it is expected that such a group will soon be formed. The state, via the University of Colorado Museum, has an excellent understanding of its native flora, and little effort would be required to pull from its computer bank a list of endangered and threatened plants.

Wyoming is the farthest behind of all five western states in this matter. Efforts to produce a list of rare and endangered or threatened vascular plants has been largely restricted to a single individual who has been denied support from the state to do the necessary work. As an official of the state's Department of Agriculture expressed it in his letter to me:

"I do not feel the State or private citizen in Wyoming should be placed in a position of having to investigate and determine whether a plant should not be listed as endangered, but rather the person who wishes to have the plant listed on the endangered species list should be responsible for conducting the investigation."

Nevada, Utah, Colorado and Wyoming all suffer from the same problem: There just are not enough professional plant taxonomists specializing in floristic botany in these states to do all the work necessary to determine what is endangered, where these plants are, and what management decisions should be made on how they can best be protected and maintained.

These decisions must be made by the specialists, and not by untrained government employees.

FEDERAL EFFORTS

At the present time, there are several federal efforts being made to study the endangered and threatened plant species in the five western states. My comments concerning them will be restricted to three federal agencies: the Forest Service, the Bureau of Land Management, and the National Park Service.

The Forest Service is actively reviewing the species of plants which are currently listed by the Smithsonian Report (1975), Powell (1974), and Welsh et al. (1976). In Nevada and Utah, the Forest Service has started a review of all plants on these lists, providing each national forest office in those two states with an original description, copies of monographs with descriptions, and such distribution data as are available. In California, the Forest Service is one of the few agencies using the data provided by Powell (1974) and CNPS via its mapping program. The Forest Service is concerned with obtaining as much data as possible for the proper management of those species which are truly endangered or threatened. They will challenge a species if the data does not support the contention of the taxonomist that the species is endangered or threatened. This attitude is commendable.

I have not reviewed the Bureau of Land Management's programs in each of the states. However, unlike the Forest Service which has taken an activist view, the Bureau has adopted a "wait and see" attitude, particularly at the district level in California and Nevada. In Utah, for example, the Bureau is gathering data from herbaria collections for each of the species listed by the Smithsonian Report (1975) and Welsh et al. (1976). It is also supporting some field studies, and just recently, contracts have been issued to various colleges and universities in Utah to do field work on these plants. A position at the national level to supervise the state's endangered species program was filled by an individual not specifically trained in floristic botany, even though the job description called for such experience. This unfortunate move may have a long-term deleterious effect upon the state's program.

The National Park Service is reviewing the flora of each park, and is attempting to determine the status of each species within its boundaries. This is long-term project and will be exceedingly useful.

RECOMMENDATIONS

As a result of this review, a number of recommendations seem necessary. The Smithsonian Report (1975) and Welsh et al. (1976) have made some proposals at the federal and local levels. I would like to expand those to include realistic suggestions for both the state governments and their botanical communities. It is essential that the states understand that the Endangered Species Act of 1973 inevitably will have a profound effect upon the management of all lands within the nation. It is equally vital that the botanical community make an effort to assist federal, state and local governments in their individual and joint efforts to protect the unique plants of the United States.

In reviewing the programs of the five states which concern us here, and in attempting to understand their realistic needs when faced with the problems of having to comply with federal law, I have reached a number of conclusions. While some of these conclusions may reflect the opinions of others with whom I have spoken, I wish to take the responsibility for their presentation here.

The Endangered Species Act of 1973 is the law of the land. As Lachenmeier (1974) has pointed out, the Act leaves some points unresolved. For example, what is meant by a "significant portion of its range"? Why does the Act not contain any provisions for reasonable actions, and to what extent does the operational Section 7 infringe upon the rights of the various states of the Union?

Section 7 of the Act is critical. It is also explicit. All federal departments and agencies shall take such actions as necessary "to insure that actions authorized, funded, or carried out by them do not jeopardize the continued existence of such endangered species and threatened species or result in the destruction or modification of habitat of such species." In the United States, there are very few things the federal government does not authorize, fund or carry out. Although the states of Utah, Wyoming, and to a much lesser degree, Colorado, are complaining about the federal government's role in the endangered species problem, the fact remains that the federal government has every right to be involved with the protection of all endangered and threatened species, especially those which occur on federal lands, or are associated with land-use management programs which are partially funded by the federal government. The states may complain, but they have no basis or grounds for complaint. However, the courts will probably ultimately decide how far the federal government may go in protecting and maintaining endangered and threatened species on state and private lands.

There are a number of critical amendments to the Endangered Species Act that must be made. They are now in the process of being proposed and need not be discussed here, except to note that "take" will have to be included if the provisions for plants are to be similar to the provisions for animals.

What exactly is meant by the term "significant portion of its range?" This is a complex problem for the states. As a basic policy decision, the Smithsonian Committee which compiled the first list (Smithsonian Report, 1975) decided to study endangered and threatened species only as they occur from a national view. Thus, if a plant is endangered in Utah but is widespread in Nevada, it is not included on the Smithsonian list. Such plants, however, should certainly be listed by the state. The Endangered Species Act does provide for state and regional application for plants to be placed on the national list, so that a plant can be listed as endangered or threatened in part of its range. I would prefer to see such listings made at the state level, but if some states refuse to act, as is likely in a few cases, then a species could be listed nationally without state action. I originally objected to this provision, feeling that since the states were more familiar with their own situation they were in a better position to act. But now that I see the reluctance on the part of some states, I can well understand the wisdom of Congress in this matter. The final or "operational" definition of what is meant by "significant" will probably be resolved only by litigation. However, a practical definition can be made for plants.

Lachenmeier's (1974) excellent review of the Act clearly indicates that the 1973 Act leaves no provisions for "reasonableness." While helping to draft the Act, I called for just such a clause with regard to plants. I argued that while it is often possible to move an endangered animal, it is often exceedingly difficult to move a plant. Such a clause, in my opinion, is still needed for plants only. The Act should provide an option which states in principle that if an essential modification of a particular habitat cannot be moved to another location, and that a minimal number of endangered or threatened plants must be destroyed, then they may be destroyed. This degree of "reasonableness" would make the

54

provisions of the Act more realistic.

The majority of plants on the Smithsonian list (1975) are highly restricted to a single, or at most, a few distinct locations. For such species any part of its range is considered "significant", for to reduce the numbers of individuals from a few thousand to a few hundred will have a profound effect upon the survival of the species. Nonetheless, there are several plants which, though widespread, are rare in each known location. It is in such instances that what is "significant" can be seriously questioned. In such situations the species may be deemed endangered in only part of its range, while not endangered in another part. This compromise may be useful but is not reflected in the current list. The individual states can play an important role in clarifying such questions.

Some states seem to be misinformed as to the need for environmental impact statements for *each* plant placed on the endangered species list. An environmental impact assessment will be made leading to a negative declaration (i.e., there is no impact by having the species listed) and a statement for all plants listed by the Secretary of the Interior will likely be written as well, but except for a few critical species, I suspect no individual impact statements will be necessary.

Listed below are a series of recommendations which, if used as a guideline, can make the relationship between the states and the botanical communities much more productive. Guidelines, I believe, can also lead to a realistic program for protecting and maintaining endangered and threatened species.

1. *The botanical community in each state should establish, formally or informally, a working group that will actively aid and support local, state, and federal agencies in the scientific matters of endangered species.*

The formation of native plant societies or organizations which individuals or employees of local, state, and federal agencies can turn to for information on endangered and threatened plants is important. Such organizations, if widely supported by professional and amateur botanists, can be the single most important source of scientific information if it takes upon itself to gather data on the native plants of the state. California's Native Plant Society is an excellent example of such an organization, and its charter and goals can be easily adapted to any state's needs.

Such an organization may be established on a regional level in those states which are either geographically small or which have only a few interested individuals. It is essential for the botanists in each area to take the initiative in this connection.

2. *The professional taxonomists, or a native plant society of the state or region, should: Undertake a detailed review of the state's flora; review those plants proposed for endangered or threatened status by the Secretary of the Smithsonian; and, determine which plants are endangered or threatened in the state.*

The task of determining what species of a state or region are endangered or threatened should be undertaken by professional plant taxonomists using the best available data. Such species should not be determined by botanically-ignorant bureaucrats, nor should such species be proposed by politicians. If the taxonomist does not apply himself to this problem, however, those are precisely the kinds of individuals who will be making the proposals. The national listing is being compiled at this time, and it is critical that professional plant taxonomists make a concerted effort to see that the species on this list are those that specifically should be included.

Problems with certain species can be handled much more efficiently at the state level rather than at the national level. The lists from both California and Utah contain plants which are endangered or threatened only within the state. This is how species which are long-range disjuncts should be handled. State lists will be used by the more responsible federal agencies (such as the Forest Service) to determine "sensitive" plants, or those plants which are endangered in a particular state or region even if they are not protected by the state or national government as a matter of policy. Thus, such lists, if assembled with reasoned judgement, will prove to be most useful, in spite of the lack of statutory sanctions.

3. *Professional taxonomists and others interested in preserving and protecting a state's unique flora, should call upon the government to pass or incorporate the concepts of the Endangered Species Act of 1973 into state legislation.*

The Department of the Interior has prepared model legislation for states wishing to accept the basic principles of the Endangered Species Act of 1973, but unless biologists and others interested in endangered species take it upon themselves to alert state legislators to the Act, it is unlikely that many states will move independently to propose such legislation.

Even if such laws are passed, as they already have been in Nevada and indirectly in California, they will prove ineffective unless there is adequate funding to enforce the legislation. If laws are passed, it is imperative that biologists and others make sure the programs are adequately funded and supported.

4.   *Detailed field studies should be undertaken via cooperative programs with the state, by professional plant taxonomists or native plant societies, to map such species considered to be endangered or threatened, or which may become threatened, and to make such data available to the appropriate land management agencies to insure proper management actions to maintain and protect the habitats of these plants.*

It is one thing to cite a species of plant as endangered or threatened, but it is equally important to know exactly where the plant is located and what its biological relationships are to other plants in the area. Proper management, control, and protection of a state's unique flora is not possible without providing sufficient information to the local, state and federal governments about the biology of the plants themselves. Ideally, this information should be gathered by plant taxonomists and plant ecologists. Such data-gathering and evaluation should not be left to untrained or poorly-trained individuals. The professional must assist the agencies in gathering information, for if he does not, the work will be done poorly or not at all and the end purpose of the Endangered Species Act will be negated.

Each species should be mapped as already has been done in California. The information should be made available to those agencies which are involved with land management. Funding to make such information available (as is now the problem in California) must be granted by the states. Such information can be obtained, in part, by extant herbarium material, but detailed field studies are critical. The biology of each species should be investigated. Many plants will require little or no direct management action except to prevent their outright destruction by man. Many species can even be grazed, burned or subjected to other stresses with little or no permanent damage. They cannot survive, however, if constantly picked by taxonomists and wild flower enthusiasts, dug up for commercial purposes, or covered over by pavement. Note please, that in the first instance the actions are done by nature, while in the latter case, the actions are done by man. We cannot control nature, but we can control what we do. What actions are to be allowed can only be determined by studying the biology of the plant. Some endangered species may actually benefit from disturbance, and it is critical to know that. Other species may be so isolated that no contemplated actions will subject these species to immediate extinction. However, if such species are not known, recorded, and studied, they unknowingly could be destroyed if they suddenly become less isolated or if land use policies of an area change.

5.   *The states should provide the necessary funding, agency support, and priorities to fully implement the provisions of the Endangered Species Act of 1973, if and when plants are listed by the Secretary of the Interior.*

The Endangered Species Act provides for cooperative programs between the states and the federal government. One problem which has developed with the Act itself — as a result of an attempt to establish a cooperative program in California dealing with plants — is the definition of the agency which should be funded in order to enable it to handle endangered plants. This will require an amendment to the Act, and while this will delay cooperative programs dealing with plants, there is no excuse to delay active state concern for endangered species when they are listed.

Most efforts made by the botanical community regarding the Endangered Species Act of 1973 have been on a volunteer basis. This cannot continue. Funding must be forthcoming from both the state and federal governments to carry out field and herbarium studies. State agencies especially must take on a responsible attitude toward the Act, particularly those provisions regarding plants. If the responsible agency is a Fish and Game Department, then that department will have to hire professional plant taxonomists and ecologists to handle the plant problems of the state. It is not realistic to merely assign someone who has little or no training in the subject to do the work. This situation is even more critical at the federal level where it is essential that the Forest Service and the Bureau of Land Management, for example, have high-caliber professionals on their staffs. These two agencies in particular have made the greatest advances in this area, especially at the national level, and to some degree even at the regional and state levels. This good record must continue; the pressure from Washington onto regional and state officials of these agencies must continue; and these agencies must continue to push their local districts to critically study and work closely with professional people to evaluate the area's specific problems. As with these two agencies, so too must the states strive to accomplish the same goals.

Lastly, the states must divorce themselves from their obstructionist policy of doing nothing about plants. I can appreciate the legal questions of states' rights, but if states object to the federal government taking the leadership in the protection and management of the states' endangered and threatened species, then the states themselves must properly protect and manage their endangered and threatened species of plants. As I see it, the federal government's function in this regard is to provide basic guidelines and policies for endangered species so that there is a uniform application of the laws throughout the nation. It is then the states' responsibility to provide local leadership and to judge and evaluate the local conditions of their floras under the guidance of professional botanists. If the states fail to assume this role, then the federal government will do so.

6. *The states should maintain and control their activities so as to prevent the destruction of "sensitive" plants until such time as these plants are formally listed by the Secretary of the Interior.*

The Smithsonian list (1975) and the various state lists (Powell, 1974; Welsh et al., 1976) are neither official nor binding upon any state or federal agency. However, these lists do represent the best available information about the nation's endangered or threatened plants. Until the Secretary of the Interior lists some of these plants as such, they belong to a category I have described as "sensitive".

Each state should establish a policy to provide the necessary protection and maintenance of all plants now listed as endangered or threatened nationally or locally. It should strive to determine the distribution, biology, and status of each listed plant. The states must review all land management programs to insure that presently contemplated actions will not jeopardize these sensitive plants. If such preventive actions are taken now rather than when plants are formally listed, the states will be well into an effective program for the protection of their endangered species of plants.

The western states are subjected to strong economic pressures, especially in the energy and recreation fields. Many of these states regard the total development of their natural resources as the only way to ensure their economic growth. The demand for nationally-owned and nationally-controlled energy sources is growing, and in these states, with their vast amounts of coal, the prospect for enormous land destruction now exists. Both the Clean Air Act and the Endangered Species Act will have a profound effect upon a state's desire to exploit its natural resources. In a small way, these two acts may already have played a role in the demise of the Kaiparowits project in southern Utah. The Endangered Species Act has certainly proven to be an effective deterrent in the eastern states in stopping several projected dams.

Unique plants in these western states often occur on unusual geological formations, such as in the vicinity of hot springs or in areas of remarkable beauty. Our national parks are set aside because of their uniqueness, and these sites are the homes of many endangered and threatened plants. The intense recreational use of some areas may well require more vigilant control if these parks are to retain their beauty and their flora. Many "sensitive" plants occur in potentially important economic locations, and state governments must make every effort to see that such areas are protected.

States must take no premature actions to circumvent the law by destroying habitats of endangered plants so that projects may proceed. Such actions have been taken by individuals, and it is important that biologists and environmentalists determine in ample time that ongoing local and state proposals do not result in the loss of irreplaceable plants.

The Secretary of the Interior will, in all likelihood, propose only those species on the revised Smithsonian list which are endangered or commercially exploited. The large number of threatened plants which are not listed should be regarded as sensitive plants, and appropriate measures should be taken to ensure their continued existence.

7. *The states should recognize, by state law, those species of plants which are endangered or threatened within the political boundaries of the state regardless of the distribution of these species in other states.*

It is one thing to have a national list which protects and maintains certain plant species which are endangered or threatened throughout their ranges, but it is another to protect a part of a species range because, in a particular locality, it is so rare that without protection, it will be lost to the state's flora. There are many such examples of long-range disjuncts in the western states, especially in Colorado, where the Rocky Mountains are the only home in all of North America for certain widespread Eurasian plants. It is critical that the states assume the responsibility for maintaining these plants when they occur on state and private lands, and also ensure that the federal agencies are instructed by state laws to protect such plants when they occur on federal lands within the state. As already noted, the Endangered Species Act provides a means of listing these plants, but it is essential that the states take on this important role. Trained, professional taxonomists can and should provide the state government with the critically-needed information to ensure the long-term existence of such plants.

8. *The states should provide funding for the preservation of critical habitats.*

The Endangered Species Act of 1973 provides for matching fund grants for the purchase of critical habitats within the states. This year marked the first time that Congress appropriated funds for this purpose. If necessary action can be taken at the state level to establish agencies for the protection of endangered plant species, then matching funds may become available upon approval of formal grants.

The protection of endangered species can be accomplished only by protection of their habitats. However, the selection, size, and nature of such sites must be carefully weighed, and this is the role of both the land manager and the professional plant ecologist. Many species require no such action, while only a very few select species will require an area to be set aside specifically for them. A more common situation will be one in which a series of rare, threatened, and endangered species can be protected within a single, critical habitat, and it is exactly those sites which should be given the highest priority.

California's Ecological Reserve Program is funded by the extra fees paid for so-called "vanity" automobile license plates. These fees are used solely to support this program. Perhaps other states may wish to adopt a similar idea to fund the purchase of critical habitat.

9. *The states should provide the leadership in the endangered plant species program within the botanical community cooperating with the state in such efforts.*

The botanical community, even if organized into a native plant society, cannot lead the state into a viable program on endangered and threatened plant species. All it can do is assist the state agencies involved with the program.

10. *The state government and the botanical community must work together to establish a mutual respect and understanding for the scientific, economic and social implications of the Endangered Species Act, and they must strive to formulate laws which will satisfy the needs of the local area.*

The Endangered Species Act of 1973 will have a profound effect upon the nation's future. It is the desire of the people of the nation, as judged by their members in Congress, to prevent, insofar as possible, the continuing loss of unique species of plants and animals. Our nation is now committed to this, and future generations will benefit from this legislative milestone.

How the provisions of this Act are carried out and administered will determine the success of this law. As easily as Congress passed it, Congress can strip the Act of its effective operational provisions. The Act must be administered with judgement and wisdom. It must be modified to provide for reasonable application of its provisions. Botanists must contribute realistic information to government officials, bearing in mind that we cannot save *all* plants, prevent *all* destruction, or curtail *all* development. The Act has grave economic implications. Preservation of a single species can stop an entire project, be it an electric power plant, a highway, or a dam. Conversely, it can also prevent the orderly growth of the nation if abused.

How we, members of the botanical community, assist the government in carrying out the provisions of this Act will largely determine its success. We must establish mutual respect for the problems the state is confronted with, and we must work with the state to resolve them. We cannot afford to aggravate even the most antagonistic state officials, but we should rather strive to educate them and those who appoint them. In other words, we must deal with facts, not personalities. If we can do this, then perhaps the attitude will be infectious. At least, let us hope so.

LITERATURE CITED

Committee on Merchant Marine and Fisheries. 1976. Hearings before the Subcommittee on Fisheries and Wildlife Conservation and the Environment of the Committee on Merchant Marine and Fisheries on implementation and administration of the Endangered Species Act and its amendments, and to review the problems and issues encountered. Series No. 94. **17**: 1-367.

Greenwalt, L. A. 1975. Endangered and threatened wildlife and plants. Federal Register **40**: 44412-44429.

Greenwalt, L. A. and J. E. Gehringer. 1975. Endangered and threatened species. Federal Register **40**: 17764-17765.

Lachenmeier, R. R. 1974. The endangered species act of 1973: Preservation or pandemonium. Environmental Law **5**: 29-83.

Powell, W. R. (ed.). 1974. Inventory of rare and endangered vascular plants of California. California Native Plant Society, Spec. Publ. **1**: 1-56.

Smithsonian Institution. 1975. Report on the endangered and threatened plant species of the United States. U.S. Congress, Committee on Merchant Marine and Fisheries, Serial No. 94-A, U.S. Government Printing Office, Washington. 200 pp.

Welsh, S. L., N. D. Atwood and J. L. Reveal. 1976. Endangered, threatened, extinct, endemic, and rare or restricted Utah vascular plants. Great Basin Naturalist **35**: 327-376.

THE SOUTHWESTERN UNITED STATES

Marshall C. Johnston

The University of Texas, Austin, Texas 78712, U.S.A.

Many kinds of threatened wild plants occur in Arizona, New Mexico, Oklahoma and Texas. At least three of these states have laws implying the desirability of minimizing rates of extinction. In some cases, these laws specifically direct state agencies to carry out programs to achieve this goal. But in spite of such legislation, very little progress has been realized.

As a beginning, it will be constructive to examine legislative goals. I shall confine my comments to statewide regulations, excluding regulations pertaining to such special areas as scenic parks and rights-of-way.

Arizona law (Senate Bill 1334, effective June, 1975) prohibits the taking of nine species of native wild plants for any purpose. It names many other plants, including entire families such as Cactaceae and Orchidaceae, that may be collected for commercial purposes only after the collector pays a permit fee for each individual plant.

New Mexico law (Article II New Mexico Statutes, 1953 annotated) cites several dozen species of "protected" plants that can be taken only on a permit basis from state lands, and from private property when permission of the owner has been obtained. But the law specifically excludes from this protection any plant growing more than 400 yards from a public highway.

Oklahoma and Texas plants have no statutory protection. Perhaps partly because so many of its plants are desperately jeopardized, Texas is the only state in this group in which there is an organized attempt to propagate threatened plants from seeds and cuttings and to try to establish them in protected areas. The Rare Plant Study Center of the University of Texas at Austin works with the state's Parks and Wildlife Department in trying to establish threatened species in state parks, especially near visitors' centers and other interpretive areas. The Rare Plant Study Center has propagated about 50 of the 500 taxa considered threatened in Texas. We consider this a major accomplishment in view of the minimal budget we have had to work with. With small private grants and some support from the University, we have constructed a greenhouse and we have been instrumental in establishing at least three botanical gardens and arboreta. We have a Bicentennial program through which threatened species are planted on the grounds of the state capitol. The major obstacle to the fulfillment of our projects is lack of adequate funding. The federal Soil Conservation Service also has plans to propagate threatened species at its Plant Materials Center in Texas.

Texas encourages its Parks and Wildlife Department to acquire areas in which rare species occur and to maintain them for educational and scientific purposes. For such acquisitions, the state levies a sales tax of one cent on each package of cigarettes. However, during the several years that such money has been available for this purpose, the Texas Parks and Wildlife Department has purchased only one so-called "natural area." There seems to be a dearth of understanding and direction.

So much for the bleak record at the state level. The slightly less disappointing record of the federal Department of the Interior has been previously outlined by Bruce MacBryde. The Department of Agriculture's Soil Conservation Service has been active in plant protection and in all four states it has helped to gather and evaluate information about threatened plants.

In fact, the gathering of information on threatened plants is the most important activity that agencies can perform at present. It is also the most difficult activity to accomplish in a meaningful and useful way. Such information is most constructive when it is based on raw data that are documented and scientifically replicable. Ideally, we want to know how many kinds of plants occur in each geographic area, the numbers and locations of populations and individuals, the populational variability, reproductive capacity, environmental relations, and so on. The adduction of such complete information is possible only after years of field and laboratory work by an army of highly trained and experienced workers. Since there is no army, but only a handful of knowledgeable people, the emerging data cannot be of an ideal quality. The information tells us only in an imprecise way where certain plants have been found in the past, and some of the places where they can still be found. Even gathering this limited amount of information requires the input of experienced botanists who are capable of determining the taxonomic identities of living and prepared specimens, using the more or less esoteric taxonomic literature and research collections of exsiccatae.

Gathering information on threatened plants in Arizona, New Mexico, Oklahoma and Texas is the function of a fragile, loose network of contributors, well known in the botanical community. The two main approaches of these workers are, obviously, the listing of taxa for which few records are available, and the listing of significant natural areas, especially those limited areas that harbor many kinds of threatened plants.

The semantics of the words "threatened" and "endangered" has long been a subject of discussion among botanists. I will confine myself to illustrations of the provincial usage to remind you of the difficulty of their application. The Smithsonian Institution's listing of threatened plants necessarily assumes a broad view and does not include plants that are rare in some states and common in others. The states' views are different. In New Mexico, the species *Limonium limbatum* is considered threatened though it is locally abundant in other states. In Texas, *Pistacia texana* is listed as endangered, though it is common for hundreds of kilometers in adjacent northern Mexico. In Oklahoma, *Sonchus oleraceus* is considered a threatened species. Examples of this kind are legion, and underscore the fact that provincial viewpoints often lead to confusion in the minds of those lacking access to adequate botanical information.

I do not want to criticize the existing lists, but I do want to emphasize the point that we desperately need to bridge the wide information gap. First, we need to flesh out our lists and their more-or-less arbitrary pronouncements of the imminence of extinction. Each taxon should be the subject of intensive study which results in the publication of information as complete as can be compiled, including microgeographic and ecological data as well as diagnostic details. The description should be largely in the form of illustrations, with details clearly drawn so that nonprofessionals can begin to participate in the work to protect these plants. This is an educational program of vast dimensions, one for which enthusiasm can be generated at the local level among all "outdoor oriented" citizens and even university administrators and politicians, if approached intelligently. Arizona, New Mexico, Oklahoma and Texas sorely need one person or a group of botanically-aware individuals to provide the impetus for concerned action to preserve threatened species and habitats in our part of the country.

PLANT CONSERVATION IN THE
UNITED STATES FISH AND WILDLIFE SERVICE

Bruce MacBryde

*Office of Endangered Species, U.S. Fish and Wildlife Service,
U.S. Department of the Interior, Washington, D.C. 20240, U.S.A.*

INTRODUCTION

Conservation, which so often uses a base in the past to emphasize the future danger of current practice, is clearly an appropriate theme for The New York Botanical Garden Symposium to commemorate the U.S. Bicentennial. The people and the Government of the United States of America are today in a reflective, historical mood, with a receptive attitude towards assessing the impact of mankind upon nature, and towards planning for a more harmonious and safer future *vis-à-vis* man and the biosphere. Photographic records of the same locale at different times, such as the bicentennial review of Trefethen (1976), and the earlier survey of Hastings and Turner (1965), dramatically document changes we have wrought.

We are fortunate in the U.S.A. to have a tradition (cf. Borland, 1965) of naturalists such as Henry David Thoreau, John Muir, John Burroughs, and even statesmen/naturalists such as Theodore Roosevelt. They all have lent perspective, and to some extent, influenced the development of this country from primarily wilderness to pervasive civilized complexity. The welfare of nature has penetrated even into some U.S. legislation (cf. Nash, 1968). Recent adverse environmental pressures have caused the U.S. Congress to develop, among other conservation laws, the National Environmental Policy Act of 1969, and the Endangered Species Act of 1973 (cf. Appendix 4). Earlier endangered species legislation (in 1966 and 1969) dealt only with animals, but since the current Act was passed on December 28, 1973, our Government has been responsible for the worldwide conservation of all endangered and threatened plants and animals, and for the conservation of the ecosystems upon which those species depend.

In addition to this unilateral U.S. commitment to its share of global ecological responsibility, the Endangered Species Act of 1973 (hereafter termed the Act) serves as the implementing legislation in the United States for certain international treaties and conventions, at least two of which may soon have significant beneficial impact upon plants. The first of these is the Convention on Nature Protection and Wildlife Preservation in the Western Hemisphere (cf. Appendix 2), which became effective on April 30, 1942, and has been ratified thus far by 17 nations, although it remained a rather quiescent Convention until recently. The second of these multinational conservation commitments is the Convention on International Trade in Endangered Species of Wild Fauna and Flora (cf. Appendix 3), which came into force on July 1, 1975 and has been ratified so far by 31 nations.

Thus there are now certain key "tools of the trade" for the discipline of plant conservation, just as the *International Code of Botanical Nomenclature* (Stafleu, 1972) serves as the mutually agreed upon regulator for a related botanical discipline.

When I was asked to participate in planning this Symposium's session on Canada and the U.S.A., my immediate reaction was to think in terms of real utility. I felt that we needed a detailed survey of the pattern of development in plant conservation in both countries and their respective provinces and states. To use our legislation wisely, both nationally and internationally, we must know items such as: 1) what already has been accomplished, and what is still unknown; 2) who are the key individuals and which are the key groups, committees and institutions to turn to for botanical conservation information in a given locale, what are their data and also what — besides funds and time — are their problems; 3) what strategy, workshop, report or format in one area could be adapted for use in another; 4) how are national, state or provincial, and local governments developing with respect to these challenges, and what (and how adequately written and enforced) are the laws either already in existence or now in preparation; and 5) how are members of professional and amateur societies and organizations, and the general public and public media, responding to the issues and problems.

In short, I had hoped to see the panel do part of my job for me: to develop more precise information that the U.S. Fish and Wildlife Service could use in responding to the Endangered Species Act of 1973. I hoped also that the questions posed would stimulate further growth toward meeting the needs of jeopardized plants and their habitats among those exchanging information for the survey. A final critical factor was that the information presented be useful to all Symposium participants and readers, so that each of us may be a more effective conservationist and act more wisely. Of course, the entire Symposium is designed to stimulate, and in part produce, useful work toward wise invocation of local, national and international laws of plant conservation, and wise practices that need no laws. Documenting the destruction cannot be enough!

The panel has done its job well in presenting us with the plant conservation activities currently under way in their regions, and it is encouraging to see so much concern. What is the role of the U.S. Government in such matters, and how can the information on jeopardized plants and areas be related to its commitments?

LISTING RESPONSIBILITIES AND PROGRESS

As indicated earlier, the U.S. Endangered Species Act of 1973 is strong, rather comprehensive legislation intended to have a significant influence in the conservation and restoration of endangered species. The 1973 Act accords plants nearly equal status with animals for the first time, in keeping with the growing recognition of their equal jeopardy. This recent comprehensive commitment to the welfare of plants has a positive aspect: we now can begin with the considerable legislative, administrative, and managerial policies, traditions and experience of those organizations that have attempted for years to cope with the problems of imperiled animals (cf. Hart, 1974; cf. Schreiner and Ruhr, 1974). We plant taxonomists should remind ourselves that the times are different than those when Carl Linnaeus flourished, and we should not look to 1753 for baseline solutions!

Since the Act is included as Appendix 4, and also has been outlined elsewhere (Williams and Baker, 1976; Baker and MacBryde, 1976), I will present here only some of its highlights and possibilities for its use. In Section 2, the U.S. Congress indicates that the federal government is responsible for the welfare of jeopardized species, and emphasizes that conservation of ecosystems is the ultimate goal. In Section 3 are the legal definitions of: "conservation," "plant," "species," "endangered species," and "threatened species." These definitions are also comprehensive, including all plants (and animals) in any significant portion of their ranges, providing for species likely to be in danger of extinction in the future (threatened species) as well as those now endangered, and allowing for full restorative management.

Section 3 also states that the Secretary of the Interior (and for certain marine and estuarine species, the Secretary of Commerce) has the major responsibility for implementation of the Act; this duty has been delegated within the Department of the Interior to the U.S. Fish and Wildlife Service. While botanists may be most familiar with the Service from such publications as Martin, Zim and Nelson (1951) and Hotchkiss (1967, 1970), the U.S. Fish and Wildlife Service in fact has extensive responsibilities in managing lands, with over 13.8 million hectares in 563 units, including some 378 National Wildlife Refuges (an area greater than that administered by the National Park Service). Furthermore, the Service has had extensive responsibilities in managing animal wildlife, including endangered vertebrates, and contains much other ecological work, so that an ecosystem approach to conservation is a continuing tradition in which plants are being readily included (cf. Murphy, 1968; cf. U.S. Fish and Wildlife Service, 1975a). For example, in the concurrent Conference on Endangered Plants in the Southeast, Gary Henry (of our Service's Regional Office in Atlanta, Georgia) has presented a review of Service-administered lands in the Southeast in relation to some candidate endangered and threatened plants (Henry, 1976).

In Section 4 of the Act the threats which can be legally considered for a species are given, and again the list is comprehensive so that no species need be excluded from consideration. This section also includes the procedures for determining and listing a species as either endangered or threatened, and this is summarized in Fig. 1. It should be noted that the first entry in the *Federal Register* that is required by the Act is a *proposed rulemaking;* consideration of the species can begin at this stage if the other appropriate factors indicated above it in Fig. 1 are complied with. Clearly the process is complex, with ample

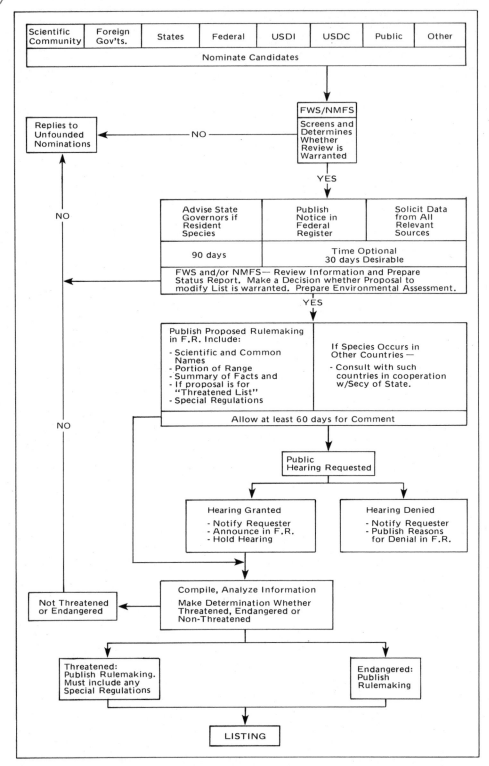

Fig. 1. General Procedures for Modifying Lists of Threatened or Endangered Species under the Endangered Species Act of 1973.
Note that this is basically a three-step process: notice (optional), proposed rulemaking, and final rulemaking. F.R.=Federal Register ; FWS=U.S. Fish and Wildlife Service; NMFS= National Marine Fisheries Service; USDC=U.S. Dept. of Commerce; USDI=U.S. Dept. of the Interior.

opportunity for comment, so it behooves the scientist and conservationist to keep up with the U.S. Government's "journal" — the *Federal Register!*

In Section 7 a further significant concept, "critical habitat," is introduced, which has been interpreted in more depth in a joint notice in the April 22, 1975 *Federal Register* (U.S. Fish and Wildlife Service and National Marine Fisheries Service, 1975). In this context, "critical habitat" is considered to be either the entire habitat, or any portion of it in which a constituent element is necessary to the normal needs or survival of the endangered or threatened species in question. To determine the legal critical habitat, the geographical area involved and autecological needs of the species require a similar process of review as in Fig. 1. Of course, certain actions may not be detrimental even within the critical habitat of such a species, so the area should not be considered inviolate. Managers of the land will often need this information on critical habitat in order to consider the welfare of the plant in their operations, and botanists should find keen scientific and intellectual challenges in providing such total life history and population data on a plant taxon. Many class papers, theses, dissertations and much research could be directed to such efforts, while still advancing pure science.

What has the U.S. Fish and Wildlife Service done with regard to implementing its responsibilities under the Act for plants? Notice was given in the April 21, 1975 *Federal Register* that, in response to a petition, four species of plants found in Canada and the midwestern and eastern United States were under review (U.S. Fish and Wildlife Service, 1975b). In Section 12 the Act directed the Smithsonian Institution to prepare a review of the situation regarding plants within one year. The report, presented to Congress in January, 1975, indicated that 3,187 vascular plant taxa in the United States were likely candidates for endangered or threatened status (or were, perhaps, extinct). Of these, 2,099, or 10.4 percent, are in the continental U.S., while the remaining 1,088 are Hawaiian — 48.9 percent of its flora (Smithsonian Institution, 1975). Some of the efforts and experiences that resulted in the above compilation are discussed in Fosberg (1972), Jenkins (1975), and DeFilipps (1975).

The U.S. Fish and Wildlife Service accepted the Smithsonian report as a petition by means of a notice in the July 1, 1975 *Federal Register,* thereby initiating the formal review of each of these plants, including seeking information regarding critical habitat for any of them. This action prompted many federal agencies, as well as certain state groups and organizations, to give increased attention to these taxa (U.S. Fish and Wildlife Service, 1975c; cf. Goldstein, 1976; cf. MacBryde, 1976). Not quite one year later, on June 16, 1976, it was proposed in the *Federal Register* that 1,779 native plant taxa might qualify as endangered species under the Act, and any further information on them and any data on critical habitats for them were again requested (U.S. Fish and Wildlife Service, 1976c). The majority of these plants were included in the two earlier notices. Eight hundred ninety-seven of them are Hawaiian, and some 347 may be extinct already, or found only in cultivation. The information on file at the Smithsonian Institution for their forthcoming revision, and the comments received by the Service, are the basis for this first proposal on native plants. Recognizing our international obligations, some of the plants are also proposed as endangered in the portion of their ranges in Canada, Cuba, Guadeloupe, México, although most are U.S. endemics.

MANAGEMENT

What are the benefits that a plant can receive once it has been determined to be an endangered or threatened species under the Act? The Nature Conservancy (1975) has indicated many of the ways in which ecosystem management could be approached, and Stover (1976) reviews some of the many options available for conserving land. If necessary, and if funds are available, U.S. land can be acquired (as indicated in Section 5 of the Act) for any endangered or threatened plant included in any of the Appendices of the Convention on International Trade in Endangered Species of Wild Fauna and Flora.

Section 7 of the Act directs all federal departments and agencies to carry out programs for the conservation of endangered and threatened species. It also enjoins all federal departments and agencies to insure that actions authorized, funded, or carried out by them do not jeopardize such a species or adversely modify its critical habitat. While the concept of critical habitat may not apply outside the U.S.A., other provisions of Section 7 do apply to U.S. actions internationally. Myers (1976) has suggested some methods for developing such global responsibility.

Section 9 of the Act not only regulates import and export of listed species, but also international and interstate commerce in them (if a change in ownership of the plant is involved). The responsibility for enforcement of import and export controls for terrestrial plants rests with the Secretary of Agriculture, as indicated in Section 3.

Obviously, not all activities involving jeopardized species are harmful. Section 10 of the Act authorizes the issuance of permits for otherwise prohibited actions which are for scientific purposes or to further the propagation[1] or survival of the affected species. To implement the Act in regard to these prohibitions and any exceptions, plant regulations were proposed in the June 7, 1976 *Federal Register* (U.S. Fish and Wildlife Service, 1976b). It might be noted here that Section 10 also allows for hardship exemptions for up to one year from the first notice of consideration, so all those plant taxa indicated in the April 21 and July 1, 1975 issues of the *Federal Register* have already enjoyed this legal countdown toward their protection.

Of course, in carrying out a conservation job of the magnitude covered by this Act, a sharing of effort is absolutely necessary. In Section 6 the Act indicates that the Secretary of the Interior may enter into an agreement with any of our States for the administration and management of any area established for the conservation of endangered or threatened species. The Secretary also is authorized to enter into a cooperative agreement with any State which establishes and maintains an adequate and active program for the conservation of such species. Financial assistance from the Secretary, within the limits appropriated by Congress, may be provided to qualifying States through the appropriate State agency, after such an agreement is negotiated with the Service.

The U.S. Fish and Wildlife Service has provided each of the States with model legislation similar to the federal Act, and also with a model cooperative agreement, to aid in their development of conservation capabilities as required by Section 6. Certain State Governments are either developing or already have developed, their capabilities in responding to the plant conservation problem. However, the Act is so worded that it may need amendment for these provisions of cooperation to apply to plants equally with fish and wildlife, and the Service is considering developing such an amendment. In any case, there is still much to be accomplished at the state level, such as the conservation of plants rare in the State but perhaps too common in another State to be protected nationally (i.e., "political artifacts"), which may still have ecologically important portions of the gene pool of that taxon. And, of course, state laws could be made even stronger than the current federal Act if desired (and in some cases already have been).

In further management efforts, the Service has developed recovery teams and plans for certain species or groups of species for which coordination and detailed evaluation are especially necessary in order to secure or restore the species as a self-sustaining member of its ecosystem. When the need becomes apparent, such teams and plans may also be developed for listed plants. In order to evaluate and compare the problems of diverse endangered and threatened species, and in order to summarize the management actions necessary and the species' potential for restoration, the Service is developing an endangered species priority system which, it is hoped, will aid in determining the most effective allocation of funds (U.S. Fish and Wildlife Service, 1976, *unpublished;* cf. U.S. Committee on Merchant Marine and Fisheries, 1976). Lovejoy (1976) has indicated the terrible dilemma such choices can involve.

As added encouragement for an individual or group to become more active in the field of conservation, Section 11(g) of the Act allows a civil suit to be instituted when necessary,

[1]The significance of cultivation in plant conservation is addressed in the conference on "The Function of Living Plant Collections in Conservation and in Conservation-Orientated Research and Public Education" which was held in September 1975 at the Royal Botanic Gardens, Kew; cf. Frankel and Hawkes (1975) for phytoagricultural perspectives. For analogous situations and experience regarding captive endangered vertebrates see Martin (1975); the concept of survival centers (or in some cases "half-way houses") applies as readily to plants. The Service has proposed a method to encourage such activity by regulation for certain animals, through captive self-sustaining populations which can relieve collecting pressure on the wild ones (U.S. Fish and Wildlife Service, 1976a). However, the limitations on survival outside the wild should be borne in mind (cf. Thompson, 1975).

in order to insure that the provisions of the Act are followed. Section 11(d) authorizes financial rewards to any private individual for information which leads to a conviction under the Act.

Since we have been reviewing the law — an activity which most botanists may consider a more strenuous exercise than a lowland rain forest field trip — it might be wise to see how lawyers view this realm of biology. Wood (1975) has presented a sympathetic review of portions of the Act, including some commentary on recent lawsuits. Lachenmeier (1974) provides an extensive analysis, including some indications of the legislative history of the Act and some of its potential conflicts.

In a pioneering essay which is fully applicable to the problems of endangered species and their habitats, and which is surprising for its lack of reference to them, Stone (1974) has indicated a convergent approach, by proposing that natural objects should have legal rights, and that the history of law will accommodate such consideration. Huffman (1974) provides a congenial review of the essay, and Stone (1976) indicates that this method of guardianship is already being explored in the courts.

INTERNATIONAL COMMITMENTS

As mentioned earlier, the Endangered Species Act of 1973, especially in Sections 2 and 8, indicates that 1) the United States will consider the welfare of jeopardized species abroad in its actions, even to the extent of offering financial assistance to other countries for programs or land acquisition, and offering training in plant management to foreign personnel; and 2) that the United States will honor its treaty commitments.

The "Pan-American" Convention

On April 13, 1976, the President of the United States signed Executive Order 11911 on the "Preservation of Endangered Species." This order directed the Secretary of the Interior to take the U.S. initiative in implementing the Convention on Nature Protection and Wildlife Preservation in the Western Hemisphere (hereafter termed the "Pan-Am." Convention). While proclaimed by the U.S. President as long ago as 1942, the Convention has not been completely ignored; for example Brower (1964) comments on its contemporary definition of wilderness, and it was addressed at the "Conferencia Especializada Interamericana para Tratar Problemas Relacionados con la Conservación de Recursos Naturales Renovables del Continente" held at Mar del Plata, Argentina in October 1965. Since the Convention is available as Appendix 2, only a few highlights of it will be noted here. The Preamble eloquently describes the desire to halt man-caused extinctions and preserve natural areas, and Article I provides some useful definitions, including the possibility of setting aside an area as a "nature monument" for a single species.

Article VIII indicates that protection of those species listed in the Pan-Am. Convention Annex is of special urgency and importance. The original lists, provided by the following countries in 1941 and 1942 to the Pan American Union, Washington, D.C. (which is the depository for the Convention), include 134 named plant species and certain other taxa: Argentina (24 spp.), Bolivia (9 spp.), Brazil (45 spp.), Guatemala (24 spp.), Haiti (7 spp.), Nicaragua (7 spp.), and Venezuela (18 spp., plus all its Orchidaceae and tree ferns). By 1965, for the Mar del Plata conference, the lists included 269 named plant species and certain other taxa (although some taxa were not repeated from the earlier lists of some countries). Those reporting at that time on jeopardized plants were Argentina (over 85 spp., using an ecosystem approach), Bolivia (over 3 spp.), Brazil (8 spp.), Costa Rica (over 36 spp.), Ecuador (21 spp.), El Salvador (10 spp.), Guatemala (24 spp.), and U.S.A. (82 spp.) (Pan American Union, 1967). Investigations are now under way to determine the complete official Annex list of plant taxa current for each country which ratified this Convention. Results of this Symposium will help immeasurably in the effort for any necessary revisions, and in developing lists for the other countries which have ratified the treaty.

Article VIII of the Pan-Am. Convention provides for the regulation of taking of listed plants, a provision not included in Section 9 of the Endangered Species Act of 1973, although restriction of such taking may be invoked through Section 7 if the U.S. federal action would jeopardize the plant species. Article IX indicates the import and export regulations that are required. The U.S. Fish and Wildlife Service and the National Park

Service are in preliminary stages of developing their responsibilities and projects for this Convention. Clearly, the Convention may become a powerful tool to use when coping with some of the problems which are the subject of this Symposium.

The International Trade Convention

The April 13, 1976 Executive Order (U.S. President, 1976) also establishes the scientific and management authorities for the Convention on International Trade in Endangered Species of Wild Fauna and Flora (hereafter termed the Trade Convention), and again placed the main U.S. responsibility with the Secretary of the Interior, who, in turn, delegated it to the U.S. Fish and Wildlife Service. This Convention is the culmination of a conference of 88 nations and some international organizations, held in Washington, D.C. in February and March 1973, in response to a resolution of the United Nations Conference on the Human Environment at Stockholm (U.S. Department of State, 1973). Of the 31 nations which already have ratified the treaty, eight are in the Americas. Since the Convention, as well as the list of nations which have ratified or acceded, is available as Appendix 3, the following will indicate only specific key points.

The introductory portion of the Trade Convention declares some of the reasons for its utility, including recognition of the ever-growing economic value of wild fauna and flora that may be threatened with extinction (cf. Myers, 1976). Article I includes definitions of: "species," "plant specimen," and "trade." Article II introduces the principles of criteria for including a species on one of the three appendices of the Convention. Appendix I includes all species threatened with extinction which are or may be affected by trade, while Appendix II provides for those species which, although not necessarily threatened with extinction at present may become so unless trade in them is subject to regulation. Appendix III includes species which any party to the Convention regulates for the purpose of preventing or restricting exploitation (and for which any party needs cooperation from other parties). Currently, only six plant species, from Nepal, are included in Appendix III.

It should be noted here that the concept for Appendix I species is rather similar to the definition in Section 3(4) and threat in Section 4(a)(1)(2) of the U.S. Endangered Species Act of 1973, and, in fact, the 45 floral taxa (about 88 species) of Appendix I were proposed (along with various animals) as endangered species in terms of the Act in the September 26, 1975 *Federal Register.* This was the first proposal under the Act for plants (U.S. Fish and Wildlife Service, 1975d).

The regulations to which a Trade Convention species are subject are indicated in Articles III-V. Basically, trade in Appendix I species will require both an export permit from the country of origin and an import permit from the country of destination, while trade in Appendix II and Appendix III species will need only the export permit. Article VI and Article VII indicate the type of permit needed and certain special provisions, including the less stringent manner in which noncommercial interchange of herbarium specimens and live plants will be regulated between scientists and scientific institutions. To implement the Trade Convention, interim regulations were proposed in the June 16, 1976 *Federal Register* (U.S. Fish and Wildlife Service, 1976d); basically it was recommended that those plant and animal regulations already proposed or finalized be adopted for this Convention as well.

Articles XV and XVI indicate the methods for amending the lists of species on the appendices: adoption (or deletion) of Appendix I and II species requires a two-thirds majority of voting parties, while Appendix III species may be submitted unilaterally. (Article XXIII allows any member nation to enter a reservation with regard to any included species, while still ratifying the Convention.) Currently the Convention provides for the regulation of trade in over 20,000 species of plants, since all Orchidaceae, all Cactaceae, and various genera with succulents are among the taxa included on Appendix I or II. As mentioned earlier, U.S. land can be acquired for any plant included in this Convention.

It is apparent that the Trade Convention and the Pan-Am. Convention both offer strong mechanisms for controlling some of the destructive impacts on plants which are being considered at this Symposium. We, as botanists, must make sure that all the plants of which we have knowledge are included under the appropriate classification, on the respective Appendices and Annex, as well as on the list of the U.S. Act, as a minimum use of our time, skills and training toward their conservation. Let us bear in mind that the majority of today's endangered plants are the direct result of man's abuse, and since the majority of plant taxonomists who have ever lived are alive today, it is up to us to become actively involved at once, while there is still time, to determine exactly which plants and areas should be conserved. The world is fortunate that Dr. Carlos Muñoz Pizzaro completed his book on endangered plants of Chile, yet who can say what critical new knowledge, indeed what endangered species, may ultimately have fallen with him?

Further international efforts include the Threatened Plants Committee of the Survival Service Commission of the International Union for Conservation of Nature and Natural Resources. This Committee is becoming an increasingly effective means for consolidating the world's knowledge of jeopardized flora (cf. Hepper, 1969; Lucas, this Symposium). The Flora North America Program, which may be reactivated, could have provided major assistance in Canada and the United States (MacBryde, 1974), judging from related activities in Europe (Lucas and Walters, 1975).

PROBLEMS AND ATTITUDES

We are all confronted with immense problems in our efforts to adequately conserve species and areas. The difficulties range from recognition and definition of the conservation problem (cf. DuMond, 1973), through documentation and education (Iltis, 1972a, 1972b), to administration and management (cf. Weinberg, 1975). With perhaps 10 percent of the biota on earth already in jeopardy, and with at best, limited funds and personnel, the problems that the U.S. Fish and Wildlife Service faces in providing for endangered species throughout the world can be considered representative. The Committee on Merchant Marine and Fisheries of the U.S. Congress House of Representatives, oversees the effects of the Endangered Species Act of 1973, and thereby the Conventions, through its Subcommittee on Fisheries and Wildlife Conservation and the Environment. In October 1975 this Subcommittee held hearings on progress thus far. Not only were governmental procedures, plans, priorities, and policies discussed, but also certain impacts on the public, were indicated and reviewed (U.S. Committee on Merchant Marine and Fisheries, 1976). The intellectual challenge to conserve wisely under such severe financial and personnel constraints, which is common to so many conservation efforts, is a problem which many of us can share. In this spirit, and in order to anticipate and avoid as many unnecessary problems as possible, the Service planned public hearings (cf. Fig. 1) with regard to three of its plant proposals (all but the interim Trade Convention regulations) for the summer of 1976, as announced in the July 2 *Federal Register* (U.S. Fish and Wildlife Service, 1976e).

Awareness, attitude, acceptance, and involvement are all basic to the success of any conservation effort, and many conservation failures are less a result of limitations on time or funds than of limited perception or action in ignorance. We may learn from such men as Eiseley (1960), Anderson (1967), and Sauer (cf. Leighley, 1963) concerning the deep relationship of mankind to the land and plants, and of the need for reflection in our actions. While botanists generally do not collect as intensively as some commercial and amateur interests do (Gosnell, 1976), we should be circumspect in our activities, lest we create a demand which causes others to exploit a species or provide knowledge so freely that destruction can result (cf. Huxley, 1974). As Kellert (1976) has shown for animals, we must be aware that attitudes and perceptions about plants in another individual or society may be vastly different than our own, with corresponding differences in actions affecting plants. We must strive to achieve conservation of plants and their habitats within the value system of the individual, society or agency involved (cf. MacBryde, 1972), rather than expecting others to adjust to our point of view. One can refer to the actual or potential economic value of jeopardized plants for such needs as food or medicine; industrial products such as wood, fiber, chemicals; geological exploration; or as ornamentals; and their other values in realms such as science, the arts, ecosystems, and the biosphere. Certainly, if plant conservation is, indeed, worthwhile, it can be justified in many cultures and philosophies. Generally, mankind will not knowingly limit its future options.

Ultimately, the degree of action each of us takes will depend primarily on the ethics each of us lives by. Albert Schweitzer has shown what a fine responsibility and burden reverence for life can bring (cf. Joy, 1950). Barbour (1973) has collected some of the diverse views that must be accommodated if we are to achieve ecological wisdom with social justice. Teilhard de Chardin (1959) has given us the vision of the "thinking layer" or *noosphere* upon which the biosphere increasingly depends for its survival.

In the field of conservation we are fortunate to have the sustaining insight of Aldo Leopold (cf. Flader, 1974; Leopold, 1949) and the examples of others who have made conservation a way of life (cf. Clepper, 1971). People such as Robert Frost and Gwen Frostic (n.d.), remind us of the value of and need for nature in the lives of all sensitive people.

The tragedy of endangered species reminds us that one species, *Homo sapiens,* has become dominant on earth, and that its power is of a magnitude new to the biosphere. The fact that mankind is rational has led to an attempt at international curtailment of its force. We recognize that we did not create the web of life in which we live. We recognize that each species on earth has a right to exist, *a right to evolve,* and that premature extinction of any species lessens both the earth's and mankind's potential.

We seek life beyond the earth, but we are too careless of it here at home. Because of his special knowledge, the botanist has a special responsibility for conservation of plants. Yet scientists too often seem predisposed to avoid taking an active role (cf. Holton, 1975). Such an approach is scarcely possible any longer, for if the scientist, or indeed anyone else, attempts what has been called the "Pontius Pilate Syndrome" (James E. Trosko, Michigan State University, discussant in the symposium on Bioethics and Accountability in Research on the Environment, A.I.B.S. Meeting, 1976), he may be accused instead of what could be termed the "Nero Syndrome;" that is, if he avoids the issue despite his knowledge, he may be so jaded that he is stimulated by destruction. Recognizing the accelerating scope of conservation concerns, each of us must now determine which is the best path to follow. The Copernican/Darwinian revolution must prevail: ecological integrity is an evolutionary imperative.

LITERATURE CITED

Anderson, E. 1967. Plants, man and life. [Revised ed.] University of California Press, Berkeley and Los Angeles. xiv + 251 pp.

Baker, G. S. and B. MacBryde. 1976. The endangered and threatened plant program of the U. S. Fish and Wildlife Service. ASB [Association of Southeastern Biologists] Bull. 23(3): 141-144.

Barbour, I. G., ed. 1973. Western man and environmental ethics: Attitudes toward nature and technology. Addison-Wesley Publishing Co., Reading, Mass. viii + 276 pp.

Borland, H., ed. 1965. Our natural world: The land and wildlife of America as seen and described by writers since the country's discovery. Doubleday and Co., Inc., Garden City, New York. xii + 849 pp.

Brower, D. 1964. Foreword. Pages 11-19 in D. Brower, ed. Wildlands in our civilization. Sierra Club, San Francisco.

Clepper, H., ed. 1971. Leaders of American conservation. Ronald Press Co., New York. x + 353 pp.

Convention on Nature Protection and Wildlife Preservation in the Western Hemisphere, October 12, 1940. 1974. Pages 115-120 in Treaties and other international agreements on fisheries, oceanographic resources, and wildlife to which the United States is party. Congressional Research Service for Committee on Commerce, U. S. Senate. U. S. Government Printing Office, Washington, D. C.

DeFilipps, R. A. 1975. How proposed lists of threatened and endangered plant species were prepared (unpublished). Symposium on Threatened and Endangered Plants. Society for Range Management and Society of American Foresters, New Mexico Sections, Albuquerque.

DuMond, D. M. 1973. A guide for the selection of rare, unique and endangered plants. Castanea 38: 387-395.

Eiseley, L. 1960 [1966]. The firmament of time. Atheneum Publishers, New York. viii + 183 pp.

Endangered Species Act of 1973. Public Law 93-205, 93rd Congress, S. 1983, December 28, 1973. 87 Stat. 884-903; 16 U. S. C. 1531-1543. 21 pp.

Flader, S. L. 1974. Thinking like a mountain: Aldo Leopold and the evolution of an ecological attitude toward deer, wolves, and forests. University of Missouri Press, Columbia, Mo. xxvi + 284 pp.

Fosberg, F. R. 1972. Our native plants: A look to the future. National Parks and Conservation Magazine 46(11): 17-21.

Frankel, O. H. and J. G. Hawkes, eds. 1975. Crop genetic resources for today and tomorrow. International Biological Programme 2. Cambridge University Press, London and New York. 492 pp.

Frostic, G. [n. d.] Beyond time. Presscraft Papers, Benzonia, Mich. [64 pp.]

Goldstein, J. 1976. How gardeners can help save endangered plants. Organic Gardening and Farming 23(2): 110-112.

Gosnell, M. 1976. Please don't pick the butterworts. National Wildlife 14(3): 32-37.

[Hart, D. 1974.] Proceedings of the symposium on endangered and threatened species of North America, June 11-14, 1974, Washington, D. C. [Wild Canid Survival and Research Center, St. Louis, Mo. ii + 339 pp.]

Hastings, J. R. and R. Turner. 1965. The changing mile: An ecological study of vegetation change with time in the lower mile of an arid and semiarid region. University of Arizona Press, Tucson. 317 pp.

Henry, V. G. 1976. Role of Fish and Wildlife Service concerning endangered flora (in press) in Conference on Endangered Plants in the Southeast. Southeastern Forest Experiment Station, Forest Service, U. S. Dept. of Agriculture and University of North Carolina at Asheville, Asheville, North Carolina.

Hepper, F. N. 1969. The conservation of rare and vanishing species of plants. Pages 352-360 in J. Fisher, N. Simon and J. Vincent. Wildlife in danger. Viking Press, New York.

Holton, G. 1975. Scientific optimism and societal concerns. Hastings Center Report 5: 39-47.

Hotchkiss, N. 1967. Underwater and floating-leaved plants of the United States and
Canada. U. S. Bureau of Sport Fisheries and Wildlife, Resource Publ. **44**. 124 pp.
_____. 1970. Common marsh plants of the United States and Canada. U. S. Bureau of
Sport Fisheries and Wildlife, Resource Publ. **93**. 99 pp.

Huffman, J. 1974. Trees as a minority. Environmental Law **5**(1): 199-202.

Huxley, A. 1974. The ethics of plant collecting. Royal Hort. Soc. J. **99**: 242-249.

Iltis, H. H. 1972a. Shepherds leading sheep to slaughter [I] : The biology teacher and
man's mad and final war on nature. Amer. Biol. Teacher **34**(3): 127-130, 137.
_____. 1972b. Shepherds leading sheep to slaughter [II] : The extinction of species and
the destruction of ecosystems. Amer. Biol. Teacher **34**(4):201-205, 221.

Jenkins, D. W. 1975. At last -- a brighter outlook for endangered plants. National Parks
and Conservation Magazine **49**(1): 13-17.

Joy, C. R., ed. 1950 [1958] . The animal world of Albert Schweitzer: Jungle insights
into reverence for life. Beacon Press, Boston. 209 pp.

Kellert, S. R. 1976. Perceptions of animals in American society. Trans. North. Amer.
Wildlife and Natural Resources Conf. **41**: *in press.*

Lachenmeier, R. R. 1974. The Endangered Species Act of 1973: Preservation or
pandemonium? Environmental Law **5**(1): 29-83.

Leighley, J. ed. 1963 [1967] . Land and life: A selection from the writings of Carl
Ortwin Sauer. University of California Press, Berkeley and Los Angeles. vi + 435 pp.

Leopold, A. 1949. A sand county almanac, and sketches here and there. Oxford University
Press, New York. xiv + 226 pp.

Lovejoy, T. E. 1976. We must decide which species will go forever. Smithsonian **7**(4):
52-59.

Lucas, G. Ll. and S. M. Walters. 1975. A preliminary draft for the list of threatened and
endemic plants for the countries of Europe. International Union for Conservation
of Nature and Natural Resources, Survival Service Commission, Threatened Plants
Committee. Royal Botanic Gardens, Kew. [194 pp.]

MacBryde, B. 1972. Set-backs to conservation in Ecuador. Biol. Conservation **4**(5):
387-388.
_____. 1974. Flora of North America Program suspended. Biol. Conservation **6**(1): 71.
_____. 1976. Adopt a plant. National Wildlife **14**(3): 38.

Martin, A. C., H. S. Zim and A. L. Nelson. 1951 [1961] . American wildlife and plants:
A guide to wildlife food habits. Dover Publications, Inc., New York. x + 500 pp.

Martin, R. D., ed. 1975. Breeding endangered species in captivity. Academic Press, Inc.,
London, New York, and San Francisco. xxvi + 420 pp.

Murphy, R. 1968. Wild sanctuaries: Our National Wildlife Refuges — a heritage restored.
E. P. Dutton and Co., Inc., New York. 288 pp.

Myers, N. 1976. An expanded approach to the problem of disappearing species. Science
(A. A. A. S.) **193**(4249): 198-202.

Nash, R., ed. 1968. The American environment: Readings in the history of conservation.
Addison-Wesley Publishing Co., Reading, Mass. xx + 236 pp.

Nature Conservancy, The. 1975. The preservation of natural diversity: A survey and
recommendations. U. S. Dept. of the Interior Contract No. CX0001-5-0110,
Final Report. The Nature Conservancy, Washington, D. C. iv + 212 pp. + 4
appendices + index.

Pan American Union. 1967. La Convención para la Protección de la Flora, de la Fauna
y de las Bellezas Escénicas Naturales de los Estados Americanos: Listas de especies
de fauna y flora en vías de extinción en los estados miembros. Organization of
American States Washington, D. C. ii + 48 pp.

Schreiner, K. M. and C. E. Ruhr. 1974. Progress in saving endangered species. Trans.
North Amer. Wildlife and Natural Resources Conf. **39**: 127-135.

Smithsonian Institution. 1975. Report on endangered and threatened plant species of
the United States. Committee on Merchant Marine and Fisheries, Serial No. 94-A.
94th Congress, 1st Session, House Document No. 94-51. U. S. Government
Printing Office, Washington, D. C. iv + 200 pp.

Stafleu, F. A., chairman. 1972. International code of botanical nomenclature, adopted by the Eleventh International Botanical Congress, Seattle, August 1969. Regnum Veg. **82.** 426 pp.

Stone, C. D. 1974. Should trees have standing?: Toward legal rights for natural objects. William Kaufmann, Inc., Los Altos, Calif. xviii + 103 pp.

_____. 1976. Toward legal rights for natural systems. Trans. North Amer. Wildlife and Natural Resources Conf. 41: *in press.*

Stover, E. J., ed. 1975. Protecting nature's estate: Techniques for saving land. Bureau of Outdoor Recreation, U. S. Dept. of the Interior, The Nature Conservancy and New York State Office of Parks and Recreation. U. S. Government Printing Office, Washington, D. C. x + 123 pp.

Teilhard de Chardin, P. 1959 [1961]. The phenomenon of man. Harper and Row, Publishers, Inc., New York. 318 pp.

Thompson, P. 1975. Should botanic gardens save rare plants? New Scientist **68**(979): 636-638.

Trefethen, J. B. 1976. The American landscape: 1776-1976, Two centuries of change. The Wildlife Management Institute, Washington, D. C. vi + 91 pp.

U. S. Committee on Merchant Marine and Fisheries. 1976. Endangered species oversight hearings . . . on implementation and administration of the Endangered Species Act [of 1973] and its amendments, and to review the problems and issues encountered: October 1, 2, 6, 1975. Subcommittee on Fisheries and Wildlife Conservation and the Environment, Committee on Merchant Marine and Fisheries, Serial No. 94—17. House of Representatives, 94th Congress, 1st Session. U. S. Government Printing Office, Washington, D. C. vi + 367 pp.

U. S. Department of State. 1973. World wildlife conference: Efforts to save endangered species. Dept. of State Publ. 8729, General Foreign Policy Ser. 279. 30 pp. [floral taxa and permit appendix IV omitted]

U. S. Fish and Wildlife Service. 1975a. Conserving our fish and wildlife heritage: Annual report—FY 1975, U. S. Fish and Wildlife Service, U. S. Dept. of the Interior. U. S. Government Printing Office, Washington, D. C. x + 96 pp.

_____. 1975b. [Notice of] Review of endangered species status [of United States plants]. Federal Register **40**(77) [April 21]:17612.

_____. 1975c. [Notice of] Review of status of over 3,000 vascular plants and determination of "Critical Habitat." Federal Register **40**(127, V) [July 1]:27823-27924.

_____. 1975d. [Proposed rules.] Proposed endangered status for 216 species appearing on Convention on International Trade. Federal Register **40**(188) [Sept. 26]: 44329-44333; correction *ibid.* **40**(205) [Oct. 22]:49347-49348; correction *ibid.* **40**(236) [Dec. 8]:57221.

_____. 1976a. [Proposed rules.] [Determinations of captive self-sustaining populations for certain species of wildlife and definition of endangered species permits for "enhancement of the survival of the species."] Proposal to determine captive self-sustaining population for: 3 large cats, 2 primates, 7 pheasants, 3 waterfowl and 1 quail; and to define endangered species permits for "enhancement of the survival of the species." Federal Register **41**(88, III) [May 5]:18617-18624.

_____. 1976b. [Proposed rules.] Proposed prohibitions on certain uses of endangered or threatened plants, permits for exceptions to such prohibitions, and related items. Federal Register **41**(110, III) [June 7]:22915-22922.

_____. 1976c. [Proposed rules.] Proposed endangered status for some 1,700 U. S. vascular plant taxa. Federal Register **41**(117, V) [June 16]: 24523-24572.

_____. 1976d. [Proposed rules.] Proposed implementation [of Convention on International Trade in Endangered Species of Wildlife (sic) Fauna and Flora]. Federal Register **41**(117) [June 16]: 24367-24378.

_____. 1976e. [Proposed rules.] Public hearings on endangered plants]. Federal Register **41**(129) [July 2]:27381.

U.S. Fish and Wildlife Service and National Marine Fisheries Service. 1975. Notice on critical habitat areas. Federal Register **40**(78) [April 22] : 17764-17765.

U.S. President. 1976. Preservation of endangered species. Executive Order 11911, April 13, 1976. Federal Register **41**(73) [April 14] : 15683-15684.

Weinberg, J. H. 1975. Botanocrats and the fading flora. Science News **108**(6): 92-95.

Williams, J. D. and G. S. Baker. 1976. A review of the Endangered Species Act of 1973. ASB [Association of Southeastern Biologists] Bull. **23**(3): 138-141.

Wood, L. D. 1975. Section 7 of the Endangered Species Act of 1973: A significant restriction for all federal activities. Environmental Law Reporter **5** (10):50189-50201.

Section 2

Mexico, Central America and the Caribbean

THE PROBLEMS OF
THREATENED AND ENDANGERED PLANT SPECIES OF MEXICO

A. P. Vovides and A. Gómez-Pompa

*Instituto de Investigaciones sobre Recursos Bióticos a.c.,
Xalapa, Veracruz, Mexico*

INTRODUCTION

Mexican territory, which covers approximately two million square kilometers, is largely mountainous and contains many high plateaus. It is geologically, topographically, and climatically very diverse. Fig. 1, for example, gives some idea of the range of climates of just one Mexican state, Veracruz. This, therefore, plus the fact that it lies approximately between 30 and 15 degrees north latitude, placing it at the tropic extremes, enables the country to support a most diversified flora. It is here that three geofloras meet: the holarctic, the autochthonous and the neotropical. The intermingling of these floras during past geological times must have contributed greatly to the present richness of the Mexican flora.

The following is a brief background of the vegetational history of Mexico. Owing to the north-south orientation of the major mountain ranges of the Americas, the southward migration of floras was easily accomplished during the Pleistocene glaciations. We are in agreement with Raven and Axelrod (1975), who state that "Opportunities for the migration of cool-temperate and montane plants and animals between North and South America were never greater than in Pleistocene and Recent Times, and such groups as the gooseberries and currants *(Ribes),* the locoweeds *(Astragalus),* willows *(Salix),* and evening primroses *(Oenothera),* now represented by many species in South America, probably arrived on that continent only during the past million years or so and evolved rapidly under the influence of the expanding and fluctuating climates characteristic of Pleistocene cycles." This would account for the boreal elements at high altitudes in present-day Mexico such as *Pinus, Liquidambar, Carpinus,* and *Acer,* to mention a few. These could be said to be the southern biotypes from temperate elements of North America, even though in the past, some of them probably were northern extensions of a tropical montane flora. But no matter what their origin may be, they are now recognized as distinct populations which should be preserved and protected.

According to Raven and Axelrod (1975), plants and animals of temperate North America spread slowly into South America. Typical of these are such genera as *Abies,*

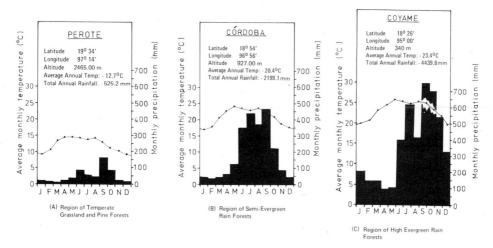

Fig. 1. *Climatographs of stations located in different vegetation areas: A, Temperate grassland and pine forests, Perote station. B, Semi-evergreen rain forests, Córdoba station. C, High evergreen rain forests, Coyame station. (Bars = precipitation; lines = temperature).*

Alnus, Liquidambar, Fagus, Juglans, and *Ulmus,* which reached the mountains of southern Mexico 16 million years ago. This accords with Graham (1973) whose findings revealed pollen of ten temperate arborescent genera in the younger Middle Miocene Paraje Solo formation of southern Veracruz. These genera include *Alnus, Abies, Celtis, Fagus, Juglans, Liquidambar, Myrica, Populus,* and *Ulmus.*

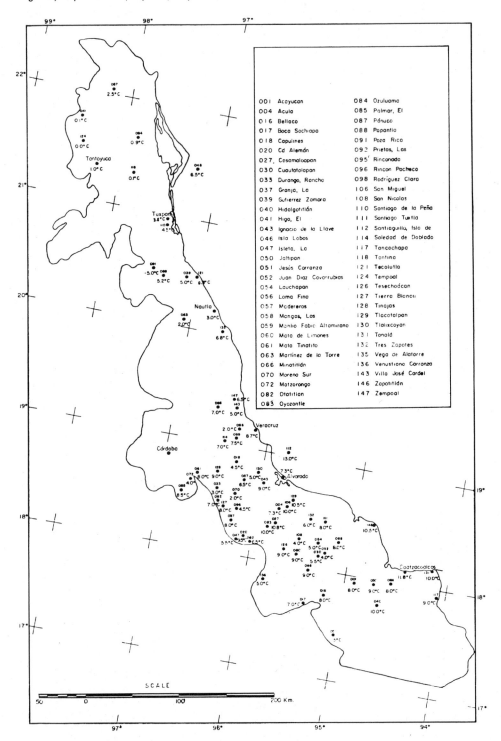

Fig. 2. Minimum extreme temperatures in the state of Veracruz, measured from 1921 to 1970. (After Gomez-Pómpa, 1973).

High altitude areas provide favorable habitats for species typical of more northern latitudes. This gives eastern Mexico an unusual structure in its humid mountain floral communities, which contain a surprising mixture of species with very diverse geographical affinities. Some of these communities are in the transition zone between *tierra caliente* and *tierra templada,* which lies at the altitude between about 1,000 and 1,200 meters. It is in these transition zones where the great geofloras intermingle that there is extreme competition and selection of populations, and it is these very same zones which are fast disappearing. Biotypes with genetic adaptations to tropic extremes are being lost before botanists can get to know them.

Temperature is an important selection factor for floras at this tropic extreme. The average values of temperature are not important factors in understanding species distribution here, but minimum extreme temperatures, even though they may occur only rarely, play a far more important role in the elimination of low-temperature sensitive species from this region. Fig. 2 shows the minimum extreme temperatures in the state of Veracruz, as measured from 1921 to 1970 in stations lower than 300 meters altitude. There is a distinct difference in the north, with minimum extreme temperatures close to 0°C, as compared to the south, with minimum extreme temperatures around 10°C.

The presence of boreal elements such as *Liquidambar* in the Gulf Coast mountains between 900 and 1,600 meters altitude, is clear evidence of a temperate flora that was probably more widely distributed during the Pleistocene, and that these elements are relicts which have found suitable niches at these altitudes (Fig. 3).

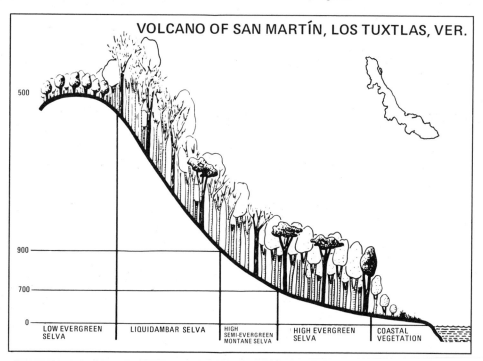

Fig. 3. *Vegetation profile of the volcano of San Martín, Los Tuxtlas, Veracruz (from Gómez-Pompa, 1973).*

When attempting to compile an inventory of threatened and endangered species, the biggest stumbling block one encounters is the lack of an inventory of the Flora of Mexico. Partial floras can help, but it is almost impossible to be absolutely certain as to which plants are rare, threatened, or endangered. The existing herbarium specimens are no indication of the plants' status in this respect. Moreover, the existence of a few individuals of a so-called "rare species" in any given locality does not necessarily indicate that the plant is rare, because it may be found to be widely dispersed or much more abundant elsewhere. On the other hand, a very rare species may be locally abundant. Lack of information — coupled with the daily destruction of habitats such as tropical rain forest and temperate cloud forests — makes the compiling of an inventory of threatened and endangered species an increasingly difficult task. With the disappearance of entire habitats and ecosystems, many species can become extinct before they even are known (Fig. 4). For example, the Misantla region of Veracruz was well-collected during the last

Fig. 4. Clear felled area in the Río Uxpanapa region.

century by botanists from Europe and many species were described from classical localities such as Colipa and Pital in that area. Today, however, no rain forest exists there at all. The problem which confronts us today, therefore, is that it is not known what has been lost irretrievably. It is assumed that there were endemics there, but two questions remain: Were they indeed endemics? Or were they products of splitter taxonomy of the last century? We do not know, and type specimens help us very little in this respect.

In the past, strong emphasis was given to species conservation, especially to animals. This awareness largely developed in temperate countries, and as in many other situations, this concern eventually reached the tropics though it has up to now contributed little to the solution of the problem.

We think it is as important to conserve the Mexican genetic pool of species populations at the tropic extreme as it is to conserve rare individual species. Species of wide distribution ranges at more southern latitudes have different biotypes from those of the tropic extremes, and we are greatly concerned because it is this important genetic diversity which is being threatened or lost. The following example illustrates this point: In an experiment to study population differentiation in a tropical rain forest species *Terminalia amazonia,* seed populations were collected from Central America and Mexico. After germination, the young seedlings were transplanted to an introduction station in Mexico. All the Central American seedlings were exterminated by predators, especially ants (Gómez-Pompa et al., 1972). This clearly indicates the existence of physiological races of *T. amazonia* in Mexico which differ from those of Central America.

The importance of genetic diversity should be stressed for both agriculturists and plant breeders. The future for conservation in Mexico, however, appears bright, as more people at the decision-making level become increasingly aware of the existing problems. But the chief stumbling block today is time — do we have enough time? Astonishing as it may seem, this concern first was taken seriously only about five or ten years ago, when increasing levels of smog began to plague the people of Mexico City. This, coupled with the smoke produced by the mass burning of vegetation in the southeast of the country which interfered with aircraft landings, alerted officials to the inherent dangers.

Some recent specific governmental action also has triggered concern for Mexico's environment: the creation of the *Subsecretaría del Mejoramiento del Ambiente* (Undersecretary for the Betterment of the Environment); the Population Law; and the Human Settlements Law. In addition, there is the creation of a National Ecological Programme and the foundation of at least four Institutes or Centers which are related to and concerned with ecology and natural resources. One of these is the Instituto de Investigaciones sobre Recursos Bióticos (INIREB - Institute for Research in Biotic Resources), to which we belong. Legal progress made during the last three years, and its subsequent increasing public awareness, makes us feel optimistic about this problem for the near future.

ENDANGERED ECOSYSTEMS

The most endangered ecosystems in Mexico are lowland tropical forests and mountain cloud forests. In these ecosystems the endemism at the generic level is very small. In fact, the relative abundance of the endemic element in the woody flora of Mexico, when analysed on the generic level, shows a clear correlation with arid climate. There is a correlation, also, between the relative scarcity of endemic elements and humid climates of Mexico (Rzedowski, 1962). At the species level, the endemism in tropical rain forests of Mexico has not been studied but it is rather high in many groups, especially herbs and shrubs.

The real problem of Mexican rain forests — and possibly of all rain forests — is their inability to regenerate under intensive use. Rain forests regenerate themselves through a process of secondary forest succession. Secondary species which are fast-growing, sun-loving, and produce seeds which have long dormancy and viability, take advantage of clearings produced by natural processes such as storms, floods, and by fires caused by lightning. However, there are always seedlings of young primary tree species on the forest floor. Under disturbed conditions these seedlings will continue to grow at an increased rate while many secondary species start growth from dormant seeds in the soil. After several years the primary trees have overtopped the secondary ones and the major step in regeneration is thus accomplished. Primary rain forest trees have large heavy seeds with short viability and natural dispersal by animals, water, and gravity. Regeneration of these species can occur only if sufficient rain forest is left undisturbed in the immediate vicinity, such as in primitive shifting cultivation systems where only minute areas are cleared and where the genetic pool of the primary trees is retained to give a supply of propagules to the disturbed areas when abandoned.

This serious problem has been discussed and emphasized in the past also. For example:

Professor E.J.H. Corner reported in his Malaysian study just after the Second World War: "There is urgent need for the preservation of the tropical forest, particularly lowland in large nature reserves in all tropical countries. The number of botanic gardens in the tropics should be increased and should develop better relations, particularly in the loan or exchange of staff" (Corner, 1946).

Professor J. Heslop-Harrison: "There are real hazards in eliminating biological diversity. Our scientific understanding of the rain forest plants now vanishing is no more than superficial; our knowledge of the economic potential of the species under threat fragmentary; and our acquaintance with their chemistry and pharmacology trivial. The high successes of modern agriculture have been bought at a cost by reducing the genetic basis on which production rests. The expansion of areas under cultivation ousts the wild races and primitive relatives of marginal lands. So the gene pool available for future breeding is drained away" (Heslop-Harrison, 1975).

"It is of imperative importance to retain pieces of the original rain forest as the only way to reconstruct future forest. With the present rate of destruction of the tropical rain forests there is great danger of mass extinction of thousands of species due to the simple fact that primary tree species from tropical rain forests are incapable of recolonizing large areas opened to intensive and extensive agriculture. Thousands of species could disappear before any aspect of their biology has been studied. This would mean the loss of millions of years of evolution. There is incomplete scientific evidence to prove this assertion, but if we wait for a generation to provide abundant evidence there will not be any rain forests left to prove it" (Gómez-Pompa et al, 1972). (Figs. 5, 6).

Fig. 5. Selective felled forest in the Balankan-Tenosique region of Tabasco.

CONSERVATION

Conservation frequently is regarded as a sentimental middle-class luxury of western

Fig. 6. Rain forest ecosystems converted into grasslands by man for grazing; Balankan-Tenosique region, Tabasco.

society which the poorer countries cannot afford. However, it should be pointed out here that conservation also forms the very essence of the great ancient ethical and religious systems of the East. But conservation must never be practised for conservation's sake alone, without including Man as an integral part of the ecosystem to be conserved. For example, it is totally wrong to earmark an area as a "Nature Reserve: Keep Out," and then have it policed, while multitudes of starving peasants in the vicinity are looking for a suitable spot where they can plant next season's crop, the latter being traditional and never varying from one or two species. This "colonialist" approach to conservation is doomed to failure, for it fails to reach the very heart of the problem and can lead only to future deep-rooted antagonism for generations to come — if the system ever manages to survive that long in spite of itself!

Conservation of natural resources has been approached in many ways and has played an increasingly important role in industrial civilization. According to Dansereau (1957) conservation, historically, can be divided into four distinct phases:

1) Legislative phase:

Protection given to wild plants and animals decreed by law and focussed almost entirely on rare species.

2) Biological phase:

As detailed inventories of flora and fauna were made, a more scientific approach was initiated; freedom of certain areas from disturbance so that individual plants and animals could be studied; protection of large populations of trees and birds.

3) Ecological phase:

With the disappearance of many species and a rising consciousness of the inter-relatedness of living things, it was claimed that no efficient protection of individual species was possible if the habitat as a whole were not free from direct or indirect disturbance.

4) Sociological phase:

The harnessing of the world's resources is being viewed in a new light and a joint attack made on the question as a whole. Conflicting forces are increasing human populations and decreasing resources. It has been said that land and available resources are an arithmetical progression, whereas human population increases in a geometrical progression.

In a bulletin it publishes based on regular worldwide reports, the International Union for the Conservation of Nature states that individual species of plants and animals and — even more important — biotypes and endemics, can receive no effective protection unless their habitat itself is spared from destruction or interference, and that ecosystems are the really vital units with which our economy deals.

What can be done to get land-hungry peasants and laymen to accept this? Basic awareness of the problem is of prime importance. This can be accomplished by a concerted educational program, and by publicity through the mass media such as radio, T.V., and newspapers. The problem must be communicated to official decision-makers as well as to the general public. The encouragement of conservation-oriented competitions and quizzes in schools also is an effective method of communicating the problem to young people in whose hands the future of countless species ultimately rests. A good example of the type of program which should be encouraged is the recent competition put forward by INIREB for primary schools of the state of Veracruz. Children were asked to submit pressed specimens, drawings and information on any useful wild plant of Veracruz. The contest has met with great enthusiasm and success and winners will receive a conducted tour of the state and a visit to the Institute's laboratories in Xalapa (Figs. 7, 8).

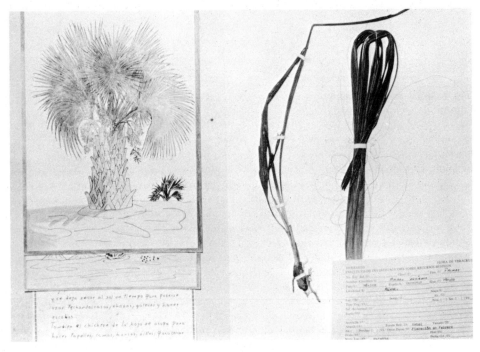

Fig. 7. Useful wild plant of Veracruz competition for primary schools. Sabal mexicana *used for making mats, chairs, roofing and bedding, prepared by Hector Díaz Hernández, age 14.*

Following are four general guidelines for a suggested National Conservation Program:

1. Technical, practical and scientific explanations must be clearly stated so that everyone will understand what is needed. Radio and T.V. documentaries can be of great help in this connection.

2. Conservation should be included in a list of sound land use practices.

3. More research biologists are urgently needed to work in the most endangered zones, such as in the Mexican lowlands and in the cloud forests of the Sierras. At present, most of the experts spend much of their time in Mexico City, rather than out in the field.

4. The possibilities of increased tourism, hunting, fishing, and other recreational activities in areas of exceptional beauty can bring untold wealth to small peasant

communities as well as a better income from more efficient farming.

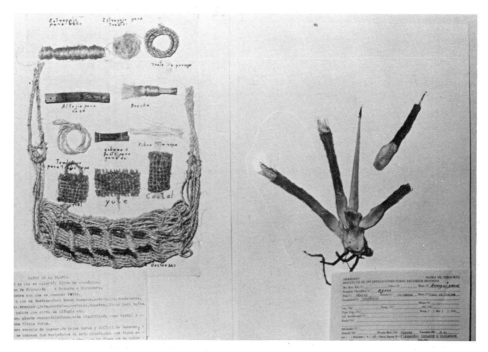

Fig. 8. Useful wild plant of Veracruz competition. Agave plant, ixtle, used for making household items, prepared by Serafín Vallejo Galán, age 12.

It should be pointed out that today's destruction of natural areas in the tropics is a symptom of a much greater problem: population growth and deficient food supply. To permanently solve the problem of conservation, it is first necessary to solve the major problems of population and food. This is by no means a defeatist attitude on our part, rather it emphasizes the need to go to the very roots of the problem. Protected National Parks are not the permanent solution. The political and socio-economic aspects of underdevelopment first have to be solved, before a definite and permanent solution for the tropics is arrived at. One possible answer to this problem in Mexico, at least, are the biosphere reserves.

BIOSPHERE RESERVES

The Man and the Biosphere (MAB) programme of UNESCO is doing much to promote international cooperation in the fields of conservation and development. Through this programme Mexico approved the general idea and adapted it to suit the country, with the result that two biosphere reserves already have been established in la Michilia and Mapimi in the state of Durango. These are desert areas, but a similar system of reserves is being planned for the Selva Lacandona of Chiapas.

The fundamental part of the programme is designed to promote a multiple agreement between scientists, landowners, the general public, and the federal and state governments. With this agreement, an association is formed with the consensus of all and signed by all parties. The obligation of the participants to the pact is to follow the agreements and to help to enforce them.

The object of biosphere reserves of Mexico is to conserve for the present and future use of man, the diversity and integrity of all biotic communities within natural ecosystems; to serve as gene-pool banks in order to safeguard the genetic diversity of animal and plant species; to provide areas for basic ecological research within the biosphere reserves as well as in adjacent areas; and to provide the means for efficient management of the reserves by teaching at different levels and training of local personnel. In addition, residents must be taught the principle of "renewable resources" in general, and how they can improve their present land use practices. For example: The Michilia reserve covers an area of 35,000 hectares and is divided into the four following zones:

(a)   Nucleus or integral reserve zone
(b)   Buffer zone
(c)   Applied reserve zone
(d)   Influence zone

Nucleus zone:

This is an area of total conservation in its truest sense, to which only research scientists have access. This zone not only serves as a germ plasm bank but also as a means of comparison of what has been done in the buffer and research zones, i.e. faunistic and floristic changes due to anthropogenic or natural causes. Valuable autecological information can be obtained from this zone, which has had practically no human disturbance.

Buffer zone:

This is a relatively large area completely enclosing the nucleus zone from easy access. It serves the important function of protecting the nucleus zone from either geochemical or microclimatic man-induced changes. The zone offers potential recreative amenities, implementation of educational programs, tourist activity, and other means, all designed to promote increased appreciation of the ecosystems.

Applied research zone:

From the point of view of economical benefits, this zone is of major importance. It is here that experiments are carried out which have long or short-term applications. The search for plant species with possible medicinal or forage use, both direct or indirect, is carried out. One of the most important factors that this zone has to offer is to channel the resolution of specific problems confronting common-land users, the general public, small landowners, and peasants, as well as the improvement of their cattle, agricultural, fructicultural, and silvicultural methods. Research into new products for local industries will give the inhabitants an alternative source of income besides the fringe benefits obtained from tourism. It is hoped that this activity also will further understanding between scientists, peasants, small landowners, and government authorities, and also will encourage amicable relations among these groups. This is of fundamental importance to the actual protection and future conservation of the nucleus and buffer zones.

Influence zone:

Technically, this zone remains a part of the reserve until such time as research results have a permanent application in the ecosystems of the buffer and research zones, but the results can and should be extended to all ecosystems with similar characteristics in the area.

As land use techniques improve and with the introduction of new products for local industries, it is hoped that there will be less need for peasants to migrate into new areas or to the big cities when traditional agricultural methods fail. With these areas the needs of all participants will be fulfilled and the protection of the area is effected by the people living there. No guards, therefore, will be needed.

PRESENT WORK ON ENDANGERED SPECIES

Owing to the above-mentioned difficulties concerning the production of a reliable list of threatened and endangered species, Instituto de Investigaciones sobre Recursos Bióticos (INIREB) has a program in collaboration with the IUCN to produce a preliminary list of endangered and threatened plant species of Mexico which offers as objective a view of the problem as possible. The main method used for achieving this is to submit a questionnaire to botanists and specialists, asking them for such information as species, locality, state, altitude, reasons, etc., for their belief that a plant species is threatened or endangered.

This, coupled with the consultation of the meagre available literature, is beginning to form a slowly-growing list of endangered plants. Interviews with amateur and commercial collectors also can be helpful, but one must be wary because these people are often those most responsible for the rapid depletion of many ornamental species such as cacti.

Tighter legislation should be enacted in order to control the activities of commercial exporters. It has been brought to our attention that a large number of Mexican cacti are exported from Mexico, for example, under the pretext of utilization for biochemical research. Cacti also are exploited ostensibly for construction of botanical gardens, but

Fig. 9. Dioon spinulosum *Dyer with female cone. One of the plants on the Mexican endangered species list seen here growing in a garden in Acatlán, Oaxaca.*

we believe they are to be resold for other uses. Mexican plants collected in the wild are exported, though such license was originally granted to cover nursery-grown stock only.

Publicity in horticultural journals should reduce demand by fostering an enlightened attitude amongst private collectors and enthusiasts. This is by no means aimed at persecuting the private collector. In fact, properly-documented private collections can be of infinite value to conservation. The desire to raise difficult plants from seeds and to propagate such plants must be fostered amongst enthusiasts. It requires far more skill on the gardener's part to raise a difficult plant from seed or cutting than to pay large sums of money for an imported specimen, which often does poorly and soon dies, due to the sudden change of environmental conditions. Flower show judges should be alert to this difficulty and should award winning points accordingly.

We wish to collect from endangered habitats samples of endangered plants for cultivation and propagation in future botanic gardens and it is hoped that in the near future, rare and endangered plants can be reintroduced into suitable habitats that have been assigned to reserves or national parks in much the same way that reindeer were reintroduced into the Highlands of Scotland, after having been extinct there for many generations.

Here are some examples of threatened or endangered plant species from our growing list:

Species	State
Astrophytum myriostigma Lem.	Coahuila
A. ornatum (DC.) Weber	San Luis Potosí and Coah.
Agave victoria-reginae Moore	Nuevo León
Ceratozamia miqueliana Wendl.	Veracruz
Dioon edule Lindl.	Veracruz
D. spinulosum Dyer	Oaxaca
Diospyros riojae Gómez-Pompa	Veracruz
Hydrangea nebulicola Nevl. & Gómez-Pompa	Veracruz
Leuchtenbergia principes Hooker	Coahuila
Zamia furfuraceae L.	Veracruz

We hope that some day soon, Man in his potential infinite wisdom will want to repair the damage done to all ecosystems, especially the tropical rain forest. We hope also that there will be enough germ plasm left to enable the reconstruction of these ecosystems as living museums for future generations of mankind.

LITERATURE CITED

Corner, E.J.H. 1946. Suggestions for Botanical Progress. New Phytol. **45** (2): 185-192.

Dansereau, P. 1957. Biogeography an Ecological Perspective. 394 pp. Ronald Press Co.

Gómez-Pompa, A. 1973. Ecology of the Vegetation of Veracruz. Vegetation and Vegetational History of Latin America. 73-148. Elsevier Publishing Co., Amsterdam.

Gómez-Pompa, A., C. Vazquez-Yanes and S. Guevara. 1972. The Tropical Rain Forest: A Nonrenewable Resource. Science **177**: 762-765.

Graham, A. 1973. History of the Arborescent Temperate Element in the Northern Latin American Biota. *In:* Vegetation and Vegetational History of Northern Latin America. Ed. Alan Graham. 301-314. Elsevier Publishing Co., Amsterdam.

Heslop-Harrison, J. 1975. Man and the Endangered Plant. IUCN International Year Book 1975. 103-106.

Raven, P.H. and D.I. Axelrod. 1975. History of the Flora nd Fauna of Latin America. Amer. Scientist **63**: 420-429. No. 4.

Rzedowski, J. 1962. Contribuciones a la Fitogeografía Histórica de México y algunas consideraciones acerca del elemento endémico en la Flora Mexicana. Boletín de la Sociedad Botánica de México. No. 27. 52-65.

ENDANGERED LANDSCAPES IN PANAMA AND CENTRAL AMERICA: THE THREAT TO PLANT SPECIES

W. G. D'Arcy

Missouri Botanical Garden, St. Louis, Missouri 63110, U.S.A.

In its broadest sense, Isthmo-America ranges from the Isthmus of Tehuantepec or Río San Juan in Mexico to the Río Atrato in Colombia (Fig 1). For the purpose of this paper, however, Central America includes only Guatemala, Belize, Honduras, El Salvador, Nicaragua, Costa Rica, and Panama. This region, which extends to about 2,000 kilometers in length, is only 600 kilometers at its widest point, while most of it is much narrower. The area covers about 530,000 square kilometers and supports approximately twenty million people[1], a density of 38 people per square kilometer or 98 people per square mile. Isthmo-America is currently ruled by nine different flags. Besides this political fragmentation, there is much division along physical, biological, and historical lines.

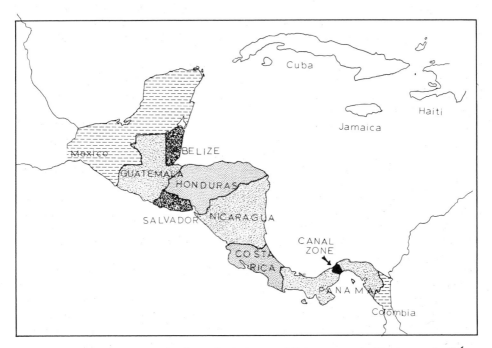

Fig. 1. Map of Isthmo-America from the Isthmus of Tehuantepec in Mexico to the Río Atrato in Colombia.

THE PLANT INVENTORIES

Central America has one of the world's richest floras, and contains about 18,000 species of vascular plants, fewer than half of which have been studied by scientists. Only about 35 percent of the land area has been covered by recent floras, and the extent of investigation is very unequal in the different countries. Panama and Guatemala have modern floras at or near completion (Woodson et al., 1943; Croat, 1976; Standley et al., 1958), while Costa Rica has a modern flora in the beginning stages (Burger, 1971). For Belize (Standley and Record, 1936), El Salvador (Standley and Calderón, 1926), and

---

[1]Vital statistics, area, and economic data are based on information from: 1975 World Data Sheet, World Population Reference Bureau, Inc., and World Almanac and Book of Facts, 1976. Newspaper Enterprise Association, 1975.

Honduras (Record, 1927; Standley, 1930, 1931; Molina, 1975) there are only annotated checklists, some of which are quite old. A new checklist for Belize will be issued shortly (Spellman et al., 1975; Dwyer and Spellman).

The modern floras of the region compare professionally with any issued anywhere in the world, but unfortunately such floras are based on incomplete sampling of the species present. For the past ten years, botanical exploration in Panama proceeded at a quite high rate, with well over 10,000 collections being made during some years. Yet for the five families he sampled, Dressler (1972) estimated an average increase of 85 percent in the number of species known for the country in the 10-30 years since publication of these families in the flora, and Gentry (1976) underlined the large understatement in even recent fascicles of the Flora of Panama. When new families are published in the Flora of Panama, several novelties often come to hand between the time the fascicle goes to press and when it is distributed to the scientific public. Since 1970, collecting in the wet forests of Panama began to reveal large numbers of new species.

A flora is a summary of the state of knowledge about a region's plants at the time it is issued. In western Europe, the Flora Europaea and Flora Mittel-Europe are thorough statements on the floristic content of the continent. But these are based on an impressive six centuries of study and publication of over 5,000 local floristic works. In Central America, botanical study goes back only about two centuries, with most of the effort having been put forth during the current century. Fewer than fifty local floristic works have been published in Central America, even though the flora there may be about three times richer than that of Europe.

All major Central American floras have been prepared by scientists working at foreign institutions. Seemann's early flora of the Isthmus of Panama (Seemann, 1852-57) and Hemsley's Biology of Tropical America (Hemsley, 1879-88) both were written in Europe. At the present time, the major floras are being written by the Field Museum of Natural History, Chicago, and the Missouri Botanical Garden, St. Louis, both of which are in the United States. Not surprisingly, the overwhelming proportion of herbarium specimens from Central America are held abroad. The Field Museum and the Missouri Botanical Garden have the richest holdings, each with upwards of a quarter million specimens from Central America. There are important holdings in several other United States institutions, notably the Smithsonian Institution, Duke University, and the University of Texas. Holdings in Europe are relatively small, but because they are often old they are of great importance. The latest inventory of herbaria (Holmgren and Keuken, 1974) lists sixteen research herbaria in Central America with a total holding of just under a quarter million specimens. It is now standard practice, however, for foreign collectors to deposit representative material from their expeditions with institutions in the countries they have visited.

Monographic studies of plant groups occurring in Central America have traditionally been done either in the United States or in Europe. United States investigators frequently have done field work in Central America. Often their studies involved groups centered in Mexico and ranging with smaller diversity into Central and South America, but there have been increasing studies centered in Central America itself. Little monographic work has been done in Central America by permanent residents of the area, although there is now more interest in this direction. The treatment of the orchids of El Salvador (Hamer, 1974) is an example of such recent work.

The journals *Turrialba, Revista Biologia Tropical, Brenesia,* and *Ceiba* are four Central American publications in which taxonomic or floristic papers appear.

There is now a firm scientific base in Central America from which plant inventories can evolve. It is admittedly small at present, but it is steadily expanding and includes some first-rate talent. None of the major floras of Central America has yet been translated into Spanish, and to my knowledge, Hamer's orchids of El Salvador is the only recent monographic treatment available with Spanish text. Until recently, there have been no texts in Spanish to facilitate teaching the residents of these countries about their native plants. The major college botany text is a Spanish translation of a North American book (Wilson and Loomis, 1968), well-illustrated with examples from the North Temperate Zone. Spanish translations of a German text (Strasburger, 1962) and several North American texts, e.g. Raven and Curtis, 1975, also are available. While these provide excellent coverage of botanical curriculum, they are not oriented specifically towards the familiar experiences of the Central American student. Several recent botanical texts published by the staff of the University of Panama (Correa, 1975; Escobar, 1976; Taylor, 1976) are most welcome.

THE FLORA

Central America is floristically rich, with 18,000-20,000 species of vascular plants. Most of the major ecological zones, such as the seasonally dry forest and tropical moist forest, are wide-ranging and extend from Mexico into South America as far as Peru and Surinam. As Gentry (1976) has stated: "The Andes and the Cordilleras of Central America may be conveniently viewed as constituting a single, interrupted montane system with major breaks in the Atrato Valley of Colombia, central and eastern Panama, and across most of Nicaragua." Over half of the 1,375 species found in the lowland forest of Barro Colorado Island, Panama, are of wide distribution, occurring in Mexico, South America, the Caribbean or other parts of the world (Croat, 1976). There are no endemic Central American families of vascular plants.

But while a large portion of the flora of Central America is of widely distributed species, there also is a large and important element of local or endemic plants. There are about 100 endemic genera and another 65 genera which are essentially Central American with only one species out of several occurring outside the region. Two hundred and sixteen species on Barro Colorado Island are confined to Panama, Costa Rica, and northern Colombia. Because the flora here is rich, this constitutes only 17 percent of the total. Many species of plants are known only from one or a few local populations. This localism is in part real and in part an artifact of insufficient knowledge.

The greatest concentrations of endemic plants are on the high mountains. Beaman has been cited (Bowers, 1970) as estimating that 70 percent of the vascular plants of the high mountains of Mexico and Guatemala are endemic, while the figure for the high mountains of Costa Rica and Chiriquí may exceed 50 percent (cf. Weber, 1958). This terrain, some of which is over 3,000 m elevation, accounts for less than two percent of the total land area of Central America (Budowski, 1965), but it is of the greatest biological importance in the area from a phytogeographical-endemism viewpoint. Floristically, the high mountains of Guatemala are allied to the uplands of Mexico while those of Costa Rica and western Panama are closely allied to floras of the Andes (Weber, 1958). The occurrence of a Mexican or an Andean species in Central America documents an amplitude of dispersal ability and ecological tolerance which would not have been suspected from examination of the central range alone. And of course we can study native populations of endemic plants only in the lands where they occur. Though intact until recently, these mountain areas are rapidly undergoing destruction at an alarming pace as a direct result of activities of sheep and charcoalers.

At lower elevations, the flora is both wide-ranging and localized. Gentry (1976) has pointed out how sampling error has led to much artificial endemism. This is particularly true for species of the lowland wet forests. But recent explorations of these relatively unknown areas do not greatly alter the overall patterns of endemism. As distributions of known species are broadened and their variability better understood, many formerly supposed endemics are united with populations from elsewhere and become wide-ranging elements. For example, the shrub *Cestrum megalophyllum* was known by nine different names in as many countries before the group was recently revised (D'Arcy, 1973). At the same time, current explorations are bringing numerous novelties to light, and these usually are considered endemic when they are first described. The long narrow shapes of much of the ecological zonation determining plant distribution in Central America has inhibited thorough sampling and encouraged parochial taxonomy.

Although a substantial proportion of the localism in the floristics of Central America is artificial, a substantial element of this localism also is real. Fragmentation of the flora is a major factor.

To summarize the geography of Central America, one may say that there is a high, dissected ridge extending most of its length, with greater rainfall on the Caribbean side and seasonal rainfall on the Pacific side. The wettest forests, mainly along the Caribbean coast, are the richest in species and the least understood botanically. The outcrop in Central America consists of sedimentary limestones and dolomites uplifted and intruded by acidic volcanic materials. The basic materials have been eroded from most of the Isthmus. Long periods of high rainfall have altered the composition and placement of overlying soils, and under forests there is advanced laterization of volcanic soil materials (Fig. 2).

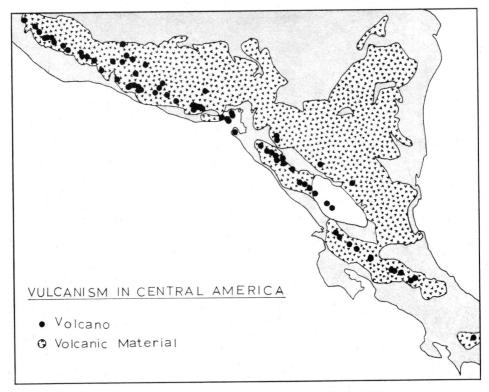

VULCANISM IN CENTRAL AMERICA

● Volcano
◉ Volcanic Material

Fig. 2. Map of vulcanism in Central America (adapted from Weyl, 1965).

Central America is an area of strong relief, with eroded peneplains in Guatemala and steep volcanic slopes on eastern areas. The combination of high relief, differing substrate, and seasonally shifting rainfall patterns has produced a reticulate physical environment. The resulting vegetation patterns are also reticulate (see Fig. 1 in Gentry's paper in this Symposium), where so-called wet and dry formations interdigitate, especially in the interior. Beard (1944, 1955), Holdridge et al. (1971), and Budowski (1965) all have described plant formations in Central America in detail and also provided systems for classifying the different ecologically determined species assemblages. The species makeup of the different plant formations is almost entirely distinctive, and one may recognize as many floras as there are vegetation types. Thus, because of the habitat fragmentation along the Central American land mass, the flora is also differentiated into numerous units.

Another physical phenomenon adding to the floristic divisions established by edaphic and climatic factors is vulcanism. Not only has vulcanism formed much of the rock substrate from which soils are formed, but in its course of activity, it has periodically covered and smothered much of the Isthmus. Fig. 2. shows to what extent the volcanic material has covered the Isthmus; most of the indicated volcanos have erupted during the comparatively recent past. These volcanic events destroy only one sector at a time, with species usually surviving in bordering areas. These survivors then repopulate the sites as soon as soil and shade conditions readjust to the plant requirements. But because some forest species spread very slowly, such recovery of range lost to vulcanism may not always have been completed.

The isolation brought about by this fragmentation of the flora continuum has provided genetic separation and has permitted greater speciation than might otherwise have taken place. During the period we are most interested in, however, the fragmentation has been multiplied by the hand of man.

THE FIRST DEFORESTATIONS

For the forests of Central America, the current felling is the second time around. In the centuries before Spanish contact, Central America was well-populated by several predominantly agricultural cultures. Permanent settlement is recorded (Lothrop, 1959) more than 6,000 years ago in a village site in central Panama. The Mayan culture reached a high development in the lowlands of Belize and the Yucatan (Puleston, 1971), as well as in the interior highlands of Guatemala (Steward, 1963; Wauchope, 1965). In Meso-America there were substantial villages and many buildings made of stone. In upland Honduras there was stone pavement in villages. Sapper (1924) estimated a population of 5-6 million for Central America and Chiapas, and Kroeber (1939) estimated a population of 3.3 million for all of Mexico and Central America. For central and eastern Panama and neighboring parts of Colombia, Oviedo (from Sauer, 1966) estimated a population of two million in the year 1513, or about the same number of people present there now. There are conflicting reports about the degree to which these early Panamanians lived in towns, some reports maintaining that there were at the time no towns at all, while others indicate towns of about 1,500 people. Espinosa (from Sauer, 1966) required cannon to subdue the fortified towns on the Azuero Peninsula. But it is not necessary to have towns in order to support a large population. In Rwanda, Central Africa, for example, there is at present a rural population of about four million people living in high density but almost entirely without towns and villages. The farmers in Rwanda have removed all the forest in the areas where they live.

One may wonder how much effect this pre-Columbian population had on the landscape. Maize cultivation is recorded as far back as 5,000 years, and this crop requires at least some clearing and tilling of the land. Columbus and other early European visitors record cultivation at various points along the Caribbean coast. At Veraguas, Columbus marched inland from the coast and found nearly 30 kilometers of "maize fields like fields of wheat" (Sauer, 1966). There are many other indications that man had altered the landscape by forest removal and other activities. Reports of large numbers of doves, white-tailed deer, and large herds of peccaries argue for woodland or somewhat open country, as do plagues of grasshoppers that "appeared and covered the sky and covered the fields of maize." The use of fire for hunting and for clearing the land of forest was common practice. Vast expanses of savanna were reported by early Spanish visitors, and there is good reason to believe these were induced by activities of man. Several visitors describe large areas without forest *(monte)* or thickets *(arcabucos)*, and Espinosa likened the terrain in what is now Pacific Panama to that of Spain. Removal of forest trees would have been necessary for cultivation, and removal of thickets was probably necessary for defense against warring neighbors. Whatever figure is accepted for the population of Central America at the time of Spanish contact, it is clear that man had by then a prolonged and drastic effect on the vegetation of the Isthmus. Forest already had been removed from large areas throughout Central America.

The Spanish contact had drastic consequences on the indigenous populations of Central America. Many inhabitants were killed outright, while many more died as a result of the social upheaval which prevented continuation of their intensive agricultural and economic systems. Still others were taken to Peru for mine labor. Malaria, smallpox, tuberculosis and yellow fever were introduced. The result was a drastic reduction in human populations and a redistribution of the remainder. Upland areas of Guatemala, which had little immediate value to the conquerors, were left with the native populations largely intact. Population recovered slowly but steadily along the seasonal lowlands of the Pacific coast and in the interior of Costa Rica, but, with the exception of the San Blas Islands which were resettled by remnants of the Cuna, malaria, yellow fever, and lack of seaports restricted populations along the Caribbean coast. In the four centuries between the arrival of Europeans and the present century, the natural vegetation of Central America regenerated under release from human activity. The estimated population (Fox, 1971) of the United Provinces of Central America at the time of their promulgation in 1823 was only 1.25 million. It must have been several times larger in pre-Colombian times, and today it is about 18 million. When the garrison at Bayano in Panama was withdrawn in 1607, most of the area reverted to forest. By 1640 the forests of Veraguas had recovered sufficiently to support a large sawmill industry based on *Cedrela* timber for export to Peru. In 1681 the entire Caribbean coast of Panama with the exception of Portobelo was reported as forest-covered (Sauer, 1966). A similar regeneration of forest occurred over most of Central America.

The effects of earlier dense populations are still evident. The distribution of palms and *Brosimum* in the Peten region of Guatemala (Puleston, 1971) reflects the slow rate with which some species recolonize earlier ranges. Many species of the tropical rain forest are dispersed by mammals and birds. Hladick and Hladick (1969) recorded 78 species of fruits eaten by the three species of monkeys they were studying on Barro Colorado Island. Croat (1976) noted 80 species of plants which seemed to have been dispersed by birds or mammals. Without the appropriate birds or mammals, these species of plants must fall back on gravity dispersal, which Gleason and Cook (1926) estimate may achieve a spread of only one kilometer in 500 years, assuming good establishment conditions. As well as farming the land, the early inhabitants hunted game, and as they reduced wildlife populations, they also reduced the ability of many plant species to recolonize felled forests. An interesting example is that of *Brunfelsia dwyeri,* a showy shrub or tree which has been found only in a lowland cloud forest near Panama City. This species has a leathery fruit the size of a large grape with a soft, custardy interior. It would seem to be mammal-dispersed, although this is discounted by Plowman (1973). Otherwise, this oval object has no means of dispersal other than falling to the ground and rolling a few meters at most, the seeds perhaps being later carried still further by torrents. Such a dispersal mechanism is slow at best, but even slower for going up hills. The closest relatives of *Brunfelsia dwyeri* are widespread in South America, occurring over a large portion of the continent in a variety of habitats, but the nearest such relative is found about 700 km to the south along the west coast of Colombia. Richards (1952) comments that this gravity system of seed dispersal is common in mature tropical forests, and Smythe (1970) reported on the relationship between seasons and mammalian seed dispersal. Hladick and Hladick (1969) planted seeds of several species of plants observed in the diet of monkeys. One of these, *Trichilia cipo* (Meliaceae), showed high germination following ingestion by monkeys and almost no germination when seeds had not passed through mammalian gut.

Because of the devastations of sporadic vulcanism and the deforestations of pre-Colombian man together with the reduction of dispersal-agent populations, it is unlikely that 400 years has been sufficient for full recovery of range by all species of Central American forests. This incomplete saturation is expressed as a high degree of floristic localism, which often makes the species composition of similar appearing tracts quite dissimilar. Many wide-ranging and well-dispersed elements are amply represented throughout the appropriate life zones, but minor elements of the forests remain geographically discontinuous.

DEFORESTATION TODAY

In the 20th century human population pressure has caught up with the recovery of natural vegetation. Coffee became an important crop for the moist uplands in the last century, and bananas replaced large areas of lowland forest in the present century. Since World War II cotton has become important to seasonally dry lowlands. And steadily the land hunger of subsistence farmers has increased. Pressed for more land because they are more numerous, and supported by higher prices from growing urban centers, these pioneers are extending the margin of cultivated lands at a rapid rate. Some of these farmers are displaced by purchasers who acquire land for speculation or agricultural purposes. Usually poor, with few possessions besides a machete, a hoe, and a cooking pot, driven by determined hope and sometimes hunger, the pioneer subsistence farmers bring an excitement to the frontier areas which commands political sympathy and often stultifies attempts at land planning and forest conservation.

There are important differences between the deforestation carried out in the past four centuries and that effected by the aboriginal people. Steel hand tools are better than anything used by the early Indians, and guns, of course, are more effective for provisioning than spears or arrows are. It is interesting to visit the San Blas coast where the Cuna inhabitants have resisted many outside influences. This is a land of small clearings, footpaths, and streams which are forded rather than bridged. Burning is on a small scale here. In fencerows, streamsides, bluff faces, and along trailsides, remnants of former vegetation persist. Forest species survive, not as parts of complex biological systems but rather as relict individuals, persistent testimony to the glorious forests of former times. As individual plants die, they are usually not replaced, yet so long as these individuals remain, there is a chance for continuity of the species. It is in such patchwork refugia that species survived the deforestation of pre-Colombian times. Every year the Cuna cut more forest for cultivation, but these people respect the forest as a source of food, medicine, and elements of folklore and mystery. It is not their enemy.

On the other hand, for the peasant, the forest is a hostile barrier to hopes and needs. When he cuts it down, he makes sure it is dead. Burning is repeated and waste places are cleared of trees and brush. This is hard work and he may not succeed in keeping his land clean, but he does manage to remove all original species. The removal of such forest species and of their chances of continuity is complete. As vehicular roads are built, roadsides are cut and burned, leaving a swath of formidable secondary species which resist even the machetes and herbicides of road gangs. The peasant precedes the roads by as much as 40 kilometers, staking out farms by squatting and cutting the woods. Removal of the forest animals and birds are part of the process. Slope makes little difference. After two or three years all species of the forest are gone, nutrients are leached from the soil, and it turns into a brick-like pavement. Viruses build up to weaken and deform his crops and vermin multiply to eat what crops he harvests. He has no chemicals to combat the destroying insects. Then he moves on to another forest site to repeat the same desperate operations. Not long ago, much of the Azuero Peninsula of Panama was cut and burned into submission in this way. Now it is losing population, and the people are moving to new forests in upland Chiriquí or to the wetlands of Darien. These two areas require techniques and crops different from those used on the seasonally-dry Azuero. Some of these emigrants go to the cities where they may become successful urban workers.

The peasant is only one element in the destruction of the natural landscape. Better-financed operations use chain saws, earthmoving equipment, insecticides and herbicides. The operations are on a large scale, sometimes altering several square kilometers on a single site. In Honduras and upland Panama some of this activity is directed to cattle raising, in Honduras and Nicaragua to increasing acreage of cotton, in many uplands for producing cabbages and other vegetables. Sisal and sugar are planted in some places. Tree crops such as coffee and cocoa also lead to plantation clearing. Working hand in hand with the plantation builders but sometimes in opposition to them are the lumbermen who select and cut the most useful trees, often long before the plantation machinery arrives and hopefully before the peasant squatter arrives. The modern deforestation proceeds at amazing rates, a modern crew being able to prepare as much land in a week as a group of peasants might prepare in a whole season. And preparation means complete sterilization of the landscape, forbidden forever to the natural plant species which are removed. It is comparable to the effects of a volcanic event, except that it is taking place over nearly the whole Central American landscape at the same time rather than as isolated cataclysms widely scattered across the Isthmus.

In addition to clearing activities, some crops command aerial insecticide spraying which destroys birds and other pollinators as well as crop pests. Extinction of organisms rather than conservation or tolerance is the goal of these anti-insect activities.

THE SCOPE OF PRESENT PRESSURES

The population of Central America is now about 20 million people. It is increasing at about 3 percent annually and will double in less than 24 years. Belize has a population density of only about 5 people per square kilometer, comparable with the neighboring Mexican States of Quinana Roo and Chiapas. El Salvador, a country of similar size, has many times this population. The density in El Salvador is 196 per square kilometer, in the same 125-200 person per square kilometer range as in Antillean countries, or as in Rwanda and Burundi, the most densely populated countries in Africa. At least until very recently, those countries with the lowest population density and where conservation is the most feasible, have unfortunately shown the least interest in conservation of natural landscapes.

Incomes in the different countries vary greatly; Panama has the highest per capita Gross National Product but it is only about 16 percent that of the United States. The Panamanian figure is slightly less than that of Greece, South Africa or Spain. Per capita GNP in El Salvador and Honduras is less than half that of Panama and compares with such countries as Korea, the Philippines or Algeria. Central America ranges from countries that are obviously poor to countries with some affluence. But in every country the peasants are mostly poor. And they form the bulk of the populations.

Central America displays birthrates only slightly lower than the unbridled natural frequencies of tropical Africa and southern Asia, and birthrates are more or less inverse to national income level. They are more than twice as high as in western Europe or the United States. Death rates are substantially lower than those of the unmitigated high levels of Central Africa or southern Asia, but they are much higher than those of western Europe. Central American birth and death rates are not directly comparable with those of Europe or North America because they have much younger populations. The Central American death rates are also nearly proportional to national income levels.

While it may be reasonable to hope or even predict that Central American countries will enjoy rising incomes, it is not reasonable to expect that they will rise rapidly enough to lead to stabilization of population growth in any of the countries before the end of the century. And long before that time, the population of Central America will have doubled to over 40 million. No large-scale emigration from the region can be foreseen. Massive birth restriction campaigns can certainly help to lower population growth, and this has already taken place in Costa Rica, but such campaigns are offset at least in part by lower death rates. The net lowering of natural increase is a long-term phenomenon. And because of the large proportion of young people in the Central American countries, reduction of family size to idealistically low levels would not prevent large increases in population during the rest of the century. There is no likelihood that population pressure will decrease or stabilize in any Central American country before the end of this century.

When one discounts aid programs, there are only small foreign exchange earnings from other than agricultural or natural product sources. Some countries, especially Panama, enjoy a good tourist trade and Panama also experiences economic stimulus from the Canal. Guatemala may be successful in finding oil in the Peten. But fundamentally, Central American countries will require greater returns from the land to feed their populations and to pay for imports of fuels, machinery and other items that keep the agricultural sphere going. And most of these countries look on the still uncut forest as their untapped legacy from which foods and export products will be obtained in the future. It is noteworthy that El Salvador, with the greatest density of population and the smallest area of remaining forests, has the highest percentage of GNP derived from manufacturing. El Salvador is one of the poorest countries, but it has been forced to emphasize other economic directions in the realization that there is no more room for expansion of agricultural lands. To other Central Americans today, the remaining forest is the basis for national expectation. It means lumber for building and furniture, fuel for cooking and warmth; it means new farms for sons and daughters in rural homes with no other promise; it means more farm produce to reduce food imports and more exports to pay for what is brought in. And the new roads which open the forests are a visible sign that the government is doing something, even if roads are expensive and perhaps not in the best long term interests of the country. This philosophy inhibits any successful establishment of conservation reserves.

Less than ten years ago the hills near Panama City were densely forested. As botanists followed the forest and road crews, Cerro Jefe, Cerro Campana and Cerro Pilón yielded several dozen species new to science and many more not elsewhere known to Panama. Now these areas are almost without forest. In upland Chiriquí, the space from Nueva California to the Costa Rican border was intact forest six years ago; now it is pasture land. The interior of Nicaragua around Matagalpa was once a vast forest of valuable furniture woods and pines. This forest is now reduced to remnants. Destruction is proceeding with amazing rapidity, and there are almost no checks in sight.

To save any natural areas, an almost complete reversal in public philosophy is required. And it must act at many levels of the society. Although government personnel may deplore destruction of watershed or forest reserves, they are powerless to reverse the course unless their superiors and important elements of the town populations they work with support their views. Fortunately, public awareness has increased in some countries.

CONSERVATION AREAS: PRESENT AND POTENTIAL

We would like to see reserves set up in each ecological zone to protect as many species as possible. But because of the fragmentation of the flora, more than one reserve is necessary for each zone. Budowski (1965) drew up 5 major vegetation types and 13 subtypes for the plant cover of Central America. Some of these already have lost their natural vegetation except for small areas. The lowlands of the Pacific coast and uplands flanking the Pacific are the most densely populated areas and have been part of the agricultural sector for a long period. Movement inwards from the Caribbean coast is now progressing rapidly at most points with the squatter vying with the major entrepreneur to claim and hold land. In a number of places deforestation has been achieved from coast to coast, and there is now similar pressure on the high mountains. There are still large tracts of forest in some places, but the situation is quite uneven. Most countries have laws regulating cutting of all trees, but such laws are frequently vaguely worded and therefore can be loosely interpreted and very variable in application. The largest areas of intact forest at present are the Darien in eastern Panama, Bocas del Toro in western Panama, the Peten region of Guatemala, and portions of the Nicaragua-Honduras interior.

At a conference held in Bogotá, Colombia, in December 1968, representatives of each Central American republic recommended an integrated system of pilot conservation areas with sites chosen for each country. Another such conference was held in San José, Costa Rica, in 1974. Some of the recommendations have become reality while others, unfortunately, already have been forgotten.

Guatemala:

The largest area yet uncut in Guatemala is the Peten and along the boundary with Belize, a vast area of lowland forest mainly on limestone substrate. Within it the Parque Nacional Tikal protects the ruins near Lake Tikal and about 1,500 square kilometers of forest around it. This is an area with almost no population. Poaching of timber became serious along the southeast boundary, but steps are being taken to prevent this and to mark the boundaries. Minerals have been found nearby and there is exploration for oil. Protection has been legislated for the Guatemala grebe, but the slopes of the lake where this unique flightless bird occurs are not protected and are now being cut off. Elsewhere there are small private preserves for the protection of the quetzal bird which lives in mature upland forests. The Bogotá conference recommended a conservation area covering the Atitlan complex in the Cuchumatanes, an upper elevation forest system of unusual physiognomy and floristic composition. Subject to charcoaling, wheat farming and sheep farming, this area has been protected in some measure by the resident Indians, who refuse permission for cutting by other Guatemalans. Guatemala has as yet no specific wildlands system, and the Peten park is administered by archeological personnel without wildland expertise.

Belize:

This is the smallest and least populated of the Central American countries, but little primary forest remains here. Small preserves have been set up to protect vestiges of former forest, e.g. at the Rio Frio Cave, and the local Audubon Society has been successful in reserving several small areas, such as that at Crooked Tree for the protection of herons. The well-known Mountain Pine Ridge is an edaphically determined pineland savanna which is subject to military gunnery practice.

Hunters come from many countries for the widely-touted bags to be had in the southern forests. Management of game kills and forest cutting here could easily maintain this forest as both a tourist resource and conservation area for the foreseeable future. Mayan ruins are scattered throughout the forest, thus adding interest for tourists and providing another argument for protection. This is the northernmost element in a tropical lowland forest formerly stretching almost without interruption into South America.

Honduras and Nicaragua:

The interior of these two countries includes a system of valleys and ridges with unusual and rich forests. These areas have been above sea level longer than any other parts of Central America, but they have not been explored botanically in any great detail. A major hardwood, *Macrohassaltia* (Flacourtiaceae), used to supply sawmills in the Matagalpa-Jinotega area of Nicaragua, was only described botanically a few years ago, yet supplies are now nearly exhausted (Williams, pers. comm.).

These two countries have much in common, and attitudes here are similar. Both countries have tree laws, but until recently neither had taken any steps to protect the natural environment. Bananas are important on the Caribbean side and cotton on the Pacific side. Coffee and other crops are planted on the uplands in both countries. Lumbering is an important occupation for both nations and provides some foreign exchange. In each country, cattle ranching often supplants the forest when the valuable timber is removed. These are the two largest countries in Central America, with low population density and low incomes, but the rates of natural increase are high. In the past ten or fifteen years forest cover has dwindled from about 70 percent to less than 50 percent. How much forest is left on the Cordillera Isabella and the Cordillera Dariense I do not know, but it will probably be gone by the end of this century unless steps are soon taken to preserve it. Neither country has a recognizable national park system, and squatters move freely throughout the areas. Neither country has implemented recommendations of the Bogotá conference.

Nicaragua recently took an encouraging major step in the direction of conservation. On March 10th, 1976, legislation was passed for the establishment of a national wildlife and forest plan. Planning is to be carried out with the help of Canadian funding, and already it has been decided to protect each of the nearly two dozen volcanos and migratory bird sites. The new law permits condemnation of land and enforcement of hunting and clearing regulations. It remains to be seen how effective this program will be.

While Honduras has no operative national park system, it is hoped that the forestry school at Siquatepeque will have an important effect on future land use planning in this country.

El Salvador:

The smallest and most densely populated of Central American countries, El Salvador now has four national parks which receive various measures of protection. For a country which has almost no remaining untamed land, these official sites come none too soon. Montecristo Park on the north is a cloud forest with strong protection. The park at Cerro Verde on the escarpment overlooking Volcán des Esalco is also a cloud forest, but it is subject to large-scale pilfering of orchids and other epiphytes, including whole tree branches which are sold primarily in the cities for Christmas trade. A tourist hotel in the park facilitates these illegal but popular activities. Establishment of the Cerro Verde park was one of the recommendations of the Bogotá conference. The Parque Deneger near the coast at La Libertad which also is new, has been subject to sporadic farming and cutting for many years. With adequate protection it is hoped that this area might recover as a good natural situation. A new park will be established next year at Thermopolis in the mountains above La Libertad. This is an old forest of Balsam de Peru, *Myroxylon perierae* (Leguminosae).

For a small, poor, and overpopulated country, the park system of El Salvador is exemplary, but it is only one step in the direction of maintaining the viability of species under threat of elimination. El Salvador is the most advanced country in Central America in terms of pressures to feed its people with diminished wild land. The parks must be better policed, and this can only be done if sufficient numbers of people are alert to the problems involved and to the importance of leaving these systems untouched in all respects. A press campaign to reduce the desirability of poached Christmas ornaments would help, and publicity on the hopelessness of trying to cultivate cloud forest orchids in seasonal lowland gardens might in the long run be of considerable value.

Panama:

Panama still has large areas of uncut forest but these are being rapidly removed for subsistence farming, plantation farming, and pasture. The denudation is proceeding on many fronts. As the Pan America Highway moves through Darién Province, peasants clear the forest and shoot the area's animals for food, in spite of laws against firearms. The Darién is floristically allied with the Chocó of bordering Colombia, and it has a unique, but as yet undescribed flora. Because human pressure on the Colombian side of this region already has cut high up the slopes (Gentry, pers. comm.), it is important that Panama set aside natural reserves in this region. Closer to Panama City, the magnificent lowland cloud forest on Cerro Jefe has been largely cut off, and forest roads provide logging trucks access to areas far north of the continental divide. This range of hills is the type locality of dozens of new species, many of them still believed to be endemic. The area now largely supports coffee plantations, chicken farms, and pasture. One landowner who wished to leave a tract of forest uncut as a nature preserve was urged to develop it or lose title to the land. Though the Cuna Indians control the San Blas coast, their slight but steady increase in population is bringing pressure on the forested ridge tops. Modern farming innovations, sponsored by outside institutions, also will cause large declines in the forest flora. No forest preserves have been proposed for Panama east of the Canal.

West of the Canal, the forests on Cerro Campana and Cerro Pilón have been a steady source of species new to science over the past 10-15 years. A new genus of Solanaceae *(Rahowardiana)* was discovered there only about 4 years ago. Cerro Campana still has forest on top, mainly due to steep rocky slopes and the present landowner's interest in land preservation. The area was recommended for protection by the Bogotá Conference and now has National Park status. Cerro Pilón is now popular with tourists from Panama City for weekend visits, and its forests may be of future tourist value if they remain intact. Moving further west, forests have been removed from all lowland Pacific Panama in recent times, except for small pockets on the Burica Peninsula near the Costa Rican border and on offshore islands. Though vigorous effort was made by enlightened citizenry of Chiriquí Province to publicize the tourist and conservation value of Volcán Chiriquí as a national park, the national government has only very recently acted on the proposals. Upland areas of Chiriquí Province are being cut off rapidly for coffee, vegetables, and pasture. The oak is being removed from the uplands and there are sheep at the highest elevations, even on Volcán Chiriquí. A new proposal has been made to construct a power dam and to establish a new 15 km long upland farming district above Gualaca. This scheme, popularly referred to as La Fortuna, involves removal of a large area of forest, but there is a bright spot in this long-range plan. The Panamanian government has authorized botanists at the University of Panama to make an ecological, forest resource and floristic survey of the area both on foot and by helicopter before actual cutting is begun. When this forest is removed, there will at least be some record of the kinds of things that grew there.

The Caribbean slopes of western Panama are still largely intact. A large forest reserve for conservation purposes would be possible here since there are comparatively few people living in the area at the present time. Since landowners in Chiriquí complain that it is now harder to obtain permits to open new lands, it would seem that the government is tightening up on unrestricted felling.

An important conservation element in Panama is now under the United States flag. In the Canal Zone, two forests, one near Arraijan and the other called the Madden Forest have long been protected, but even with the assistance of armed U.S. troops dropped at night by helicopter, poaching of timber and animals persists. The forest near Arraijan seems to have been abandoned as a forest preserve.

The most important nature preserve in Central America is Barro Colorado Island, situated in the Canal itself on the northern side of the Isthmus. Though this little island is only about six square miles in size, it has been protected for many years and has been the site of numerous scientific studies. It hosts some 1,400 vascular plant species known to date. Another good forest in the Canal Zone is the area bordering the north half of the east side of the Canal. Except for the oil pipeline which runs through it and the pipeline road which serves it, there is no sign of man. It includes both tropical moist and premontane wet forest. This is good forest with rich faunal life, and it is officially protected and even given some policing. A ride on the Panama Railroad past this forest is an exciting tourist event.

Under present political arrangements, these Canal Zone resources seem secure, and steps have been taken to protect Barro Colorado Island in the event of a shift of administration in the Panamanian government. It is highly desirable that the area served by the pipeline road be given similar international scientific status. The institution involved here is the Smithsonian Tropical Research Institute, which has good rapport with both the Panamanian government and the Panamanian population at large. Any steps which can be taken in the near future to strengthen this rapport would be of value, especially in view of a possible change in political administration of the Canal.

Costa Rica:

Costa Rica has a national park system with a number of parks already officially recognized as such. These are of different sizes and represent different sorts of vegetation, and the degree of disturbance and policing varies greatly in each. A small area at the top of Cerro Poas is protected, as is the summit of Cerro Chirripo, which contains a much larger park. The Cerro Poas summit park was a recommendation of the Bogotá meeting. A small area is protected in the Pacific lowlands at Santa Rosa, while another small lowland area at Cahuita now protects an already-disturbed coastal swamp forest. There is also the extensive Cabo Blanco dry forest reserve on the Nicoya peninsula. The watershed area on the Río Grande de Orosi at Tapinti is also well-policed. The Organization for Tropical Studies operates an integrated field biology facility in several parts of the country. This organization, which has operated in Costa Rica for a number of years, is funded mainly from the United States with supplemental income from use charges. Although the business arrangements are less than perfect and there are recurring financial crises, the story of the organization is largely one of success. For example, La Selva in northeastern Costa Rica, an area of about 4,320 hectares, is held in pristine condition, most of which is mature lowland wet forest. This station is not only protected but it is actively studied, with policing of both poaching and overuse by scientists.

In 1975 a major step was taken to place a substantial tract of prime, mature Costa Rican landscape under government control and to develop mechanisms for its protection and study. This is the 290km^2 Corcovado Park which covers about one-third of the Osa Peninsula. It is almost the only remaining area of approximately equivalent size of uncut, lowland tropical wet forest in Central America. There are also inclusions here of many other vegetation types. Steps have been taken to set up policing policies and to remove squatters who have intruded.

The conservation picture in Costa Rica is strong, and it is supported by an alert scientific community which embraces both professional and amateur groups.

POSSIBILITIES AND FEASIBILITIES

The preceding is a brief summary of the state of destruction of the Central American landscape and species complex. A few suggestions on areas to save have been made. But the difficulties in achieving any sort of conservation in Central America are numerous and commanding. In many cases, cutting of the forest for pasture or subsistence farming are a short-term gain, and governments of poor countries find it understandably difficult to deny peasants this possibility for their economic betterment, even if the long-term results of their actions work against them. Under pressure from abroad, governments may set up official parks, but without a firm conservation philosophy, the parks more often than not are unpoliced and violated. The case for active conservation must rest on better grounds than tokenism for international appeasement. It is quite plain that continuance of present activities in much of Central America will inevitably result in the extinction of hundreds or thousands of plant organisms, but even this does not seem to be sufficient justification for the peasant to refrain from obtaining his living in the ways he has been traditionally using for many generations. Other reasons for protecting forests should be coordinated to the fullest extent. Watershed protection is a solid reason for prohibiting forest felling and the costs of neglecting this are so great that it should have top national priority. This does not always mean saving the forest, however. For example, the banks of the Bayano River in Panama were cut off before the lake behind the new Bayano power dam was filled several months ago. But if the forest is cut, it means landscaping will be required which may be, in the long run, much more expensive than letting the natural forest control erosion. Tourism is a lucrative source of revenue, especially of foreign exchange, and where tourism benefits from forest, government may couple the two activities in a combined program. Thus, protection of the game forests in southern Belize and of the woods bordering the Panama Railway in Panama have dual benefit. The forest at Cerro Verde in El Salvador improves the attractiveness of the hotel, but the tourist resource must be protected from the tourists themselves. In other places game management requires restraint on overkill, and forest-tourist management requires that tourists be enjoined from removing epiphytes and other plants from the woods. Military reserves are another place where forests can persist, but there are several disadvantages in this form of conservation area. Military personnel are hired to use firearms and they often use them indiscriminately on any animal or tree that comes into view. It is a well-known fact that armoured vehicles and artillery ranges can make desert out of almost any vegetation. And with changes in system of government, military reservations are sometimes discontinued.

To be successful, nature reserves must receive the attention and concern of both an enlightened public and an active scientific echelon. Being told they are squandering their landscapes does not make for an aware or receptive public, especially when such criticism is levied by foreigners. But when a vocal element of the resident population campaigns to restrain unbridled destruction, the possibilities for creating and maintaining natural systems improve enormously. If we can point to any single positive factor in the Costa Rican success in establishing national parks, we must single out the alert and interested elements of the population who have had the foresight to realize the importance of removing these irreplaceable areas from the immediate economic sphere. And in Central America as a whole, the extent to which governments have moved towards protecting landscapes is proportional to the size and vitality of the interested scientific communities. This professional base in Central America, unfortunately, is small, and so it has not always been able to translate the need for conservation into political action. The success in Costa Rica has resulted at least in part from the presence of OTS, which has been instrumental in channelling foreign assistance into local scientific undertakings. A similar but smaller beginning has now been made in Guatemala. In Panama, the university and the forestry agency both have become increasingly effective in educating the population at large, and the amateur Chiriquí park group has drawn attention to the importance of its goals through a film documenting the value of protecting this unique national resource. Canadian assistance to the government of Nicaragua has encouraged the important new legal steps taken there.

The writing of a flora is an important step in the botanical investigation of a country, and to be done properly it requires large herbarium resources. But this is only one step in understanding the botany of the country. Though the 1/4 million specimens now deposited in Central American institutions is an impressive holding, it will be augmented as local studies continue. It is encouraging to realize that Central American countries no longer need to send abroad for routine determination of specimens. There are well-trained curators and teachers throughout the area, and several of the universities also make extensive use of visiting scientists. Extension instruction now includes botany and other nature sciences in Panama and Costa Rica. When President Oduber of Costa Rica is lauded by foreign dignitaries and scientists in the national press for his conservation efforts, it gives the average Costa Rican a feeling of personal pride in his country and his park, even though he may be opposed to Oduber's political party and candidacy. And botanical teaching in the schools based on local examples goes a long way towards helping people respect their environment. In essence, what is required in many parts of Central America, is a complete reversal of past attitudes. From a belief that the forest is an unlimited hostile resource to be cut down and annihilated to the understanding that the forest is a very limited and unique resource, worthy of careful management rather than destruction is a revolution in thought for the populations of Central America. Yet without such a change in thinking, no amount of money, pressure, or propaganda from outside Central America will result in a single hectare being given more than nominal protection.

Efforts to protect the remaining biota of Central America must move swiftly. But in order to be effective, such efforts must move either through or in close cooperation with local institutions. And all efforts must also be made to strengthen the local scientific base work towards the goal of preserving a modicum of the area wildlife. In some countries this is easy to do, because there already are well-developed institutions and channels for cooperating with local scientists. In other countries, however, biological studies are less advanced, and it is in these countries that it is even more necessary that help be given to improve the influence of local biologists. This may involve keeping agriculture out of certain valuable habitat areas, and channeling the land requirements of an impoverished and politically vocal peasantry to other areas, but it is worth the effort because much is at stake.

ACKNOWLEDGEMENTS

The author thanks the many people who assisted with data used in this report, especially G. Budowski, Dr. Celedon, M. Correa, H. Cutler, J. Dwyer, A. Gentry, R. Keating, G. Pilz, D. Puleston, D. Spellman and L. Williams.

LITERATURE CITED

Beard, J. S. 1955. The classification of tropical American vegetation types. Ecology **36:** 89-100.

_____. 1944. Climax vegetation in tropical America. Ecology **25:** 125-58.

Bowers, F. D. 1970. High elevation mosses of Costa Rica. Jour. Hattori Bot. Lab **33:** 7-35.

Beaman, J. 1968. Some statistics on the alpine flora of Mexico and Guatemala. Unpublished mimeographed leaflet. Seminar at the Univ. of Tenn., Knoxville, Tenn. February 8. (not seen).

Budowski, G. 1965. The choice and classification of natural habitats in need of preservation in Central America. Turrialba **15:** 238-246.

Burger, W. 1971. Flora Costaricensis. Fieldiana Botany.

Correa A. and D. Mireya. 1975. Manual Laboratio de Botánica General. Panama.

Croat, T. 1976. Flora of Barro Colorado Island. Stanford Univ. Press. in press.

D'Arcy, W. G. 1973. (1974) Solanaceae *in* R. E. Woodson et al. eds. Flora of Panama. Ann. Missouri Bot. Gard. **60:** 573-780.

Dressler, R. L. 1972. Terrestrial plants of Panama. Bull. Biol. Soc. Wash. **2:** 179-186.

Dwyer, J. D. and D. L. Spellman. A list of the dicotyledons of Belize. in prep.

Escobar, N. 1976. Angiospermas, decciones de Botánica Sistemática. Panama.

Fox, D. J. 1971. Central America, including Panama. pp. 121-178 *In* H. Blakemore and C. T. Smith eds. Latin America: Geographical Perspectives. Methuen and Co. London.

Gentry, A. L. 1976. Floristic needs in Pacific tropical America. Allertonia. in press.

Gleason, H. A. and M. T. Cook. 1926 (1927). Plant ecology of Porto Rico, Vol. 8 *in* Scientific Survey of Porto Rico and the Virgin Islands.

Hamer, F. 1974. Las orquideas de El Salvador. San Salvador.

Hemsley, W. B. 1879-88. Botany *In* F. D. Godman and O. Slavin eds. Biologia centrali-americana. London.

Hladick, A. and C. M. Hladick. 1969. Rapports trophiques entre vegetation et primates dans la forêt de Barro Colorado (Panama). La Terre et la Vie 23: 25-117.

Holdridge, L. R., W. C. Grenke, W. H. Hatheway, T. Liang and J. A. Tosi, Jr. 1971. Forest environments in Tropical Life Zones, A Pilot Study. Pergamon Press, Oxford. 747 pp.

Holmgren, P. K. and W. Keuken. 1974. Index Herbariorum Part 1. The Herbaria of the World. ed. 6.

Kroeber, A. L. 1939. Cultural and natural areas of native North America. Univ. Calif. Publ. Amer. Archaeol. and Ethnol. **38.** (not seen). Cited by McBryde, F. W. 1947. Cultural and historical geography of southwest Guatemala. Smithsonian Inst. Soc. Anthrop. Publ. **4:** 184.

Lothrop, S. K. 1959. "A re-appraisal of Isthmian Archeology", Mitt. Mus. Volkerkunde Hamburg 25: 87-91. Cited in C. F. Bennett 1968. Human influences on the biogeography of Panama. Ibero-Americana **51:** 1-112. Univ. of California Press.

Molina, A. 1975. Enumeracion de las plantas de Honduras. Cieba **19:** 1-118.

Plowman, T. 1973. The South American Species of *Brunfelsia* (Solanaceae). Ph.D. thesis, Harvard University.

Puleston, D. C. 1971. An experimental approach to the function of classic Maya chultuns. Amer. Antiq. **36:** 322-335.

Raven, P. and H. Curtis. 1975. Biologia Vegetal. p. 716. Omega. Barcelona (Spanish Translation of Biology of Plants. (1970). Worth, New York)

Record, S. J. 1927. Trees of Honduras. Trop. Woods **10:** 10-47.

Richards, P. W. 1952. The Tropical Rain Forest. Cambridge Univ. Press. 450 pp.

Sauer, C. O. 1966. The Early Spanish Main. Univ. Calif. Press. 306 pp.

Sapper, K. T. 1924. Die Zahl und Volksdichte der indianischen Bevolkerung in Amerika vor der Conquista und in der Gegenwart. Int. Congr. Amer. **21:** 95-104. The Hague (not seen). cited by McBryde, F. W. 1947. Cultural and historical geography of southwest Guatemala. Smithsonian Inst. Inst. Soc. Anthrop. Publ. **4:** 184.

Seemann, B. C. 1852-57. The botany of the voyage of H. M. S. Herald. London.

Smythe, N. 1970. Relationships between fruiting seasons and seed dispersal methods in a neotropical forest. Amer. Nat. **104:** 24-35.

Spellman, D. L., J. D. Dwyer, and G. Davidse. 1975. A list of the monocotyledons of Belize including a historical introduction to plant collecting in Belize. Rhodora **77:** 105-140.

Standley, P. C. 1930. A second list of the trees of Honduras. Trop. Woods **21:** 9-41.

_____. 1931. Flora of the Lancetilla Valley. Publ. Field Mus. Bot. **10:** 1-418.

_____. and S. J. Record. 1936. The forest and flora of British Honduras. Publ. Field Mus. Bot. **12:** 1-432.

_____. and S. Calderón. 1926 (1925). Lista preliminar de la flora de El Salvador, San Salvador.

_____. et al. (varies) 1958 . . . Flora of Guatemala. Fieldiana Botany.

Stevens, R. L. 1964. The soils of Middle America and their relation to Indian peoples and cultures. _In_ R. Wauchope, Ed. Handbook of the Middle American Indians vol. **1:** 265-315.

Steward, J. H. 1963. ed. The Circum Caribbean Tribes, Vol. 4. Handbook of the South American Indians. Cooper Square Publishers. New York.

Strasburger, E. 1962. Lehrbuch der Botanik für Hochschulen. ed. 28. 832 pp. Fischer, Stuttgart.

Taylor, B. 1976. Las Leguminosas Papilionaceas Herbaceas y Arbustives da Panama. Editorial Universitaria — Panama.

Wauchope, R. 1965. Southern Meso America, _In_ J. D. Jennings and E. Norbeck eds. Prehistoric man in the New World. pp. 331-386. Univ. Chicago Press.

Weber, H. 1958. Die Paramos von Costa Rica. Akad. Wissensch. Lit. Mainz. 1958(3): 1-78.

Weyl, R. 1965. Erdgeschichte und Landschaftsbild in Mittelamerika. Kramer, Frankfurt 175 pp.

Wilson, C. L. and W. E. Loomis. 1968. Botánica. First edition in Spanish. Uteha, Mexico.

Woodson, R. E. et al. 1943 . . . Flora of Panama. Ann. Missouri Bot. Gard.

CONSERVATION AND THE ENDANGERED SPECIES OF PLANTS IN THE CARIBBEAN ISLANDS

Richard A. Howard

The Arnold Arboretum, Cambridge, Massachusetts 02138, U.S.A.

The Caribbean Islands have been a magnet for tourists for many years. Travellers of all ages and descriptions, from all walks of life, never cease to be fascinated by the Islands' small, isolated sanctuaries, fine climate, clear water, sandy beaches, hazy mountains, and dramatic spewing volcanoes. More recently, gambling casinos and quick divorces have added appeal for a few people. Such exotics as sugarcane and rum, cacao and coffee, bananas and plantains, nutmeg and ginger — all introduced plants — have been some of the products conjured up in the popular image of the Caribbean Islands. In addition to these superficial aspects, the island archipelago has a scientific interest and appeal. Island land masses stretch over 1,700 miles east and west from Barbados to the Pinar del Rio tip of Cuba, and 1,200 miles north and south from Grenada to the tip of the Bahamas. The nearly one thousand islands range in size from a scant few square yards of sandbar or rock, which are pinpointed on the maps and possess a few plants, to the island of Cuba with 44,250 square miles and an impressive diversity of vegetation numbering over 6,000 species, of which more than fifty percent may be endemic. Soil types vary in their productivity and vegetation types from pure salt rock, volcanic ejecta, dogtooth limestone through serpentine and bauxite-rich soils to fertile lands heavily used for agriculture. Each soil type may contain few to many local endemic species. Land occurs below sea level in the Enriquillo basin and at 9,700 feet above sea level in the Cordillera Central on Hispaniola. Graben thrusts, fault lines, elevated beaches and volcanism in many forms indicate the recent instability of the land.

In the past, populations have been controlled largely by emigration to many different homelands. Though heavily populated today, many islands supported even larger numbers of residents in past years. Agricultural lands, which were abandoned in the last century, have now been revegetated in natural succession. Old roads are now so overgrown that they are hard to follow. The substantial Great Houses of successful planters — with their broad expanses and huge ceilings so reminiscent of a long-vanished prosperous era — have now fallen into crumbly, bleak ruins. In other areas, by contrast, cities have grown and encroached on natural areas. New oil refineries have replaced the old sugar mills in contaminating the air and water. Strip mining, a comparatively recent development, alters land profiles on a large scale. Natural beaches have been cleared of their typical vegetation for the "enjoyment" of tourists, while in the hills squatters and milpa (migratory, cut, burn, plant) agriculture present images of gross destruction and cause for concern.

In view of all of the above, consideration of the endangered species in the Caribbean Islands is, obviously, not a simple matter. An emotional reaction to a potential loss is no more satisfactory than a laissez-faire attitude. A serious analysis, therefore, is in order as part of this Symposium.

THE NATURE OF THE VEGETATION

The plant life of the Caribbean Islands cannot be regarded as unknown or needing immediate study or a massive collecting program. Botanical observations have been made in the New World since 1492 when Columbus reported on existing plants in the area, even though he did not collect or preserve herbarium specimens. Subsequently, through the activities of botanists interested in colonial lands, or more recently through an international approach, collections have been made by Spanish, British, French, Dutch, German, Norwegian, Swedish, Canadian, American and Soviet bloc botanists whose dried specimens are, at present, distributed in the herbaria of each country.

A very few floras have been prepared by local botanists, such as: Duss (1897): *Flore Phanérogamique des Antilles Françaises;* León and Alain (1946-1969): *Flora de Cuba;* Moscoso (1943): *Catalogus Florae Domingensis;* Adams et al. (1972): *Flowering Plants of Jamaica;* Gooding et al. (1965): *Flora of Barbados.* Non-resident botanists contributed classic studies if not floras, viz., Swartz (1797): *Flora Indiae Occidentalis;* Urban (1898-1928): *Symbolae Antillanae;* Fawcett and Rendle (1910-1936): *Flora of Jamaica;* Grisebach (1859-1864): *Flora of the British West Indian Islands;* Britton and Wilson (1923-1930): *Botany of Puerto Rico and the Virgin Islands;* and Britton and Millspaugh (1920): *Bahama Flora.*

Unfortunately, the multinational interests in the Caribbean Islands have led to a plethora of uncoordinated, scientific studies. Grisebach's Flora of the British West Indian Islands, for example, did not consider the islands of Martinique and Guadeloupe, while Duss's Flore Phanérogamique des Antilles Françaises covered Martinique and Guadeloupe but neglected to include the island of Dominica, which is situated immediately between them. Major monographic treatments of families have been relatively few, with Hitchcock's Manual of the Grasses of the West Indies, perhaps, the only large, inclusive family treatment. It is the monographic approach, hopefully — including northern South America, Central America and Mexico — which will eventually determine the true nature, identity, delimitation and distribution of the plants of the Caribbean Islands. Yet, every botanist should be aware that the typification of many of the Caribbean periphery plants will be based on the early collections of botanists who may have visited several areas but whose published findings are only associated with the Antillean islands.

A consideration of the existing floras and lists of species reported from the various islands would suggest a total flora of the Antilles in the range of 12,000-15,000 species. About 25 percent of these represent introduced or widely distributed tropical species; about 50 percent of the species are restricted to a Caribbean-wide area, and about 25 percent of the species are of localized distribution, ranging from single island endemics to those with an Antillean delimitation. León and Alain estimated the vascular plants of Cuba included 50 percent endemic species; however, Duek recently indicated that only 6.2 percent of its ferns and allied plants were endemic. Hispaniola has specific endemism of 33 percent, Jamaica 20 percent, Puerto Rico 4 percent, the Bahamas 13 percent and the Lesser Antilles 12 percent. The nationalist or single-island approach has led to the over-description of species, as recent monographs clearly show. In all such studies, more endemic species are being reduced than new species described.

The compilers of the existing floras of the Caribbean have had to consider species represented not only by single and old collections but often with incomplete data. The entry is frequently considered an endemic species. In reality, many of these old records have proven to be misidentifications, either because of improper localization or because an introduced plant was improperly recognized as being part of the native flora. Disparate areas were visited by such early collectors as Surian and Plumier, Jacquin, Houstoun, Richard, and Sessé and Mociño, and their collections are often wrongly attributed to the Caribbean islands. For example, several species collected by Houstoun in Veracruz, Mexico, were named "barbadensis", misleading subsequent workers to attribute the species to the Antilles. To this day, Barbuda and Barbados, Dominica and the Dominican Republic, Grenada and Nouvelle Grenada are confused in literature as well as the herbarium. Localities cited in the early literature as India Occidentale may not apply to the Antilles at all but rather to coastal South America or Central America. Richard's printed label of Guyanensi-Antillanum may be found on specimens from northern South America or some island of the Antilles. "Montserrat" has proven to be a troublesome "locality" applying to European species, wrongly attributed to the West Indies, to plantations of nostalgic local owners or to the actual island of Montserrat. Ossa's collections from "Habana" are commonly of plants cultivated in a former Botanic Garden in that city, though the plants originated in other islands or countries. Specimens by Guilding, Anderson or Caley may have been cultivated in the St. Vincent Garden under their supervision or gathered during their travels from plants growing elsewhere.

However, some caution must be used in excluding all old records from a modern flora without examining a valid specimen. A plant named *Maytenus vincentinus,* described in the Celastraceae by Turczaninoff in 1863, could not be identified by subsequent monographers. A specimen was finally located in Leningrad and proved to be common *Zanthoxylum monophyllum* (Lam.) Wils. of the Rutaceae. The identity is still unknown of *Celastrus grenadensis,* described by Urban in 1904 (Vol. 5:51) and based on an Egger's 6,222 from Grenada. The genus *Celastrus* is not to be expected in the West Indies, unless cultivated. The description does not permit identification, although Urban was a careful observer. Egger's collections are widely distributed but this number cannot be located and the Berlin specimen was destroyed in World War II. It is possible that someone subsequently identified the plant correctly and refiled it properly without any annotation in the Celastraceae.

Jacquin's plant, described as *Brunfelsia spinosa* but without location, could not be identified by a recent monographer who even questions the application of the specific name in *Brunfelsia*. No material has subsequently been found for this or such other species as *Cordia juglandifolia, Phaseolus sclareodes, Rondeletia disperma,* or *Vicia disticha.* When correctly identified, such names dating from 1760 almost surely will displace names in current use.

Occasionally, modern workers have described old collections from well-known areas where the plant has not been seen for many years. Thus, Box and Philipson (1951) described *Mastichodendron sloaneanum,* designating as the holotype a collection made by Sir Hans Sloane on Barbados in 1687. Although it is unusual for a plant to be considered 264 years after having first been collected, Box and Philipson were able to trace accounts of this species by common name and usage through the literature of the 18th century. Today, a few trees designated as Turner's Hall Woods on Barbados are all that remain of this once large and more diverse forest.

Perhaps the most famous plant on the island of St. Vincent is *Spachea perforata.* In traditional stories, this plant, which has been called the Soufrière tree, once existed on the slopes of the Soufrière volcano, and all native specimens were destroyed by the eruption in 1902. Even the plant in the Botanic Garden on the island "disappeared" for a decade when it was not recognized because the label had been removed. When this visitor pointed out the tree in 1950, the publicity about it was revived. Ironically, there is no evidence to suggest that *Spachea* was ever a native of the mountains of St. Vincent; the existing plant is probably an introduced species from South America. Although Guilding's list of the plants in the Botanic Garden of St. Vincent published in 1825 uses only common names, it has been useful in verifying the fact that certain plants now established in the local vegetation originally were introduced species.

The St. Pierre Botanical Garden on Martinique (1803-1902) was destroyed in the eruption of Mt. Pelée. The early curators, Hahn and Bélanger, did collect herbarium specimens of plants grown there. Duss (1897) promised a publication on the cultivated plants of Martinique and Guadeloupe, but this project never materialized, and not even a manuscript has ever been found. Recently, Dr. Alicia Lourteig of the Muséum d'Histoire Naturelle in Paris located some of the Hahn and Bélanger specimens which were identified. Thus, 100 years after their collection, these specimens have become available for study. An occasional label indicates the date of introduction of a species commonly cultivated in the area today.

The recognition of endemic or rare plants requires careful study. Automatically adding these plants to lists of endangered species can be a scientific error. The endemic species may not be endangered at all, but instead may be extremely abundant in a localized area. *Pilea krugii, Psilogramme portoricensis, Psychotria grosourdyana, Wallenia yunguensis* and *Cyathea pubescens* of the Puerto Rican flora represent examples of the many facets of the problem. The first four all have been considered endemic to Puerto Rico. *Pilea krugii,* however, is represented by literally hundreds of thousands of living plants in the mountains throughout Puerto Rico. Perhaps it is endemic, although the genus needs a monographic treatment. It certainly is not endangered nor will it be in the foreseeable future. Though *Psilogramme portoricensis* is an endemic species, Maxon in Britton and Wilson (1926, vol. 6:439), it is now recognized as *Eriosorus hispidulus* (Kunze) Vareschi (A. Tryon, 1970) and occurs in South America, Central America, and in one locality on El Yunque in Puerto Rico. Having searched personally for material, I can verify that the plants are exceedingly rare in accessible locations. It is possible that more specimens may be found in a rope descent of the treacherous vertical rockface of that mountain. Thus, the species is endangered locally but not throughout its range.

Psychotria grosourdyana was considered an endemic species, but it has now been included in the very common *Psychotria guadalupensis* (DC.) Howard, of wide range in the Antilles. Maxon in Britton and Wilson (1926, vol. 6:386), gives *Cyathea pubescens* Mett. a range of Jamaica, Cuba, Hispaniola, and Puerto Rico. In a recent monograph, Gastony (1973), indicated how a serious misidentification could be clarified by a detailed study. The Jamaican plants called *Cyathea pubescens* are *Nephelea pubescens* (Kuhn) Tryon and are endemic to Jamaica. The Puerto Rican plant called *Cyathea pubescens* is now known as *Alsophila bryophila* Tryon (1972), and is endemic to Puerto Rico. The Cuban and Hispaniolan plants are *Alsophila minor* (D.C. Eaton) Tryon (1972). Not one of these segregates is endangered since all are abundant locally.

Wallenia yunquensis (Urb.) Mez. also is listed as an endemic species localized on the summit of El Yunque. Although it was poorly collected for many years, a recent ecological study (Howard, 1968) revealed that the plant was very abundant throughout the Luquillo range, which includes El Yunque. This plant is known to contain a chemical constituent showing value for its biological activity against certain types of cancer. The analysis was done originally on 40 pounds of dried material. Later requests for more material were filled, but when the request was for "several tons" for commercial production, it was obvious that the wild population could never supply a commercial need. This species clearly could be listed as endangered, and an attempt should be made to synthesize the chemicals rather than depend on a native source.

THE LOCATION OF ENDANGERED SPECIES

The diverse types of vegetation of the Antilles have been variously classified by ecologists. Beard (1955) and Stehlé (1945) followed a Clemensian approach in describing the various units. Howard (1973) has suggested a much simpler classification, utilizing geographic location as well as the environmental features of the area:

Coastal formations:
   beach
   strand
   rock pavement
   mangrove

Lowland formation:
   thorn scrub
   savanna
   marsh or swamp
   alluvium

Montane formations:
   wet or dry forests on limestone
   montane sclerophyll
   palm brakes
   tree fern communities
   pine forests
   cloud forests
   volcanic and soufrière communities
   crater lakes
   elfin thickets

Most of these areas contain endemic and often rare species, and each may be "endangered" in a different way. Thus, a consideration of the reasons for concern may be appropriate.

THE NATURE OF THE PRESENT THREAT

Nature and man are running a close race in the Caribbean Islands on the extent of destruction of the vegetation.

The frequency of hurricanes has been well-documented. Evidence of their destructive influence are clearly visible at the time of the turbulence and for many years afterward. While hurricanes may decimate an area, it is doubtful if any species have actually been lost through these dramatic storms. The aftereffects are striking, however. Forest destruction may be followed by the regrowth of different rapid-growing, shade-producing species, and the effects of the different cover are difficult to evaluate. A mixed native vegetation canopy has a significantly different effect from that produced by *Cecropia* or *Prestoea* on the type of undergrowth which develops or survives. Landslides may seem to be spontaneous alterations of surface, yet their frequency can be associated with the supersaturation of the land following the passage of tropical storms. Such areas may either remain barren for many years or they may be dominated rather quickly by solid stands of ferns such as: *Nephrolepis, Dicranopteris* or *Odontosoria,* or even by such fast-growing shrubs as *Pluchea, Eupatorium* or *Clibadium.* Local populations of rare species can be eliminated by landslides.

Volcanism has been a more spectacular element in the devastation of large areas of the Caribbean. The slopes of Mt. Pelée on Martinique, including the coastal town of St. Pierre, and the slopes of the Soufrière in St. Vincent each have been studied for the succession and timing of the reestablishment of vegetation. Fumaroles and soufrière fumes continue to play a role in altering vegetation near or downwind from their orifices. The recent eruption of the Soufrière on St. Vincent produced a large cone of cinders within the crater lake but it did not overflow the caldera. The potential destruction was sufficient to alert several zoologists and botanists to examine the area for collections and basic distribution data in case the eruption continued (Howard, 1975). The Soufrières on Montserrat, which continue to increase in size, pose a major threat in the potential pollution of watershed areas with sulfurous water. The Boiling Lake on Dominica shows current activity of unusual strength, while an underwater ejection of lava is occurring near Kick 'em Jenny. Volcanic activity may have eliminated some unusual species but the evidence for such conclusions is difficult to obtain. This type of destruction, however, is beyond the control of man.

The effects of man on the vegetation of the Antilles have been many. The animals he keeps, the crops he cultivates, his activities in mining, forestry, housing and tourism — all have played a role in the past and continue to do so today.

Controlled grazing of large animals has had a limited destructive role in the Caribbean, for herds have not been large here nor have there been extensive acreages used for pasture. Destruction by goats in uncontrolled grazing still occurs in many of the drier areas on both large and small islands. These animals have outlasted the human occupation of Beata, Alta Vela and Redonda islands, and their browsing impact is obvious to the botanist. Although goats have been allowed to run free for decades, the herds, surprisingly, have not overdeveloped. Perhaps the occasional hunter has kept the herd under control or, possibly, the population has attained a natural balance. In any case, the damage to vegetation by goats is more extensive when herds are controlled with periodic roundups for milking or moving to new areas. This concentration of grazing has inevitably resulted in many overgrazed areas. The demand for goat meat is less at present than in the recent past, and the number of goats may actually be smaller now than before the era of imported meat and of refrigeration. Today, destructive browsing by goats is less of a factor endangering the vegetation than it has been in the past.

Though agriculture still plays a major role in the use of land in the Antilles, it had its greatest effect in the early decades of the 19th century. In sugarcane production, for example, before the development of machinery for harvesting and transport, and before the introduction of oil furnaces, a larger labor force utilized hand labor for cutting cane and native timber for fuel. Fields for food crop production and livestock grazing also utilized manual labor. The agricultural industry was actually lost on St. Lucia during World War II when the principal acreage was converted to a military base and airport, and the sole remaining sugar mill was removed from the island. In recent years, higher labor costs in Puerto Rico have forced the marginally profitable areas out of business, and throughout the Antilles many fields, which once were in cane, have now reverted to the invading vegetation or have been converted to housing.

Citrus production on several islands of the Lesser Antilles diminished because of the inability to control diseases of the crop. The breakup of the banana monopoly and the large-scale company production in Central America, by contrast, has made banana growing profitable in the Antilles for the European Common Market. In many cases, the areas now devoted to bananas represent encroachment on new lands in the mountains. It is also in these same areas that a milpa type of agriculture still prevails. The local effects of such practices can be serious when they occur in a watershed, for the protection given to these former "crown lands" has been removed by the local independent governments. It cannot be determined how often such squatters destroy rare or endangered species. However, the previously-mentioned case of *Mastichodendron sloaneanum* indicates that that species, at least, was in all likelihood eliminated from the wild by man's overuse of the vegetation.

More recently, the late Dr. Caroline Allen described a new species in the Lauraceae, citing one of my own collections from the Dominican Republic. The area in which I made the collection was a native, undisturbed, mountainous woodland. When it was decided that a new species was involved, Dr. A. Liogier attempted to follow a route I could describe from field notes in order to re-collect the material. On his arrival, he found the area was now a coffee plantation with patches of cassava, corn and dasheen. The plant was found again several miles away, but the location of the original collection was altered by man in milpa agriculture.

Mining for minerals may currently be at an all-time high in the acreage involved and in the destruction of the native vegetation. Gold was never abundant enough for commercially-viable mining activities in the Antilles, although Columbus and subsequent explorers avidly sought it. The botanist, Padre Fuertes, worked at a vein of copper in Hispaniola, but neither he nor subsequent explorations have ever located particularly significant quantities. Phosphates were briefly mined on the now uninhabited island of Redonda, and yet no early description of the defaced landscape is available to determine what preceded the largely introduced weedy flora which now is present there. The same is true for the guano mining operation on Alta Vela during the Grant administration. The land contours on both islands today show the obvious effects of these mining operations.

In Cuba, large areas of red-colored soil in the Oriente province were held in reserve as iron deposits by American steel companies, while in the southern Oriente area nickel ore was mined. However, in the period since World War II, there has been active strip mining of large quantities of bauxite-rich soils, the source of aluminum, with removal for processing and export. As a result, large areas have been excavated in several sections of Jamaica and the Dominican Republic. Where local processing of the ore to alumina has taken place, quantities of fines or altered residues have been dumped in a few mined-out pits or behind newly-created dams. These areas are as slow to recover as the untreated mined-out pits.

Another minor and secondary effect of the ore-drying operation has been the appearance of a red dust which coated the nearby vegetation and discolored the coastal waters. These effects brought much criticism from local conservationists, but there is no indication that the dust has eliminated or even affected any unusual or rare plants in the area.

The initial mapping for bauxite deposits in the Dominican Republic permitted a few correlated studies of the local vegetation. It was determined that some species of endemic plants occurred primarily, if not exclusively, on ore deposits of commercial significance. Obviously, some of these endemic species would have to be removed with the excavation of large deposits, but there is little likelihood that all of the rare species will be eliminated as a result of these mining operations.

The laws in Jamaica initially called for replacing the topsoil and restoring the "original vegetation" after strip mining was completed. Since mining already had been under way for several years when it became necessary to obey the law, a survey was conducted on areas to be mined in order to determine exactly what the "original vegetation" consisted of. An inventory of these areas showed no endemic or rare species, but rather a commercially worthless growth of secondary vegetation of an area from which all valuable trees had been cut. A recommendation was made to the effect that a reconditioning of the soil with the establishment of useful grasses or fruit or lumber trees was a sounder procedure than replacing the original component species in an artificial mixture. This recommendation, by and large, has been followed.

The commercial production of oil has not been successful in the Caribbean Islands in spite of repeated, often expensive prospecting and actual drilling operations. The recently-completed drilling operations in the Enriquillo Valley have had less deleterious effect on the vegetation than have the fumes and waste products generated by the fractionation plants newly-established on St. Croix, Aruba and Curaçao.

Cutting of native forests for commercial timber production has not been extensive in the Antilles except in Cuba and Hispaniola. Successful lumber mills did operate in eastern Cuba, with *Pinus* the principal resource. No attempt at reforestation in cut areas has been attempted on a large scale. By contrast, under the administration of Rafael

Trujillo in the Dominican Republic, lumbering concessions were granted, and the regulation that ten trees had to be planted for each one that was cut was enforced on some concessionaires. Plantations of useful species of *Swietenia, Cedrela* and *Eucalyptus* have been made on Guadeloupe and Martinique. Plantings of several species of *Pinus, Cassia siamea, Hibiscus tiliaceus* and other species have been attempted with varying degrees of success on mined-out bauxite lands on Jamaica. However, the most conspicuous reforestation attempt throughout Jamaica has been with *Hibiscus tiliaceus,* the mahoo. This tree has been planted after hillsides have been cleared of native vegetation or strips have been created through the forests for such plantings. Any long-term effect on the rare species may well have been in altering conditions of shade, moisture or drainage, rather than in elimination through clearing. Attempts have been made to establish plantations of *Tectona* in various areas in the archipelago but they have been unsuccessful. A slightly greater success was experienced by the Forestry Department in Puerto Rico, where combinations of species and types of land preparation have been tried experimentally. It is worth noting here that for educational purposes the name of the Luquillo National Forest was changed to Luquillo Experimental Forest.

On Dominica, a Canadian company tried to establish sawmill operations, but this proved to be an economic failure. Though the concession was protested vigorously by conservationists, the commercial operation ceased only because of the local problems of harvesting and milling, combined with the lack of a sufficient market for the product.

For many years, the local Antillean governments prohibited the introduction and use of mechanical equipment for felling or processing trees. Trees were cut with machetes or axes or burned at the base to topple them. Pit-sawing with long, two-man saws produced crude boards. Such operations had only selective effect on the forests, but they reduced the number of large trees of the Sapotaceae, commonly called "bullet." These are the large timbers found so characteristically in old Great Houses, windmill foundations, and associated buildings.

Tourism as an industry has had a variety of effects on the vegetation of the Antilles. Beach areas are regarded as private property to be bought and sold at the whim of the investor. In order to develop hotels west of San Juan, Puerto Rico, large areas of a wet, sandy savanna, with such unusual plants as *Drosera, Pinguicula, Osmunda* and *Piriquita,* were drained and filled. The coastal strand vegetation, containing *Scaevola, Canella, Zizyphus* and *Tournefortia,* was removed, and the area now is landscaped with exotic introduced species. Grand Anse Beach on Grenada was a classic collecting locality for Broadway and H.H. and G.W. Smith. Today, it is a long line of expensive hotels, and the plants previously reported there cannot be found. Among the first plants to be removed when tourists come to an area are *Hippomane mancinella,* the poisonous manchineel. In fact, it is this plant, along with the coastal species of *Comocladia* and *Metopium,* which causes dermatitis and should be included on lists of the endangered species in the Antilles.

The Conquistador Hotel in the Fajardo area of Puerto Rico was once a small, privately-owned hostelry. The manager asked that plants along a trail to the beach be labeled, and a nature walk was established for the guests. Several unusual plants grew there, including a probable new species of the *Myrtaceae* found in fruit. Unfortunately, the owner died and the hotel changed hands. It subsequently was tripled in size and a casino represented the interests of the new management. The nature trail was replaced by a mechanical people-transport system, and in the process, the *Eugenia* relative was destroyed. I have been unable to locate the species anywhere else on the peninsula, and the plant remains undescribed for lack of flowering material. A 36-hole irrigated golf course has now replaced a thorn scrub vegetation, which was used for many years as a study area to show classes the Caribbean type of dryland vegetation. Gone, too, is one of the few stands of *Anacardium occidentale* believed to be native in that area and the northernmost extension of the range. A breakwater was created to protect a new harbor, and a cement rim and floor now surrounds a deep saltwater coral depression where Marshall Howe once described the local algal flora of Puerto Rico.

West of San Juan in the area of Bayamon, there are isolated mogotes of ancient limestone rock. Many of these have been excavated; the material has been removed for use in road construction and landfill, and industrial buildings now occupy the leveled site. One of the lost mogotes was the type location of *Daphnopsis helleriana.* Britton

111

and Wilson (1924, vol. 5:620) noted that the plant was "collected only by Heller in 1900, Endemic." Fortunately, the species occurs on some of the remaining mogotes.

Dunn's River Fall in Jamaica is another classic collecting locality equally well-known for the unusual development of travertine rock over living plants. Today, the area is featured in advertisements for the Playboy Club of Jamaica, and the unusual herbaceous species along the stream have been decimated.

Cacti and orchids of the Antilles are not the most attractive species, but nonetheless, both groups are subject to exploitation by commercial and private collectors. The globular forms of *Melocactus* have been exported in quantity from St. Eustatius and Grenada. A commercial tour to Jamaica featured the opportunity to acquire orchids from native locations in the Cockpit country, and the tour even promised participants help in packaging and bringing these specimens back to the United States. The rare species to be found were listed by name in some of the publicity. Letters of protest to the government of Jamaica, the Jamaica Tourist Commission and Jamaican newspapers went unacknowledged and unheeded. One of the routes the group followed is now almost without visible orchid plants along a trail well-known botanically to members of the Jamaican Natural History Society.

ATTEMPTS AT CONSERVATION

Though threats to rare species in the Antilles are many and diverse, the attempts at conservation have been few and far between. International symposia on the subject of conservation have been held in the West Indies, and the reports have been published, but generally these have been unheeded by local governments, although the efforts of individuals and societies continue. The Bahama National Trust has designated a few islands or areas as reserves, but for the protection of birds, not plants. A few areas in Jamaica have been purchased and set aside under the direction of the Institute of Jamaica. In Puerto Rico, the National Parks and Forests are operated as multiple-purpose areas with emphasis, it seems, on recreation rather than on preservation. However, in Puerto Rico, several excellent examples of mangrove forests, siliceous savannas, mogotes and thorn scrub are within military reservations. Although they are heavily used for military maneuvers or artillery practice, at least the public is excluded and development is postponed. Under the administration of Trujillo, the island of Beata in the Dominican Republic, one of the few locations of the endemic palm *Zombia,* and the island of Alta Vela were both under military control and visitors were excluded. Through government — even military — ownership, there is at least a temporary semblance of protection against massive destruction.

The small size of many of the islands of the Antilles combined with the long period of human occupancy, also provides a degree of protection. In general, areas that are amenable to agricultural development have already been cultivated. The practical difficulties of opening new areas, therefore, may in fact protect them. Watersheds may be invaded, but in general, their need is recognized, and such areas will probably not be despoiled. Nevertheless, the need exists for the designation and protection of conservation areas throughout the Antilles. The emphasis, I believe, should be placed on the preservation of types of vegetation, rather than on the protection of so-called endangered species, a view I share with Peter Thompson (1975). The publicity following the designation of individual species has called unnecessary attention to a rare species and has increased the threat to *Orothamnus,* the Marsh Rose Protea (van der Merwe, 1975).

The resolutions passed at the 1975 meeting of the International Association of Botanical Gardens suggested a role for arboreta and botanical gardens in the permanent preservation of endangered species within their living collections. This calls for a dedicated director and a scientific program for each botanical garden. Neither is found within the Antilles. With very few exceptions, the botanical gardens of the Caribbean region are recreational areas featuring colorful displays only. By contrast, it is interesting to note that during the 19th century, the botanical gardens of the Lesser Antilles were scientific stations managed by European-trained personnel and financed outside of the local economy. There was a scientific and intellectual association between them and research organizations abroad. The Royal Botanic Gardens at Kew and the Muséum d'Histoire Naturelle of Paris, for example, cooperated closely with British and French gardens in the Antilles. In the 20th century, with the sole exception of the former Harvard Botanical Garden in Cuba, these contacts were broken. Although collaboration is still possible, there is a lack of qualified managerial personnel and a lack of support on the part of local governments for scientifically active botanic gardens. The controlled culture and maintenance of stock of endangered species within botanical gardens of the Antillean area, therefore, is not likely in the immediate future.

The designation of endangered species and the publication of a list of such species would be a futile exercise for the Caribbean area. An environmental impact statement might be enforced within Puerto Rico and the Virgin Islands due to their association with the United States. But for other areas of the West Indies, such ideas are incomprehensible or premature at the present time. We do applaud the establishment of the Osa Forest Preserve in Costa Rica, however, and even the legislation of Mexico and Brazil regarding plant collecting is noteworthy. We hope similar control measures soon will be adopted elsewhere and especially for some of the unusual vegetation types found within the Antilles.

LITERATURE CITED

Adams, C. D., et al. 1972. Flowering Plants of Jamaica. Mona, Jamaica: University of the West Indies. 848 pp.

Beard, G. S. 1955. The classification of tropical American vegetation-types. Ecology **36**: 89-100.

Box, H. E. and W. R. Philipson. 1951. An undescribed species of *Mastichodendron* (Sapotaceae) from Barbados and Antigua. Bull. Brit. Mus. Nat. Hist. **1**: 21-23.

Britton, N. L. and C. F. Millspaugh. 1920. The Bahama Flora. Lancaster: New Era Printing Co. 695 + viii pp.

Britton, N. L. and P. Wilson. 1923-1930. Botany of Puerto Rico and the Virgin Islands. Vol. V (626 pp.) — VI (663 pp.). *In:* Scientific Survey of Puerto Rico and the Virgin Islands. New York: New York Academy of Science.

Duss, A. 1897. Flore. Phanérogamique des Antilles Françaises. Macon: Protat Frères. 656 + xxviii pp.

Fawcett, W. and A. B. Rendle. 1910-1936. Flora of Jamaica. London: William Clowes and Sons, Ltd. Vol. 1-7.

Gastony, G. J. 1973. A revision of the fern genus *Nephelea*. Contr. Gray Herb. **203**: 81-148.

Gooding, E. G. B., et al. 1965. Flora of Barbados. London: Henry Blacklock and Co., Ltd. 486 + xvi pp.

Grisebach, A. H. R. 1859-1864. Flora of the British West Indian Islands. London: 789 + xvi pp.

Guilding, L. 1825. An Account of the Botanic Garden in the Island of St. Vincent. Glasgow: Richard Griffin and Co. 47 pp.

Howard, R. A. 1968. The Ecology of an Elfin Forest in Puerto Rico, I. Introduction and Composition Studies. Jour. Arnold Arbor. **49**: 381-418.

_____. 1973. The Vegetation of the Antilles. *In:* Graham, A. ed. Vegetation and Vegetational History of Northern Latin America. Amsterdam: Elsevier Scientific Publ. Co. 393 + xiii pp.

_____. 1975. *Lindernia brucei,* A new West Indian species of the Asian section *Tittmannia.* Jour. Arnold Arbor. **56**: 449-455.

León, H. and H. Alain. 1946-1969. Flora de Cuba. Published variously. Vol. 1-5 + suppl.

Moscoso, R. M.   1943. Catalogus Florae Domingensis. New York: L. and S. Printing Co., Inc. 732 + xlviii pp.

Stehlé, H.   1945. Forest types of the Caribbean Islands. Caribb. For. 6 (suppl.): 274-408.

Swartz, O. 1797. Flora Indiae Occidentalis. London: Benj. White and Son. 2018 + tab. xxix pp.

Thompson, P. 1975. Should botanical gardens save rare plants? New Scientist **68**: 636-639.

Tryon, R.   1970. The classification of the Cyatheaceae. Contr. Gray Herb. **200**:1-32.

_____. 1972. Taxonomic fern notes, VI — New Species of American Cyatheaceae. Rhodora **74**: 441-450.

Turczaninoff, N. 1863. Animadversiones ad Catalogum primum et secundum herbarii Universitates Charkoviensis. Bull. Soc. Nat. Moscow **36**(1): 600.

Urban, I. 1898-1928. Symbolae Antillanae. Lipsiae: Fratres Borntraeger. Vol. 1-9 + Index (Compiled by E. Carroll and S. Sutton, 1965, Jamaica Plain: Arnold Arboretum. 272 pp.).

van der Merwe, P. 1975. Impossible to Save the Marsh Rose Protea? Veld and Flora **1**: 4, 5.

Section 3

South America

THE PREPARATION OF
THE ENDANGERED SPECIES LIST OF COLOMBIA

Alvaro Fernández-Pérez

*Instituto de Ciencias Naturales, Universidad Nacional,
Apartado Aéreo 7495, Bogotá, Colombia*

The major part of the Colombian population, since before the discovery of America, is localized in the Andean and Interandean regions. The diversity and notable endemism of the flora of this area allowed pre-Columbian cultures to discover and improve numerous food plants, such as potatoes, which today are utilized throughout the world. As a result of population pressures, modern civilization has increased the production of these food plants and has even introduced many new crops from other countries. However, we do not know how many potential food plants may have been lost. I do not doubt that the introduction into Colombia of the forage grass *pasto quicuyo (Pennisetum clandestinum),* a vigorous colonizer, has also totally destroyed numerous herbaceous plants.

Species such as the insectivorous plants *Pinguicula elongata* and *Utricularia alpina,* which were once frequently observed in areas around Bogotá, have disappeared locally as a result of the introduction of this grass. Examples of plants which are gradually disappearing due to the disturbance of man are timber trees such as *nogal (Juglans), roble (Quercus), pino romeron (Podocarpus),* and *cedro (Cedrela).* Herbs such as *ipecacuana (Cephaelis),* of great value in the pharmaceutical industry, are also in danger of extinction because of man's interference.

On various occasions I have searched in vain for samples of plants for chemical and pharmacological analysis. As long ago as 1955 I was unable to locate *Prestonia amazonica* in the Department of Valle though collections were made there in 1927. I was able to locate just one tree of *Rauvolfia leptophylla,* known only from the Sierra Nevada de Santa Marta, when I searched for it in 1957 with Prof. Romero Castañeda, who had discovered it there a few years earlier. *Rauvolfia sanctorum,* discovered by E. P. Killip and A. C. Smith in 1926 in Mesa de los Santos in the Department of Santander at 1500 m elevation is today very rare in the type locality. Although this species has also been found in the Amazon lowlands at Mishuayacu near Iquitos, Peru, we cannot be certain the species could be reintroduced into highland Colombia with living material of the Peruvian populations because of the probability of ecotopic differentiation to the different altitudes. We know that the species produces alkaloids in the plants at Mesa de los Santos but the alkaloid content, if any, of the Peruvian collections is not known.

The endangered species data cards that we are using in Colombia for compiling the available information are shown in Fig. 1.

The advance of progress is, at best, a slow and laborious process, and this is particularly true if data that is compiled locally become modified when larger foreign herbaria and specialists are consulted. According to Dr. D. B. Lellinger, "*Ctenitis squamosissima* is apparently very rare in Colombia. We have at the Smithsonian only one Triana collection, but it was collected several times by Sodiro in Ecuador." According to Dr. Thomas Soderstrom, "*Rhipidocladum racemiflorum* is extremely common in Central America and northern South America. It flowers infrequently so is often not collected. Although it is seldom collected and appears to be rare in Colombia it is not really an endangered species."

Strict adherence to the information presented in various publications on systematic botany can often be misleading. For example, one of the commonest plants in Colombia is the orchid *Epidendrum paniculatum,* according to the systematic treatment by Ames, Hubbard and Schweinfurth (1934). Twenty-six different taxa were united in that study. Recently, my colleague Leslie Garay informed me that in preparing the Orchid Flora of Colombia he has found that *E. paniculatum* is not only distinct from its so-called twenty-five synonyms, but it does not even occur in Colombia. In addition to the type from Peru, he has seen only five collections of this attractive plant in herbaria.

Classical botanical works, latest taxonomic monographs and revision of sundry genera are being consulted for the compilation of the Colombian list. Modern equivalents of all of the names used in Jacquin's *Enumeratio Systematica Plantarum* (1760) are given by

PLANTAS COLOMBIANAS EN PELIGRO DE EXTINCION: Familia _____	
Nombre científico	Nombre vernáculo
Localidad original	Otras localidades
Hábitat	Hábito
Donde, cuando y quién la coleccionó por última vez	
Extinguida ☐ En peligro ☐ Muy rara ☐ Rara ☐ Dudosa ☐	
Otros informes:	
Acopió estos datos:_____ Fecha _____	

Fig. 1. Data collection card for the endangered list of Colombia.

Howard (1973). Since a large number of the species published in the *Enumeratio* were collected in Colombia, mainly in Cartagena and along the Magdalena river, Howard's work is very useful also in the preparation of the endangered species list of Colombia. Similar studies on the publications of Jacquin and Humboldt were initiated by my former professor of botany, Dr. Armando Dugand, whose generous help is appropriately acknowledged by Howard: "In compiling the list of equivalents which follows, I have been aided significantly by the work of the late Armando Dugand of Colombia." In 1966, Dugand wrote: *"Mas tarde publicaré, Dios mediante, un catálogo de todas las especies cartageneras de Jacquin, citando en primer lugar el nombre con que él las dio a conocer al mundo científico; y dando ademas el nombre técnico que hoy tienen o deben llevar conforme a los estudios taxonómicos que se han publicado desde la época de Jacquin hasta la nuestra . . ."* [1] Regrettably, Dr. Dugand did not live to complete his studies, so it remains for a botanist who is familiar with the coastal flora of Colombia to locate the plants of the Cartagena region that Jacquin collected and named, which were beyond the scope of the present study. A preliminary study of Jacquin's plants from Colombia indicates that few are endangered, except the orchid, *Brassavola nodosa,* which is becoming rare in the wild, due to excessive collecting by amateur orchid fanciers; *Swietenia mahogoni* shares the fate of becoming extinct with other valuable timber trees of the Caribbean area.

The second classical botanical collection of my country is that of the José Celestino Mutis Botanical Expedition to Nueva Granada (1783-1819). The so-called Mutis herbarium contains about 20,000 specimens, including 8,000 watercolor drawings, representing ca. 9,000 species. This herbarium is in Madrid, with a considerable number of duplicates at the Smithsonian Institution and a few in Bogotá. E. P. Killip named most

[1]"Later with God's help I hope to publish a catalog of all of Jacquin's species from the vicinity of Cartagena, citing both the original Jacquin names and the modern scientific names based on all taxonomic studies made since Jacquin's time."

of the plants of this collection as well as described numerous new species based on that material. Unfortunately, the specimens lack specific locations and therefore are useless in studying endangered and threatened species in Colombia. I highly recommend that neotropical monographers do not omit such information when it is available.

In analyzing the texts I have written for the orchid books of Mutis' Botanical Expedition, the following, among 151 drawings, are without citation of supporting Colombian specimens; some have not yet been named because of lack of documenting samples; others have not been re-collected in Colombia during the present century:

Habenaria corydophora: the drawing[2] represents the only citation for Colombia.
Pseudocentrum sylvicola: not collected in this century.
Ponthieva disema: the drawing represents first citation for Colombia.
Stelis maxima: not collected in this century.
Masdevallia cuculata: not collected in this century.
M. fertilis: not collected in this century.
M. simulatrix: the drawing represents the first citation for Colombia.
Restrepia guttulata: not collected in this century.
Lepanthes aquila-boroussiae: the drawing represents the first citation for Colombia.
L. nubicula: not collected in this century.
L. polygonoides: the drawing represents the first citation for Colombia.
L. rhombipetala: not collected in this century.
L. astrophora: not collected in this century.
Pleurothallis cardiophyllax: not collected in this century. The original description was based on material introduced in London gardens.
Pl. decurrens: the drawing represents first citation for Colombia.
Pl. lanceana: the drawing represents first citation for Colombia; the species has a large distribution in tropical America.
Pl. lancipetala: not collected in this century.
Pl. monocardia: not collected in this century.
Malaxis parthonii: not collected in this century; this species has a large distribution in tropical America.
Epidendrum armeniacum: the drawing represents first citation for Colombia.
E. attenuatum: I have seen only a photograph of the type specimen at Ames Herbarium; perhaps the drawing represents the first citation for Colombia.
E. bivalve: not collected in this century.
E. floribundum: formerly named as *E. paniculatum,* which is a complex under study by Dr. Garay.
E. inornatum: the drawing represents first citation for Colombia.
E. laeve: not collected in this century.
E. ottonis: no citation of Colombian material, but the species is common in both Peru and Venezuela.
E. pittieri: known only from the type collected by H. Pittier in Colombia in 1906.
E. smaragdinum: the drawing represents first citation for Colombia.
E. stramineum: not collected in this century.
E. tipuloideum: no citation of Colombian material; also Venezuela.

A partial analysis of the type collection of J. J. Triana at the herbarium in Bogotá reveals the following results:

	Locality	Year of first collection	Year of most recent collection
ACANTHACEAE			
Aphelandra botanodes	Chocó, Nóvita	1853	
diachyla	Chocó, Sn. Pablo	1853	

[2]Drawing throughout this list refers to the Mutis drawings and not illustrations in the present work.

		Locality	Year of first collection	Year of most recent collection
Aphelandra	lasiophylla	Antioquia	1852	
	serichantha	Antioquia	1852	
	sericophylla	Bogotá	1852	1952
Dicliptera	bogotensis	Bogotá	1855	
	conformis	Popayán	1853	
	inamoena	Pasto	1853	
	trianae	Cauca	1853	
Justicia	novogranatensis	Popayán	1853	
	pectoralis var.			
	stenophylla	San Martín	1856	1952
Pseuderanthemum				
	micranthum	Buenaventura	1853	

ACTINIDIACEAE (Dillen.)

Saurauia	floccifera	Bogotá, Ubalá	1855	1939
	parviflora	Cauca	1853	1962
	peduncularis	Túquerres	1853	1953
	ursina	Antioquia	1852	1963

ANACARDIACEAE

Tapirira	myriantha	Buenaventura	1853	1955

ANONACEAE

Anona	ionophylla	Antioquia	1852	
Guatteria	cargadero	Cauca	1853	
	cestrifolia	Villavicencio	1856	
	longipes	Bogotá	1855	
	platyphylla	Bogotá	1855	
Raimondia	quinduensis var.			
	latifolia	Bogotá	1855	

APOCYNACEAE

Mandevilla	trianae	Chocó	1853	1949

AQUIFOLIACEAE

Ilex	laureola	Bogotá	1855	1956
	micrantha	Popayán	1853	
	nervosa	Bogotá	1855	
	pustulosa	Fusagasugá	1855	

ARISTOLOCHIACEAE

Aristolochia	tenera	San Martín	1856	
	trianae	Barbacoas	1853	

BEGONIACEAE

Begonia	trianae	Bogotá, Ubalá	1855	1941

	Locality	Year of first collection	Year of most recent collection
BIXACEAE			
Bixa sphaerocarpa	Villavicencio	1856	1964

BOMBACACEAE (Material of *Matisia* was all on loan when this list was prepared).

BORAGINACEAE

Cordia trianae	Mariquita	1854	
Cynoglossum trianae	Antioquia	1852	
Tournefortia stenoloba	Bogotá	1853	1944

BRUNELLIACEAE

Brunellia trianae	Antioquia	1852	1958

BURSERACEAE

Protium sagotianum	San Martín	1856	1957

CAMPANULACEAE

Siphocamphylus lactus	Túquerres	1853	

CARICACEAE

Carica manihot	San Martín	1856	1956

CARYOPHYLLACEAE

Arenaria musciformis	Túquerres	1853	1957
Cerastium caespitosum	Bogotá	1855	1952
Paronychia bogotense	Bogotá	1855	1944

CELASTRACEAE

Maytenus buxifolius	Popayán	1855	1934
laxiflorus	Barbacoas	1855	

CLUSIACEAE

Balboa membranaceae	Barbacoas	1853	
Calophyllum mariae	Bogotá	1855	
Clusiella elegans	Valle	1855	1944
Rheedia pulvinata	Bogotá	1856	
Tovomita turbinata	Buenaventura	1853	1939

COMPOSITAE

Espeletia glandulosa var.			
scaberrima	Pamplona	1851	1967
trianae	Pamplona	1851	

	Locality	Year of first collection	Year of most recent collection
CUNONIACEAE			
Weinmannia trianae	Bogotá	1855	1967
DILLENIACEAE			
Davilla densiflora	San Martín	1856	1964
ERICACEAE			
Anthopterus cuneatus	Barbacoas	1853	
Cavendishia coccinea	Barbacoas	1853	1946
compacta	Cauca	1853	1944
hispida	Chocó	1853	
oligantha	Túquerres	1853	
rhychophylla	Buenaventura	1853	
Thibaudia pachyantha	Barbacoas	1853	1948
Psammisia caloneura	Barbacoas	1853	
EUPHORBIACEAE			
Sapium bogotense	Bogotá	1855	1945
ELAEOCARPACEAE			
Sloanea castanocarpa	Villavicencio	1856	
FLACOURTIACEAE			
Casearia lasiosperma	Chocó	1853	
rufidula	Mariquita	1853	
Mayna suaveolens	Magdalena	1852	
Ryania chocoensis	Chocó	1853	1944
HYPERICACEAE			
Hypericum lycopodioides	Bogotá	1853	
Vismia laevis	Antioquia	1852	1952
MALPIGHIACEAE			
Hiraea brachyptera	Chocó	1853	1944
cephalotes	Pasto	1853	1948
ternifolia	Mariquita	1853	1947
Banisteriopsis martiniana	Cord. Occident.	1853	1951
Bunchosia retusa	La Mesa	1853	1955
Byrsonima adenophylla	Barbacoas	1853	
Mascagnia hippocrateoides	Anapoima, Tena	1854	1945
macrodisca	Villavicencio	1856	1969
violacea	Anapoima, Tena	1853	1950
Stigmaphyllon alternans	Villavicencio	1856	1935
bogotense	Bogotá, Popayán	1853	1963
columbicum	Bogotá	1853	1963
brachiatum	Villavicencio	1856	1965

	Locality	Year of first collection	Year of most recent collection
Tetrapteris benthamii	Magdalena	1853	1952

MALVACEAE

	Locality	Year of first collection	Year of most recent collection
Malvaviscus leucocarpus		1851	1948
velutinus	Mariquita	1854	1946

MARCGRAVIACEAE (five types out on loan when list was prepared).

MELIACEAE

		Locality	Year of first collection	Year of most recent collection
Guarea	*gigantea*	San Martın	1855	1963
	glauca	Bogotá	1856	
Trichilia	*goudotiana*	Bogotá, La Mesa	1853	1963

MELASTOMATACEAE

		Locality	Year of first collection	Year of most recent collection
Adelobotrys	*fuscesens*	Antioquia	1852	1969
Blakea	*caudata*	Bogotá	1855	1967
	podagrica	Barbacoas	1853	1969
Centronia	*haemantha*	Ocaña	1851	1957
Conostegia	*trianae*	Chocó	1853	1939
Diplarpea	*paleacea*	Barbacoas	1853	1967
Diolena	*agrimonioides*	Buenaventura	1853	1946
Meriania	*nobilis*	Antioquia	1952	1963
Miconia	*anisophylla*	Chocó	1853	1967
	haematostemon	Pasto	1853	1959
	lamprophylla	Barbacoas	1853	1953
	notabilis	Caldas	1852	1969
	ochrecea	Popayán	1853	1962
	pterocaulon	Villavicencio	1856	1969
	reduscens	Buenaventura	1853	1962
	reticulata	Cauca	1853	
	subnodosa	Buenaventura	1853	
Monolema	*cordifolia*	Chocó	1853	
Ossaea	*rufibarbis*	Barbacoas	1853	1959
Tococa	*spadiciflora*	Barbacoas	1853	1963
	symphyandra	Barbacoas	1853	1946
Topobea	*subscaberula*	Barbacoas	1853	1944
	trianaei	Barbacoas	1853	1968
	setosa	Barbacoas	1853	1968

MENISPERMACEAE

		Locality	Year of first collection	Year of most recent collection
Cissampelos acuta		Quindío	1853	1946
	grandifolia	Quindío	1853	1945
	acutigera	Bogotá	1856	

MIMOSACEAE

		Locality	Year of first collection	Year of most recent collection
Mimosa	*trianae*	Villavicencio	1856	

	Locality	Year of first collection	Year of most recent collection

MORACEAE

Cecropia	*radlkoferiana*	Choco	1853	
Coussapoa	*oligoneura*	Barbacoas	1853	
Ficus	*trianae*	Barbacoas	1853	1947

MYRTACEAE

Eugenia	*variareolata*	Villavicencio	1856	

NYCTAGINACEAE

Cephalotomandra fragrans	Tequendama	1854	1958

OCHNACEAE

Ouratea	*magdalenae*	Rio Magdalena	1853	

OLACACEAE

Heisteria	*celastrinae*	Ocaña	1851	1941

OXALIDACEAE

Oxalis	*trianae*	Bogotá; La Mesa	1854	1956

PIPERACEAE

Peperomia	*ciliaris*	Buenaventura	1853	1939
	ciliosa	Barbacoas	1853	
	trianae	Antioquia	1852	
Piper	*androgynum*	Pasto	1853	
	bullosum	Barbacoas	1853	
	calceolarium var.			
	magnifolium	Antioquia	1852	
	multinervium	Barbacoas	1853	1946
	novogranatense	Barbacoas	1853	1947
	ottoniaefolium	Chocó	1853	1944
	petiolare	Pasto	1853	
	pulchrum	Pasto	1853	1948
	trigonum	Buenaventura	1853	1944
	trianae	Pasto	1853	
	villosum	Barbacoas	1853	

POLYGALACEAE

Monnina	*speciosa*	Barbacoas	1853	1949

RHAMNACEAE

Gouania	*podocephala*	Mariquita	1853	
	rumicina	Villavicencio	1856	1939

	Locality	Year of first collection	Year of most recent collection
RUBIACEAE			
Remigia trianae	Villavicencio	1856	
RUTACEAE			
Esenbeckia alata	Bogotá; Tocaima	1854	1963
SAPINDACEAE			
Allophyllus angustatus	Bogotá	1852	1948
excelsus	Bogotá	1856	
nitidulus	Bogotá	1853	1939
Cupania triloba	San Martín	1856	
Paullinia eriocarpa	Villavicencio	1856	
pterocarpa	San Martín	1856	
pterophylla	San Martín	1856	1946
serjaniaefolia	San Martín	1856	
triptera	Mariquita	1853	
Serjania clematidea	Bogotá	1856	1946
Talisia stricta	Mariquita	1853	1952
SOLANACEAE			
Cestrum granadense	Fuzagasugá	1855	
STERCULIACEAE			
Ayenia stipularis	Mariquita	1854	
Melochia kerriifolia	Mariquita	1854	
STYRACACEAE			
Styrax bogotensis	San Martín?	1856	
macrocalyx	Bogotá	1855	
trichocalyx	Tequendama	1853	
SYMPLOCACEAE			
Symplocos trianae	San Martín	1856	
THEACEAE			
Freziera arbutifolia	Antioquia	1852	
calophylla	Antioquia	1852	
Pelliceria rhizophorae	Buenaventura	1852	
Laplacea fruticosa	Bogotá	1855	1966
URTICACEAE			
Hemistylus velutina	Bogotá	1853	1947
Pilea fasciata	Chocó	1853	
pteropogon	Barbacoas	1853	1940
Urera simplex	Bogotá	1856	1964

Among the most recent monographic studies pertaining to tropical plants are the seventeen different treatments published by leading authorities in *Flora Neotropica*. I would strongly recommend these studies to botanists for the preparation of national lists of endangered and threatened species. With such information, working groups or cooperative programs for mapping and protecting species can be initiated throughout tropical America.

The following list of Colombian plants known only from type collection is derived from several monographs published in *Flora Neotropica:*

From Monograph No. 1 — SWARTZIA Cowan (1968).
Swartzia magdalena
S. macrophylla

From Monograph No. 2 — BRUNELLIACEAE Cuatrecasas (1970).
Brunellia almaguerensis

From Monograph No. 9 — CHRYSOBALANACEAE Prance (1972a).
Licania maritima	*Licania Velata*
L. cuspidata	*L. salicifolia*
L. calvescens	*L. caldasiana*
L. fuchsii	

From Monograph No. 10 — DICHAPETALACEAE Prance (1972b).
Dichapetalum nervatum
Tapura bullata
T. colombiana

From Monograph No. 12 — CARYOCARACEAE Prance and Silva (1973).
Anthodiscus montanus

From *The Bromeliaceae of Colombia,* Smith (1957) and *Flora Neotropica,* Monograph No. 14 —

PITCAIRNIOIDEAE Smith and Downs (1974).
Puya thomasiane	*T. rariflora*	*T. inconspicua*
Pitcairnia macrobotrys	*T. humboldtii*	*T. adpresa*
P. laxissima	*T. rhomboidea*	*Vriesia simplex*
P. arcuata	*T. platyrhachis*	*Guzmania lehmanniana*
P. andreana	*T. palacea*	*G. andreana*
Tillandsia lajensis	*T. trapexiformis*	*G. palustris*
		G. pearcei

In studying the distribution of rare and little known plants, reliance on herbarium material can be very misleading. In herbaria where specimens have not been recently revised by specialists, several distinct species of plants may be included under the same scientific name. It frequently will be necessary, therefore, to search for certain species in their original habitat in order to ascertain if the species is truly extinct or is in danger of extinction. Furthermore, certain populations of the same species will become adapted to different ecological conditions, presenting major obstacles in saving the species from extinction by transplanting to different areas.

ACKNOWLEDGMENTS

I express my appreciation to Dr. F. R. Fosberg for his helpful advice and assistance in the preparation of the data cards used for the Colombian list of endangered species; to Pedro Rodriguez, student of biology, for helping me compile herbaria information; and many more friends and co-workers than can be acknowledged in this small space.

Financial support has been given by Fondo Colombiano de Investigaciones Cientificas (COLCIENCIAS), Project No. c-2-73. This work is being carried out in collaboration with the Jardin Botanico of Medellin, Colombia.

LITERATURE CITED

Ames, O., F. T. Hubbard and C. Schweinfurth. 1936. The Genus *Epidendrum* in the United States and Middle America. Bot. Museum, Cambridge, Mass.

Cowan, R. S. 1968. Monograph of *Swartzia* (Leguminosae). Flora Neotropica no. **1.** Hafner, New York. 228 pp.

Cuatrecasas, J. 1970. Monograph of Brunelliaceae. Flora Neotropica no. **2.** Hafner, New York. 189 pp.

Dugand, A. 1966. Asclepiadaceae nuevas o interesantes de Colombia y paises vecinos. Caldasia **9** (45): 400.

Howard, R. A. 1973. The Enumeratio and Selectarum of Nicolaus von Jacquin. Jour. Arnold Arb. **54:** 435-470.

Jacquin, N. 1760. Enumeratio Systematica Plantarum. 44 pp. Leiden.

Prance, G. T. 1972a. Monograph of Chrysobalanaceae. Flora Neotropica **9,** Hafner, New York. 410 pp.

_____. 1972b. Monograph of Dichapetalaceae. Flora Neotropica **10** Hafner, New York. 84 pp.

_____. **and M. F. da Silva.** 1973. Monograph of Caryocaraceae. Flora Neotropica **12.** Hafner, New York. 75 pp.

Smith, L. B. 1957. The Bromeliaceae of Colombia. Contr. U.S. Nat. Herb. **33:** 311 pp.

_____. **and R. J. Downs.** 1974. Monograph of Pitcairnioideae (Bromeliaceae). Flora Neotropica **14,** Hafner, New York. 658 pp.

FUTURE OUTLOOK FOR
THREATENED AND ENDANGERED SPECIES IN VENEZUELA

Julian A. Steyermark

*Instituto Botanico, Ministerio de Agricultura,
Apartado 2156, Caracas, Venezuela*

Introduction

Venezuela has long been considered a botanical paradise, but the status of this paradise beyond the immediate future is problematical, despite the fact that many Venezuelans are far-sighted men and women who are greatly concerned about the country's preservation of the flora and natural areas. Many conferences and symposia based on protection and preservation of the landscape are frequently held. Seven National Monuments, eighteen National Parks, and several wildlife refuges have been established by acts of law and set aside for posterity. Numerous individuals, agencies and institutions are continually at work conducting surveys, inventories, ecological research, and generally devoting their efforts to detailed studies of natural areas. The present president of the Republic, Carlos Andres Perez, even received the prestigious Earthcare Award from the Sierra Club in 1975. He also was recently named an honorary member of the Sociedad Venezolana de Ciencias Naturales (The Venezuelan Society of Natural Sciences) and was instrumental in instigating a number of controls on hunting, timber removal, and water pollution. Local conservation groups, such as the Sociedad Venezolana de Ciencias Naturales, Sociedad de Ciencias Naturales de La Salle (La Salle Society of Natural Sciences), Academia de Ciencias Fisicas, Matematicas y Naturales (The Academy of Physical, Mathematical and Natural Sciences), Consejo Bienestar Rural, (The Council of Rural Welfare), Audubon Society, and Sociedad Conservacionista del Estado Aragua (The Conservation Society of the State of Aragua), are dedicating programs to the conservation and preservation of the native flora and fauna. These encouraging signs of growing public awareness, emphasized by the press, radio and television, on the dangers of destruction and loss of the natural biota and resources of Venezuela, all would seem to be a strong indication of a positive trend toward protection of our natural heritage. Consequently, the future outlook for threatened and endangered species of Venezuela should appear to be on the bright side.

Unfortunately, this optimistic attitude is largely misleading, and before discussing the reasons for such false optimism, let us examine in detail the actual state of the flora of Venezuela at the present time.

The flora of Venezuela, like other neotropical countries, is rich in endemic species and in those of restricted phytogeographical distribution. As a result of its particular geological and orogenic history, Venezuela is framed by the Coastal Cordillera, Andes, and Perijá Mountains on its western and northern borders, the Guayana Highlands with their mountains and intervening forests and savannas in the southern half, and the *llanos* (or savannas) lying between these extremes.

So far as endemism is concerned, there is only a low representation of such in the llanos, typified by *Hymenocallis venezuelensis* and *Limnosipanea ternifolia*. The highest percentage of endemism occurs in the Guayana Highlands, where 75 percent or more of the total flora has been estimated to be endemic (Maguire, 1970). This is especially true on the summits and on the more elevated portions of the sandstone mountains. The figure varies from one mountain to another, however, and in some cases may even drop to as low as 25 percent endemism. A large number of genera are also endemic to the Guayana Highlands. Likewise, in the areas of low altitude savannas and forests of Territorio Federal Amazonas and Estado Bolívar of the Venezuelan Guayana, one finds many endemic species and some genera.

Smaller but significant percentages of endemism are found in the cloud forests and paramos of the Andes and Perijá Mountains, as well as in the cloud forests of the Coastal Cordillera. Noteworthy areas of endemism in the Coastal Cordillera are situated in the cloud forests of the Peninsula of Paria in the State of Sucre, the forests of Avila and

Colonia Tovar in Distrito Federal and State of Aragua, Pittier National Park in the State of Aragua, Sierra de Aroa and Nirgua Mountains in the State of Yaracuy, and in the mountains south of Borburata in the State of Carabobo. Examples of genera endemic to the Coastal Cordillera are *Croizatia* (Euphorbiaceae), *Caracasia* (Marcgraviaceae), *Llewelynia* (Melastomataceae), and *Tammsia* and *Neoblakea* (both Rubiaceae).

While a large number of species are endemic to the cloud forests and páramos of the Venezuelan Andes and of the Sierra de Perijá, no endemic genera are known, although a number of genera, such as *Lagenanthus* (Gentianaceae), *Castratella* (Melastomataceae), *Ochoterenaea* (Anacardiaceae), are restricted to the Andean portions of Colombia and Venezuela. Within the Andean States of Lara, Trujillo, Mérida, Táchira, and in portions of Barinas and Portuguesa, local pockets of endemism can be found. One particularly interesting area is that bordering Colombia in the vicinity of the Páramo de Tamá and Páramo de Judio. In Venezuela this encompasses the terminus of the northeastern prolongation of the Cordillera Oriental of the Andes, and is separated from the rest of the Venezuelan Andes to the east by a geosynclinic depression, known as the Táchira Depression (Steyermark, 1975). The flora here is unique, with not only numerous endemic species but also containing many other species reaching their known northeastern limit of phytogeographical distribution and elsewhere occurring in the Andes of Colombia, Ecuador, and Peru or Bolivia. Among the twelve genera known in Venezuela exclusively from this zone are *Oreobolus* (Cyperaceae), *Porroglossum, Caucaea, Pityphyllum,* and *Sertifera* (all Orchidaceae), *Tovomitopsis* (Guttiferae), *Lythrum* (Lythraceae), *Bucquetia* and *Castratella* (Melastomataceae), *Lagenanthus* (Gentianaceae), *Desfontainea* (Loganiaceae), and *Delostoma* (Bignoniaceae).

Although harboring few endemic species, another area of great phytogeographical interest is that comprising the Altiplanicie de Nuria and Sierra Imataca of the State of Bolívar and adjacent Territorio Federal Delta Amacuro of northeastern Venezuela. The flora here comprises 24 genera and 317 species found nowhere else in Venezuela, and represents a western extension of the flora of the Guianas to the east, as well as a northern extension of the flora of Pará and Amazonas States of Brazil to the south (Steyermark, 1968).

Other patterns of phytogeographical distribution in Venezuela reveal species of rare or limited occurrence, such as in the low altitude wet forests of the San Camilo forest reserve of the State of Apure of western Venezuela. It is here that numerous species are found, whose center of distribution lies westward in Amazonian Peru, Brazil and Colombia, and which reach their known eastern limit in this part of Venezuela.

Within the vast territory comprising southern Venezuela, study of the geographical patterns of distribution reveals a complex kaleidoscope instead of uniformity. Not only are the genera and species of the high mountains of the Guayana Highlands often separated into their eastern components of the State of Bolívar and western components of Territorio Federal Amazonas, but other centers or nuclei of dispersal are found within these eastern and western divisions. Similarly, white sand savannas at low altitudes in the western part of Territorio Federal Amazonas, such as Yapacana Savanna and the area of Maroa, harbor unique floras not found elsewhere. Additionally, igneous outcrops, such as those in the vicinity of Puerto Ayacucho in extreme western Venezuela, comprise elements of both endemic and restricted flora. Moreover, within the vast forested continuity of southern Venezuela, several floristic patterns are manifest. Thus, in the region of San Carlos de Río Negro, Yavita, Pimichin, and Cocuy in extreme southwestern Venezuela are found floras of often very limited geographical range as well as floras containing many endemic elements. At the forested bases of the tabular sandstone mountains of the Venezuelan Guayana occur not only endemic species, but many others also which otherwise are only known from portions of Amazonian Brazil or Colombia. In southeastern Venezuela in the State of Bolívar are found elements of the flora known elsewhere from northern Brazil, Guyana, or Surinam. Especially noteworthy in this respect is the flora of the Cuyuni drainage in extreme eastern Venezuelan Guayana, where a concentration occurs of species known elsewhere only from Guyana or Surinam, and are at their westernmost stations here in Venezuela. In the forests of southeastern Venezuela lying between the Brazilian border and the headwaters of the Caroni, Paragua and Caura Rivers, the forest flora is different from that to the west, and consists of floral elements of northern Brazil.

This general review of the extent and occurrence of endemism and floristic relationships in Venezuela is a good basis for an understanding of the subject of threatened and endangered species.

Present evaluation of threatened and endangered species

As far as can be determined at the present time, no species of Venezuelan flora has been exterminated. On the contrary, with every new exploration into botanically unexplored areas, new stations are being found for species previously known only from one locality. Moreover, as a result of these new explorations, hundreds of species new to the flora of Venezuela and to science generally are being discovered and added to the flora. Botanists familiar with Venezuelan flora estimate that the eventual total number of species of vascular plants in Venezuela will be between 20,000 to as many as 35,000. However, the higher of these two estimates probably will not be reached until many more decades of plant exploration. Of course, it remains to be seen whether the present natural areas in which plant explorations need to be carried out, will still be available for such studies 50 to 100 years or more from now.

Even though no known Venezuelan species has yet been exterminated, some of them are certainly either endangered and threatened, or on the verge of extermination. This is especially true of some species of orchids, such as *Oncidium papilio* (butterfly orchid), *Cycnoches loddigesii, Masdevallia tovarensis,* and all *Cattleya* species, except *C. violacea.* In the case of *Oncidium papilio,* it is avidly sought by commercial orchid collectors and relentlessly taken from its forested haunts. The showy *Masdevallia tovarensis,* a Venezuelan endemic of the cool cloud forests of the central portion of the Coastal Cordillera, is brought down from its cooler higher altitudes and sold in quantities on street corners of the drier and warmer ambiance of Caracas, eventually to die. *Cattleya violacea,* because it is so common along river margins of southern Venezuela, is at present in no danger, but other species in Venezuela, such as *C. gaskelliana, C. mossieae, C. lueddemanniana, C. lawrenceana, C. jenmanii,* and *C. percivaliana,* must be included among the threatened and endangered species.

Each year, hundreds of the showy *C. mossiae* are loaded onto trucks or moved on muleback from their native forest habitats to be sold in Venezuelan cities or even exported abroad, despite laws to the contrary. With the recent completion of the El Dorado-Santa Elena Highway in the eastern part of the Venezuelan Guayana, avid orchid collectors and the native Indians whom they employ, are already making encroachments into the natural habitats of *C. lawrenceana* and the very rare *C. jenmanii,* the latter until recently known only from adjacent Guyana. The rare *C. percivaliana* is being taken from its Andean localities to be sold in Mérida and other cities or openly by the side of Andean roads. It is hard to believe, but sad, that even some members of local orchid societies — the very people who should preach and practice orchid conservation and protection — are among the worst offenders. They are the ones who are responsible for the endangerment and obliteration of these native species. Nor does the government, with its regulations prohibiting the collecting of *Cattleya* and other orchids from national territory, make any attempt to enforce these laws or to protect the orchids from this wanton pilfering. Another orchid that is undoubtedly being threatened is the spectacular *Phragmipedium caudatum.* This is such a rare orchid that it is known only from one or two localities in the Venezuelan Andes. Even a minimal amount of collecting can endanger its survival.

The orchid mania is an incredible phenomenon. Persons who otherwise show no particular interest in plants, become infatuated and imbued with a feverish pitch of enthusiasm when orchids are mentioned. To these people, the orchid, any orchid, is something so prized that it must be obtained at any cost. Airplane pilots and their personnel who stop even very briefly on flights in portions of the Venezuelan Guayana, attempt to obtain quantities of orchids to take back with them. Regardless of how ill-adapted the orchids may be for transporting from cooler to hot climes or from special edaphic soil types to any other soil type, the orchids, nevertheless, are brought back. Witness the quantities of *Zygosepalum tatei, Mendoncella burkei, Eriopsis biloba, Sobralia liliastrum,* and other miscellaneous associated species from the Venezuelan Guayana, uprooted from their special acid soils of cooler areas and brought back to hotter and drier localities, where they are hastily planted and eventually die. It is difficult to estimate how many hundreds or even thousands of orchids, coming from special edaphic habitats, suffer this fate each year as a result of such uncontrolled depredation.

Other plants in immediate danger are the many kinds of Venezuelan tree ferns, whose trunks are in great demand for growing orchids or for lining hanging baskets. Quantities of these trunks are sold in all garden supply shops of Caracas as well as in many other cities. Especially in danger are such tree ferns as *Dicksonia karsteniana,* and now the very rare *Cyathea flaccida* and *Trichipteris tryonorum.* However, many additional Venezuelan species belonging to the genera *Cyathea, Trichipteris, Sphaeropteris,* and *Cnemidaria,* are threatened with exploitation during the coming years, especially those species occurring in the cloud forests of the Coastal Cordillera and the Andes.

The completion of the Yacambú dam in the State of Lara will inundate and destroy three unusually rare species: 1) *Simira lezamae,* an endemic rubiaceous tree known from only one other locality in the State of Barinas; 2) *Begonia williamsii,* identified by Dr. Lyman B. Smith as identical with the Bolivian species, but with no known intermediate stations; and 3) *Lafoensia punicifolia,* a handsome lythraceous tree, known elsewhere in Venezuela only from the Lake Maracaibo area in the State of Zulia. There is a recent report of another station for *Begonia williamsii* in the State of Lara, but thus far no substantiating specimen has come to hand.

Present practices leading to the destruction of the flora

Various factors presently at work are leading to a drastic reduction of the flora and threatening the destruction of many habitats, including rare and endemic species. Alteration and obliteration of such habitats include areas within the different kinds of floristic regions previously referred to.

Among these unique threatened areas are the cloud forest summits of the Peninsula of Paria in the State of Sucre (Steyermark, 1973) opposite Trinidad. In addition to the area's concentration of species known heretofore only from Trinidad (Steyermark and Agostini, 1966), there are associated with the flora numerous endemic species *(Piper pariense, Heliconia steyermarkii, Elvasia steyermarkii, Topobea steyermarkii, Besleria hirsutissima, B. mortoniana,* and *Ixora agostiniana).* Moreover, associated with this flora are relict species of the Amazonas-Guayana region, such as *Platycentrum clidemioides,* isolated on these cloud forest summits (Steyermark, 1974). In the past five years, uncontrolled cutting and clearing of portions of this forested area are threatening the flora.

The second of these unique areas centers around the zone bordering the Páramo de Tamá and Páramo de Judío by the Venezuelan-Colombian frontier in the State of Táchira. This area is the only part of Venezuela where a number of Andean species and twelve Andean genera reach their northeastern limits of dispersal, and in addition, is the locale of numerous endemic species (Steyermark, 1975). Unfortunately, burning and cutting of the forest in order to increase local pasturage, especially in the rich forests below the Páramo de Judío, is leading to obliteration of some of the unique flora. For example, on one of the trees already destroyed in this area was the epiphytic orchid, *Pityphyllum amesianum,* known elsewhere only from Colombia. As in the case of *Phragmipedium caudatum,* even a minimal amount of collecting could endanger the continued existence of *Pityphyllum* in Venezuela. Additionally, the orchid, *Odontoglossum cordatum,* also was found on one of the felled trees of this area, but unless and until additional plants of this species are discovered, it may well have to be considered as a species exterminated in Venezuela.

A serious threat to change both in habitat and ecological balance is to be observed in the Páramo of Mucubají near Lake Mucubají, in the State of Mérida, at the elevation of 3,750 meters. Although this area lies well within the Sierra Nevada National Park, an area presumably to be left, by legal decree, in its natural state, an attempt already has been made to introduce hundreds of pine saplings belonging to an exotic species. The pines have been planted here in a natural páramo where many endemic species occur, as well as ones of restricted phytogeographical range. Many plants can also be found here of the characteristic páramo species of frailejon *(Espeletia schultzii).* If the plantings thrive, the result would be an obliteration of the páramo flora, which eventually would be shaded out by an arborescent plant succession. These plantings were started by foresters without any previous consultation with experts in ecology or with botanists who were qualified to advise on the subject. There has been general disapproval by botanists of this project, which eventually is expected to fail. The important point is that the project has set a dangerous precedent, insofar as plantings such as these have been carried out on nationally-owned terrain within national parks in projects that could well upset the ecological balance of the finely-adjusted natural areas of the páramos with their special floras.

Following the construction of the Pan American Highway and the development of large cattle enterprises, the once magnificent tall rain forest previously had occupied an enormous area south and southwest of Lake Maracaibo in the State of Zulia, and had harbored a rich variety of species of trees reaching 50 meters in height. This area has now been reduced through clearing to a paltry remnant of isolated individuals surrounded by extensive grazing and farm land. No intensive plant survey was ever carried out before this destruction occurred. Since this was one of the large rain forest areas north of the Orinoco River, the loss is especially regrettable.

At present, sawmills are so actively logging such commercial tree species as mahogany (Swietenia macrophylla) cedro (Cedrela mexicana), mijao (Anacardium excelsum), saquisaqui (Bombacopsis quinata), and apamate (Tabebuia rosea), that the remaining numbers of these species are being rapidly reduced in their natural habitats.

The forest reserves of Venezuela include ten units, originally consisting of some 11,707,050 hectares. Four of these (Imataca, La Paragua, El Caura, Sipapo) lie in southern Venezuela south of the Orinoco River, and include 10,334,750 hectares or 91 percent of the total forest reserve area. The other six reserves (San Camilo, Capáro, Ticoporo, Turén, Guarapiche and Río Tocuyo) lie north of the Orinoco River and originally comprised the remaining nine percent. Although these reserves were originally set aside for their valuable timber and for forest production and utilization purposes, some of the reserves north of the Orinoco River (Turén, Ticoporo, and San Camilo) suffered badly from illegal penetration by "squatters" or by families who converted many hectares into grazing and farming plots, with the consequent disappearance of valuable timber. It has been estimated that the Turén reserve has nearly disappeared, and that the San Camilo and Ticoporo reserves already have been reduced by as much as one-third of their original area, due to such type of farming activities. Since the San Camilo reserve contains many species not found elsewhere in Venezuela, and comprises taxa found elsewhere only in Amazonian Peru, Colombia, or northern Brazil, it is apparent that many species from this portion of Venezuela are faced with imminent destruction.

Likewise, south of the Orinoco River, the Imataca forest reserve contains a special flora not found elsewhere in Venezuela. It comprises numerous species of the low Guianas and northern Brazil which are at their northern and western limits of distribution there. In addition, this reserve also harbors species of the Cuyuni drainage, otherwise known only from adjacent Guyana. Despite the rarity of the flora, with many species of trees found nowhere else in Venezuela (Erisma uncinatum, Qualea dinizii, Systemodaphne mezii, Pausandra martinii, Lecointea amazonica, Parahancornia amapa, Inga calantha, and numerous others), portions of this forest already are being converted to farming practices by various settlers (Steyermark, 1968). Where rare and unique floras occur within these forest reserves, it may be possible to set aside a portion of the area for permanent protection, as is being considered for the Imataca forest preserve.

Studies carried out by Dr. Veillon (maps 3 and 5 accompanying the report of the Sierra Club-Consejo Bienestar Rural, Hamilton, 1976) in various parts of the Venezuelan Andes at ten-year intervals since 1950 show the extent of deforestation over the past 25 years. The data from Dr. Veillon's work, covering an area of 88,518 km^2, reveals a 33 percent reduction in forest area during the period 1950-75 alone. Of the total area which was originally in forest, only about 30 percent remains.

Unfortunately, the huge forest reserve of El Caura in southern Venezuela, which comprises virgin forests of a highly restricted flora, is now the scene of preliminary studies that will lead to the construction of a large dam and hydroelectric plant. When this project materializes, not only will one of Venezuela's most beautiful waterfall areas be destroyed, but thousands of hectares of forest in all likelihood also will be destroyed.

The Guri Dam, south of the Orinoco River, already has destroyed a substantial area of forest bordering the Caroni River. Now an additional enlarged dam on the Caroni River above the Guri reservoir is being planned. This new dam is expected to flood another 400,000 hectares, comprising two million cubic meters of wood, with the inevitable result that rich forest flora can never be replaced.

The El Guapo Dam in the State of Miranda, when completed, will flood a unique area of rich virgin forest which has never been botanized. It is anyone's guess as to how many species of great rarity or entirely new to science will be affected by the construction of this dam. Judging by the known flora of the nearby Guatopo National Park, however, it is safe to say that there will be considerable obliteration of many rare and unusual species.

It appears at present that there is an increasing emphasis on the construction of large dams. How many more will be planned in the future can only be conjectured. Unfortunately, not only have there been no previous botanical inventories made in the dam-threatened areas to determine the potential loss of rare or important tree species, but those officials involved in planning the dams do not consult with ecologists or other experts in order to evaluate site alternatives or the amount of destruction to be incurred. Natural habitats, beautiful scenery, and forest resources are viewed by the dam planners as unimportant or secondary to their water-storage projects. As in the United States, hydroelectric dams are frequently constructed because of a combination of influential political pressures collaborating with a powerful Corps of Engineers or public works officials. In the United States, environmental battles have been fought and won by many opposition groups combatting the construction of unnecessary dams. In Venezuela, the government determines the dam sites and projects, but opposition groups, even when present, are unable to win approval. Recently, the agency responsible for supplying water to Caracas (INOS) announced plans to construct a dam in the newly-created Macarao National Park near Caracas. Although plans for construction of this dam were objected to vociferously by various well-known conservationists, the opposition was not strong enough to prevent the project.

The already well-developed highway system of Venezuela is being further "advanced" by an increase in the construction of farm-to-market and penetration roads. These roads are encroaching upon previously virgin or slightly altered terrain, much of which is in primary rich forest or undisturbed savanna. The roads are opening up many new areas to logging, burning, cultivation of small farm plots, mining and other forms of human activity. While the country's need to increase its domestic food production is fully recognized, the various roads now under construction are penetrating national park territory, such as the El Manteco-Canaima road, while the road being constructed from Caicara to San Juan de Manapiare penetrates virgin and botanically unexplored territory. Such roads, bringing with them modified habitats and human settlement, will eventually lead to the inevitable destruction of the native flora, unless strict policing of anti-conservation activities along the course of such roads is initiated. This could easily be accomplished by frequent patrolling by a very small force of vehicle-mounted forest guards empowered to act at the first signs of encroachment.

In its national park system, Venezuela has legally set aside many unusual areas which harbor a rich and diversified flora. For example, in the area of the Gran Sabana (Canaima) National Park, the largest in Venezuela and the one including a portion of the Guayana Highland flora, there is absolutely no official protection of its unique flora. Anyone may climb to the summit of the famous Auyan-tepui, situated within the park, and there he may set up his own campsite for an indefinite period, leaving exposed to view trash and garbage litter dispersed over the spectacular landscape for any subsequent visitor to see. No park guards or rangers are stationed at the summit or on any other part of the mountain to protect it against such depredation or against uncontrolled fires, defacing of rock formations with graffiti, or pilfering of plant material. The Indian inhabitants in this park also are permitted to kindle the savannas at the base of the mountains at will, a continuation of an age-old tradition. Tourists are allowed to leave the area with their collection of pilfered orchids, so that instead of an "Orchid Island" above Canaima where a goodly variety of orchids used to thrive, there is now, it is sad to report, nearly no orchid flora at all.

Lack of personnel to police and protect the national parks against vandalism and destructive activities of all kinds is a common complaint in most Venezuelan national parks. Thus, without any supervised controls, individuals entering the parks are allowed unlimited freedom to carry on activities that would be prohibited elsewhere.

Future outlook of threatened and endangered species

In view of the preceding facts, what are the future prospects in Venezuela, as far as threatened and endangered species are concerned? On the basis of what has already been stated, it is more realistic to be a pessimist rather than an optimist, particularly because of the nature of the numerous ideas and planning schemes now being considered for developing and changing the face of the country. These schemes include: 1) large-scale drainage of the unusual swamp forests and other terrain of the enormous Territorio Federal Delta Amacuro for agricultural development; 2) similar plans to convert the swampy savannas of the gigantic State of Apure into a system of canals for agriculture and grazing development; 3) dynamiting the famous Rapids of the Atures and Maipures on the Orinoco River south of Puerto Ayacucho — the classical type locale for many species described by Humboldt and Bonpland — for the purpose of converting this beautiful sector into a deepwater passage to allow larger boat navigation in a proposed throughway to unite the Orinoco, Rio Negro and Amazon river systems with Brazil and Colombia; 4) a proposed canal to unite the Orinoco River with the Caribbean Sea; and 5) construction of various hydroelectric dams, such as El Guapo, El Caura, enlargement of the Guri Dam area, and others.

Other discouraging factors are the lack of policing of national parks, the lack of law enforcement personnel in national parks and wildlife refuges, the illegal entry into national forest reserves for farming activities, the official tendency to open up more and more primary and previously-inaccessible forested areas by increased road construction, and the tendency to permit the conversion of forested areas into agricultural lands, as is currently happening just south of Barquisimeto, for example, where many new potato fields are rapidly replacing what was once primary forest.

Such foolish and ill-advised schemes as changing natural páramo habitats by the planting of exotic pine trees, are instances of ill-conceived plans carried out without the advice of professional botanists. These actions on the part of officials bode ill for the future of the country if similar projects are allowed to be carried out without adequate government controls.

Furthermore, even though present and recent administrations have decreed the establishment of new national parks and have formulated protective laws for the benefit of forest conservation, there is no guarantee that future administrations, if influenced by increased population or political pressure, will continue to uphold and respect the decrees for the preservation of the natural flora.

On the optimistic side of the picture, one must take into account that Venezuela is fortunate in having many farsighted, alert, and knowledgeable citizens and groups of persons who are highly concerned about the future manner in which our natural resources will be protected and managed. Only time will reveal how strong and effective their voices and their influence will be. Moreover, we should be encouraged by the fact that the present administration has established a training school for forest rangers and national guards, with the object of placing an increased number of such trained personnel within the national parks and forest reserves of the country. Also, a new Ministry of Natural Resources is being planned, and the primary schools are including courses in conservation in their curricula.

All of these steps point to an awakening of "conservation consciousness" in Venezuela, and hopefully this new trend will become a powerful stimulus for the future conservation and preservation of our country's natural plant resources. There now also is an awareness of the necessity to set aside biological preserves or natural areas as refuges for rare and unusual species or habitats so that they can be saved for posterity. Such a system of botanical and other preserves was first proposed by the author, and won the approval of the First Botanical Congress of Venezuela in 1971. The idea was then elaborated upon in 1975 for the Technical Advisory Group of the Sierra Club, in conjunction with its work which is carried out in Venezuela with the Consejo Bienestar Rural, and is also sponsored by the Academia de Ciencias Fisicas, Matematicas y Naturales, together with the International Union for the Conservation of Nature and Natural Resources (I.U.C.N.) and the World Wildlife Fund.

In conclusion, I would like to emphasize that although a definite trend exists to develop the natural resources of Venezuela as quickly as possible in order to boost its agricultural and industrial programs, such trends inevitably will result in the destruction of many areas through drainage projects, dam construction, road penetration, and farming activities. Yet there may well be a positive control in the future on the part of its citizenry and government to protect and set aside natural areas for future generations, which should give us much cause for hope.

Although predictions have been made by some biologists (Raven, 1974a, 1974b) that no virgin or primary forest will be found in either the neotropics or in the world's remaining tropics by the year 2000, the author is confident that in Venezuela, at least, some virgin or primary forest will continue to exist indefinitely. It is anticipated that many areas in the Venezuelan Guayana of the State of Bolívar and Territorio Federal Amazonas, as well as in some areas north of the Orinoco River, by virtue of their being included within the national park system or in other preserves, will be permanently maintained and protected against such activities as cutting. The unique flora of the mountains of the Guayana Highlands, with their spectacular scenery and the vast continuous wilderness dominating much of the lowland terrain of the Territorio Federal Amazonas, undoubtedly also will be preserved and protected for future generations. In addition, the increased recognition of the extreme poverty of the forest soils of the Venezuelan Guayana will be an added incentive for preserving and conserving the area.

The future outlook for threatened and endangered species in Venezuela, therefore, may well be on the positive side. Let us hope that growing public and governmental awareness for conservation and protection of Venezuela's natural areas and resources will prevail.

LITERATURE CITED

Hamilton, L. S. 1976. Tropical rain forest use and preservation: A study of Problems and Practices in Venezuela. Sierra Club Special Publication International Series No. 4, 72 pp.

Maguire, B. 1970. On the flora of the Guayana Highland. Biotropica 2 (2): 85-100.

Raven, P. 1974a. Trends, priorities, and needs in systematics and evolutionary biology. Systematic Zoology 23 (3): 417, 418, 435.

_____. 1974b. Plant systematics. Ann. Missouri Bot. Gard. 61: 75.

Steyermark, J. 1968. Contribuciones a la flora de la Sierra de Imataca, Altiplanicie de Nuria y region adyacente del Territorio Federal Delta Amacuro al sur del Rio Orinoco. Acta Bot. Venez. 3: 49-175.

_____. 1973. Preservamos las cumbres de la Peninsula de Paria. Defensa de la Naturaleza. Ano 2 (6): 33-35.

_____. 1974. Relacion floristica entre la Cordillera de la Costa y la Zona de Guayana y Amazonas. Acta Bot. Venez. 9: 245-252.

_____. 1975. La region del Tamá debe ser preservada. Natura 57: 5-8. Nov.

Steyermark, J. and G. Agostini. 1966. Exploracion botanica del Cerro Patao y zonas adyacented a Puerto Hierro, en la Peninsula de Paria, Edo. Sucre. Acta Bot. Venez. 1: 7-80.

ENDANGERED PLANT SPECIES AND HABITATS OF ECUADOR AND AMAZONIAN PERU

Alwyn H. Gentry

Missouri Botanical Garden, St. Louis, Missouri 63110, U.S.A.

INTRODUCTION

Ecuador and Amazonian Peru may be considered as a reasonably natural geographic unit despite the political boundary between them. In fact, the political boundary involved is not universally agreed upon — maps prepared in Ecuador still show the bulk of Amazonian Peru as part of Ecuador! More important, the area encompassed under this heading includes several rather distinct and well-demarcated phytogeographic regions which transcend political boundaries. Indeed, Amazonian Ecuador is so little known botanically that extrapolation from adjacent Peru, itself poorly known, provides the main source of such information for this region. Thus, the Flora of Amazonian Peru must be consulted extensively in all evaluation of Ecuador's flora, and the geographic circumscription adopted here is a convenient one.

Though this region as a whole is floristically very rich, it is extremely poorly known, as has been generally recognized: "Botanically, Ecuador is least known of the countries that lie along the Andes" (Standley, 1931); "Ecuador is botanically one of the least known, though one of the richest countries of South America" (Svenson, 1945). Recent estimates of the number of plant species in Ecuador range from 10,000 (Sparre, 1968) to 20,000 (Gentry, 1976b) while inclusion of those in Amazonian Peru should add an additional 5,000 to 10,000 species. Thus, between 15,000 and 30,000 plant species probably occur in this region, as compared to estimates of 20,000 species for all of Central America (including southern Mexico) (Gentry, 1976b) and 25,000-30,000 for all the vast expanse of Brazilian Amazonia (Prance, pers. comm.). Steere (1950) has suggested that Ecuador may have the greatest number of plant species per unit area in all of South America.

Despite its great botanical diversity, most of the region is not covered by any published Flora. Publication of a Sweden-based Flora of Ecuador was begun during the last two years under the direction of Benkt Sparre and Gunnar Harling, though only four small families have been treated thus far. But even these treatments are hampered by lack of adequate collections to provide a source of floristic knowledge. The Flora of Peru provides some information about Ecuadorean plants, though it is lamentably incomplete in its coverage, even for Peru, and is only partially completed. While the plants of Amazonian Peru are included in that country's Flora, they are treated most inadequately. This continues to be the least known region of the country, and in many earlier familial treatments probably less than half of the species actually occurring in Amazonian Peru were included.

The few other sources of floristic information available are certainly useful but even less complete. The most significant of these is Llewelyn Williams' (1936) *Woods of Northeastern Peru,* which covers primarily the region around Iquitos and provides a wealth of ecological and ethnobotanical information. Little (1969) has published a nicely-illustrated compilation of the *Arboles Comunes de la Provincia de Esmeraldas,* which includes similar information for a limited region in northwestern Ecuador. Both of these works are restricted entirely to trees. Svenson's (1945) study of the dry coastal vegetation of Pacific Ecuador and Peru is critical and scholarly but it considers only the relatively few species actually collected by the author. Diels' important earlier work (1937, 1938-42) on the Ecuadorean Andes hardly does more than list and describe new species. Weberbauer's (1936) classical phytogeographic studies of Peru expressly exclude Amazonia and are of only slight relevance to coastal and upland Ecuador. Even at the local flora level, little or nothing is available. The Flora of the Río Palenque biological station in lowland western Ecuador, which Calaway Dodson and I are now compiling (Dodson and Gentry, 1977), will be the first such listing for any site in this region.

While lack of floristic knowledge precludes direct assessment of endangered species, sufficient information is available for comparison of habitats. Since plant species distributions are closely correlated with habitats, and since it is impossible to evaluate which species of this area are rarest and most endangered, an assessment of threatened and endangered habitats offers the best available information on threatened plant species.

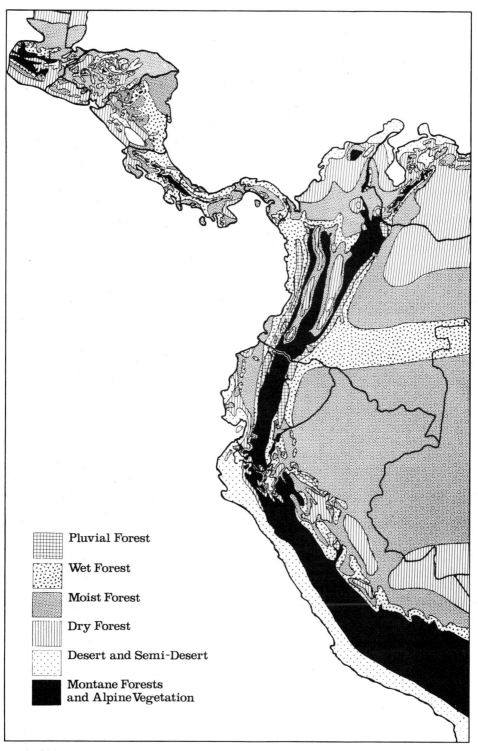

Fig. 1. *Vegetational types of northwest South America and Central America modified from the Holdridge Life Zone system of classification (from Gentry, 1976b).*

A brief phytogeographic survey is a useful preliminary to evaluation of the region's various habitats. Several noteworthy distributional and diversity patterns are beginning to be apparent. It now appears that most tropical plant species have geographically ample but ecologically restricted distributions. In most of tropical America, these distributions are much more highly correlated with the Holdridge Life Zone System (Fig. 1) than with any other system of ecological classification. Only in climatically relatively uniform Amazonia, where edaphic factors become major determinants of distribution, does this phytogeographic correlation break down. The wide ranges and ecological fidelity of most plant species, coupled with the availability of Life Zone maps for most of Latin America, make possible a reasonably accurate extrapolation of species distributions.

The extent of this ecological specificity has been inadequately appreciated. Even between physiognomically rather similar lowland moist and wet forest, which are lumped together under non-Holdridge classifications, there is surprisingly little species overlap. Thus, only 32 percent of the tree species listed by Little for Esmeraldas Province (mostly from tropical moist forest) are at Río Palenque (wet forest), and most of the tree species which do occur at Río Palenque — including many of the commoner species — are not included by Little. Only a few weeds occur both at Río Palenque and the dry part of Pacific Ecuador, which is just a few kilometers away. In vivid contrast, 80 percent of the Río Palenque species reach wet forest habitats in Central America and 65 percent reach Amazonia. A second example of wide ranges but narrow ecological tolerance may be taken from my specialty group, the Bignoniaceae (Gentry, 1973, 1976a). Ninety-four percent of the dry forest bignon species of Costa Rica and Panama reach Venezuela, and 94 percent reach Guatemala and southern Mexico; 97 percent of the moist forest species (excluding mammal-dispersed and mangrove species) reach Venezuela and 80 percent Guatemala; for wet forest species, these figures are 94 percent and 73 percent. Yet these same species show little overlap between different life zones within Panama and Costa Rica.

That most species are wide-ranging despite the sharp ecological limits of their distributions greatly complicates evaluation of potential endangered status. A species which is in danger of extinction in Amazonian Peru might be common in Brazil; a rare species of coastal Ecuador could be common in adjacent Colombia or on the opposite side of the Andes. Its total range and the level of habitat destruction throughout its range must be considered in estimating the status of any species as a whole.

A second key geographical consideration is the presence of the Andes as a major barrier for lowland plant species. Thus, Gilmartin (1973) notes that only 17 of 249 Ecuadorean Bromeliaceae species have a trans-Andean distribution. A preliminary[1] distributional analysis for Río Palenque species emphasizes the significance of such geographical factors. One-quarter of the 80 percent of these species which reach Central America (19 percent of the total flora) do not occur in Amazonia. In contrast, only 3 percent of the Río Palenque species occur in Amazonia but not in Central America. Thus, we have a somewhat paradoxical situation in which level of destruction in faraway Central America is much more relevant to likelihood of species survival of the wet forest flora of coastal Ecuador than what happens in Amazonian Ecuador, which is geographically much closer. For montane species, of course, the obverse is true, though most of these species also occur in Peru or Colombia or both.

Another important general consideration is that species diversity, at least for lowland plant communities, seems directly related to the amount of precipitation in any given area. Thus, a 1,000 m^2 sample at wet forest Río Palenque contained 118 species of woody plants representing 49 families. A similar sample of Panamanian wet forest included 151 species, while moist and dry forest samples from a variety of sites in South and Central America contained progressively decreasing species diversities. The almost 600 species per km^2 recorded from Río Palenque is probably the highest recorded plant diversity in the world. However, even tropical dry forest, the most depauperate tropical lowland habitat known, is much more diverse than rich temperate plant communities (Table I). In general, the wetter a lowland tropical area, the richer its flora, and destruction of tropical wet forest habitats threatens many more plant species

[1] Distributional data are currently compiled for only half the species.

than does comparable destruction of moist or dry forest habitats.

Table I. Relative diversity of plant species. For individuals over 1" (2.54 cm) dbh in 1,000 m^2 sample area. See Gentry, 1976a, for details.

	No. of families	No. of Species
Dry Forest		
Guanacaste, Costa Rica (upland)	18	41
Guanacaste, Costa Rica (riparian)	33	64
Venezuelan Llanos (Calabozo) (500 m^2)	20 (+ 1 indet)	41
Coastal Venezuela (Boca de Uchire)	20 (+ 3 indet)	67
Moist Forest		
Curundu, Pacific Canal Zone	38	88
Manaus, Brazil	29	91
Madden Forest, Central Canal Zone	40	125
Wet Forest		
Pipeline Road, Canal Zone	50	151
Río Palenque, Ecuador	49	118

Comparable data are not available for documentation, but diversity generally seems to decrease with increasing altitude. As a result, high Andean plant communities consist of many fewer species than do lowland tropical ones. There is some conflicting evidence that for certain animal groups (Scott, 1976) diversity is greatest at middle altitudes (ca. 1,200 m). Whether this pattern is also applicable to plants is unclear but if so, it is likely to prove a reflection of the more mesic nature of many intermediate elevation sites: Holdridge system "rain forest" is found only at middle elevations in Ecuador and Amazonian Peru.

Ecuador and Amazonian Peru may be conveniently subdivided into several phytogeographic regions, each with a distinctive flora posing different problems for preservation. These include coastal dry forest, coastal wet forest, montane forest, dry montane valleys, paramo, and the Amazonian lowlands, the latter constituting over two-thirds of the total area under consideration.

Lowland Amazonia

Of all the above-named regions, lowland Amazonia is, by and large, the least modified by man, and its species are, accordingly, the least threatened. Amazonia is extremely diverse edaphically and may be divided into several subregions, each with characteristic vegetations and preservation problems. These include:

Upland tropical moist forest

This vegetation covers by far the greatest portion of the Amazonian region. It is largely undisturbed because of the lack of roads in this region and its relative inaccessibility from rivers. There is no apparent present danger to any plant species of this habitat. However, in the long-term even the species of the forested vastnesses of Amazonia may face a grim future. Several disturbing trends already are evident. New roads are being pushed into this region in both Ecuador and Peru. In each case, the new road has resulted in immediate (within 2-3 years) and complete destruction of strips of natural forests extending both along the road and for a few kilometers on either side of it. That this practice has had no appreciable effect as yet on the total extent of the forest is obvious from the air, where the narrow roadside ribbons pale to insignificance. But past experience tells us that rapid settlement invariably takes place along new roads, and this bodes ill for the future as additional roads penetrate Amazonia. In Amazonian Ecuador, the discovery of oil has resulted in intense road-building activity since that time, with the pace of destruction rapidly and continually accelerating. On the other hand, since petroleum exploration has been less productive in Peru, relatively little road-building is now in progress there. However, destruction along the ten-year-old road from Tingo Maria to Pucallpa and the new Carretera Marginal along the base of the Andes (which will soon be completed along the entire length of the Huallaga Valley) indicate that the same irreversible forces are operating. Construction of a new oil pipeline in the Marañon Valley will soon provide additional access and resultant destruction of vegetation. While none of these activities currently threatens whole plant species or habitats, they are the first step which elsewhere has inevitably led to rampant destruction (cf. Pacific Ecuador, below).

Even though there is no imminent danger to upland Amazonian habitats, it can be assumed that certain individual species could be threatened through commercial exploitation there. It is well-known that many of the primates and game animals like tapirs have been reduced to virtual extinction over large parts of this region from hunting pressure (e.g. Freese, 1975). Mahogany *(Swietenia macrophylla),* the most intensively exploited plant species in upper Amazonia, already has been locally eliminated around Iquitos. But even mahogany is not seriously threatened in the less accessible regions (Schunke, pers. comm.) in contrast to its situation in Panama, for example. Orchids, too, are often excessively gathered, but no Amazonian orchid is seriously threatened at present. Our conclusion as to the non-endangered status of upland Amazonian plant species remains intact, although elimination of important mammalian seed dispersal agents introduces potentially serious disturbances in the plant community as well.

Seasonally inundated forest (Tahuampal)

Large areas of upper Amazonia, especially along the many rivers, are seasonally inundated. The vegetation of these areas is floristically quite distinct from that of adjacent upland forest. In the absence of roads, these seasonal swamp forests are more easily accessible than the upland forest is, with the result that human disturbance there is somewhat more intense, although generally negligible in terms of being a threat to the survival of any plant species. Perhaps the most likely candidates for endangered Amazonian plant species are the two iriarteoid palms, which are extensively used for house construction and are ecologically restricted to these lowland areas; both species are becoming very rare at least along well-traveled rivers (Revilla, pers. comm.).

White sand areas

It is well-known that in Amazonian Brazil a white sand substrate with a very distinctive flora underlies most of the north side drainage of the Amazon River. Less well-known is the occurrence of similar but less extensive white sand areas in northern Amazonian Peru. These low productivity ecosystems are extremely slow to regenerate (Janzen, 1974), and even a relatively low intensity of disturbance has destroyed most of their natural vegetation in the Iquitos vicinity. There seems to be a certain amount of plant endemism in the Peruvian white sand areas, although most of the species are widespread ones characteristic of similar regions in southern Venezuela and northern Brazil. While not in any immediate danger, the totally distinct biota of these regions — even some mammals are endemic to them — makes them important from the conservation viewpoint. Special care must be taken to include such areas among those eventually to be designated as preserves or sanctuaries.

Dry forest regions

The southern part of Amazonian Peru contains outliers of dry forest vegetation similar to that of the Brazilian cerrado. The region is unknown botanically. Thus, *Bertholletia excelsa* (Brazil nut), the most common canopy tree around Puerto Maldonado where it is even economically important, has not previously been collected in Peru (Prance, pers. comm.). In all likelihood, the species of this region do not differ markedly from the wide-ranging ones of the cerrado, but there is at least a possibility that they may represent, in part, a distinct western South American dry forest assemblage. Once again, there is no immediate threat to this vegetation, though it should be included if a comprehensive system of natural preserves ever becomes a reality. Several river valleys farther north, most notably the Río Huallaga near Tarapoto, contain disjunct lowland dry forest areas which have been much more seriously disturbed by rather extensive agriculture. Some of the plant species described from these areas may be seriously threatened if they are not also found to be present in the more extensive southern dry forests; we have inadequate data at the present time to even begin to assess this relationship.

Wet forest regions

In general, the parts of Amazonian Peru and Ecuador which lie nearest to the base of the Andes are the wettest. Some of these regions fall into Holdridge-system wet forest and might be expected to have both a distinctive and more diverse flora. The meager floristic data available strongly support this possibility. For example, the recent collections of Brent Berlin and associates from the wet Río Cenepa region on the Peru/Ecuador border have been unusually rich in new species, as were the collections of T. Dudley from the

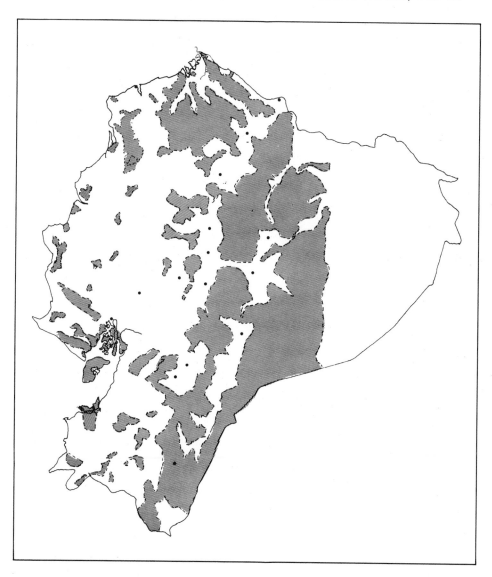

Fig. 2. Remaining natural vegetation in Ecuador. Shaded area indicates natural vegetation; white area indicates vegetation destroyed (except in eastern Amazonia which was not surveyed). Black dots are major towns. Note vertical white strip running the length of the country along the interandean valley (from Putney, 1976).

ecologically similar Vilcabamba region further south. Isolated areas of even wetter Holdridge-system rain forest near Aguaytia and on similar ridge tops north of Tarapoto and Moyobamba — all completely unexplored botanically — are also likely to contain a diverse and unique flora which needs preservation.

Major sections of the lowland wet forest habitat, much more restricted in area than the Amazonian moist forest, are now being threatened by construction of the Carretera Marginal between Tingo Maria and Tarapoto in Peru and by several new roads penetrating Ecuadorean Amazonia. Conservation officials in Ecuador (Putney, pers. comm.) consider this region the third most endangered area of the country (after the interandean valley and coastal dry forest). Nevertheless, it is unlikely that any plant species of this region is presently endangered. Preservation *now* is necessary, however, to forestall present and future destruction. This preservation needs to be better funded in order to be more effective than is now the case in Parque Nacional de Tingo Maria, the only lowland wet forest currently designated a preserve in either country.

ANDEAN UPLANDS

At the other extreme from the relatively well-preserved vegetation of the Amazonian lowlands, the natural vegetation of the upland Andean region has been heavily modified and even largely destroyed, at the hand of man. This is especially true of the fertile interandean valley which runs the length of Ecuador and in which not a single wildland area remains (Fig. 2). The largest native forest on the valley floor (Bosque de Arrayán) is only six hectares in size (Putney, pers. comm.), while on the inner slopes on either side of the central valley, a few larger but highly disturbed cloud forest remnants still exist. The FAO National Parks project has virtually eliminated the interandean valley as having any natural areas worth potential preservation (Putney, pers. comm.). Despite the lack of natural areas, however, it is doubtful if many plant species of this region are in danger of extinction. Most of the extant species are fairly widespread and have adapted to such available habitats as the traditional fencerows and roadsides which have changed very little during the past centuries. A state of relative equilibrium seems to have been reached a long time ago, and any restricted local endemics unable to cope with human settlement probably became extinct in pre-Colombian times. The likely exceptions to this generalization are in Azuay and Loja Provinces in the south, where some endemism is known in the extant flora (Gilmartin, 1972 and pers. comm.; Olga MacBryde, pers. comm.). Endemic plants of this region, such as *Puya pygmaea* L. B. Smith and *P. aequatorialis* var. *albiflora* André, are probably threatened (Gilmartin, pers. comm.).

The higher mountains (above ca. 3,500 m) and the upper parts of the numerous volcanoes have the very different above-timberline natural vegetation called páramo in Ecuador. Wet páramo with *Espeletia,* similar to that of Colombia, occurs only in the extreme northern section of Ecuador, and also reputedly in the isolated and botanically unknown "Llanganate" region. Most Ecuadorean páramo contains bunch grass and various low shrubs and herbs (especially Compositae). This region has been extensively modified by human activity, especially the omnipresent seasonal burning, but probably few, if any, of its plant species are truly endangered. They are mostly quite widespread, while the Ecuadorean páramo flora as a whole is neither so rich nor so prone to endemism as that of the Colombian páramos. In addition, the chief agricultural use for this area is grazing, which is relatively undestructive to the natural vegetation.

Above the páramo on the higher volcanoes is an alpine zone characterized by scattered vegetation composed mostly of cushion and rosette plants. Since this region has not been too affected by man, its depauperate but interesting flora is not threatened.

The interandean valley is also interrupted by numerous canyons and ridge systems which cause local rain-shadow effects. These, in turn, give rise to a high altitude desert vegetation that is very different from the montane forest formerly native to most of the valley. Compared to adjacent, more mesic regions, these upland deserts are only slightly disturbed. These regions merit preservation for their phytogeographic and ecologic interest, but none of their plant species seem especially endangered in Ecuador.

Below the Andean crests on the outer slopes of the Andes, approximately between 1,000 or 1,500 and 3,500 m, the natural vegetation is montane cloud forest similar to that which presumably once covered most of the interandean valley. Like the interandean forest, large areas of this steep and erosion-prone region have been extensively modified by man. Except in the extreme north, there are few large areas of montane forest on the Pacific slope which have not been substantially altered (Putney, 1976; Fig. 2). The eastern slopes are as yet relatively undisturbed, except along several new roads to the "oriente." But, undoubtedly, they await a fate similar to that which has already befallen their western equivalent, ie, population pressure in the uplands continually forces more and more movement of settlers into the montane forest and adjacent oriente.

The Ecuadorean cloud forest is botanically famous as the home of *Cinchona officinalis*, whose bark is the source of quinine, the famous anti-malaria drug. *Cinchona*, first discovered near Loja in southern Ecuador, occurs between 1,000 and 3,500m and demarcates the montane forest region. At one time *Cinchona* itself was considered in danger of extinction because of the bark removal for quinine production (Standley, 1931), but commercial quinine now comes largely from trees cultivated in Asia. However, the native Ecuadorean tree is once again endangered, this time by the rampant destruction of its habitat.

Ecuador is also famous for the epiphytic plants of its montane forests. Thus, Gilmartin (1973) notes that Ecuador is a prime distributional area for bromeliad species. It also is justly famous for orchids, and, in fact, Dodson and Gillespie (1967) consider the montane forest area along the Río Pastaza canyon between Baños and Puyo "perhaps the finest in the world" for orchids. These beautiful and showy flowers are especially susceptible to selective extinction through overcollecting by enthusiasts, but even rare orchid species may be in relatively little danger of extinction as long as their habitat remains intact. However, continuation of unchecked destruction of the montane forest makes all the species of this rich habitat endangered. Most of the montane forest species of Ecuador, including the orchids and bromeliads, are fairly wide-ranging, so preservation of even a few representative areas of montane vegetation should preserve most of them.

In summary, we may conclude that the montane forest and its associated species are rapidly being destroyed and therefore are in urgent need of protection, even though surprisingly few species of this habitat have yet become extinct. On the other hand, current patterns of development, exploitation, and population increase may well extinguish this whole habitat and all its species unless immediate conservation measures are taken.

Coastal Dry Forest

The Pacific coast tropical dry forest vegetation is judged by Ecuadorean conservation officials to be the country's second most endangered habitat (Putney, pers. comm.). Most of the region has been devastated for centuries by such activities as lumbering, wood-gathering for fuel, and overgrazing, especially by goats. This region (including the adjacent corner of Peru) contrasts sharply with other phytogeographic regions of Ecuador and with the rather homogeneous western Central America — northern South America dry forest vegetation in its high degree of endemism. In fact, many of the most characteristic and dominant species of the coastal dry forest are endemic, some at the generic level. For example, *Macranthisiphon*, one of the most distinctive genera of Bignoniaceae (a family generally noted for wide-ranging species and genera) is common throughout this region but found nowhere else in the world. The common, grotesque-trunked "ceibo" of the hilly regions, *Ceiba trischistandra*, is also endemic here (and incidentally almost unrepresented in the world's herbaria). Its lesser bombacaceous relative *Eriotheca ruiziana* is also common and endemic. This region shares with northern Venezuela the distinction of being the site of the most prolific speciation of the family Capparidaceae and a number of endemic *Capparis* species are characteristic and important components of its vegetation. Such conspicuous dry area species as *Hymenocallis quitoensis (Amaryllidaceae)* which covers the ground in February with its large, white flowers, and *Carica paniculata* (Caricaceae), whose shockingly red inflorescences and flowers make a vivid contrast to the brown, dry-season landscape, are likewise found nowhere else in the world.

With such unusually high levels of endemism, the coastal dry forest clearly merits efforts at preservation. The need for preservation of this unique plant community is acute, although probably few or none of its species have yet become extinct. This relatively heavily-populated region has been settled far longer than have the other lowland areas of

Ecuador and Amazonian Peru. For example, the now barren and desolate hills near Salinas were covered by magnificent dry forest as recently as the 1940's (A. Marmol, pers. comm.), but the trees and even the shrubs were destroyed first for lumber and later for firewood, charcoal, and by overgrazing of goats. Moreover, the surviving species have undergone a kind of negative selection with species like *Maytenus octagona,* which are least valuable for timber or charcoal, becoming increasingly prevalent (Gilmartin, pers. comm.). Actually, protection from goats may be as important as preservation from direct human destruction. For example, in the hilly region near Jipijapa, where fairly extensive areas of physiognomically well-developed dry forest still remain, there appears to be no regeneration, presumably due to destruction of seedlings and undergrowth by goats.

A typical example of plant species destruction directly resulting from human exploitation is that of "guayacan" *(Tabebuia chrysantha),* the most important timber of the region and the material from which the area's famous and economically important wood carvings are made. Though this species once was a dominant one throughout large areas, and mature "guayacan" forest is reported still to exist in such remote regions as the Cordillera de Chongon, I have been unable to locate any even relatively undisturbed areas of "guayacan" forest, and only isolated mature trees. Personnel at various lumber yards and carpenter shops tell me that logs of this species, which formerly were harvested from nearby forests, now come from much greater distances (mostly near Julio Moreno). In addition, the trees have been decreasing in both size and quality in recent years. Throughout much of the region, this commercially and esthetically important species has been either wiped out or reduced to small and mostly pre-reproductive juvenile plants. While the species as such is not threatened, its population structure and economic potential have been substantially altered.

Colonial accounts indicate that about 200 years ago much of the coastal region was rich savanna pastureland (Dodson, pers. comm.), though the area is now covered by thorn scrub and cactus alternating with vegetationless desert. Numerous 200-year-old "ghost towns" scattered throughout this now-desert region bear mute testimony to its climatic deterioration (Dodson, pers. comm.). This dramatic change in historical times from savanna to desert, presumably caused at least in part by human misuse, may bode ill for the long-term survival of the interesting flora of the region, even in the event that remaining forested areas are preserved. Without preservation, the last vestiges of this highly endemic, natural vegetation undoubtedly will disappear in a very few years.

COASTAL MOIST AND WET FORESTS

The vegetation of this region as a whole has not been considered by local conservation officials to be especially threatened (Putney, pers. comm.) because fairly extensive, though rapidly-diminishing, forested areas still occur near the Colombian border (Putney, 1976; Fig. 2). Actually, the rich Pacific lowland forest is not the single vegetational type which is generally recognized (e.g., Anonymous, IUCN, 1974) but rather it is comprised of two floristically very different vegetational units — moist forest and wet forest (see Fig. 1). From the viewpoint of endangered plant species, these must therefore be considered separately. One of them — the lowland wet forest — probably contains more acutely threatened plant species than any other part of Ecuador and Amazonian Peru, and possibly even all South America.

Unfortunately, both coastal moist and wet forests are poorly known floristically. The Chocó region of Pacific Colombia is even more poorly known, and is perhaps botanically the least known region of South America (Gentry, 1976b). Although both moist and wet forest regions in Ecuador have apparently endemic species, there are many more apparent endemics in wet than moist forest. On the whole, the moist forest flora of Pacific Ecuador is almost identical in species composition to similar regions of Panama. This flora also covers a greater area in Ecuador than does the wet forest vegetation, and fairly extensive areas of it still exist in the northern region, which has not been penetrated by roads. All these factors indicate that the coastal moist forest contains relatively few endangered plant species. The wet forest may be very similar to the virtually unknown Chocó flora. However, on the basis of the many species known only from this region, one is forced to conclude that it has a very strong endemic element. The flora of this region is especially rich in epiphytes. Thus, five of the eleven largest families at Río Palenque are entirely or predominantly epiphytic — Orchidaceae, Araceae, Piperaceae, Bromeliaceae, Gesneriaceae — and epiphytic ferns are also very well-represented. Moraceae is the most important family of canopy trees.

Moreover, the narrow strip of wet forest extending south along the base of the Andes is currently the country's fastest-disappearing habitat. It has been opened up within the last decade by road construction and road improvement, and in the last five or six years formerly extensive wet forest vegetation has been converted to almost solid banana and oil palm plantations. Perhaps the only unthreatened area of this vegetation in the southern seven-eighths of Ecuador is the minuscule (less than 2 km^2) Río Palenque Field Station. At least thirty species of plants are known only from this field station (Table II). Many of these, which are only now being formally described, quite probably are otherwise already extinct. The northern part of this wet forest strip is poorly known, but the rapidity with which southern and central regions of wet forest have disappeared renders the whole habitat an acutely threatened one. Even if these Ecuadorean wet forest endemics prove to extend north into the Chocó, they would have to be classified as extremely threatened.

Table II. New Species from the Río Palenque Biological Station coastal lowland wet forest. Those species marked with asterisk have only been found inside the 1.7 km^2 of the field station and most of these may already be on the verge of extinction (see text).

Acanthaceae
  *Dicliptera dodsonii Wasshausen
  Justicia ianthina Wasshausen
  J. riopalenquensis Wasshausen
  *J. pectoralis var. ovata Wasshausen
  *Ruellia riopalenquensis Wasshausen
Aristolochiaceae
  Aristolochia pichinchensis Pfeiffer
Asclepiadaceae
  Fischeria aequatorialis Spellman
Bombacaceae
  *Quararibea palenquiana A. Robyns
Burseraceae
  *Protium sp.
Capparidaceae
  Podandrogyne brevipedunculata Cochrane
Gesneriaceae
  *Alloplectus dodsonii Wiehler
  Columnea dodsonii Wiehler
  *Dalbergaria rubriacuta Wiehler
  Drymonia ecuadorensis Wiehler
  D. laciniata Wiehler
  D. rhodoloma Wiehler
  Gloxinia dodsonii Wiehler
  Pentadenia occidentalis Wiehler
  P. zapotalana Wiehler
Lauraceae
  *Persea theobromifolia A. Gentry
  *Ocotea sp.

Lecythidaceae
  *Gustavia dodsonii Mori
Leguminosae
  *Inga riopalenquensis A. Gentry
Meliaceae
  *Guarea sp.
Menispermaceae
  *Odontocarya perforata Barneby
Monimiaceae
  Siparuna domatiata A. Gentry
Moraceae
  *Maquira sp.
Myrsinaceae
  Stylogyne gentryi Lundell
Orchidaceae
  Dichaea richei Dodson
  *D. riopalenquensis Dodson
  Epidendrum mininocturnum Dodson
  Trigonidium riopalenquensis Dodson
Sapotaceae
  Pouteria capaciflora Pilz
Solanaceae
  *Solanum palenquense D'Arcy
  *S. sp.
Thymelaeaceae
  *Daphnopsis sp.
Urticaceae
  *Pilea selbyanorum Dodson

Pteridophytes
  Diplazium striatastrum Lellinger

In summation, there are two distinct vegetation types which have usually been lumped together as humid lowland coastal forest. Both are fast disappearing and both are in need

145

of protection, though the wet forest is especially endangered. This rich wet forest vegetation may contain more plant species endemic to Ecuador than any other habitat and all species restricted to it are threatened. It is especially unfortunate that the most species-rich vegetation (along with the ecologically similar transandean region) in Ecuador or Amazonian Peru is currently also the most endangered.

CURRENT CONSERVATION EFFORTS

We have now completed our survey of the major vegetational units of Ecuador and Amazonian Peru. We have also noted those which are being destroyed most rapidly. It is now appropriate to consider current conservation efforts and to suggest those additional priorities which will most effectively preserve threatened plant species and habitats.

Peru and Ecuador are obviously both aware of the need for natural area preservation. Both countries have taken commendable steps toward the establishment of national parks, national forests and ecological reserves. They both have active land use planning programs which also include preservation of natural areas. Unfortunately, as past experience has frequently shown, designation of preserved areas on a map is still a long way from creating an effectively protected sanctuary. Both countries need additional money and manpower

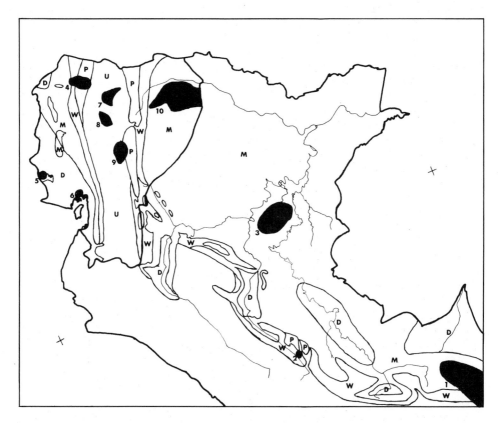

Fig. 3. Preserved areas in Ecuador and Amazonian Peru. Peruvian preserves (all extant) are: 1. Parque Nacional Manu; 2. Parque Nacional Tingo Maria; and 3. Reserva Nacional Pacaya y Samiria. Ecuadorean preserves (mostly proposed) include: 4. Reserva Ecológica Cotacachi-Cayapas; 5. Parque Nacional Puerto Lopez; 6. Reserva Ecológica Manglares de Río Guayas; 7. Parque Nacional Cayambe-Coca (systema ampliado); 8. Parque Nacional Cotopaxi; 9. Parque Nacional Volcán Sangay; 10. Parque Nacional Río Yasuni. Black dots indicate towns. Letters indicate ecological classification: P = pluvial forest; W = wet forest; M = moist forest; D = dry forest and desert; U = upland (montane) vegetations (see also Fig. 1).

to transfer into preserved realities the designated protected areas which are included in their developing land-use plans.

At this point in time, Peru is well ahead of Ecuador in actual establishment of monitored and protected reserves. Of the three major preserves in Amazonian Peru (Fig. 3), for example, the Parque Nacional de Manu (over 1.5 million hectares) in Madre de Dios Department is very effectively protected by the fact that its only access via the Manu River is monitored. Interestingly, the first guard post at Manu was financed by the contributions of British schoolchildren (F. Benavida, pers. comm.), an indication of one way outside interest can be effectively used to assist local conservation efforts. Similarly, the Reserva Nacional Pacaya y Samiria (almost 1.4 million hectares) in Loreto is protected, since access is restricted along the two rivers for which it is named. This reserve, established primarily for fish protection, has also been effective in protecting plants and other animals. An interesting managerial innovation gives the guards the exclusive right to fish within the reserve and this emolument, undoubtedly, makes them much more zealous in protecting it. Similar compromises with total preservation may be the most effective answer for achieving workable conservation in this part of the world. The third major conservation unit in Amazonian Peru is the smaller Parque Nacional de Tingo Maria (18,000 hectares) in Huanuco Department, which is not adequately protected and as a result is losing much of its natural vegetation to squatters. The Tingo Maria site is especially important insofar as it is the only conservation unit in the tropical wet forest region except for a strip much farther south in Manu. Peru also has plans for a still-undesignated Parque Nacional de Loreto in Amazonia.

In addition to national parks and reserves, Peru has a fairly extensive system of national forests in Amazonia. These national forests are designated for controlled and sustained yield of timber, including reforestation programs. In some of these forests, profit and timber yield are said to be emphasized at the expense of conservation and reforestation. I have visited only one such national forest, the Bosque Nactional de von Humboldt, between Pucallpa and Tingo Maria. It appeared to be very effectively managed, with experimental forestry being emphasized, though the program there stressed a practical view of forest conservation through controlled harvest of timber resources rather than preservation in an untouched state. About four million hectares of Amazonian Peru have been designated as National Forests.

Ecuador has a functional National Park in the Galapagos Islands, but mainland conservation units are basically in the proposal stage, although enabling legislation for Cotopaxi (National Park), Cotacachi-Cayapas (Reserve), Cayambe-Coca (Reserve), and Sangay (Reserve) has very recently been approved, and each of these areas now has the beginnings of an administrative structure. An ambitious program to survey the country for wildland areas appropriate for preservation has been carried out and a strategy for preservation of selected areas has already been established (Putney, 1976). Highest priority has been given to several proposed national parks and reserves which are exceedingly well chosen to represent the different vegetation types of Ecuador. Parque Nacional de Volcán Sangay spans a large expanse of mostly virgin montane forest stretching from the Andean páramo to the lowland rain forest along the Río Pastaza. On the opposite side of the Andes, the proposed Reserva Ecológica Cotacachi-Cayapas would preserve most of the largest extant area of west-facing montane forest and includes diverse ecological regions ranging from snow-covered Volcán Cotacachi to the lowland forests of Esmeraldas. The proposed Parque Nacional near Puerto Lopez would preserve part of one of the few remaining areas of coastal dry forest. The Parque Nacional del Río Yasuni would preserve a significant portion of lowland Amazonian forest, including two entire river valleys. These four suggested parks, already accorded the highest developmental priority, could hardly have been better chosen from the viewpoint of preservation of plant species. However, from the botanical perspective, the Cotacachi-Cayapas reserve should perhaps be accorded the highest priority of them all because it includes the greatest number of vegetational types and a portion of the especially endangered coastal wet forest.

Three other areas considered of exceptionally high priority by Ecuadorean conservation officials are Parque Nacional de Cotopaxi, south of Quito in the uplands, a Río Cuyabeno faunistic reserve in the Amazonian lowlands near Lago Agrio, a Reserva Ecológica Manglares del Río Guayas near Guayaquil, and Los Lagos del nudo Las Cajas near Cuenca, proposed as a recreation area. A much more ambitious "sistema ampliado" would include a substantial part of the remaining wildland areas of Ecuador under some form of protection. Among the thirty such proposed areas, two near Quito — Cayambe-Coca and Pichincha — are proposed as national parks; the others are classified as ecological reserves, faunistic reserves, or recreation areas. Among these, the Cordillera de Cutucu region near Macas might be especially important in preserving a portion of the otherwise unprotected Amazonian lowland tropical wet forest.

The existing Ecuadorean plan could hardly be improved upon for preservation of endangered or potentially endangered plant species. One can only hope that the proposed preserves — especially Cotacachi-Cayapas, Sangay, and Puerto Lopez — will become realities before the vegetation they contain is destroyed.

The generally more advanced Peruvian conservation effort might well be expanded in accordance with potential for protecting portions of additional vegetation types. For example, a dry forest park on the Alto Río Purus or near Ibarra, and a wet forest/rain forest park in the low mountains north of Moyobamba or in the vicinity of Aguaytia, would both be excellent additions to the existing system. Preservation of one of the incredibly abrupt transition zones coming down from the mountains toward Amazonia would also be of great botanical interest.

CONCLUSION

Although habitat destruction is accelerating in Ecuador and Amazonian Peru, probably relatively few plant species of this region have become extinct or are on the verge of extinction, with the exception of those endemic to the strip of lowland wet forest at the west base of the Andes. However, the general trend of rapidly increasing population, accompanied by the opening of new areas to settlement, threatens most of the plant species and habitats of the region. Necessary conservation steps must be put into effect now if representative samples of the rich and varied floras of this region are to survive into the next century. Clearly, these steps must be undertaken by the governments and people involved. Both Peru and Ecuador have already initiated important preservational efforts. Appropriate support from the international scientific and conservationist communities might include such steps as moral and financial support for parks and reserves, cooperation with conservation-inspired regulations, and assistance with basic floristic surveys and exploration in order to pinpoint and define areas of special biological interest.

ACKNOWLEDGEMENTS

The author's study of the plants of Ecuador and Peru has been supported by National Science Foundation grants GB-40103 and DEB 75-20325. I thank Bruce and Olga MacBryde, Amy Jean Gilmartin, Allen Putney, Henry Rodriguez, Franklin Ayala, and Fernando Ortiz for reviewing drafts of this manuscript and C. Dodson, J. Gilmartin, J. Revilla, and especially A. Putney for providing various relevant information.

LITERATURE CITED

Anonymous. 1974. Biotic Provinces of the World. IUCN Occasional paper no. 8 Morges, Switzerland.

Diels, L. 1937. Beitrage Zur kenntnis der Vegetation und Flora von Ecuador. Bibl. Bot. **29** (116): 1-190.

_____. 1938-1942. Neue Arten aus Ecuador I-V. Notizbl. 14: 25-44; 323-341. Notizbl. **15**: 23-58; 366-393; 783-787.

Dodson, C. and A. Gentry. 1977. Flora of the Río Palenque Biological Station. Selbyana **2**.

_____.and **R. V. Gillespie.** 1967. The Biology of the Orchids. Mid-America Orchid Congress, Inc.

Freese, C. 1975. A census of non-human primates in Peru. pp. 17-41 in Primate Censusing Studies in Peru and Colombia. Report to the NAS on the activities of project AMRO-0719, Pan-American Health Organization, Washington, D.C.

Gentry, A. H. 1973. Generic delimitations of Central American Bignoniaceae. Brittonia **25**: 226-242.

_____. 1976a. Bignoniaceae of southern Central America: Distribution and ecological specificity. Biotropica **8**(2):117-131.

_____. 1976b. Floristic needs in Pacific Tropical America. Allertonia (in press).

Gilmartin, A. J. 1972. The Bromeliaceae of Ecuador. Phanerogamarum Monographiae Tomus 4: 1-255.

_____. 1973. Transandean distributions of Bromeliaceae in Ecuador. Ecology **54**: 1389-1393.

Janzen, D. H. 1974. Tropical blackwater rivers, animals and mast fruiting by the Dipterocarpaceae. Biotropica **6**: 69-103.

Little, E. 1969. Arboles Comunes de la Provincia de Esmeraldas. FAO/SF: 76/ECU 13, Rome. 536 pp.

Putney, A. D. 1976. Estratagia preliminar para la conservación de areas silvestres sobresalientes del Ecuador. UNDP/FAO-ECU/71/527. 61 pp.

Scott, N. J. 1976. The abundance and diversity of herpetofaunas of tropical forest litter. Biotropica **8**: 41-58.

Sparre, B. 1968. Ecuadors flora som svensk intressesfär. Fauna och Flora **63**:110-114.

Standley, P. 1931. The Rubiaceae of Ecuador. Field Mus. Nat. Hist. Publ., Bot. Ser. **7**: 179-251.

Steere, W. 1950. The phytogeography of Ecuador, pp. 83-86. In E. Ferndon, Studies in Ecuadorean geography. Univ. S. Calif. Monogr. Sch. Am. Res. **15**: 1-86.

Svenson, H. K. 1945. Vegetation of the coast of Ecuador and Peru and its relation to the Galapagos Islands. Amer. Jour. Bot. **33**: 394-498.

Weberbauer, A. 1936. Phytogeography of the Peruvian Andes. Field. Mus. Nat. Hist., Bot. Ser. **13**: 13-81.

Williams, Ll. 1936. Woods of northeastern Peru. Field Mus. Nat. Hist., Bot. Ser. **15**: 1-587.

ENDANGERED SPECIES AND PLANT COMMUNITIES IN ANDEAN AND COASTAL PERU

Ramon Ferreyra

Museo de Historia Natural, Universidad Nacional Mayor de San Marcos, Lima, Peru

Peruvian territory, which comprises 1,300,000 square kilometers, is characterized by very broken topography because the Andean Cordilleras divide the country into three major ecosystems: Costa, Sierra, and Selva (Fig. 1). During the last 30 years, these ecosystems, particularly that of the Coastal area, have been greatly disturbed by man.

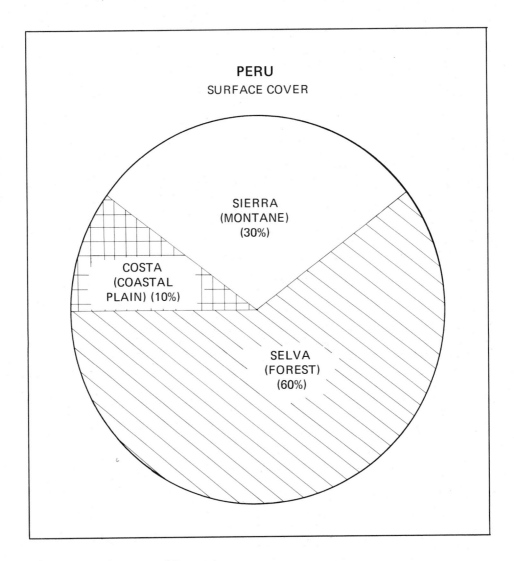

Fig. 1. Vegetation cover of Peru.

COSTA (10 percent of the territory) — This is a belt more than 2,800 km long, quite narrow in the south but up to 100 km wide in the north. There are two main Plant Formations: the *Lomas,* which occupy part of the northern, central, and southern part, and the *Algarrobal* of the extreme north.

The most remarkable fact about the Loma Formation is that it contains an unusually large number of endemic plants. The Lomas are seasonally covered by luxuriant vegetation, and extend along the Cordillera Occidental and the Pacific shore in an interrupted belt near the sea.

Climatic factors throughout the area depend upon the cold Peruvian Current. As the coast is cooled by this current during winter, the wet sea fog drifts landward and approaches the hills. This brings the fine precipitation, commonly referred to as *Garúa* in Peru and *Camanchaca* in Chile. The Garúa averages less than 80 mm a year but its effect is enhanced by the cloud cover. The mean annual temperature ranges from 20.6° C to 15° C (Fig. 2).

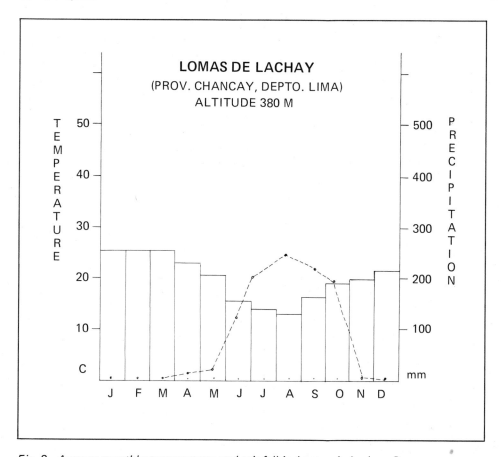

Fig. 2. Average monthly temperatures and rainfall in Lomas de Lachay, Peru.

Floristic investigation reveals that more than 600 species of Phanerogams occur in this area. The most conspicuous group is the Nolanaceae, a natural plant family characteristic of the southern coast of Peru and northern Chile. In addition, several species of *Palaua* (Malvaceae) are also typical of these lomas, as are endemic species of many other families.

Some important loma localities, such as those of Lachay, Chancay, Amancaes, Agustino, Atocongo, Lurin, Atiquipa, Chala, Atico, etc., are greatly influenced by pollution.

The isolated relicts of the woody strata, although containing few species *(Prosopis chilensis, P. limensis, Acacia macracantha, Caesalpinia spinosa, Capparis prisca,* and so on) also are being exterminated by the natives.

The herbaceous stratum has many species which are adapted to the high temperature and very light precipitation of the desert plains. Many of these species are more or less succulent. Typical of these are *Nolana* with more than 80 ephemeral species, of which at least 40 percent are threatened because of the severe pollution from the Toquepala, Ilo, and Marcona mining companies.

Several type localities similar to the previously mentioned lomas no longer exist, having been replaced by new towns and villages. As a result, some loma species like *Solanum neoweberbaueri* are severely endangered.

Annotated list of the endangered species of the lomas:

ACANTHACEAE:	*Dicliptera tomentosa* Nees
AIZOACEAE:	*Tetragonia crystallina* L'Hérit.
	T. expansa Murr.
	T. macrocarpa Phil.
	T. maritima Barnh.
	T. ovata Phil.
	T. pedunculata Phil.
	T. vestita Johnston
AMARYLLIDACEAE:	*Hymenocallis amancaes* (R. & P.) Nichols
BEGONIACEAE:	*Begonia octopetala* L'Hérit.
	B. geraniifolia Hook.
BIGNONIACEAE:	*Tourretia lappacea* (L'Hérit) Willd. = *T. volubilis* J. F. Gmel.
	Argylia feuillei DC.
BORAGINACEAE:	*Coldenia conspicua* Johnston
	C. ferreyrae Johnston
	Heliotropium pilosum R. & P.
BROMELIACEAE:	*Tillandsia paleacea* Presl
	T. purpurea R. & P.
	T. werdermanii Harms
CAPPARIDACEAE:	*Capparis prisca* Macbride
CARYOPHYLLACEAE:	*Spergularia collina* Johnston
	S. congestifolia Johnston
	S. laciniata Baehni & Macbride
COMMELINACEAE:	*Commelina hispida* R. & P.
	Tinantia erecta (Jacq.) Schlecht.
COMPOSITAE:	*Ambrosia parviflora* Payne
	Polyachyrus annuus Johnston
	Senecio subcandidus Gray
	S. smithianus Cabrera
	Viguiera weberbaueri Blake
CONVOLVULACEAE:	*Ipomoea nationis* Nich.
	Jacquemontia unilateralis (R. & S.) O'Donell
CUCURBITACEAE:	*Apodanthera ferreyrana* Martinez
	Cyclanthera mathewsii Arn.
DIOSCOREACEAE:	*Dioscorea chancayensis* Knuth
EUPHORBIACEAE:	*Euphorbia ruiziana* (Kl. & Gke.) Boiss.
	E. tacnensis Phil.
GENTIANACEAE:	*Erythraea lomae* Gilg
GERANIACEAE:	*Geranium limae* Knuth
	G. mollendinense Knuth
GRAMINEAE:	*Bromus striatus* Hitchc.
	Eragrostis attenuata Hitchc.
	Luziola peruviana Gmel.
GUTTIFERAE:	*Hypericum thesiifolium* H. B. K.
HYDROPHYLLACEAE:	*Nama dichotomum* (R. & P.) Choisy
IRIDACEAE:	*Tigridia lutea* Link.
LABIATAE:	*Salvia rhombifolia* R. & P.
LEGUMINOSAE:	*Coursetia weberbaueri* Harms
	Caesalpinia ternata (Phil.) Macbride
	Calliandra prostrata Benth.
	Astragalus triflorus (DC.) Gray
	Hoffmanseggia miranda Sandw.
	H. stipulata (Sandw.) Macbride
	Lupinus arequipensis C. P. Smith
	L. lorenzensis C. P. Smith

152

LEGUMINOSAE (cont'd.):	*Lupinus mollendoensis* Ulbrich
	Vicia lomensis Macbride
	Weberbauerella brongniartioides Ulbrich
	W. raimondiana Ferreyra
LILIACEAE:	*Pasithea coerulea* (R. & P.) D. Don
	Fortunatia biflora (R. & P.) Macbride
LINACEAE:	*Linum parvum* Johnston
	L. prostratum Lam.
LOASACEAE:	*Loasa fulva* Urb. & Gilg
	L. nitida Desr.
MALVACEAE:	*Malvastrum sandemanii* Sandw.
	Cristaria multifida Cav.
	Palaua dissecta Benth.
	P. quentheri Bruns.
	P. rhombifolia Graham
	P. tomentosa Hochr.
	P. trisepala Hochr.
	P. velutina Ulbr. & Hill
MYRTACEAE:	*Myrcianthes ferreyrae* (McVaugh) McVaugh
	Eugenia minimifolia McVaugh
NOLANACEAE:	*Nolana aticoana* Ferreyra
	N. adansoni (R. & S.) Johnston
	N. arenicola Johnston
	N. confinis Johnston
	N. cerrateana Ferreyra
	N. coronata R. & P.
	N. amplexicaulis Ferreyra
	N. inflata R. & P.
	N. gracillima (Johns.) Johnston
	N. ivaniana Ferreyra
	N. latipes Johnston
	N. lycioides Johnston
	N. guentheri Johnston
	N. plicata Johnston
	N. mariarosea Ferreyra
	N. pilosa Johnston
	N. johnstonii Vargas
	N. pallida Johnston
	N. minor Ferreyra
	N. pallidula Johnston
	N. platyphylla (Johns.) Johnston
	N. spergularioides Ferreyra
	N. spathulata R. & P.
	N. scaposa Ferreyra
	N. tomentella Ferreyra
	N. thinophila Johnston
	N. tovariana Ferreyra
	N. volcanica Ferreyra
	N. weissiana Ferreyra
	N. willeana Ferreyra
OXALIDACEAE:	*Oxalis lomana* Diels
	O. bulbigera Knuth
	O. sepalosa Diels
PIPERACEAE:	*Peperomia atocongona* Trelease
	P. crystallina R. & P.
	P. limaensis Trelease
	P. non-hispidula Trelease
	P. seleri C. DC.

PIPERACEAE (cont'd):	*Peperomia umbelliformis* C. DC.
	P. pseudo-galapagensis Trelease
POLEMONIACEAE:	*Gilia laciniata* R. & P.
POLYGALACEAE:	*Monnina arenicola* Ferreyra
	M. weberbaueri Chodat
PORTULACACEAE:	*Portulaca pilosissima* Hook.
	Calandrinia alba (R. & P.) DC.
	C. crenata (R. & P.) Macbride
	C. paniculata (R. & P.) DC.
	C. ruizii Macbride
RUBIACEAE:	*Richardia lomensis* (Krause) Standl.
SANTALACEAE:	*Quinchamalium lomae* Pilger
	Q. brevistaminatum Pilger
SCROPHULARIACEAE:	*Calceolaria anagalloides* Kranzl.
	C. pinnata L.
	C. dichotoma Lam.
	C. verticillata R. & P.
	Alonsoa caulialata R. & P.
UMBELLIFERAE:	*Domeykoa amplexicaulis* (Wolff) M. & C.
	D. saniculifolia M. & C.
	Spananthe paniculata Jacq.
	Eremocharis ferreyrae M. & C.
	E. piscoensis M. & C.
	E. longiramea (Wolff) Johnston
URTICACEAE:	*Pilea lamioides* Wedd.
VALERIANACEAE:	*Valeriana pinnatifidia* R. & P.
VERBENACEAE:	*Verbena clavata* R. & P. var. *casmensis* Moldenke

ALGARROBAL FORMATION

The major climatic influence on this area is the Equatorial Niño Current, which runs southward against the cold Peruvian Current, and results in heavy rains from December through March. The area affected by these seasonal rains stretches from the west to the base of the Cordillera Occidental.

As a whole, the Algarrobal contains two dominant species: *Prosopis chilensis* and *Capparis angulata, algarrobo* and *sapote,* respectively. Other characteristic woody species of these communities include: *Caesalpinia pai-pai, Bursera graveolens, Loxopterygium huasango, Acacia macracantha,* and *Tabebuia billbergii.* The abundant herbaceous plants include: *Monnina pterocarpa, Oxalis dombeii, Jacquemontia unilateralis, Heliotropium ferreyrae, Onoseris odorata,* etc.

A list of the threatened species follows:

ACANTHACEAE:	*Dicliptera peruviana* (Lam.) Juss.
AMARANTHACEAE:	*Amaranthus haughtii* Standl.
ANACARDIACEAE:	*Loxopterygium huasango* Spruce.
BIGNONIACEAE:	*Tecoma billbergii* (Bur. & K. Schum.) Standl.
BORAGINACEAE:	*Heliotropium ferreyrae* Johnston
BURSERACEAE:	*Bursera graveolens* (H. B. K.) Tr. & Pl.
CAPPARIDACEAE:	*Capparis ovalifolia* R. & P.
	C. angulata R. & P.
COMPOSITAE:	*Pectis linifolia* L.
	P. arenaria Benth.
	P. ciliaris L.
	Fulcaldea laurifolia (H. B. K.) Poir.
	Wedelia latifolia DC.
CONVOLVULACEAE:	*Jacquemontia unilateralis* (R. & S.) O'Donell
CUCURBITACEAE:	*Sicyos chaetocephalus* Harms
	S. weberbaueri Harms

154

LEGUMINOSAE:	Aeschynomene indica L.
	Caesalpinia pal-pai R. & P.
	Dalea microphylla H. B. K.
	Stylosanthes psammophila Harms
MALVACEAE:	Gossypium raimondii Ulbr.
	Abutilon piurensis Ulbr.
SOLANACEAE:	Solanum amotapense Svenson

SIERRA (30 percent of the territory) — The three Cordilleras divide this territory into different regions. The first of these is the Puna or plateau (3,500-4,500 m altitude), with approximately 20 million hectares of natural pasture. The dominant species here are grasses called "Ichu" *(Stipa ichu, Festuca dolichophylla, Calamagrostis vicunarum).* The second is the Inter-Andean Valleys (2,500-3,000 m), including their quebradas with quite a high level of endemism.

Man and camelids (llama, vicuña, alpaca, paco-vicuña, and huanacos) have been disturbing this area for hundreds of years. In addition to erosion, landslides *(huaicos)* and other geologic events have caused much disturbance. Another factor threatening this type of vegetation is indiscriminate burning. Many annual and ephemeral plants such as *Agrostis breviculmis, A. tolucensis, Dissanthelium peruvianum*, etc., have become rare as a result of these factors.

Until recently, the forest of *queñoa (Polylepis* spp.) was relatively abundant in the higher mountains. Today, only relicts remain and these are confined to slopes too steep for the natives to approach them.

The *Puya raimondii,* the famous giant Bromeliaceae, also grows here and nowhere else in the world except in Bolivia. It, too, is in serious danger due to burning.

An annotated list of the endangered species of this region follows:

ANACARDIACEAE:	Haplorhus peruviana Engler Hjasi
BROMELIACEAE:	Puya raimondii Harms
CAPRIFOLIACEAE:	Sambucus peruviana H. B. K.
COMPOSITAE:	Arnaldoa macbrideana Ferreyra
JUGLANDACEAE:	Juglans neotropica Diels
JULIANACEAE:	Orthopterygium huancui (Gray) Hemsley
LEGUMINOSAE:	Prosopis atacamensis
LOGANIACEAE:	Buddleia incana R. & P.
	B. longifolia H. B. K.
MYRTACEAE:	Myrcianthes quinqueloba (McVaugh) McVaugh
ROSACEAE:	Polylepis albicans Pilger
	P. incana H. B. K.
	P. racemosa R. & P.
	P. villosa H. B. K.
	Kageneckia lanceolata R. & P.
SAXIFRAGACEAE:	Escallonia myrtilloidos L. f. tasta
	E. Piurensis Mattf.

SELVA (60 percent of the territory) — For the purposes of this Symposium, we will only consider the Cordillera Oriental at the present time. In Peru, we call this area *Ceja de la Montaña* (the eyebrow of the mountains).

The climate of the *Ceja de la Montaña* is so diverse that it supports many types of vegetation. These plant communities are very rich in species, and there are an especially large number of epiphytic plants of the Bromeliaceae, Piperaceae, and Orchidaceae here.

One of the most interesting valleys insofar as speciation is concerned is the Huallaga, where we have been successfully botanizing for many years, with the result that a number of new species have been discovered there. Unfortunately, many of the type localities have disappeared since the Carretera Marginal was started in 1957. The increase of settlers sent to the Alto Huallaga in order to cultivate various local resources, has been the main cause for destruction of this biota, including: *Monnina amplibracteata, M. divaristachya, Sanchezia dasia, S. ferreyrae,* etc. The damage to this ecosystem is evident between 6° and 10° south latitude.

List of the Endangered Species of the Selva:

ACANTHACEAE:	*Aphelandra mucronata* (R. & P.) Nees
	A. hapala Wasshausen
	A. latibracteata Wasshausen
	A. ferreyrae Wasshausen
	Sanchezia aurea Leonard & Smith
	S. aurantiaca Leonard & Smith
	S. dasia Leonard & Smith
	S. klugii Leonard & Smith
	S. lasia Leonard & Smith
	S. ferreyrae Leonard & Smith
	S. villosa Leonard & Smith
	Ruellia tarapotana Lindau
AMARYLLIDACEAE:	*Bomarea engleriana* Kranzl
	B. filicaulis Kranzl
	B. ferreyrae Vargas
	B. klugii Killip.
	Amaryllis traubii Moldenke
	A. ferreyrae Traub
	Eucharis ferreyrae Traub
BROMELIACEAE:	*Pitcairnia casapensis* Mez
	P. cyanopetala Ule
	P. ferreyrae L. B. Smith
	P. tarapotensis Baker
	Guzmania tarapotina Ule
	Tillandsia brevilingua Mez
	T. harmsiana L. B. Smith
COMPOSITAE:	*Liabum angustum* Blake *(=Munnozia angosta* R. & P.)
	L. amplexicaule Poeppig
	Senecio carpishensis Cuatr.
ERICACEAE:	*Gaultheria weberbaueriana* Sleumer
	Disterigma ulei Sleumer
	Orthaea ferreyrae A. C. Smith
EUPHORBIACEAE:	*Croton olivaceus* M. Arg.
	C. pilgeri Ule
	Acalypha fulva Johnston
GENTIANACEAE:	*Lisianthius calygonus* R. & P.
	L. corymbosus R. & P.
LEGUMINOSAE:	*Calliandra bombycina* Spruce
	Pithecellobium chazutense Standl.
MELASTOMACEAE:	*Miconia adrieni* Macbride
	M. barbeyana Cogn.
	M. ferreyrae Wurdack
MYRTACEAE:	*Myrciaria paraensis* Berg
ORCHIDACEAE:	*Stelis gracilispica* Schweinfurth
PALMAE:	*Euterpe precatoria* Mart.
	Iriartea ventricosa Mart.
	Jessenia polycarpa Mart.
PIPERACEAE:	*Piper monzonense* C. DC.
	P. yurimaguasanum Trelease
	P. tocacheanum C. DC.

POLYGALACEAE: *Monnina amplibracteata* Ferreyra
M. huallagensis Chodat
M. polystachya R. & P.
M. ruiziana Chodat

DISCUSSION

Dr. Gerardo Budowski: "Does the removal of the tree cover from Lomas de Lachay explain the changes in fog and precipitation? If so, this is a subject that can be used as a very interesting and strong argument to defend retention of existing vegetation."

Dr. R. Ferreyra: "Yes indeed, in the Lomas de Lachay, it was humid ten or fifteen years ago, but during the last few years the climate has changed, and so far as I know during the past two years there has been very little precipitation. However, the Minister of Agriculture is entrusted with forestry development, and his Department is now planting *Eucalyptus* from Australia. We wonder what is going to happen with this experiment. We know, at least, that it is a good place to reforest, especially with the native species instead of *Eucalyptus*. For instance, Lachay used to be very rich in the two native trees, *Caesalpinia spinosa* and *Capparis prisca*. Now only two or three specimens can be seen and only one tree of *Capparis prisca* remained a few days ago. Perhaps it will be necessary to suggest to the government that native species be cultivated here instead, since these are adapted to this particular climate. I remember when I was botanizing with Professor Ellenberg some years ago. We were collecting data about this climatic factor and we found that the trees managed to obtain some moisture. Probably species like *Prosopis*, for example, have some particularly long roots which can take advantage of the humidity deep in the soil."

LITERATURE CITED

Cerrate, E. 1957. Notas sobre la vegetación del Valle de Chiquian, Folia Biol. And. 1: 9-39.

Ferreyra, R. 1961. Las Lomas Costaneras del Extremo Sur del Peru, Bol. Soc. Arg. Bot. 9: 87-120.

_____. 1961. Revisión de las especies peruanas del género *Nolana,* Mem. Mus. Hist. "Jav. Pra.," No. 12.

_____. 1970. Flora invasora de los cultivos de Pucallpa y Tingo Maria, Lima.

_____. 1974. Una nueva especie de *Nolana* para el Peru. Bol. Soc. Per. Bot. 7:1-2.

Johnston, I. M. 1936. A Study of the Nolanaceae, Contr. Gray Herb. 112:1-83.

Macbride, J. F. 1936-1961. Flora of Peru. Field Museum, Chicago.

Petersen, G. 1935. Estudios Climatológicos en el Norte del Peru, Bol. Soc. Geo. Per. 7, Lima.

Tovar, O. 1960. Revisión de las especies peruanas del género *Calamagrostis,* Mem. Mus. Hist. Nat. No. 11.

_____. 1972. Revisión de las especies peruanas del género *Festuca* Mem. Mus. Hist. Nat. No. 16.

Weberbauer, A. 1945. El Mundo Vegetal de los Andes Peruanos, Lima.

THE AMAZON FOREST: A NATURAL HERITAGE
TO BE PRESERVED

João Murça Pires

Museu Paraense Emílio Goeldi, Belém, Pará, Brazil

and

Ghillean T. Prance

The New York Botanical Garden, Bronx, New York 10458, U.S.A.

INTRODUCTION

Amazonia is the last large area of natural forest left in the world. It covers an area of about six million square kilometers of which more than half (3.5 million sq. km) fall within the boundaries of Brazil. In view of its vast size, man's interference with this natural forest is still comparatively little and is mostly confined to the margins of rivers and roads which cross this wilderness. However, there already are indications that this situation will soon change. Twentieth-century living, with all its concomitant problems, is leaving its indelible mark on this last remaining unspoiled great wilderness area. Such factors as accelerated highway construction programs, increasing population pressures, and economic and political problems all have forced the governments of the different Amazonian countries to intensify development of the region.

A realistic appraisal of the situation forces us to recognize the fact that at this late date it is impossible to turn the whole area into one large biological reserve. We do believe, however, that an equilibrium between man and nature can still be established in Amazonia. We also believe that representative samples of the natural forest can still be preserved while at the same time allowing man to utilize the forest in a more rational manner than he does at present. Obviously, the study of Amazonia has become a matter of extreme urgency which can no longer be postponed.

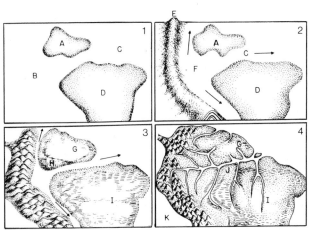
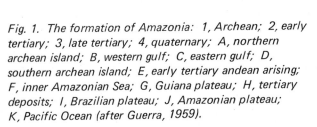
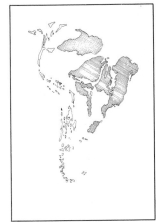

Fig. 1. The formation of Amazonia: 1, Archean; 2, early tertiary; 3, late tertiary; 4, quaternary; A, northern archean island; B, western gulf; C, eastern gulf; D, southern archean island; E, early tertiary andean arising; F, inner Amazonian Sea; G, Guiana plateau; H, tertiary deposits; I, Brazilian plateau; J, Amazonian plateau; K, Pacific Ocean (after Guerra, 1959).

Fig. 2. South American areas which were under water in the permo-carboniferous (after Roxo, 1943).

In order to fully understand how best to preserve a natural balance between man and the forest, intensive study is urgently necessary in the following areas:

1. The identification of and intensive research in those areas most threatened by the interference of man.

2. The collection of vital documentary data such as herbarium material, zoological, and mineralogical specimens, which can then be meticulously studied in the laboratory.

3. The selection and official designation, on the basis of scientific information, of representative areas of forest as biological reserves, the collection of living material for propagation in botanical gardens and zoos (preferably to be located in the tropics), and the establishment of more Indian reservations.

It should be obvious from the above three points that the preservation of threatened areas in Amazonia is at present much more important than that of threatened species. Consequently, in this paper we will present some basic information and observations which we hope will increase our knowledge of the complex Amazonian ecosystem.

AMAZONIA — SOME BASIC FACTS

In order to better understand the current problems of Amazonia, it is necessary to briefly relate the geographical history of the area. The Amazon region originated from two archean massifs, the Guayana and the Brazilian shields. Later in the Miocene epoch the Andes arose and created a large inland mediterranean sea which gradually filled with

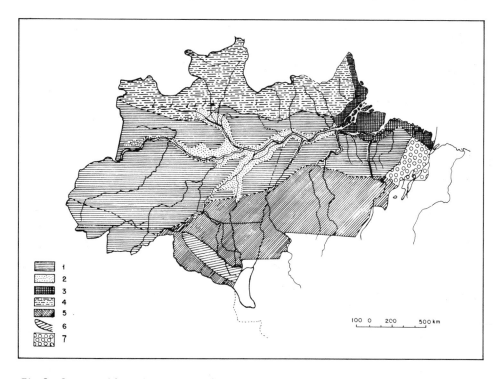

Fig. 3. Geomorphic regions of Brazilian Amazonia: 1, lowland plateau of tertiary lake bed mostly lower than 100 m in altitude and rarely above 200 m; 2, flood plain of quaternary origin, the most recent formation, covered by várzea forests, swamp forests and swamp savannas. A large part of this area is covered by water of lakes and watercourses; 3, lowland coastal plateau covered by forest, upland savanna; 4, the Guayana shield of Precambrian origin which together with the Brazilian shield (5) forms the most ancient part of Amazonia and is where rapids and falls begin on the rivers; 6, Pareceis Mountains, a sandstone ridge rarely over 600 m altitude with a characteristic lower shrubby vegetation; 7, sedimentary plateau of the Maranhão uplands (based on Guerra, 1959).

sediments. In some places today these sediments are extremely deep. For example, on the island of Marajó in the Amazon estuary, PETROBRAS (the Brazilian Petroleum Company) found sediments of over 4,000 m in depth. The area has also been changed by tectonic movement in more recent times and by the fluctuation in sea level during the Pleistocene era (see Figs. 1, 2, and 3).

The flora was profoundly affected by these changes as new areas continued to rise above sea level and were colonized by plants. Furthermore, there were earlier natural barriers of water which made plant migration more difficult. The two crystalline massifs, the Brazilian and Guayana Shields, were among the oldest centers of dispersion of plants, and in many cases the network of water acted as a barrier while in others it served as a means of dispersal of plant species. Today, the main types of vegetation are much influenced by these different characteristics and elevations of Amazonia's surface areas. The vegetation of *terra firme* (upland non-flooded ground), the flood plain vegetation *(várzea* and *igapó),* and the various transitional vegetation types such as *campinas* (open vegetation on white sand) or flooded *caatingas* (low campina forest) of the Rio Negro and Rio Branco, are all the result of the relief of the land. The influence of relief can further be seen by comparing Figures 4 and 5. The non-forested parts of South America are mostly above 500 m altitude, and the Amazon forest region is almost all under 200 m above sea level.

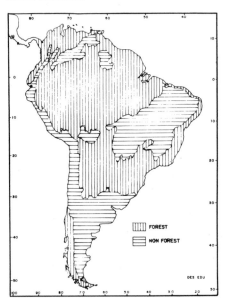

Fig. 4. Forest and non-forest areas of South America (after Hueck, 1972).

Those who are unfamiliar with Amazonia tend to think of it as a very wet and inundated region. This is not true, even though the climate is very wet. About 90 percent of Amazonia is on upland or terra firme which is slightly elevated, well-drained, plateau above the flood level. This includes large areas of ancient Cambrian and Precambrian origin. A traveller on any tributary of the Amazon will eventually come to rapids. The point at which these rapids begin indicates the boundary of the crystalline shields (Fig. 3). Mountains occupy a relatively small surface area of the crystalline shield area of Amazonia. Nevertheless, they contribute a special type of montane vegetation which is of extraordinary historical interest and with much species endemism. The two highest points in Brazil are within Amazonia, Pico da Neblina (3,014 m high) and Pico 31 de Março (2,992 m high). Both are on the Brazil-Venezuela frontier.

The types of Amazonian vegetation are complex, little known, and poorly defined. Summaries of the principal types of vegetation are given in Pires (1973) and Prance in Goodland and Irwin (1975). In a broad sense, the vegetation types are classed as: forests on terra firme, várzea forest (slightly-flooded), igapó or swamp forest (heavily-flooded), caatinga or campina forest, liana forest (with an extraordinarily high quantity of woody lianas), savanna on terra firme, open savanna (almost without woody plants), closed savanna (with some woody plants), várzea or floodplain savannas, and shrubby savannas and caatingas. A recent aerial survey of Amazonia using radar photography (Projeto RADAM) delimited an area of 85,000 square km of bamboo forest in the State of Acre. This forest is dominated by various species of bamboo.

The forest areas of Amazonia appear quite uniform to the untrained eye, but this uniformity is deceiving because the similarity is only in the physiognomy. When a floristic analysis is made of the area's species composition, the greater variation from one locality to another is quickly apparent. There are even differences in physiognomy, but our knowledge of this is still only rudimentary.

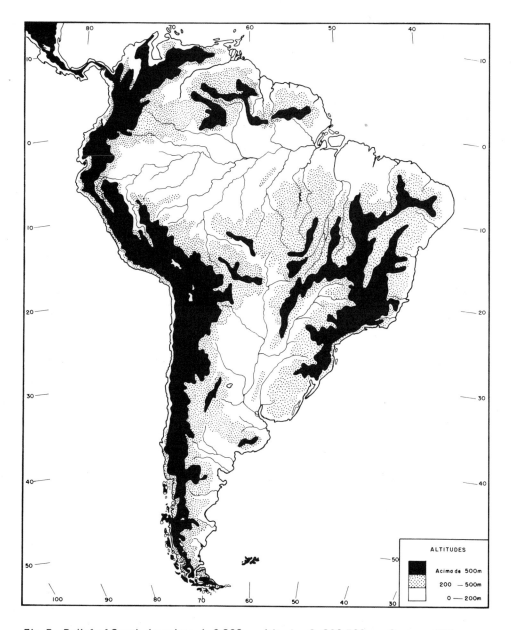

Fig. 5. Relief of South America: 1, 0-200 m altitude; 2, 200-500 m; 3, above 500 m.

The Brazilian government, with its ongoing RADAM project, is compiling an inventory of many aspects of the region, including the vegetation, minerals, land use potential, etc., and a parallel study is also underway in Venezuela. The radar images which have already been made have contributed immeasurably to our knowledge of the Amazon region. We now have, for example, a much better idea of the exact course of rivers and of the distribution of the major types of vegetation there. However, site visits on the ground are an enormous task, and much more groundwork must be done before we can accurately interpret the radar images in terms of soil and vegetation types. Correlation between these is still not possible because existing surveys group the soils in large categories that do not appear to have any correlation with the vegetation types. In order to make the soil-vegetation correlation much more detailed and precise, soil surveys are necessary, including studies of soil fertility.

There are large areas of recent quaternary origin where the vegetation has not yet reached a climax formation. Such areas are still distinctly seral, and contain pioneer plants in the colonization phase in some areas, while in others conditions have improved and the biological succession has progressed further. These areas have been greatly influenced by climate changes and variations in sea level over a long period. The best example of this type of vegetation is the *caatinga* of the Rio Negro and the Rio Branco. In those regions, there has been complete successional transition from herbaceous *campinas* through shrubby scrub to tall forest. An analysis of the pioneer stages of these vegetation formations reveals that the heretofore accepted theory that these areas are climax must now be changed to one of a continual succession. If the Rio Negro region is considered in its entirety, there appears to have been a cyclic process of development in the *caatinga* areas. The flora of that region is certainly not recent in the evolutionary sense, and it is rich in plant endemism. For example, it contains many primitive genera such as *Zamia* (Cycadaceae).

As a general rule, the environment of any equatorial tropical region with high temperature, abundant rainfall, and little seasonal variation, leads to the development of dense rain forest with a large biomass. One of the most important factors in the evolution of species adapted to this forest type with large biomass is either the degree of light or the strategy and adaptations of species to obtain light.

Climate is also an important factor influencing the type of vegetation. When a small restricted area is under study, climate is not very important because any climatic variations are too small to affect the floristic composition of the vegetation. It usually is the result of such other factors as the soil types, for example, which account for the striking local changes in the floristic composition of the forest. Climate is an important consideration in the definition of major vegetation types such as the boundary between the Amazon forest and the *cerrado* of the central Brazilian plateau. The change from forest to cerrado is closely linked with the change from a short to a long dry season, i.e., the number of days without rain (see Fig. 6). When the dry days are expressed in terms of air humidity as well as rainfall, the difference of this climate factor is even more apparent (see Fig. 7). Climate also influences the formation of the different major phytogeographic regions of Amazonia, as defined by Prance (1977) in this Symposium.

With this brief background introduction to the very complex Amazonian ecosystem, we will now discuss a few specific aspects of the forest which we have studied, i.e., horizontal and vertical distribution of its trees and their growth over a 15-year period. These data indicate the complexity of the forest and show the amount of useful information that will inevitably and irrevocably be destroyed if representative samples are not preserved soon.

THE HORIZONTAL DISTRIBUTION OF PLANTS IN THE FOREST

Tropical rain forest is characterized by the large number of tree species per unit area. For example, a hectare of forest near Manaus studied by the junior author contained 235 species of trees of 5 cm or more in diameter (Prance et al., 1976); see also the figures of diversity given by Kubitzki (1977) in this Symposium.

These species have each adapted to the environment in a variety of ways. There are many biological niches within the forest ecosystem and each species has evolved different survival mechanisms. These mechanisms are usually related to the factor of light in rain forest. Under normal conditions where there are enough nutrients and water, the natural vegetation of the Amazon region is forest with a large biomass. The non-forest or low forest with low biomass vegetation types occur where there is some limiting factor, such as extreme shortage of nutrients or either too much or too little water. The amount of nutrients is barely sufficient in most cases because of the extremely closed nutrient cycle where there is no leakage of nutrients by leaching or other processes. There is no loss of nutrients in germination, growth, the fall of dead parts, the death of plants, decomposition, and in primary and secondary production. This closed nutrient cycle makes the utilization of the region by a modern industrialized society extremely difficult. When the forest is felled, the equilibrium is lost, bringing with it a concomitant series of complications which still have to be solved before a sustained yield is achieved. The intensity of the closed nutrient cycle is in direct proportion to the genetic variability of the individuals which have to compete among themselves to survive.

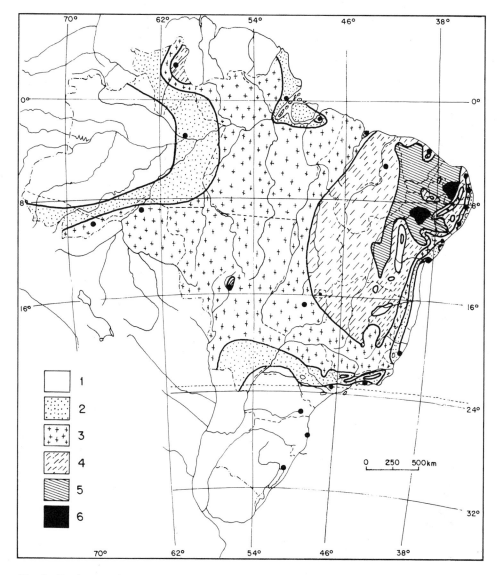

Fig. 6. Dry season length: 1, without dry season; 2, very short dry season (1-2 months); 3, short dry season (3-4 months); 4, medium dry season (5-6 months); 5, long dry season (7-8 months); 6, extended dry season (8 months or more) (after Nimer, 1966).

In regions where there is an extreme environmental factor, such as cold in the temperate regions, plants there have had to adapt to this adverse factor. The degree of specialization or adaptation for some type of stress in forest-covered areas is inversely proportional to the number of species which occur there. Thus, dominance or even the occurrence of only a single species in forests occurs where there is stress caused by one or more environmental factors. Tropical rain forest contains many species, and absolute dominance of a single species does not occur where there is the suitable combination of rainfall, nutrients, and temperature. In rain forest, no single species dominates either in number of individuals (density) or in ground cover (biomass or shade cover). Dominance in Amazonia only occurs in areas where there is obvious stress, such as in the white sand campinas or on some rocky mountain tops.

If one studies trees of 30 cm circumference or more (breast height) in any part of the Amazon rain forest, one usually finds that five to fifteen species account for fifty percent of the individuals present. Some species are more abundant than others in mixed forest. It is these more frequent species which can be considered characteristic of the area because

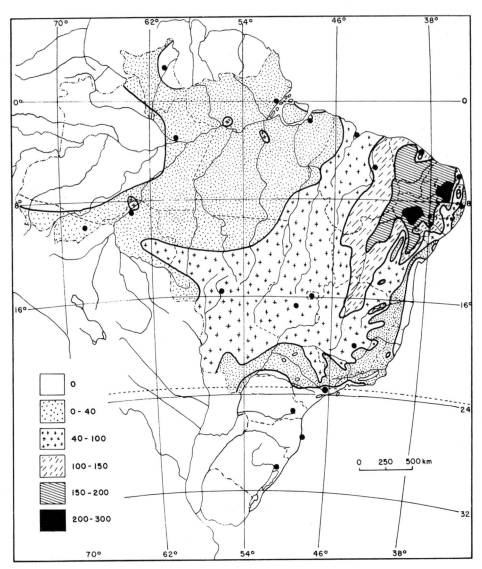

Fig. 7. Xerothermic regions, i.e. the dry season regions of Fig. 6 corrected by measurement of relative humidity of the air as well, expressed in number of days. The shorter dry season areas correspond with the areas covered with forest (after Nimer, 1966).

they have greater effect on the forest. For example, in the hectare studied near Manaus, the most common eleven species accounted for 25.43 percent of the total number of trees measuring 15 cm diameter or more. These, therefore, can be considered characteristic species of that hectare (see Table I).

However, characterization of the forest by means of the characteristic species is extremely difficult because they differ from area to area, often not even very far apart spatially. Forest inventories made by both authors throughout Amazonia (for example, those of Pires from Belém and Rio Jarí discussed below, and Prance et al., 1976) have shown that the characteristic species of one patch of forest are often rare species in another. Each species varies considerably in its abundance from place to place within its range of dispersion.

Each species has a range which is either continuous or disjunct. Each locality has a slightly different floristic composition and it is impossible to find two areas with identical species because of the variations in species density and the number of species involved. Variation between areas increases as distance between areas increases, but there are also distinct correlations between variation and the direction of the distance. For example,

Table I. The 11 most common species (5 or more individuals) in a hectare of forest near Manaus which contained 350 trees and 179 species of 15 cm diameter or more.

Species	No. of Individuals 15 cm diam.	Percent of Total
Eschweilera odora (Poepp.) Miers	26	7.43
Scleronema micranthum Ducke	9	2.57
Oenocarpus bacaba Mart.	8	2.29
Eperua	7	2.00
Protium sp.	6	1.71
Duckeodendron cestroides Kuhlm.	6	1.71
Corythophora rimosa W. Rodr.	6	1.71
Sloanea guianensis (Aubl.) Benth.	6	1.71
Neea cf. *altissima* Poepp. & Endl.	5	1.43
Neea sp.	5	1.43
Swartzia reticulata Ducke	5	1.43
	89	25.42

the flora to the north and the south of the Amazon River differs for historical geographical reasons based on the barriers that isolated the two areas and the barrier of the river itself.

Figures 9-17 illustrate this horizontal distribution of species. Figures 9-16 are distributions of individuals of 7 species in a small area near Belém discussed below, and Figure 17 represents the general distribution of the ten species of the genus *Hevea* or rubber, a relatively well-known genus. Figures 9-16 represent 3 areas within two biological reserves of EMBRAPA in Belém. These two reserves, Catú and Aurá, are located near the Rio Guamá and are about 1.5 km apart. They are shown in Figure 8, which is a map of the study area.

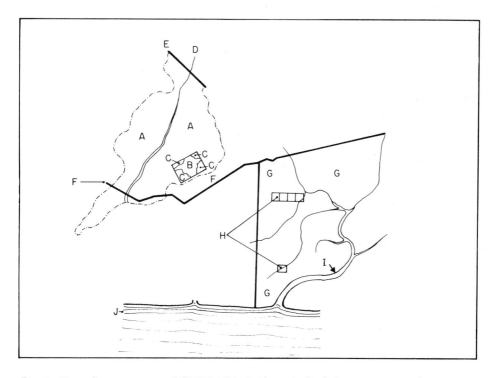

Fig. 8. Map of reserve areas of EMBRAPA, Belém: A, Catú Reserve, swamp forest; B, Mocambo terra firme (5.7 hectares); C, study areas of Catú Reserve; D, Catú stream; E, Utinga Dam; F, access roads; G, Aurá Reserve; H, study areas of Aurá Reserve; I, Aurá stream; J, Rio Guamá.

The Catú Reserve consists of about 100 hectares which are mostly swamp or *igapó* forest. The swamp here is caused by the daily inundation from the Catú stream. The flow of this stream is blocked each day by the high tide in the Rio Guamá. The forest is not flooded by the muddy water of the Rio Guamá but by the backup of the Catú itself. The other area that was studied, the Aurá Reserve, is covered by about 400 hectares of *várzea* forest. The Aurá Reserve is also swampy but less so than the Catú Reserve. Flooding at Aurá is caused by the muddy waters of the Rio Guamá.

Within the Catú Reserve, there is a small area of 5.7 hectares of forest on terra firme called Mocambo Reserve, which has been the subject of many biological studies, for example, Hatheway, 1971, Elton, 1973, and Lovejoy, 1975. This 5.7 hectares is a relatively small sample of terra firme forest, but it represents a very important area because it is, unfortunately, the largest surviving remnant of this forest type in the vicinity of Belém. The Catú reserve consists of a rectangular area of 420 x 250 m (10.5 hectares), containing the terra firme Mocambo within it. The Catú area is shown on the left side of Figures 9-16, where the four corners of the rectangle are swamp forest and the center area is the terra firme forest. The whole area was divided into 1,050 quadrats of 10 x 10 m. The frequency of individuals of all tree species was recorded on the maps for each quadrat.

Two areas from the várzea forest of Aurá Reserve were similarly selected and divided into quadrats. The upper right-hand area of Figs. 9-16 represents an area of 100 x 100 m divided into 100 quadrats. The lower right-hand area of the figures depicts 400 x 100 m likewise divided into 400 quadrats.

Trees of 30 cm circumference or greater were mapped and measured, species by species, in all three areas. These measurements included the trunk length and circumference (breast height), total height, and crown diameter. The soil of these areas was also studied. It was found that the terra firme contains various types of clay latosol, whereas the flooded areas of swamp forest (the far corners of Catú) and the várzea and Aurá have a slightly humid Gley, a hydromorphic soil. The maps in Figures 9-16 depict the eight most common species in the area summarized in Table II.

Table II. Occurrence of 8 common species in the three areas studied.

Figure No.	Species	Mocambo Terra Firme	Catú Igapó	Aurá Várzea	Total
9	*Mauritia martiana* Spruce (Palmae)	0	29	0	29
10	*Eschweilera odora* (Poepp.) Miers (Lecythidaceae)	287	70	72	429
11	*Eschweilera amara* Ndz. (Lecythidaceae)	252	60	50	362
12	*Theobroma subincanum* Mart. (Sterculiaceae)	44	49	98	191
13	*Caraipa grandiflora* Mart. (Clusiaceae)	0	230	14	244
14	*Euterpe oleracea* Mart. (Palmae)	0	267	702	969
15	*Virola surinamensis* Warb. (Myristicaceae)	1	203	56	260
16	*Goupia glabra* Aubl. (Celastraceae)	69	10	5	84

These maps clearly show the correlation between soil type and distribution pattern of the species studied. They reveal also that the amount of flooding is important for most species. A more precise and detailed soil study immediately surrounding every tree would be necessary to further correlate soil type with species distribution. The maps indicate that certain species have either a distinct ecological preference or are adapted to a specific habitat. Some species such as *Euterpe oleracea* Mart. (Fig. 14) occur exclusively in the flooded areas, while others, such as *Eschweilera odora* (Poepp.) Miers (Fig. 10), occur only on terra firme or on very lightly-flooded areas. Some species have complementary distribution. There are also many pairs of closely related species in which one occurs on

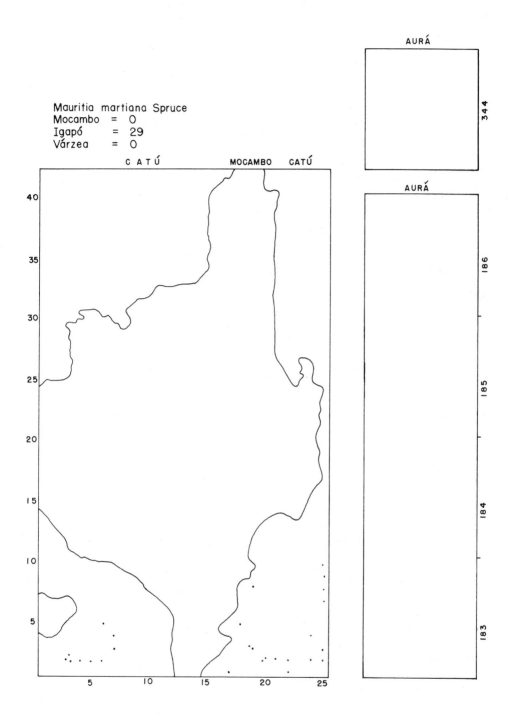

Fig. 9. Distribution of Mauritia martiana *Spruce in the study area in Belém. The dots represent presence of trunks of 30 cm circumference or more in each 100m^2 quadrat.*

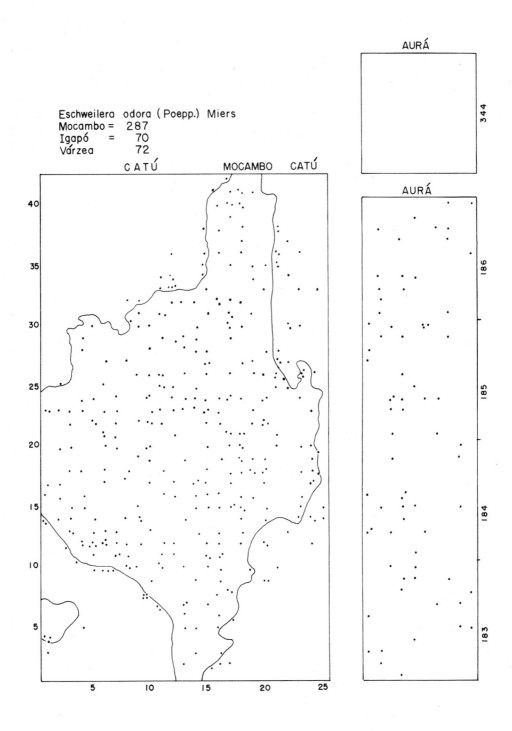

Fig. 10. Distribution of Eschweilera odora *(Poepp.) Miers, with 287 trees on terra firme, 70 in swamp forest, and 72 in várzea forest. This species was the commonest species on the terra firme.*

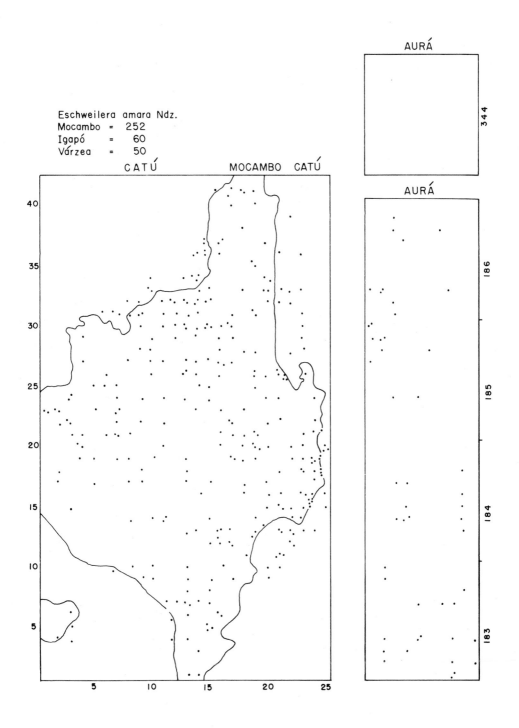

Fig. 11. *Distribution of* Eschweilera amara *Ndz., with 252 trees on terra firme, 60 on swamp and 50 on várzea. This was the second most common species on terra firme.*

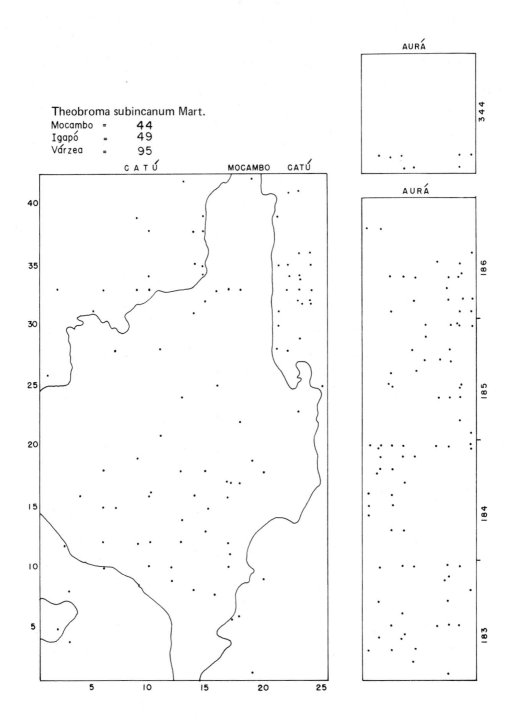

Fig. 12. Distribution of Theobroma subincanum *Mart., a species without a definite preference for terra firme or várzea.*

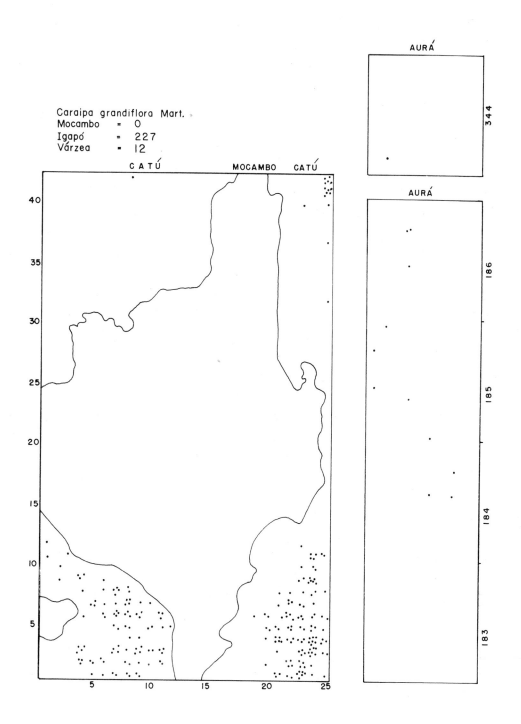

Caraipa grandiflora Mart.
Mocambo = 0
Igapó = 227
Várzea = 12

Fig. 13. Distribution of Caraipa grandiflora Mart., which does not have a single tree growing in the terra firme area and has 227 trees in the swamp forest and 12 in the várzea. In this species, there is a distinct grouping into certain areas probably related to soil differences which were not discussed.

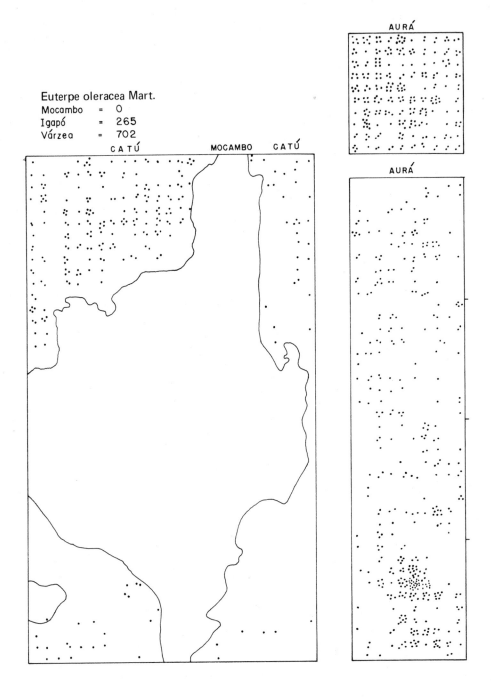

Euterpe oleracea Mart.
Mocambo = 0
Igapó = 265
Várzea = 702

Fig. 14. Distribution of Euterpe oleracea *Mart.*, with 265 in the swamp and 702 in the várzea, showing a species confined to the flooded areas. In this species, each dot represents a clump as individuals are formed of 1-11 trunks by sprouting. This was the commonest species in the várzea with 23.8 percent of the total number of trees, and also in the igapó with 9.28 percent.

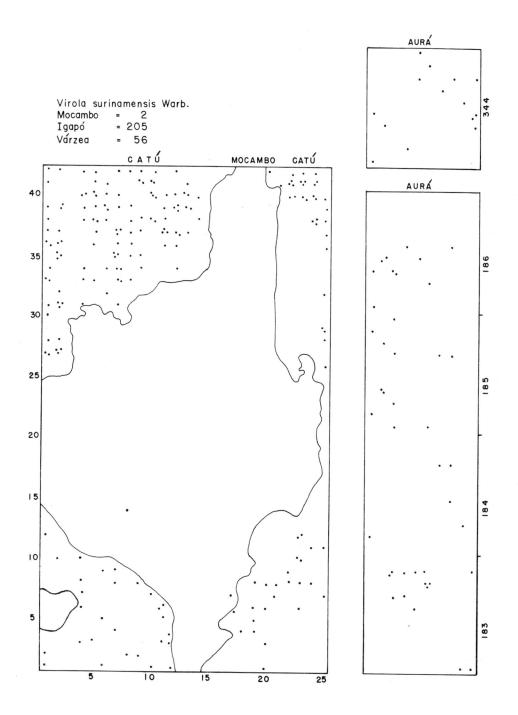

Fig. 15. Distribution of Virola surinamensis *Warb., another species almost confined to the flooded areas.*

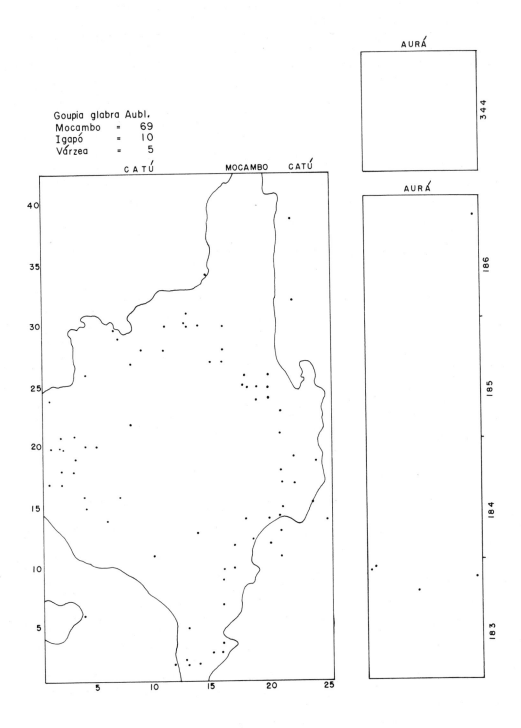

Fig. 16. Distribution of Goupia glabra *Aubl.*, a terra firme species with a distinctly zonal distribution.

the várzea and the other on terra firme, for example, *Licania macrophylla* Benth. (várzea) and *L. oblongifolia* Standl. (terra firme).

Only eight species are delineated in the accompanying maps, but the three areas studied contained 345 species of 30 cm girth or more and are represented by 8,996 trees (in a total area of 15.5 hectares or 856 individuals per hectare). Eighty-three species (24 percent) occur in all three areas. The occurrence of these species is summarized in Tables III and IV.

Table III. Number of species over 30 cm girth in the three study areas:

Locality Vegetation Type	Mocambo Terra Firme (A)	Catú Igapó (B)	Aurá Várzea (C)
No. of species	224	180	196
Size of area	5.7 ha	4.8 ha	5 ha
Density	39.3/ha	37.5/ha	39.2/ha

Table IV. Species in common between the 3 areas: A, Mocambo, 5.7 ha; B, Catú, 4.8 ha; C, Aurá, 5 ha.

	No. of species	No. of species in common	Percent species in common
A + B	282	122	43.26
A + C	305	115	37.70
B + C	269	107	39.78
A + B + C	342	83	24.27

Since areas A (Mocambo) and B (Catú igapó) are contiguous, dispersal between them is facilitated. As expected, these two areas have the highest percentage of species in common although the environmental conditions of each are quite different (A is terra firme, B is igapó), and várzea has more species in common with igapó than with terra firme.

This study consisted of only individuals which were 30 cm circumference or more. If a smaller girth is used as the determining factor, the number of individuals would rise rapidly above the 8,996 studied. For example, a more detailed study of quadrat 15-1 of 10 x 10 m in the Mocambo terra firme contained 440 individual plants of 5 cm girth or more — an equivalent of 44,000 plants per hectare, in contrast to the average of 856 of over 30 cm girth.

These smaller trees have two types of growth potential: (1) those species that can grow larger and eventually exceed the 30 cm circumference category, and (2) those species of smaller stature which form part of the lower layer of the forest and are never expected to attain 30 cm circumference. In the case of quadrat number 15-1 mentioned above, there were only 7 species with a total of 8 individuals of over 30 cm circumference. The remaining 432 individuals in the quadrat were less than 30 cm in circumference, and consisted of 68 species with the potential to exceed the 30 cm limit, and 31 species which normally remain small and occupy the lower and more shaded layer of the forest (this included 3 species of epiphytes). Although lianas usually do not have very large trunks, they can reach the canopy of the forest to obtain light. The 100m^2 quadrat contained 37 lianas as well as 52 epiphytes, 25 herbs and small shrubs, and 349 additional individuals of species which could eventually grow to more than 30 cm in circumference.

THE VERTICAL DISTRIBUTION OF PLANTS IN THE FOREST

No distinct stratification could be detected in the area we studied, though some workers such as Richards (1952) emphasize stratification. It is probable that stratification does not really exist in the Amazonian rain forest, and that it is, rather, a distorted interpretation of profile diagrams. Figure 28 shows the total number of plants in each height class for all plants over 2 m tall in an area of 5,000 m^2 in the three areas studied.

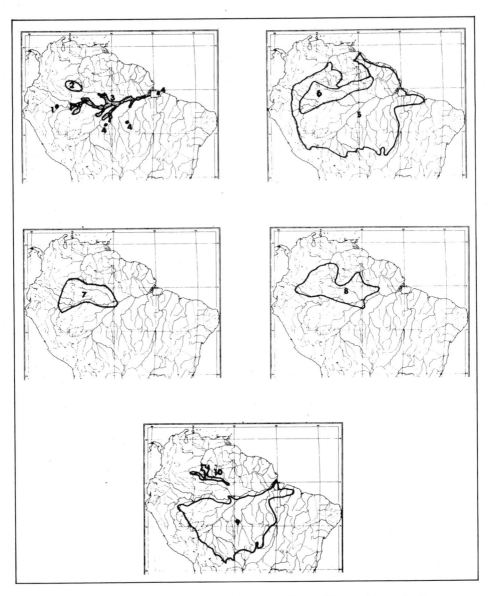

Fig. 17. Distribution of the ten species of the genus Hevea (Euphorbiaceae). The range of the genus corresponds with the limits of Amazonia. 1, H. paludosa Ule; 2, H. rigidifolia Muell. Arg.; 3, H. spruceana Muell. Arg.; 4, H. camporum Ducke; 5, H. guianensis Aubl.; 6, H. nitida Muell. Arg.; 7, H. pauciflora Muell. Arg.; 8, H. benthamiana Muell. Arg.; 9, H. brasiliensis Muell. Arg.; 10, H. microphylla Ule.

The nature of the 3 curves reveals that all heights are represented and that no distinct strata are evident.

The branching of the crowns is quite irregular and variable, and although the basic architectural patterns of Hallé and Oldeman (1970) are easily seen in young trees, the crowns of the mature trees are quite irregular because they develop in the way which makes the best possible use of available light. Crown shape is adaptive for light as well as being a genetic trait. It is very difficult to draw crowns of trees faithfully because of these irregularities in their development. By contrast, in open types of vegetation such as savannas, the arrangement of branches in relation to light is not so important because light there is available in abundance. Consequently, branches can grow freely in any direction and the trees have a distinct and more regular architecture. This is one of the reasons for the characteristic landscape of the savannas where the common trees repeatedly have the same genetically-controlled crown shape.

Although increases in girth and height indicate an increase in the biomass in an individual tree, the biomass of an area of mature forest should remain constant. An understanding of the distribution of individual plants of an area among the different size classes is most important for an understanding of their adaptive strategy towards light. In tropical rain forest, natural selection is closely linked with the factor of light economy. The increase in size of plants (an increase in biomass) means that other plants or parts of plants must be dying, either of old age or for other reasons, such as disease or insect attack, in order for the biomass to remain constant. In forest vegetation with a large biomass, the biological cycle of organic material is efficient and rapid (primary and secondary productivity). There is the continuous process of individuals dying and others growing as substitutes for those which are eliminated.

Nutrients are important in the forest ecosystem, but they are not generally the most important controlling factor for the distribution of individual trees within a small area of forest. The forest plants have evolved many specialized mechanisms to reduce loss of

Figs. 18-25. Graphs of tree height plotted against circumference for 8 species in the Jarí forests. The shade-tolerant trees have a large number of small individuals (Figs. 18 and 23) whose number diminishes as size increases. The light-demanding species do not have many individuals of small size in mature forests because their regeneration depends upon natural clearings. These graphs also show that in the larger individuals there is an increase in trunk growth in comparison to height. All individuals over 15 cm tall were studied.

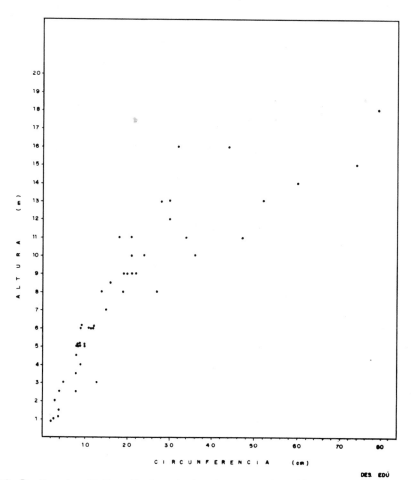

Fig. 18. Protium tenuifolium *Engl., a shade-tolerant species with a large number of small plants.*

Fig. 19. Iryanthera juruensis *Warb., with a large number of small plants. The trees rarely exceed 80 cm circumference. Growth of trunk and height are rather closely correlated as shown by the linear pattern of distribution of the dots on the graph.*

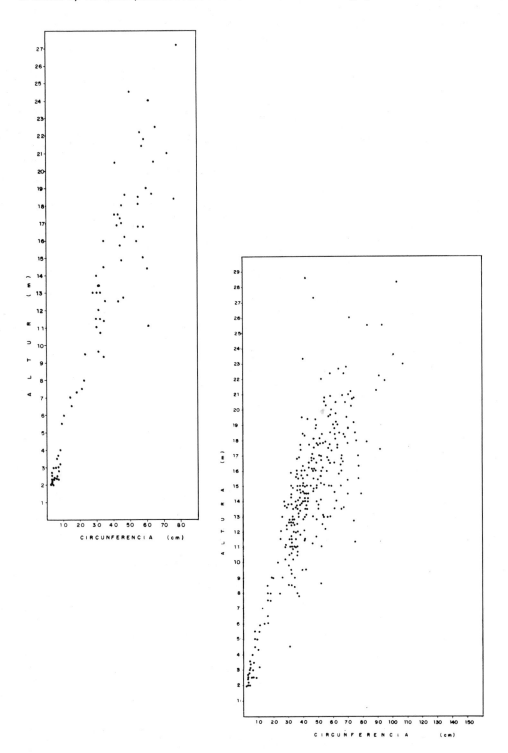

Fig. 20. Eschweilera amara *Ndz., a very common species in the Jarí forest.*

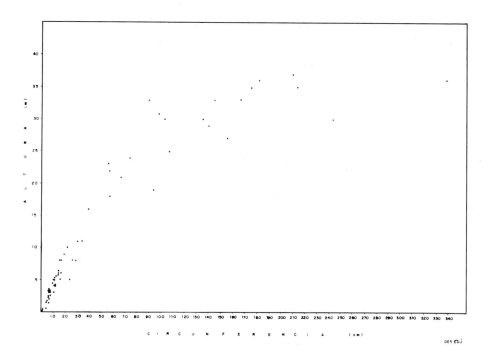

Fig. 21. Manilkara huberi *Standl., one of the largest trees of the forest, but it is shade-tolerant as evidenced by the large number of small plants.*

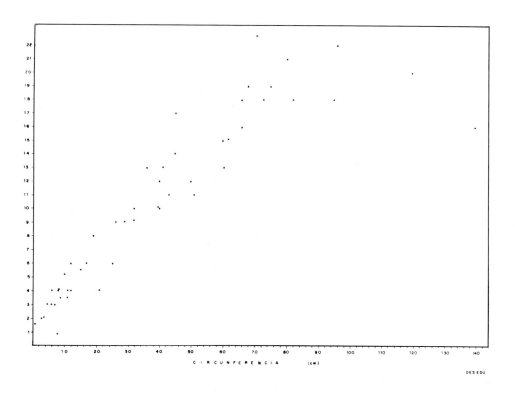

Fig. 22. Eschweilera odora *(Poepp.) Miers.*

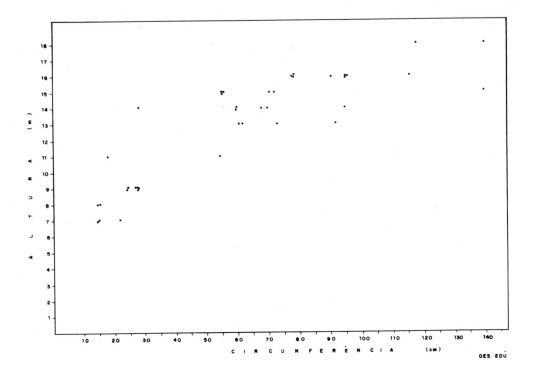

Fig. 23. Geissospermum sericeum *(Sagot) Benth. & Hook.*, a light-demanding species where individuals under 15 cm circumference are rare.

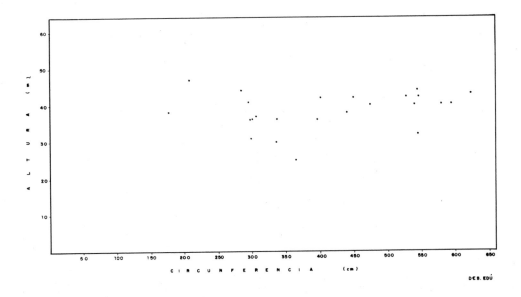

Fig. 24. Bertholletia excelsa *Humb. & Bonpl.*, in which there are no small individuals at all in mature forest; also, all individual trees fall in the same height class regardless of trunk size.

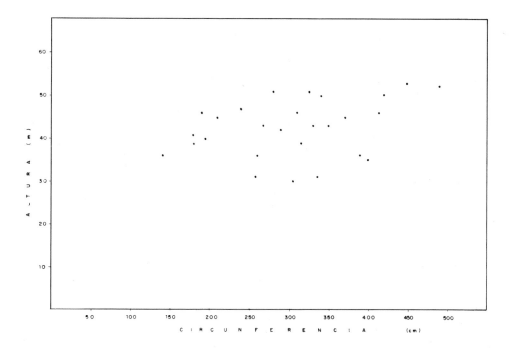

Fig. 25. Dinizia excelsa *Ducke, a species with no small individuals in mature* forest; like Bertholletia *(Fig. 24) all trees are of the same height regardless of trunk size.*

nutrients by leaching and other ways. Several workers such as Sioli (1968) and Schmidt (1972) have shown the lack of nutrients in various rivers and streams. They are not washed out of the forest. The forest functions as a closed circuit without leakage where the greater part of the nutrients are stored in the plants themselves rather than in the soil. When a leaf either drops or is still on the tree, it is attacked by fungi and micro-organisms, and when it is on the ground it also is soon attacked by small roots. It decomposes rapidly as a result of these external factors. This happens with all dead parts of plants. The layer of organic matter on the ground is a major part of the forest and functions like a sponge which is difficult to wash out. However, most of the available nutrients are in the vegetation itself and the luxuriance of tropical rain forest is not usually linked directly with a richness in soil nutrients. In fact, open types of vegetation can occur on much richer soils than those of the dense tropical forests.

The availability of light in the forest at different levels above the ground is directly linked to the adaptive strategy for survival developed by each species. Some species are *shade tolerant,* and therefore are able to develop in the shade of the forest. These species most frequently are low or medium in height and never reach the crown of the forest. However, there are some notable exceptions, such as the shade tolerant *Maçaranduba (Manilkara huberi* Standl. — Sapotaceae) which is one of the larger trees of the forest. Other species are light-demanding in order to grow, and consequently they can only thrive in clearings. In the forest, clearings evolve naturally without the influence of man. They are caused by storms, by the falling of dead trees, or by defoliation. The species which need light to grow are usually the large ones which reach the canopy of the forest (emergent species).

Examination of the occurrence of different diameter classes of a tree species in a forest is all that is necessary to determine whether it is a shade-tolerant or light-demanding species. To illustrate this, Figures 18-25 depict tree height plotted against circumference for 8 species. These data are taken from a 6,000 x 10 m transect in the Rio Jarí region of Pará in which all trees above 15 cm in height were measured. If a species is shade-tolerant, most of the individuals should be small plants, as in *Protium tenuifolium* Engl. (Fig. 18) or *Eschweilera amara* Ndz. (Fig. 20). In the graphs for these species, the number of individuals drops as the size class increases. In the case of the few larger individuals, there is also a greater variation in height for the same circumference, and the rate of growth of the trunk increases proportionately. This is because the trees need less light as they become taller. At the same time, they also become thicker in order to better withstand a new factor, wind, which is not as important for the smaller and shorter trees. Two other shade-tolerant trees are illustrated in Fig. 19 (*Iryanthera juruensis* Warb.) and Fig. 21 (*Manilkara huberi* Standl.). In these cases, there are many individuals in the smaller classes.

In contrast to the above examples, there are some interesting cases, such as those of *Bertholletia excelsa* Humb. & Bonpl. (Fig. 24) and *Dinizia excelsa* Ducke (Fig. 25), in which almost all the individuals, regardless of circumference, are of the same height. These are both light-demanding species. In the Rio Jarí forests, the majority of individuals of *Bertholletia excelsa* were approximately 4.5 m in circumference and 34-40 m high.

A summary of the differences between shade-tolerant and light-demanding species is given in Fig. 26.

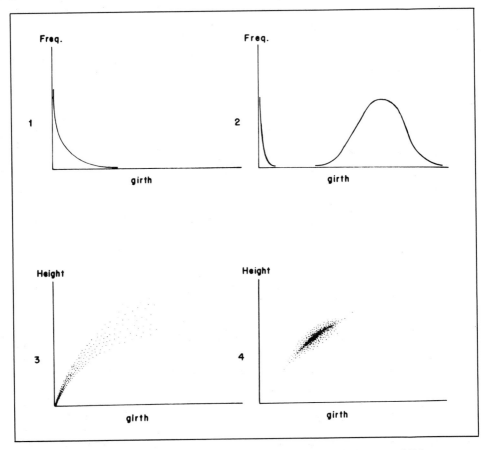

Fig. 26. *Illustrating the difference in behaviour between shade-tolerant and light-demanding species, by showing the frequency plotted against girth which drops rapidly in shade-tolerant trees (1), and increases in light-demanding species (2). Below, tree height is plotted against girth showing: 3, shade-tolerant species with a large number of smaller individuals, and 4, light-demanding species with the majority of trees of a large size. Mature forest always contains a mixture of these two types of trees.*

Another aspect of the vertical distribution of trees is the frequency of different tree heights, and this is shown in Fig. 27 for the Mocambo terra firme forest. Considering trees of over 30 cm in circumference, the most frequent trunk height in this area is about 15 m in 7-8 percent of the individuals present there. This graph reveals no evidence of stratification in the forest.

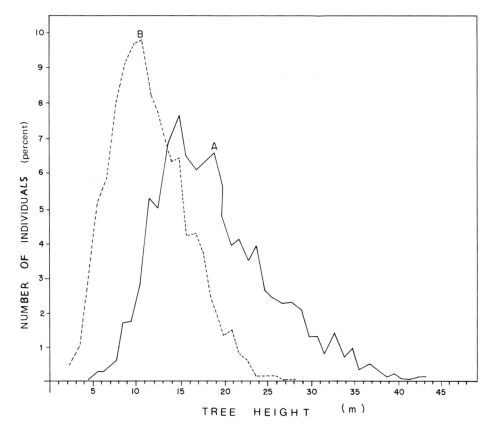

Fig. 27. The distribution of tree heights in the Mocambo terra firme forest expressed in terms of percentage of the total number of trees over 30 cm in circumference: A, total height, B, trunk height.

GROWTH RATE OF TRUNKS IN THE FOREST

In the Mocambo Reserve described above, an area of 100 x 200 m of terra firme forest (see Fig. 30) was chosen in which the circumference of all trees over 30 cm was measured. Six measurements were made during the fifteen-year period 1956 to 1971. The results of this study produced some interesting data which are summarized below:

1. 1,144 trees were measured in 1956. By 1971, 868 of these trees were still alive and 276 had died (for the distribution of the dead trees see Fig. 30). In 1956, the total basal area of all 1,144 trees was 553,984 cm^2. In 1971, the basal area of the remaining 868 trees plus an additional 112 trees that then had newly surpassed the 30 cm girth category was 554,612 cm^2. The basal area of the trees that had died (based on their 1956 circumference) was 128,618 cm^2. Thus, the living basal area (and, therefore, presumably biomass) of these 2 hectares remained relatively constant during the 15-year period. There was, however, a reduction of 164 in the number of trees.

2. When the forest as a whole is considered with its mixture of different species, it is apparent that the growth in trunk size is very uneven. There is no uniform growth pattern in the forest. In any circumference class, there were some trees whose girth increased more, others where it increased a little, and some which did not change at all during the period of study (see Figs. 31 to 34).

3. The irregular growth pattern of the forest as a whole was also true between the different individuals of the same species. For example, there were individuals of a particular species with zero increase in girth in each different circumference class (see Figs. 31-34 which show girth growth rates of all the individuals of 4 species).

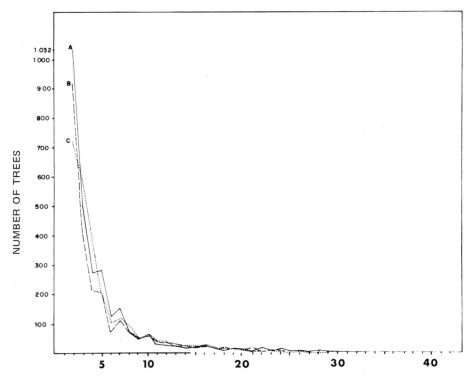

Fig. 28. Height of all trees over 2 m tall in 5,000 m^2 quarters: A, Mocambo terra firme; B, Aurá várzea; C, Catú, igapó. The distribution of tree heights is similar in all three forest types and there is no indication of strata when 1 m increments in tree height are plotted as in this graph.

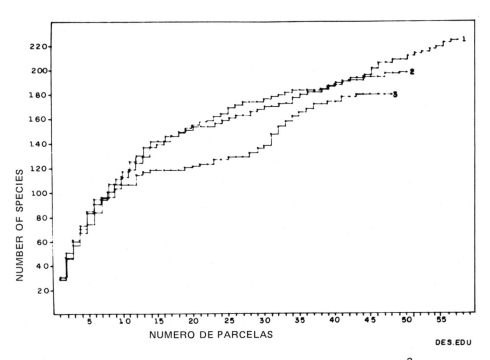

NUMERO DE PARCELAS

DES.EDU

Fig. 29. Number of additional species found by adding an additional 100 m^2 quadrat to the sample size: 1, Mocambo terra firme; 2, Aurá várzea; 3, Catú, igapó. It reflects the greater diversity of the terra firme, and that várzea is richer in species than the igapó.

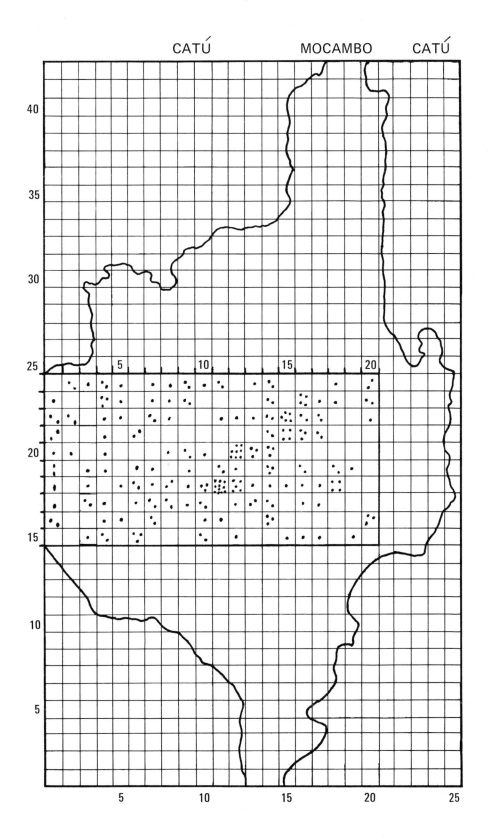

Fig. 30. Localization of trees which died over a 15-year period (1956-1971) in an area of 200 x 100 m of the Mocambo terra firme. In some quadrats there is a cluster of several dead trees caused by the effect of storms.

The most unusual and interesting observation which resulted from this study is the different behavior in growth of the trunks of different trees of the same species. This can be explained by the location of the individual plants in the forest. A certain tree will receive a greater or lesser quantity of light, depending upon its position in relation to the other trees. The growth pattern will depend upon whether it is a shade-tolerant or light-demanding species. Individuals of light-requiring species will often have zero growth for many years until such time as a natural clearing in the forest occurs which produces the light necessary for the tree to grow.

Basal area is a useful figure for an estimate of biomass. Although basal area does not represent the exact biomass, it is a figure that can be easily calculated. Data on the volume and weight of the vegetation are more precise but their calculation involves measurement of tree height, crown, etc., measurements which are difficult to make in tropical rain forest. Since basal area remained relatively constant during the 15-year study

Figs. 31-34. Increase in trunk diameter of four species over a 15-year period for the different circumference sizes, showing the lack of uniformity in growth pattern within the same species or same circumference size.

ESCHWEILERA AMARA

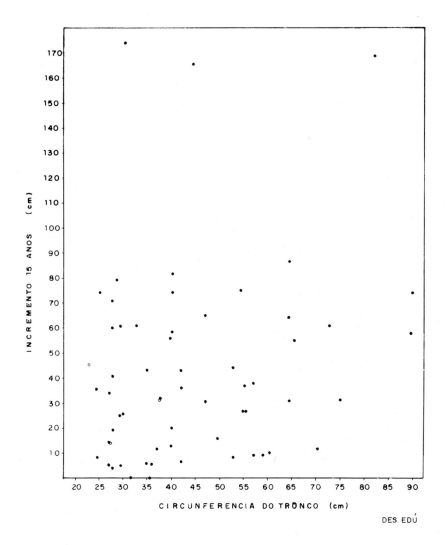

Fig. 31. *Increase in trunk size of individuals of* Eschweilera amara *Ndz.*

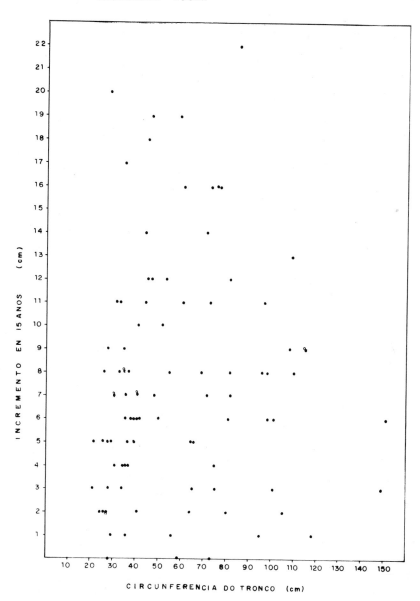

Fig. 32. *Increase in trunk size of individuals of* Eschweilera odora *(Poepp.) Miers.*

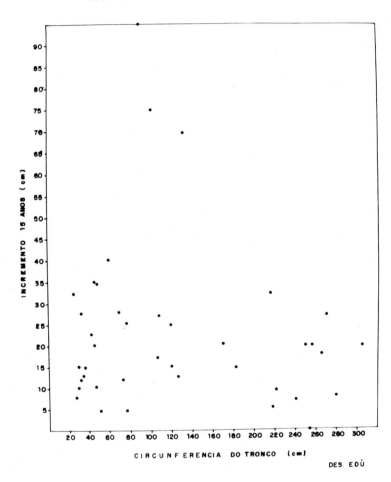

Fig. 33. *Increase in trunk size of individuals of* Vochysia guianensis *Aubl.*

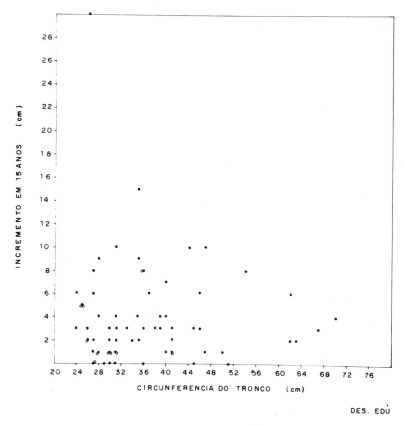

Fig. 34. Increase in trunk size of individuals of Tetragastris trifoliolata *(Engl.) Cuatr.*

period, we can assume that biomass also remained constant despite the reduction in the number of trees.

In general, trees of the tropical rain forest are comparatively short-lived. Heinsdijk (1965) estimated that in Amazonian rain forest a typical tree in the 25-35 cm diameter class has an annual growth in trunk size of 0.8 cm, and in the 25-155 cm diameter class, an annual growth of 0.37-0.8 cm. He calculated theoretically that the ages of the trees studied varied between 27 and 418 years. His data were based on measurement of 120,000 trees measured in a series of FAO inventories. These age calculations are theoretical and based on averages and must therefore not be taken literally, especially in view of our observation in the Mocambo Reserve, where zero growth over a fifteen-year period is a common feature. The lack of growth uniformity even within individual trees of the same species makes it impossible to calculate age on the basis of growth rate. Growth rings do not necessarily help in tropical forest studies because they are frequently either absent or hardly apparent in most species. When present, rings also often are either not continuous, or are eccentric or broken up in such a way that one counts a different number in different radial directions. In general, trees of the tropical rain forest are comparatively short-lived, rarely exceeding 400 years old. Occasionally, one finds impressive giant trees which are probably more than a thousand years old. For example, the tree illustrated in the drawing in Martius' "Flora Brasiliensis" in 1841 required 13 Indians with outstretched arms to extend around the trunk. We also can cite other examples of giant trees, such as three large trees in the forest of the Rio Jarí which were measured for us by Nilo T. Silva: 1. *Bertholletia excelsa* Humb. & Bonpl. — 1,400 cm in circumference; 2. *Vochysia maxima* Ducke — 425 cm in circumference and 62 m tall; 3. *Dinizia excelsa* Ducke — 320 cm in circumference and 66 m tall. Each of these giants had hollow trunks, which is quite usual among the more ancient trees of the forest. Janzen (1976) commented on the frequency of hollow trunks in tropical trees and hypothesized that this has a selective advantage for trapping nitrogen and minerals especially from animal nests and animal defecation.

In the two hectares studied for growth measurements, the positions of the trees that died during the 15-year study period were mapped (see Fig. 30). The groups of dead trees are also in the largest natural clearings where light penetrates to the lower layers of the forest, enabling the light-demanding trees to grow.

CONSERVATION AND DEVELOPMENT OF AMAZONIA

The three aspects of the forest discussed above — horizontal and vertical distribution of trees, and growth increments — serve to show the complexity of the Amazon ecosystem. There are many other aspects which could be mentioned but which we personally have not studied, such as nutrient cycles, mycorrhizal fungi, pollinators, dispersal of diaspores, etc. The complexity of this forest indicates that it is much more important to conserve ecosystems rather than individual species in Amazonia. Both conservation and utilization of the Amazon rain forest can only be carried out rationally through an understanding of this unique ecosystem. It is essential that methods be developed which are better adapted to the ecology of the region than the present wasteful destruction of the forest, especially by the prevalence of slash and burn agriculture.

The governments of the countries within whose boundaries the vast Amazon forest lies are facing complex problems in how best to achieve a balance between man and conservation. There is no easy solution. Of all the countries which adjoin Amazonia, Brazil is confronted with the greatest opportunity because more than half of the huge forest lies within its territorial limits. But the future of Amazonia is infinitely more complex than at first meets the eye. Its problems are political, social, economic, and scientific. How can such a vast part of the country be conserved and utilized at one and the same time? What are the wisest uses of its enormous natural resources? How can the country's national security be protected in view of these physical problems? It seems unlikely that any of these problems will be resolved without some irreversible mistakes being made. For example, we feel that already a grave judgemental error is being perpetuated by the degree of burning and defoliation that is now being practiced on such a wide scale. Such errors can only result in irrevocable long-range damage to Amazonia's delicate ecosystem.

Brazil has given much thought to the situation, and a number of different regional plans already have been proposed specifically for Amazonia. In fact, a considerable part of Brazil's GNP has been used for the various Amazonian programs. Inevitably, these programs have a profound effect on the future of the forest, regardless of whether they are designed for development or for conservation. Among the many government programs in Amazonia, two deserve special mention: The Humid Tropics Program and the Polamazônia Program. Both belong to the National Council for Scientific and Technological Development (CNP_q).

The Humid Tropics Program was formed to establish centers for the study of basic and applied research problems involving the region's development. Its program was discussed in this Symposium by Dr. Dantas Machado. The Polamazônia program was designed to select areas or nuclei which will receive greater financial resources to enable them to serve as focal points for development of agriculture and mineral exploitation. The program also includes provisions for the designation of reserves. We hope that choice of these reserves will be based on sound scientific data and guidance, rather than because they offer no potential for either agriculture or mining, and therefore could be "dispensed" with by designating them as reserves.

The Superintendency for the Development of Amazonia (SUDAM) has at the present time a similar project designed to establish permanent production forest areas for the sustained yield of timber and other forest products. These will be in priority areas for managed forests. We hope that these areas will avoid monocultures. SUDAM also has a project underway to intensify agriculture and cattle production in the várzea regions along the margins of the major rivers of Amazonia. This has the advantage of diverting the pressure of development from the interior forests on the plateau or terra firme areas.

In addition, problems of colonization along the rivers are less complex. It is to be hoped, however, that development of the várzea will include the conservation of areas large enough to preserve the natural gene pool of várzea species.

The Brazilian Organization of Agricultural Research (EMBRAPA) functions throughout Brazil, though it has only two study centers in all of Amazonia: The recently created Center for the Study of Rubber Trees and Rubber in Manaus, and the Center for Agricultural Research in the Humid Tropics in Belém. The EMBRAPA Center for the Study of the Cerrados located in Brasília is already increasing the agricultural production of the cerrado region of central Brazil. This type of program should certainly be encouraged, since the greater use of the savannas where physical difficulties are fewer and environmental destruction is less could considerably alleviate the pressure on Amazonia. More time would then be available for research into the ecology of the terra firme forests which cover most of the region. This has been well-expounded in this Symposium by Dr. Robert Goodland.

Figures 35-37 offer an idea of the Brazilian government's development and conservation programs. Figure 35 shows reserves of four different types for indigenous populations:

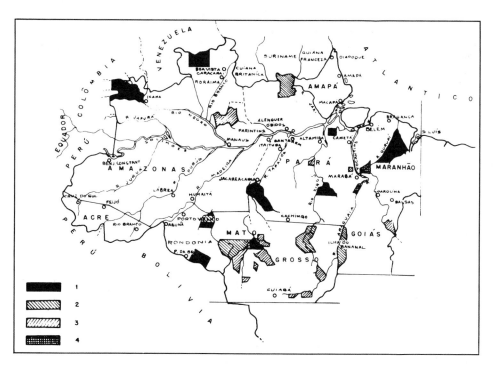

Fig. 35. Forest reserves for indigenous populations: 1, Forest Reserves; 2, Indian Parks; 3, Indian Reservations; 4, Controlled areas for future pacification of Indian groups. (after Pandolpho, 1974).

forest reserves, Indian parks, Indian Reserves, and areas set aside for Indian tribes known to exist but not yet contacted. Figure 36 indicates forest reserves (conservation areas) which are either already officially decreed or proposed. Figure 37 pinpoints the foci of the Polamazônia development program. Besides the official reserves depicted in these maps, there are numerous other areas set aside both for development projects, and as reserves which belong to other government or to private organizations. For example, a small 9 km^2 reserve belonging to the National Amazon Research Institute and the Superintendency of the Free Port of Manaus was recently increased in size to over 300 km^2, enabling it now to include several different types of vegetation. On the other hand, all large agricultural projects receive governmental stimulus such as tax incentives and financing, which is controlled by SUDAM, the development superintendency. One of the requirements for SUDAM's approval of projects is that half of any property be left as a forest reserve. The control of such reserves, however, is very difficult to administer

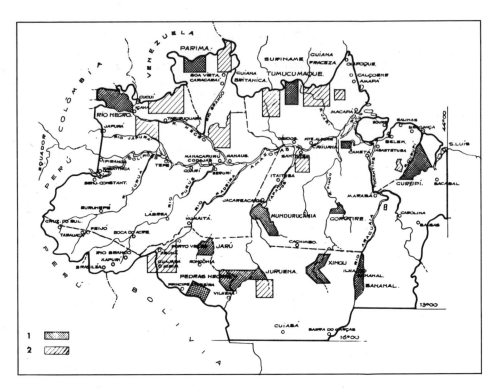

Fig. 36. Forest reserves of Amazonia: 1, areas already decreed; 2, proposed areas (after Pandolpho, 1974).

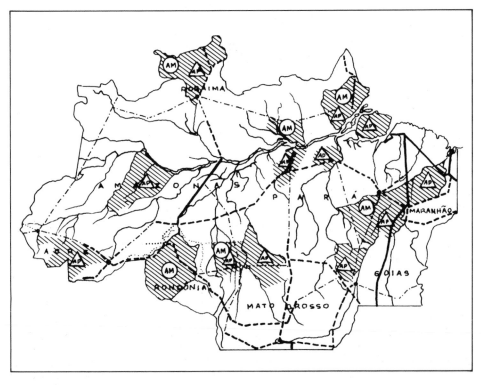

Fig. 37. Map of the areas selected by the Polamazônia program as foci for development of the region: AM, mineral and mining development; AP, agricultural development (after Pandolpho, 1974).

192

in an area as vast as Amazonia, and the legal requirements are not always observed, despite considerable government efforts to control them.

In addition to these problems, the main danger in SUDAM's policy of fifty percent conservation of an area is the chessboard effect that it will inevitably have on the forest. This is particularly noticeable already in smaller colonization areas. These small conserved areas are interspersed with exploited areas, and they are illogical in terms of the biology of the forest ecosystem. Separate large isolated reserve areas are needed, especially to protect many of the animals.

The protection and regulation of the fauna and flora of Brazilian Amazonia is under the auspices of the Institute of Forestry Development (IBDF). In addition, the recently-created Environmental Secretariat (SEMA) is now in the process of establishing ecological reserves in Amazonia. SEMA's charter is to study the problems of the balance between Man and Nature, in addition to examining the environmental conditions and the problems of pollution in the country as a whole.

Recently, the National Council for Scientific and Technological Development (CNP_q) initiated *Programa Flora,* which aims at accelerating the process of basic inventory of the Brazilian flora. The program is creating a computer data bank based on herbarium and library information, and is in the process of organizing intensive botanical collections in the region. The program has begun with the five-year Amazon phase called *Projeto Flora Amazônica.* The National Science Foundation of the United States has been invited to participate in this project.

The above brief outline of the most important Brazilian government projects in Amazonia serves to show that the Nation is fully aware and cognizant of the extraordinarily complex problems resulting from human interference in Amazonia. Although Man is fallible and many mistakes are inevitable, it is to be hoped that research will be supported to such a degree that sufficient data will soon become available which will contribute greatly towards wiser use and better protection of the forest. This is in the interest of all Humanity, rather than of Brazil alone.

ACKNOWLEDGEMENTS

The joint work of the two authors was made possible by grants from the National Science Foundation (INT75-19282) and the Conselho Nacional de Desenvolvimento Científico e Tecnológico (CNP_q) as part of their Cooperative Science Programs in Latin America. We thank Dr. Robert A. Goodland for reviewing the manuscript. We are grateful to the Smithsonian Institution who collaborated with the formation of the study area under their program APEG (Área de Pesquisas Ecológicas de Guamá).

LITERATURE CITED

Elton, C. S. 1973. The structure of invertebrate populations inside neotropical rain forest. Jour. Animal Ecol. **42:** 55-104.

Guerra, A. T. 1959. Estrutura geologica, relevo e litoral. In: Grande Região Norte, IBGE, Rio de Janeiro: 17-60.

Hallé and R.A.A. Oldeman. 1970. Essai sur l'architecture et la dynamique de la croissance des abres tropicaux. Paris. 178 pp.

Hatheway, W. H. 1971. Contingency table analysis of rain forest vegetation. In: Many species populations, ecosystems, and systems analyses eds. G. P. Patil, E. C. Pielou and W. Z. Waters, Pennsylvania State Univ. Press.

Heinsdijk, D. 1965. A distribuição dos diâmetros nas florestas brasileiras. Min. Agric., Dept. Rec. Nat. Renov., Boletim **11:** 1-56.

Hueck, K. 1972. As florestas da América do Sul. 466 pp. Polígono, São Paulo.

Janzen, D. H. 1976. Why tropical trees have rotten cores. Biotropica **8:** 110.

Kubitzki, K. 1977. The problem of rare and of frequent species: the monographer's view. In this symposium.

Lovejoy, T. E. 1975. Bird diversity and abundance in Amazon forest communities. The Living Bird **13:** 127-191.

Martius, C. P. F. 1841. Tabulae Physiognomicae Brasiliae Regiones Iconibus Expressas. Martius, Fl. Bras. **1** (1): 22-29, tab. 9.

Nimer, E. 1966. Regime xerotérmico. In: Atlas Nacional do Brasil. Plate II-8. IBGE, Rio de Janeiro.

Pandolpho, C. 1974. Estudos básicos para o estabelecimento de uma política de desenvolvimento dos recursos florestais e uso racional das terras da Amazônia. Ed. 2, 54 pp. SUDAM, Belém.

Pires, J. M. 1973. Tipos de vegetação da Amazônia. Museu Goeldi, Belém, Publ. Avulsa **20:** 179-202.

Prance, G. T. 1975. Flora and vegetation. In: Goodland, R. J. A. and H. S. Irwin, Amazon jungle: green hell to red desert? Pages 101-111. Elsevier, New York.

_____ 1977. The phytogeographic subdivisions of Amazonia and their influence on the selection of biological reserves. In this Symposium.

_____. W. A. Rodrigues and M. F. da Silva. 1976. Inventário florestal de um hectare de mata de terra firme km 30 da Estrada Manaus-Itacoatiara. Acta Amazonica **6** (1): 9-35.

Richards, P. W. 1952. The Tropical rain forest. Cambridge University Press, 450 pp.

Roxo, M. G. D. 1943. Terras sulamericanas emersas nos tempos permo-carboníferos. Rev. Bras. Geog. **5** (1): 43.

Sioli, H. 1968. Hydrochemistry and geology in the Brazilian Amazon region. Amazoniana **1** (3) : 267-277.

Schmidt, G. W. 1972. Chemical properties of some waters in the tropical rain forest region of Central Amazonia along the new road Manaus-Caracarai. Amazoniana **3**(2):199-207.

THE PHYTOGEOGRAPHIC SUBDIVISIONS OF AMAZONIA AND THEIR INFLUENCE ON THE SELECTION OF BIOLOGICAL RESERVES

Ghillean T. Prance

The New York Botanical Garden, Bronx, New York 10458, U.S.A.

INTRODUCTION

In terms of conservation, it is a dangerous tendency to consider the entire Amazonian rain forest as a uniform mass of vegetation. This leads to the idea (prevalent among certain South American politicians) that it is possible to conserve only one single contiguous area of forest anywhere in the vast region to preserve the gene pool of Amazonia. In reality, however, Amazonia is extremely varied in terms of the species composition of its vegetation. In this paper I shall deal only with the forest on high (non-flooded) ground, the vegetation which covers about 90 percent of the area. It is physiognomically uniform throughout the region, but botanically very varied. Several other vegetation types also exist in Amazonia, each of which is more restricted in occurrence, but equally important from the standpoint of conserving representative areas. Some of these other vegetation types are: the savannas, Amazonian campinas on white sand, campina (caatinga) forests of the upper Rio Negro, swamp forest, transition forest, and montane forest. These other vegetation types are described in Pires (1973) and Prance (1975).

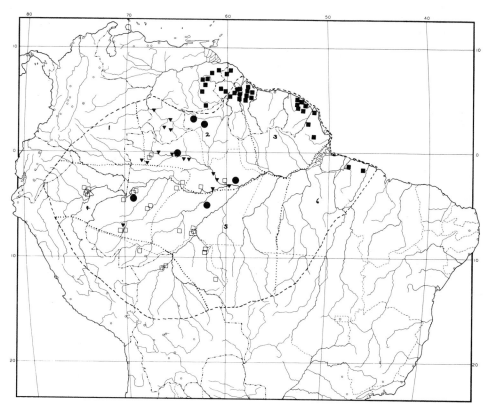

Fig. 1. Map of the 7 phytogeographic regions of Ducke and Black (1953) and the distribution of four species of Rourea *(Connaraceae):* □ R. camptoneura *Radlk;* ● R. krukovii *Steyerm;* ▼ R. cuspidata *Benth. ex Bak. var.* cuspidata; ■ R. frutescens *Aubl.*

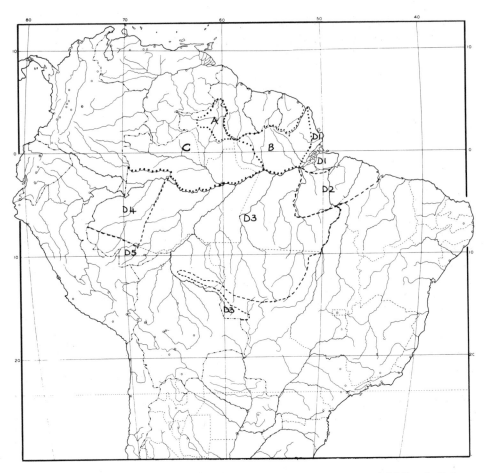

Fig. 2. Map of the Phytogeographic divisions of Amazonia of Rizzini (1963): A, Subprovince upper Rio Branco; B, Subprovince Jari-Trombetas; C, Subprovince Rio Negro;
D, Subprovince tertiary plane; D_1, Oceanic region; D_2, Southeast region; D_3, Southern
region; D_4, Western region; D_5, Southwestern or Acre region.

The aim of this paper is to identify the major phytogeographic divisions of the high-
ground forests of Amazonia and to compare them with areas scheduled for development
because, unfortunately, some of the most interesting areas of vegetation correspond with
those destined for intensified development in the foreseeable future.

AMAZONIAN PHYTOGEOGRAPHY — GENERAL CONSIDERATIONS

It has long been known by both phytogeographers and zoogeographers alike that the
Amazonian forest is not one uniform mass. Various attempts to explain this and to sub-
divide the area have been made. The best known subdivision of Amazonia into phytogeographic regions is that of Ducke and Black (1953, 1954), shown in Fig. 1. This will be
discussed and slightly modified later in this paper. Rizzini (1963) divided Amazonia into
the eight regions shown in Fig. 2. The purpose here is to study the phytogeographic
subdivisions of Amazonia in terms of the conservation needs of the area. In order to
understand the situation, however, we must also consider why such variations exist in the
vegetation cover of the Amazon rain forest. There are three basic reasons for these
variations:

Climate: Amazonia does not have a uniform rainfall. Considerable variation occurs,
with a wet region around the Atlantic coast, a drier area in the center, and the wettest
region nearer to the Andes (see Fig. 3). The rainfall variation is delineated well in the

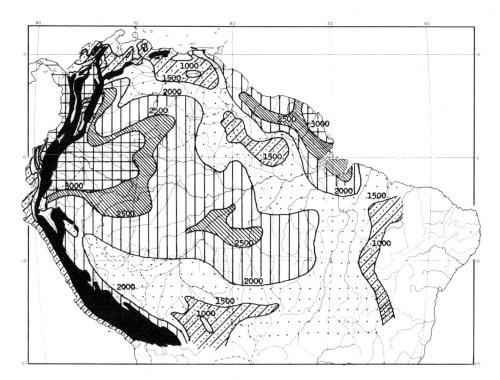

Fig. 3. Rainfall map of northern South America showing annual rainfall in millimeters. (adapted from R. Rienke, Das Klima Amazoniens, thesis. University of Tubingen, 1962).

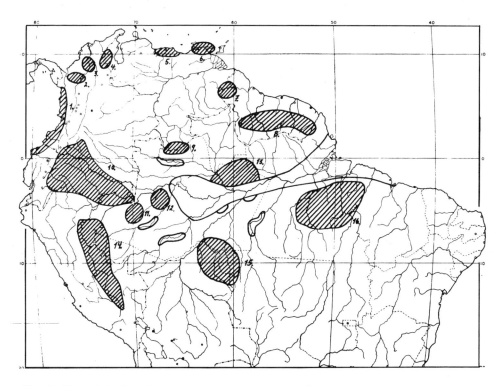

Fig. 4. Map of the Pleistocene forest refuges proposed in Prance (1974): 1, Chocó; 2, Nechi; 3, Santa Marta; 4, Catatumbo; 5, Rancho Grande; 6, Paria; 7, Imataca; 8, Guiana; 9, Imerí; 10, Napo; 11, Olivença; 12, Tefé; 13, Manaus; 14, East Peru; 15, Rondônia-Aripuanã; 16, Belêm-Xingu.

197

rainfall maps of Walter et al. (1975). The distribution of some species of plants is correlated with the present-day rainfall variation within Amazonia.

History: The history of the Amazon vegetation has been closely linked to climate changes in the past. Drier periods accompanied the Pleistocene glaciations. At that time, the area of rain forest in Amazonia diminished considerably, and savanna predominated. The forest was reduced to a number of small areas or refuges from which it later recolonized the savanna areas. This theory was first proposed by Haffer (1969) who studied the geography of the subspecies of various Amazonian birds. The evidence for Pleistocene climatic changes and their consequences on the vegetation of Amazonia is summarized in Haffer (1974). Van der Hammen (1972) also provided some palynological evidence for Pleistocene variations. I proposed some modifications of Haffer's refuges (Prance, 1974), based on distribution patterns of four of the woody plant families discussed below as evidence of phytogeographic regions of Amazonia. Fig. 4 shows the locations of the refuges proposed in Prance (1974).

The origin of present-day forest from a number of small areas has many interesting implications in terms of conservation. It is important to conserve these refuges because they are also areas of high endemism. Much of the forest was reduced to savanna by natural means during these drier periods, and the trees successfully recolonized the entire Amazon basin when the wetter climate returned. A greater understanding of this migration and recolonization should enable us to plan better reserves of the gene-pool of the Amazon forest, since these are the areas from which the present-day forest was formed. They are also probably the areas most able to provide the material for the recolonization of areas disturbed by modern man. The refuges are reflected in the present-day distribution of woody plant species and the phytogeographic subdivisions discussed below.

Geology and Soil: Geology and soil are both varied within the Amazon basin. Four main types of geological formations occur in Amazonia: (1) flood plain or alluvial plain which is covered by *várzea* and *igapó* forest; (2) the dry land area adjacent to the flood plain, which consists of forest-covered plains of Quaternary sediments; (3) the high plain (planalto) with forest cover over the greater part of Tertiary origin; and (4) the crystalline shield which is covered by high forest and begins above the waterfall line to both the north and south. Many plant species are confined to one of these areas. The forest on white sand (regosol) is different from forest on the clay laterite (latosol). Soil types in Amazonia have been documented and described by Falesi (1967, 1971, 1972).

THE PLANT FAMILIES USED AS A SOURCE OF PHYTOGEOGRAPHIC DATA

The phytogeographic areas defined below are based primarily on the monographic study of five woody plant families, all common and widespread in Amazonia. Though only a few examples are offered here, the conclusions are drawn from a detailed study of all five families.

1. CHRYSOBALANACEAE: The center of distribution of this family is Amazonia. A complete phytogeographic breakdown of the family is listed in Table I. Amazonian distributions for the genera *Licania* and *Couepia* are given in Tables II—VII. The genus *Licania* has 160 species, with 75 represented in Amazonia. The genus *Couepia* has 58 species, with 29 represented in Amazonia.

2. CARYOCARACEAE: This is a small, predominantly Amazonian family with 23 species belonging to two genera.

3. DICHAPETALACEAE: Two of the three genera in this family are represented in Amazonia. One is *Tapura* with 17 American species, 10 of which occur in Amazonia; the other is *Dichapetalum* containing 15 American species, with eight occurring in Amazonia.

4. CONNARACEAE: This is a pantropical family represented in Amazonia. Data from Forero (1976).

5. LECYTHIDACEAE: This is a large woody family whose distribution center is Amazonia, and in which many species are abundant and important in the ecology of

the Amazonian forest. For example, a single hectare of forest studied by Prance et al. (1976) contains 74 trees of Lecythidaceae, or 21.14 percent of the total of 350 trees over 15 cm diameter.

During the course of monographic studies in these five families of woody plants, I have plotted the distribution of a large number of species. The mapping of plant distribution may seem routine, but it provides considerable descriptive information about the phytogeography of the region. Botanists working on the African flora, which is much better collected than that of South America, have recently started to publish a series of plant distribution maps of individual species in order to further their knowledge of African phytogeography (Bruxelles 1969—present). Through international collaboration, these botanists have produced an extremely accurate vegetation map of Africa, under the auspices of the Association pour l'Etude Taxonomique de la Flore d'Afrique Tropicale. A similar project is urgently required for tropical South America. For Brazilian Amazonia, the lack of detailed vegetation maps is being rectified gradually through the work of Projeto RADAM which is producing vegetation maps based on radar photography of the region (Projeto RADAM 1973-75). As of this writing, these maps cover primarily the eastern part of Amazonia where the vegetation is the most highly disturbed.

PHYTOGEOGRAPHIC SUBDIVISIONS OF AMAZONIA

Ducke and Black (1953, 1954) recognized seven sectors (Fig. 1) of the Amazonian flora. These may be summarized as follows:

Northern Hylaea sector: The basin of the Rio Negro and west to the Rio Japurá.
Northeastern sector: From the Essequibo in Guyana through the Guianas south to the Rio Amazonas, between the Rios Jarí to the east and Trombetas to the west.
Atlantic sector: From the coast west to the Rio Jarí and Rio Xingu, roughly the area affected by tides. (See Fig. 5 for location of rivers discussed here.)

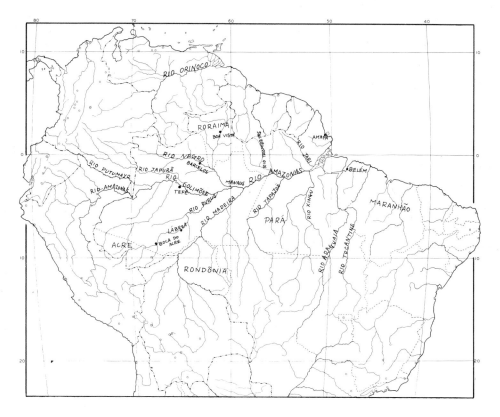

Fig. 5. Map showing principal towns and rivers mentioned in text.

199

Western Amazonia sector: The limits to the southeast are the Rio Tefé, to the northwest the Rio Japurá, extending north into Colombian Amazonia and south to include the State of Acre, and also extending west into Peru as far as the Pongo de Manseriche.

Southern Hylaea sector: The basin of the Rio Tapajóz, Madeira, the Rio Purus south to Bôca do Acre, the tributaries of the Rio Solimões west to Tefé.

Southeastern sector: The basin of the Rio Tocantins as far south as the Rio Araguaia.

Sub-Andean zone: This area lies entirely within Peru against the Andes.

Ducke and Black listed a number of species characteristic for each of the vegetation zones given above. My study of the five woody plant families shows a similar division of Amazonia, with only slight modifications of the areas proposed by Ducke and Black. Rizzini in 1963 divided his vegetation category of the province of Amazonia into four subprovinces, as shown on Fig. 2.

COMMON DISTRIBUTION PATTERNS

The distribution patterns of a few typical tree species — mainly of species of the Caryocaraceae — are presented here by distribution maps. To give an idea of the frequency of each distribution pattern, Tables II—VII list the distribution of the species of two genera of Chrysobalanaceae, *Licania* and *Couepia.* Many peculiar cases of plants with unusual distributions differ from the categories listed below, but from my studies I recognize eight frequently occurring distribution patterns involving Amazonian plant species. Also superimposed upon these patterns are the distributions of widespread species which have become redispersed over a wide area because they either are adaptable or have a good dispersal mechanism.

1. *Widespread:* This usually includes the Guianas and may extend as far north as Central America and as far south as central Brazil, though more frequently it includes only the Guianas and Venezuela. Examples: *Caryocar microcarpum* — Fig. 6, *C. glabrum* — Fig. 7, species of *Licania* and *Couepia* — Table II.

2. *Widespread but confined to Amazonia:* Examples: species of *Licania* and *Couepia,* listed in Table II.

3. *The Guianas and eastern Amazonia:* This is a common distribution which extends from the Orinoco delta eastward to the Brazilian States of Pará and Maranhão. Examples: *Rourea frutescens* — Fig. 1, species of *Licania* and *Couepia* — Table III.

In addition, a few species occur in the Guianas and in the northern part of Roraima Territory, Brazil. Examples: *Licania discolor, Caryocar montanum* — Fig. 10, and some extend west to the upper Rio Negro region, such as *Anthodiscus mazarunensis* — Fig. 8.

4. *The Guianas and central or western Amazonia:* Examples: *Couepia canomensis* and *C. parillo* — Fig. 11, species of *Licania* and *Couepia* — Table IV.

5. *Confined to eastern Amazonia:* Examples: *Couepia caryophylloides* and *Couepia excelsa* — Fig. 11, species of *Licania* and *Couepia* — Table V.

6. *Confined to central Amazonia:* Examples: *Rourea krukovii* — Fig. 1, species of *Licania* and *Couepia* — Table VI, *Cariniana micrantha.*

7. *Confined to western Amazonia:* Examples: *Anthodiscus amazonicus, A. pilosus* — Fig. 8, *Caryocar gracile* — Fig. 10, species of *Licania* and *Couepia* — Table VII.

8. *Confined to western and southern Amazonia:* Examples: *Rourea camptoneura, R. amazonica* — Fig. 1.

Table I. Phytogeographic breakdown of the distribution of Neotropical Chrysobalanaceae, excluding very widespread species.

	Chrysobalanus	Licania	Parinari	Exellodendron	Maranthes	Couepia	Hirtella	Acioa
Number of Species in Genus	2	152	16	5	1	55	80	3
U.S.A. Gulf States	–	1	–	–	–	–	–	–
Central America	–	7	–	–	1	1	3+i	–
Caribbean Islands	1	1	–	–	–	–	2	–
Pacific coastal Colombia and Ecuador	–	9	1	–	–	–	3	–
Northern Colombia and Venezuela	–	2	1	–	–	–	–	–
Central Colombia	–	2	–	–	–	1	2+i	–
Guayana Highlands (over 1,000 m)	–	1	–	–	–	1	2	–
Guianas incl. Venezuela-Bolívar and D. Amacuro, Brazil-Amapá and N. Terr. Roraima	–	23+i	2	–	–	7+iii	10	–
Amazonia widespread	–	12+i	–	–	–	6+i	6	–
western	–	14	3	–	–	6	18	1
central	–	10+ii	1	–	–	8	4	–
eastern	–	2	–	–	–	3+i	4	–
Bolivia	–	1	–	–	–	–	1	–
Planalto of Central Brazil	–	4	1	2	–	2	4	–
Northeastern Brazil	–	3	–	–	–	3	–	–
Eastern Brazil	–	11+i	1	1	–	7	9+i	–
Widespread in two or more of above, Central America, Colombia, Guianas and Amazon	–	i	–	–	–	–	ii	–
Caribbean and/or Trinidad and northern coastal Venezuela	–	3	–	–	–	–		
Colombia, Peru and Guianas, Venezuela-Bolívar	–	1	–	–	–	1	–	–
Guianas and eastern Amazonia	–	11	1	1	–	–	3	–
Guianas and western or central Amazonia or widespread Amazonia	–	19+i	2	1	–	5	8+ii	1
Guianas, Amazon to S. Brazil	–	1+iii	1	–	–	–	2	–
Amazon and E. central Brazil	–	1	–	–	–	–	–	–
Amazon and Bolivia	–	1	1	–	–	–	–	–
Eastern and northeast Brazil	–	i	–	–	–	1	–	–
S. coastal U.S.A., Caribbean, Central America to S. Brazil	1	–	–	–	–	–	–	–
Central America, Cuba, Colombia and Venezuela	–	–	–	–	–	–	1	–
Amazon and northeastern Brazil	–	–	–	–	–	–	i	–

Arabic Numerals represent species
Roman Numerals represent subspecific taxa

Table II. Species of Licania *and* Couepia *with widespread distribution, Categories 1 and 2.*

Species	Distribution	Species	Distribution
Licania apetala	V,G,A,ECB	*Licania canescens*	G,EA,CA
L. egleri	EA,SA,ECB	*L. fritschii*	EA,SA,ECB
L. guianensis	G,EA,CA	*L. heteromorpha*	
L. heteromorpha		var. *glabra*	WA,EA
var. *heteromorpha*	V,G,A	,*L. hypoleuca*	V,G,A,ECB
L. impressa	CA,EA	*L. intrapetiolaris*	V,G,CA
L. kunthiana	G,A,ECB	*L. latifolia*	G,A
L. licaneaeflora	G,A	*L. longistyla*	G,A
L. micrantha	G,WA,CA,EA	*L. minutiflora*	G,CA,EA
L. octandra subsp.		*L. pallida*	V,CA,EA
octandra	V,G,A,ECB	*L. parviflora*	G,A
L. paraensis	EA,SA	*L. polita*	G,A
L. parvifructa	G,A	*L. sclerophylla*	EA,SA,ECB
L. rodriguesii	CA,EA	*Couepia bracteosa*	G,CA,SA,EA
L. silvae	CA,EA	*C. chrysocalyx*	A
Couepia canomensis	G,CA,WA	*C. leptostachya*	WA,CA,EA
C. glandulosa	G,WA,CA,EA	*C. robusta*	CA,EA
C. paraensis subsp.			
paraensis	V,WA,CA		

Key to distribution

A — Widespread within Amazonia   G — Guianas

CA — Central Amazon   SA — Southern Amazon

EA — Eastern Amazon   WA — Western Amazon

ECB — Eastern Central Brazil   V — Venezuela

Table III. Species of Licania *and* Couepia *which occur in the Guianas and Eastern Amazonia or Roraima Territory, Category 3.*

Guianas and Eastern Amazonia   Guianas and Northern Roraima Terr.

Guianas and Eastern Amazonia	Guianas and Northern Roraima Terr.
Licania affinis	*Licania discolor*
L. alba	*Couepia multiflora*
L. davillaefolia	
L. glabriflora	
L. incana	
L. leptostachya	
L. macrophylla	
L. membranacea	
L. pruinosa	
L. robusta	
Couepia caryophylloides	

Table IV. Species of Licania *and* Couepia *which occur in the Guianas and Central and/or Western Amazonia, Category 4.*

Licania arachnoidea
L. coriacea
L. elliptica
L. laxiflora
L. parvifolia
L. sprucei

Couepia habrantha
C. obovata
C. parillo

Table V. Species of Licania *and* Couepia *confined to Eastern Amazonia, Category 5.*

Licania amapaensis
L. blackii
L. maxima
L. piresii

Couepia cataractae
C. excelsa
C. froesii
C. paraensis subsp. paraensis
C. reflexa

Table VI. Species of Licania *and* Couepia *confined to Central Amazonia, Category 6.*

Licania apiculata
L. bellingtonii
L. bracteata
L. caudata
L. gracilipes
L. heteromorpha var.
  subcordata
L. hirsuta
L. hypoleuca var. foveolata
L. longipedicellata
L. maguirei
L. mollis
L. niloi
L. oblongifolia

Couepia elata
C. eriantha
C. krukovii
C. latifolia
C. longipendula
C. magnoliifolia
C. spicata
C. stipularis
C. subcordata

Table VII. Species of Licania *and* Couepia *confined to Western Amazonia, Category 7.*

Licania angustata
L. boliviensis
L. britteniana
L. crassivenia
L. emarginata
L. hebantha
L. klugii
L. krukovii
L. lata
L. longipetala
L. octandra subsp. pallida
L. reticulata

Licania savannarum
L. triandra
L. trigonioides
L. unguiculata
L. urceolaris
L. vaupesiana
L. wurdackii
Couepia macrophylla
C. racemosa
C. trapezioana
C. ulei
C. williamsii

Tables 1-7 give examples solely from the Chrysobalanaceae, but these distribution patterns occur repeatedly in each plant family studied. For example, Table VIII presents a phytogeographic breakdown of the Dichapetalaceae, a family containing lianas rather than the trees and shrubs of the Chrysobalanaceae. The 18 Amazonian species of Dichapetalaceae show all eight distribution patterns discussed above. Further distribution maps are given in Prance (1974) and are therefore not repeated here.

Table VIII. Phytogeographic breakdown of the distribution of Amazonian species of Dichapetalaceae. The numbers refer to the distribution patterns enumerated in the text.

1. Widespread

   Dichapetalum latifolium — Venezuela, Western Amazonia
   D. pedunculatum — Trinidad, Venezuela, Guianas, Amazonia
   D. rugosum — Venezuela, Guianas, Western, Central and Eastern Amazonia
   Tapura amazonica — Guianas, Western, Central, Eastern and Southern Amazonia, Eastern-Central Brazil
   T. capitulifera — Venezuela, Guianas, Amazonia
   T. guianensis — Guianas, Western Amazonia

2. Widespread within Amazonia

   Dichapetalum pauper

5. Eastern Amazonia

   Tapura singularis

6. Central Amazonia

   Dichapetalum coelhoi
   Tapura lanceolata

7. Western Amazonia

   Dichapetalum froesii   *Tapura coriacea*
   D. odoratum   *T. peruviana*
   D. spruceanum   *T. tessmannii*

8. Western and Southern Amazonia

   Tapura acreana
   T. juruana

In addition to the various continuous distributions discussed above, there are a number of disjunctions common in Amazonia; for example, *Couepia parillo* (Fig. 10) which is found in the Guianas and western Amazonia. These are best explained by separation which occurred during the drier periods of the post-Pleistocene.

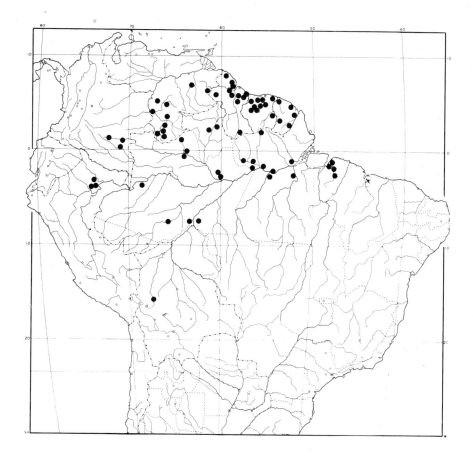

Fig. 6. Distribution map of Caryocar microcarpum *Ducke.*

Based on the plant distributions of the five families studied, Amazonia may be divided into the seven sectors shown in Fig. 12.

1. *Atlantic Coastal:* This sector begins in the west in the Orinoco delta, stretches through the northern half of the Guianas to the eastern limits of Amazonia in Maranhão, Brazil, and west to the Rio Xingu.

2. *Jari-Trombetas:* The sector is bordered by the Rio Jari to the east, the boundary of the States of Amazonas and Pará to the west, and the Rio Amazonas to the south. It is a region containing many hills, savannas, and savanna forests with a high degree of endemism. This region is also distinguished by Ducke and Black (1954) and Rizzini (1963).

3. *Xingu-Madeira:* The area bounded to the north by the Rios Solimões and Amazonas east of Tefé, to the south by the planalto of central Brazil, to the east by the Rio Xingu, and to the west by the basin of the Rio Purus south to Lábrea.

4. *Roraima-Manaus:* This region is subdivided in order to distinguish the large savanna in Roraima that extends into Guyana.

    a.   Roraima: The forest bordering the savannas is different from that farther south, though many species in the rain forest are related to the Manaus region.

    b.   Manaus: This is an area of high endemism which contains many species confined solely to it.

Fig. 7. Distribution map of Caryocar glabrum *(Aubl.) Pers.* ◆ C. glabrum *var.* glabrum; ◇ C. glabrum *var.* parviflorum *(A.C. Smith) Prance & Silva.*

    5. *Northwest-Upper Rio Negro:* This region consists of the basin of the Rio Negro west of Barcelos to the limits of Amazonia in Colombia and Venezuela.

    6. *Solimões-Amazonas west:* This sector consists of the basins of the Rios Solimões, Japurá and Putamayo.

    7. *Southwest:* This sector includes the Rio Amazonas of Peru west to the Andes, the upper reaches of the basins of the Rios Juruá and Purus, the State of Acre, and the Territory of Rondônia west to the Rio Roosevelt.

    Phytogeographical divisions cannot be based on the distribution of any single species, but on the comparison of the distributions of a large number of species and subspecies. The distribution of the species of a given genus does not depend on the region itself, but rather on such varied factors as ecology, dispersal, pollination, etc. Perhaps dispersal mechanisms are the most important factors of all. After defining the phytogeographic divisions of a region, one will find many transgressive species which cross the boundary from one area to another. In the African flora, where phytogeography and ecology are much better known, species which cross phytogeographical or ecological boundaries were termed, respectively, chorological (= phytogeographical) transgressors and ecological transgressors (see White 1976). Thus, while the areas defined here are based on distribution of different species, I have also presented distributions of some widespread species or chorological transgressors such as *Caryocar glabrum* (Aubl.) Pers.

    An important aspect in determining phytogeographic regions based on distribution patterns is the recognition of a phenomenon that occurs in most woody groups, including the five discussed in this paper: that is, the variation between species of any genus in

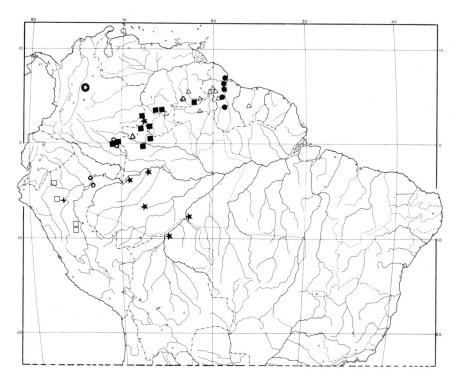

Fig. 8. Distribution map of species of Anthodiscus (Caryocaraceae).   +   A. klugii *Standl. ex Prance;*   ●   A. trifoliatus *G.W.F. Meyer;*   ○   A. pilosus *Ducke;*   ✪ A. montanus *Gleason;*   ■   A. obovatus *Benth. ex Wittm.;*   △   A. mazarunensis *Gilly;*   ★   A. amazonicus *Gleason & A.C. Smith;*   □   A. peruanus *Baill.*

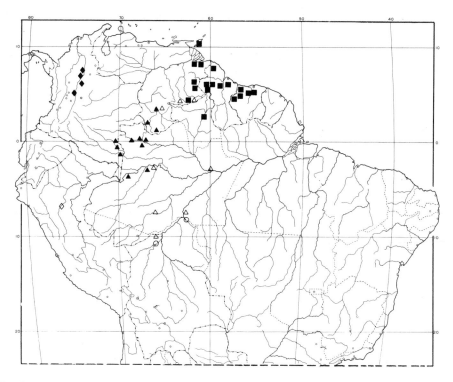

Fig. 9. Distribution map of species of Caryocar: ○   C. dentatum *Gleason;*   ■ C. nuciferum *L.;*   ▲   C. gracile *Wittm.;*   ◇ C. amydaliforme *G. Don;*   ◆ C. amygdaliferum *Mutis;*   △ C. pallidum *A.C. Smith.*

207

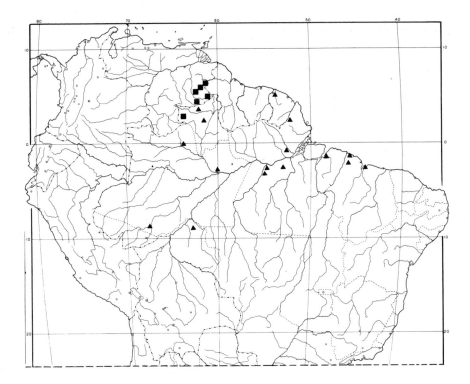

Fig. 10. Distribution map of species of Caryocar: ■ C. montanum *Prance;*
▲ C. villosum *(Aubl.) Pers.*

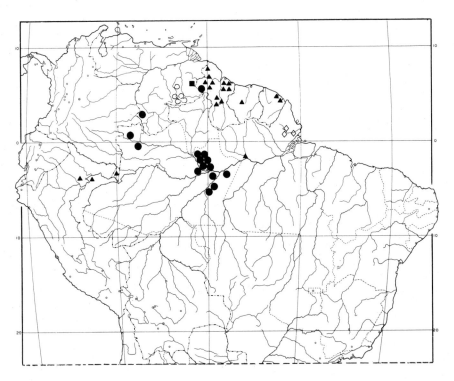

Fig. 11. Distribution map of species of Couepia *(Chrysobalanaceae):* ▲ C. parillo *DC.;*
○ *C.* foveolata *Prance;* ■ C. steyermarkii *Maguire;* ● C. canomensis *(Mart.)*
Benth. ex Hook.f.; ◇ C. excelsa *Ducke.*

208

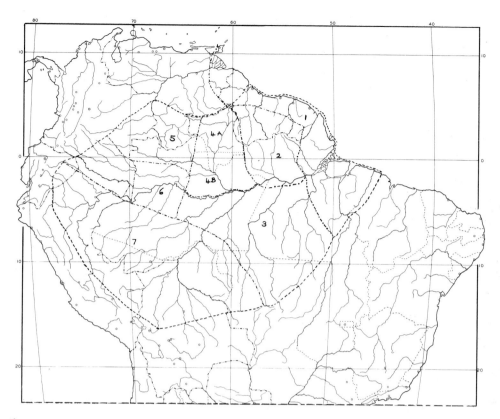

Fig. 12. Map of the seven major phytogeographic regions of Amazonia proposed in this paper: 1, Atlantic coastal; 2, Jari-Trombetas; 3, Xingu-Madeira; 4, Roraima-Manaus; 5, Northwest-upper Rio Negro; 6, Solimões-Amazonas west; 7, Southwest.

their frequency and breadth of distribution. It was this phenomenon which led Willis (1922) to propose his "age and area" theory which was soon refuted by many other botanists. This problem was discussed recently by Kubitzki (1975) who analyzed distributions and primitiveness in species of Hernandiaceae and Dilleniaceae. He found that species which possess primitive characters generally have a more restricted distribution. The widely distributed species are the most derived species. He concluded that progressive evolution tends to bring about an improvement in the ability for dispersal and seedling establishment. This is paralleled by the loss of primitive characters.

The ability for dispersal and seedling establishment must have been important factors in the redistribution of the Amazon forest after the dry periods. The distribution of areas of endemism helps to indicate the areas from which the present-day forest has spread, but dispersal mechanisms and seedling establishment cannot be ignored.

THE PHYTOGEOGRAPHICAL REGIONS AND CONSERVATION OF THE AMAZON FOREST

The purpose of defining and discussing phytogeographic sectors at a Symposium on Threatened and Endangered Species of Plants is to discuss them in terms of the conservation of the Amazonian ecosystem. Consequently, I have been more concerned in defining the present-day plant distributions than in considering the historic plant aspects of Amazonian phytogeography. A knowledge of climatic changes, geology, migration, and evolution is vital for the understanding of the Amazonian ecosystem; however, I have concentrated on a description of actual distribution patterns to show areas in need of protection.

If any one of the seven regions of forest shown on Fig. 12 is completely destroyed, a large number of species will definitely become extinct. Fig. 12 shows the seven phytogeographical areas defined here, and Fig. 4 shows the forest refuges proposed in Prance

(1974). These refuges are also the centers of high endemism. Both the refuges and the phytogeographical divisions must be taken into account in formulating conservation programs. Fig. 13 shows the areas called poles — the focus points for development — of *Projeto Polamazônia,* a government program designed to stimulate development of the

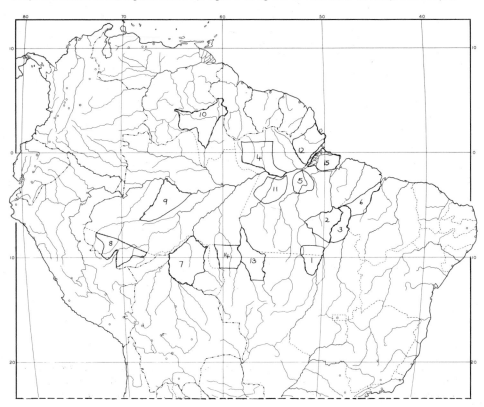

Fig. 13. Map showing areas of development under the Polamazônia Project. 1, Xingú-Araguaia; 2, Carajás; 3, Araguaia-Tocantins; 4, Trombetas; 5, Altamira; 6, Pre-Amazonia-Maranhão; 7, Rondônia; 8, Acre; 9, Juruá-Solimões; 10, Roraima; 1, Tapajóz; 12, Amapá; 13, Juruena; 14, Aripuanã; 15, Marajó. Officially approved by Decreto (law) 74607 of November 28, 1974.

Brazilian Amazon. It can be seen that some of the refuge areas fall entirely within these poles. For example, Poles 2, 3, and 6 — Carajás, Araguaia-Tocantins, and Pre-Amazônia-Maranhão — include most of the Belém refuge (Fig. 4, refuge 16); Poles 13 and 14 — Juruena and Aripuanã — cover the refuge Rondônia-Aripuanã (Fig. 4, refuge 15); Pole 9 — Juruá-Solimões — includes the refuges Olivença and Tefé (Fig. 4, refuges 11 and 12). In other words, a considerable number of Amazonian species of plants could be threatened by the development plan of Projeto Polamazônia, unless their reserves are selected carefully and early in the program.

Much has already been written about general threats to the Amazon ecosystem, and this is summarized in Goodland and Irwin (1975). The danger to individual species, however, is not so much from the highways as from the projects which cover large areas, such as some of the enormous cattle ranch operations which have been sponsored by various international companies.

I trust that the information given here about phytogeography will not be construed as criticism of the development of Amazonia, but rather as an indication of the most interesting areas in terms of the gene pool of future plant resources. The data are being offered with the hope that they will contribute to an enlightened program of setting up reserves. Though the Polamazônia project itself contains provisions for reservations, I hope that we can now draw attention to those areas most urgently in need of reserves.

There has been a tendency to place reserves in areas of low economic potential rather

210

than to consider the distribution of the species of plants to be conserved. It is essential to consider the phytogeographic regions and the centers of endemism for any conservation program to be effective. Botanists have a responsibility to point this out to legislators, rather than to allow the information to be concealed in learned taxonomic monographs, interesting as this data may be to one's fellow researchers!

From a botanical point of view, the prime need is to create reserves which will represent samples of each of the sectors of the Amazon forest, each of the suggested refuge areas, and adequate examples of all the other vegetation types. At present, the most crucial areas for action in Brazil are the Southwest sector (Fig. 12-7) and the Xingu-Madeira sector (Fig. 12-3). The greatest amount of destruction is now taking place in these areas because of their accessibility from the south of Brazil and also because these two regions are traversed by several of the recently constructed highways, i.e., the Trans-Amazon and the Cuiabá-Santarém highways. Large areas of the most interesting transition forest from the botanical viewpoint, between Amazonia and the cerrado of central Brazil, already have been destroyed. The recently-established Araguaia National Park, however, will help to conserve some of this area.

The need for a large number of reserves to preserve the rain forest adequately was emphasized by Hamilton (1976) in a study of the preservation of the Venezuela rain forest. Hamilton's report suggested 64 rain forest areas in need of conservation in Venezuela alone, of which 28 were in the lowland forest areas of that country. The basic selection of these areas was based on the botanical data of Julian A. Steyermark, a speaker at this Symposium. If a comparatively small area the size of Venezuela needs 64 reserves to conserve the forest diversity, then it would seem logical that an area the size of Brazilian Amazonia warrants a proportionately larger number of reserves.

Botanists and ecologists assembled at a symposium of this type face a second challenge. We will not save threatened and endangered species by concentrating all our efforts toward the creation of reserves. We must also use our knowledge to discover ways to use the tropical rain forest without either destroying it or interrupting its delicate nutrient cycle.

The majority of people attending this Symposium are North Americans. You may ask what you can do for Amazonia which, after all, is geographically really so far from here. In fact, much of the rape of Amazonia is due to foreign investments, rather than to Brazilian national programs. For example, two major international motor firms recently have diversified their interests by acquiring vast areas of Amazonia for cattle ranches. A recent survey of the government agency in charge of land distribution in only one part of Amazonia revealed that the available purchasing capital there was 30 percent U.S., 20 percent West German, 10 percent Japanese, 10 percent other foreigners, 20 percent São Paulo residents, and only 10 percent residents of Amazonian Brazil. This is a typical and accurate indication of what is happening today in Amazonia; foreign capital represents as much as 70 percent of the investment which is destroying much of the vegetation of the region.

Though a number of reserves have been proposed — on paper — for Brazilian Amazonia, the few which actually have been created by legislation lack adequate vigilance and control because of lack of funds. One of the main problems is the absence of coordination among the many organizations which have proposed reserves, often in competition or without knowledge of each other. An urgent need therefore exists for a single government entity endowed with the authority to control not only all reserves, but all reserve policy. At present, reserves are being proposed by such diverse groups as the Forestry Development Institute, the Environmental Secretariat, the Indian Protection Service, RADAM project, and various research institutes, to cite just a few. A recent survey of this situation was conducted by Dr. Gary B. Wetterberg (pers. comm.) under the sponsorship of the FAO of the United Nations. It is to be hoped that FAO recommendations in this regard will lead to a greater unity of purpose in the reserve policy of Brazil for Amazonia.

I have discussed the conservation situation primarily in Brazilian terms because there are other speakers at this Symposium representing other countries whose territory also includes part of Amazonia.

Much of the information given for Brazil applies equally to neighboring countries. Policies toward reserves, however, vary greatly from one country to another. For example, Venezuela is quite conservation-conscious and has already set up large national parks such as Canaima. Peru has established Manu, an equally important reserve of Amazon forest. The speaker from Surinam who was invited but unfortunately was unable to attend this Symposium (Dr. Jop Schulz, pers. comm.) has written that species in his country are not actually threatened or endangered because of the establishment of large forest reserves covering all major vegetation types in Surinam. This is heartening news indeed, in contrast to the threat to some of the most interesting areas of Brazilian Amazonia.

ACKNOWLEDGMENTS

Field work in Amazonia was carried out while I was B. A. Krukoff Curator of Amazonian Botany at the New York Botanical Garden, with the support of National Science Foundation Grants GB-32575X3 and BMS-75-03724, which are gratefully acknowledged. I am grateful to the director and staff of the Instituto Nacional de Pesquisas da Amazônia, Manaus, Brazil, for continuing collaboration; to Dr. Gary Wetterberg, FAO IBRA/71/545, Manaus, for helpful discussion of the forest reserve situation in Amazonia; to Dr. João Murça Pires of the Museu Goeldi Belém for continued help with my Amazonian field work; to Anne E. Prance for preparing the maps, and to Frances Maroncelli for typing the manuscript.

LITERATURE CITED

Bruxelles. 1969–. Distributiones Plantarum Africanarum Bruxelles, Jard. Bot. Nat. Belg.

Ducke, A., and G. Black. 1953. Phytogeographical notes on the Brazilian Amazon. Anais da Acad. Brasil. de Ciências **25**(1):1-46.

_____. 1954. Notas sobre a fitogeografia da Amazônia Brasileira. Bol. Técn. Inst. Agron. do Norte **29**:1-62.

Falesi, I. C. 1967. O estado atual dos conhecimentos sobre o solos da Amazônia Brasileira. Atas do Simpósio Sôbre a Biota Amazônica **1**: 151-168. Rio de Janeiro.

_____. 1971. Solos do distrito Agropecuário da SUFRAMA. Publ. IPEAAOC Solos **1**(1):1-99. Manaus.

_____. 1972. Solos da Rodovia Transamazônica. Inst. Pesq. Agropec. Norte (IPEAN). Belém.

Forero, E. 1976. A revision of the American species of *Rourea* subgenus *Rourea* (Connaraceae). Mem. N. Y. Bot. Gard. **26**(1): 1-119.

Goodland, R. J. A., and H. S. Irwin. 1975. Amazon jungle: Green hell to red desert? Elsevier. Amsterdam and New York.

Haffer, J. 1969. Speciation in Amazonian forest birds. Science **165**:131-137.

_____. 1974. Avian speciation in tropical South America. Cambridge, Mass.

Hamilton, L. S. 1976. Tropical rain forest use and preservation: A study of problems and practices in Venezuela. Sierra Club Special Publication, International Series No. 4.

Kubitzki, K. 1975. Relationships between distribution and evolution in some heterobathmic tropical groups. Bot. Jahrb. Syst. **96**:212-230.

Pires, J. M. 1973. Tipos de vegetação da Amazônia. Publ. Avulsas Museu Goeldi. Belém **20**:179-202.

Prance, G. T. 1974. Phytogeographic support for the theory of Pleistocene forest refuges in the Amazon Basin. Acta Amazonica **3**(3):5-28.

_____. 1975. Flora and vegetation. *In:* Goodland, R. J. A., and H. S. Irwin. Amazon jungle: Green hell to red desert? 101-111. Elsevier. Amsterdam and New York.

_____**, W. A. Rodrigues, and M. F. da Silva.** 1976. Inventário florestal de um hectare de mata da terra firme km 30 da estrada Manaus-Itacoatiara. Acta Amazonica **6**(1):9-35.

Projeto RADAM. 1973-1975. Levantamento de recursos naturais. Volumes 1-7. Ministério das Minas e Energia, Rio de Janeiro.

Rizzini, C. T. 1963. Nota prévia sôbre a divisão fitogeografica do Brasil. Rev. Bras. Geog. **25**(1):1-64.

Van der Hammen, T. 1972. Changes in vegetation and climate in the Amazon Basin and surrounding areas during the Pleistocene. Geologie en Mijnbouw **51**: 641-643.

Walter, H., E. Harnickell, and D. Mueller-Dombois. 1975. Klimadiagramm-Karten der einzeln Kontinente und de ökologische Klimagliederung der Erde. Vegetations monographien der einzelnen Grossaum 10. Gustav. Fischer Verlag, Stuttgart.

White, F. 1976. The taxonomy, ecology and chorology of African Chrysobalanaceae (excluding *Acioa*). Bull. Jard. Bot. Brux. **46**: 265-350.

Willis, J. C. 1922. Age and area. Cambridge.

AMAZONIAN FOREST AND CERRADO:
DEVELOPMENT AND ENVIRONMENTAL CONSERVATION

R. J. Goodland and H. S. Irwin

*The Cary Arboretum of the New York Botanical Garden,
Millbrook, New York 12545, U.S.A.*

INTRODUCTION: TROPICAL FORESTRY DEVELOPMENT

This book and the symposium from which it grew concerns endangered species of plants and their protection in the Americas. This paper concerns the most extensive and diverse ecosystem in the Americas — the Amazon forest. Diverging from the almost total taxonomic stance of the book, this chapter presents the ecological viewpoint that protection of individual rare species in Amazonia is either impossible, ineffective, or inappropriate. After reviewing development in tropical forests in general, this paper proposes that current practices are responsible for increasingly severe and possibly irreversible environmental destruction — including endangering species — in Amazonia, and that rational, sustained yield methods have not yet been fully elucidated. Until they are, this paper advocates deflecting agricultural colonization to the vast cerrado region to the South, thus alleviating pressure on Amazonia and buying time to determine less destructive courses of action. Proposals to achieve successful sustainable development of the cerrado based on ecological principles are included.

Tropical moist forests are being destroyed throughout the world on an unprecedented scale, primarily in the name of 'development'. All exploitation of tropical forests, therefore, must be considered together with the development that spurs it. In this paper, an extreme viewpoint is presented that no practice yet devised for the use of tropical wet forests is ecologically sustainable and economically justifiable in the long term. We question whether development is best served through prevailing tropical forest activities and we maintain that the present course is self-defeating, that clear-cutting of tropical wet forest is so perilous that it should be supplanted by more rational alternatives until sustained yield can be achieved.

Objectives and goals for development include desires to improve the lot of the poorer segment of society, to eliminate slums, to increase food production, to utilize a natural resource (forest), to augment exports, and to earn foreign exchange. We propose that clear-cutting tropical wet forest may assuage such desires for a brief period, but that ultimately it culminates in a worse situation than obtained before development.

Conversion of tropical wet forest to oligocultures of oil palm, rubber, cocoa, bananas, coconuts or, worse, to annuals or monocarpic crops such as rice, corn, millet, manihot, beans, or tobacco is an immediate realization of capital assets, a spending of capital for derisory returns lasting a few generations in the case of perennial tree crops, or only a few years in the case of annuals. Most conventional tropical agriculture fails the moment investment of fertilizer, pesticides, or other energy-intensive inputs cease.

A Statement of the Value of Tropical Wet Forest Ecosystem

Opposing groups differ in the value they attach to the tropical wet forest ecosystem. Developers assert that it is necessary to exploit the forest now to enable profit to be derived from it. The value of conserved forest is assumed to be great by the opposing faction, but this value is rarely, if ever, formally stated. The following is an attempt at stating this value.

The tropical wet forest ecosystem is not readily appreciated by owner-governments, largely because its value is esoteric, long-term, and not readily converted into profit. Further, the value is not restricted solely to the owner, and the benefits may be conferred on all humanity. Much of the value is like an insurance policy that may pay off handsomely in the future. Meanwhile the premiums — in this case, the pressures to cut — are onerous. Some of the value is assumed rather than readily proven. Part of the value is subjective while part will only become valuable if sociological trends continue as projected. Finally, part of the value reposes in the minds of people in other countries, and is not appreciated by the owners themselves. Although the capital value of the ecosystem is immense, the sustainable withdrawable interest is minimal.

Primal diversity embodies much of the value of the wet forest ecosystem. This ecosystem is biologically the richest on earth, a huge genetic pool of unknown dimensions which contains more interlinked organisms than any other ecosystem. As we have heard, many of these organisms have not even been named and inventoried; far less are their roles and unique properties known. Scientists reason that some of these organisms could suddenly become transcendentally important, and may provide a cure for such disorders as cancer or schistosomiasis perhaps, or a productive pest-free source of protein. Possibly the value is more remote but no less vital — an organism capable of controlling a future major world pest or a genetic aberrant.

Primal diversity also is of paramount importance in evolutionary terms: as the environment changes, new needs arise. Crops on which we are overdependent today may not thrive in future changes. Maintaining diversity preserves options for posterity, although action and investment now for future generations is not politically attractive. Once extinct, an organism with a specific spectrum of unique properties can never be recreated. As evidence that it is at last becoming widely recognized that dependence on a few crops is dangerous, crop diversification is healthily accelerating. This ecosystem is a cornucopia of genetic diversity as yet scarcely tapped. Within 50 years, oil-palm plantations have become one of the most productive sources of edible-industrial oil in the world. One hundred years ago there were no rubber plantations; now our dependence on them is profound and will increase commensurate with petroleum-based synthetic rubber prices!

Along with diversity and the genetic repository value of the tropical wet forest ecosystem, the other values are esthetic — a value indulged in largely by the replete segments of the population. The labyrinthine complexity and immemorial antiquity of this ecosystem contrasts with its fragility and the ease with which it is irrevocably shattered. *It must be taken as a cardinal responsibility that we do not destroy what we cannot recreate and do not yet comprehend.*[1] Increasing financial affluence of people forced to live and work in cities and deprived of the least contact with the natural world is raising the value of forest for such activities as recreation and tourism. This is a use of tropical wet forest potentially capable of turning a large and sustainable revenue for owner-nations if it is planned with foresight.

Possibly the greatest value of forest is the hidden service of environmental protection: enrichment and protection of soil, attenuation of climatic extremes, moderation of water flows, buffering of the atmosphere, purification of air and water, and suppression of large fluctuations of plant and animal populations. Although provided and maintained free to society, disruption of this service can be unimaginably expensive and damaging. Substitution of tiny components of this environmental protection service — such as flood and erosion control, agricultural fertilizers and pesticides, water supply, fuel, and subsidies for afflicted communities — consumes inordinate quantities of resources and human energy better applied elsewhere. This waste is entirely and cheaply avoidable by leaving protective forest intact. Cures of environmental ills, at best, are ineffective, expensive, or nonexistent. Thus prevention of environmental abuse, in the long run, is not only essential for future generations, but the most economical and wisest course to pursue. Yet as one sees the great scale of forested Amazonia, the question inevitably arises: must all of it be saved? How much loss is tolerable?

ENDANGERED SPECIES AND HABITAT IN AMAZONIA

The extinction of unknown species may exceed the extinction rate of rare but known species. In this volume, Moore mentions that he maintains a personal inventory of undescribed *genera* of palms. Considering that most palm genera are well-localized and that there is only a modest number — 200 or so — this fact reveals how little we know of rare taxa. Similarly, at the specific level, our ignorance reaches such depths that a sizeable proportion of a collection may consist of undescribed species. Prance,

[1] Author's italics.

215

for example, has made collections of plants which were subsequently discovered to consist of several percent new species (Prance, 1976, pers. comm.; Prance, 1975, *in* Goodland and Irwin, 1975). *Corythophora rimosa,* one of the commonest species of trees occurring in one of the most thoroughly collected tropical wet forests in the world — that around Manaus — was only described as recently as 1974. These three examples support the contention that extinction of unknown species may exceed that of known species. If this is so, then the preservation of rare and endangered species appears a futile exercise in tropical forests, or at least less appropriate than in temperate ecosystems.

The preservation of a particular rare species of plant, while often feasible in temperate ecosystems, may be improbable or exceptionally difficult in the case of tropical species, which appear to be more narrowly specific in requirements than are temperate species. Tropical pollinators may be oligolectic and dispersal mechanisms may be tightly dependent on specific situations, while mutualisms are commonplace and frequently obligate. Environmental or edaphic amplitude of tolerance is more circumscribed in the tropics. These circumstances mean that tropical species are more intimately linked to each other than they are to their temperate counterparts, so that preservation of a rare species in isolation is likely to be less successful or meaningful. Furthermore, the rhizosphere may be more specific and critical to the survival of individual tropical species than to temperate species.

Species differ in importance to society and there are both many more species and many more closely-related species in the tropics. This means that it may not be as worthwhile to preserve all the rare tropical species as it is to preserve the rare temperate species. Inventory of new tropical species contributes significantly to evolutionary knowledge and floral morphology, but this may not be adequate justification for the money spent without some concomitant evaluation of their potential uses, properties, or relevance in the ecosystem. Herbarium specimens, the major and largely the sole result of plant inventory can, in Devil's Advocacy, be likened to collections of bricks or building blocks although a collection of bricks is arguably more valuable. Such collections of components are interesting and revealing to botanists, but without some knowledge of where they fit in the structure or what their role is in the whole, their functional value is much less.

Nearly all tropical wet lowland forest will have been cut within one generation, predicts Dr. Warwick Kerr, Director of INPA, the National Amazonian Research Institute in Manaus. Similarly, the Agricultural Commission of the Federal Senate deposes that within 27 years ". . . there will exist no more trees (in Amazonia) . . ." should clearing continue (as reported in *Correio Brasiliense,* 5 June 1976, p. 9). This severe limit should serve as a warning that the brief time still available should be used as wisely as possible. In view of stringently limited resources of human capability and financial support, the question then arises: is the inventory of plants the most judicious allocation? The description of new species may not be the most effective deployment of finite resources.

The preservation of rare and endangered species themselves, therefore, is even now acknowledged to be less effective than the preservation of habitat. But the preservation of tropical habitats or ecosystems is more complicated and possibly less effective than habitat preservation would be in temperate regions. Since many tropical taxa occur within severely circumscribed ranges, any one tract will contain only a small proportion of threatened species. Decisions will have to be made to preserve a known area of high endemism, or of unusually great species diversity, or completely unknown tract. There is not yet adequate knowledge today to enable us to select judiciously such areas for preservation, although the best available selection is better than none. Furthermore, it is not known how large a tract must be in order for it to attain a measure of biological self-sustainability. The critical minimum number of individuals of a population necessary for self-perpetuation also is unknown. Fortunately, some of these problems are now being addressed (e.g. Terborgh, 1974, 1975) and establishment of UNESCO's biosphere reserves is encouraging.

The preservation of both habitat and of rare species is, therefore, expensive and unlikely to be totally effective. While such efforts should be encouraged, society's demands indicate that a more effective approach would be to avoid situations in which habitats will be destroyed. It appears inevitable that the wet forest ecosystem will be extirpated unless effective alternatives are provided. This paper proposes the case that the present widespread substitution of wet forest for cattle pasture is irrational and it goes on to suggest more rational alternatives.

At this late stage, there should no longer be any doubt that soaring human populations exert increasing pressure on the entire globe — its atmosphere, oceans, resources, infrastructure, and ecosystems — as well as on forest. Demographic trends absorb resources and social energy that could be otherwise devoted to qualitative improvement. Though this problem is acknowledged to be crucial to environmental conservation, it is outside the scope of this paper.

Amazonia is in an unusually fortunate situation in that its human population is not at present excessive. The environmentally rational course engenders societal circumstances in which couples will neither feel they need nor elect to create large families. Stable, reliable conditions and continuity of community organizations should be fostered which will promote small families. Under such conditions the natural feedback loops of self-regulation function well. Overdependence on foreign markets (e.g. banana export) and on foreign products (e.g. pesticides, machinery) and sudden transfusion of antibiotics into a society are examples of how feedback loops are removed.

CURRENT UTILIZATION OF TROPICAL WET FOREST

Conversions of tropical wet forest into an agricultural commodity maximizes short-term returns while at the same time it irrevocably precludes even the possibility of sustained yield in the future, since much of the attendant deterioration is irreversible. With a small human population and almost unlimited unpopulated regions, sustained yield was not so important. Now this era has ended and as we reach the limits to growth, we must learn how to make the best use of finite resources. The familiar litany of such degradation — reduced yields, depleted soil fertility, proliferation of weeds and pests, damaging erosion — is described elsewhere (e.g. Goodland and Irwin, 1975).

The reasons for such failure are summarized in Figure 1. The stupendous volumes of water boiled off by the evapo-transpiration of intact forest speedily leaches the soil as soon as the forest is cut. Most tropical soils are notoriously infertile, even under virgin forest. Warm tropical temperatures accelerate decomposition of organic matter and soil chemical reactions in general. Nutrients are rapidly and thoroughly leached by the warm and torrential rains — a process vastly accelerated when the enormous force of transpiration ceases. Where these processes have been operating through long geological time, the soils become senile and lixiviated. For many years it was thought, erroneously, that the unparalleled luxuriance of tropical forests indicated rich soil; biomass was equated with productivity. A more realistic point of view is the hyperbole, "tropical forest is a desert covered by trees!"

The introduction of a crop into such an environment aggravates the infertility. The crop demands the same nutrients, in the same proportion, and at the same time as such rapid nutrient depletion intensifies in the unprotected soil. As predators retreat in the face of felling, pests burgeon uncontrolled. The crop planted for human food is attacked by pests which need not wait for harvest. The plot is swiftly colonized by large numbers of weeds — plants much better adapted than the crop to the local environment or to stringent conditions. After one or two harvests, crop yields plummet as soil nutrients are depleted; competition with weeds intensifies and pests proliferate.

In the case of perennial tree crops, this demise is briefly delayed. The tree canopy affords some protection for the ground; weeds are retarded by shade and leaching is mitigated. The bulk of the tree is generally large in comparison with the harvestable portion and the tree is less palatable than is a succulent crop. In addition, latex is a natural protection against pests. Even so, without heavy fertilization, expensive pest control and energy-consuming attention, the yield gradually diminishes.

In sum, conversion of tropical forest does not yield sustained increases in food. Rubber and palm oil are used to earn foreign exchange, which then must largely be spent on necessary imports of pesticides, fertilizers, and foreign machinery for more forest clearance. In other words, forest is exported in the form of a cheap product such as cocoa or lumber, in exchange for expensive technological products needed to perpetuate the process itself and the economic society dependent upon it. Though some employment is provided, the prime beneficiary in this vicious cycle is the manufacturer.

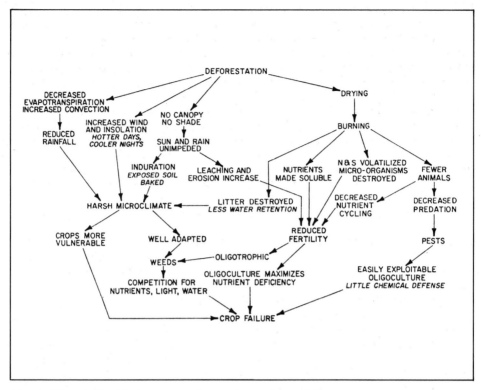

Fig. 1. The relationship between deforestation and crop failure.

Many plantations are not labor-intensive. The entire process of creating such plantations — from tropical forest conversion and clearance through harvesting of crops — usually is mechanized, with the result that fewer people are supported on the same area of land than could support themselves by subsistence agriculture. As settlement schemes and for slum clearance, these modern plantations also are less effective than subsistence plots. From a development point of view, it would therefore be preferable to export more finished products — cabinets and furniture instead of logs; soap, candles, and cooking oil rather than palm oil; tires instead of rubber; paper instead of wood chips. In this way, it is the labor of the people which becomes a renewable resource, added to the value of the product which can then be exported for greater profit to the owner-nation. Development should be more effectively aimed at enabling the people to export their work, that is, to transmute their labor and skills into worked products and commodities, and not merely to export crude, unprocessed raw materials.

Conversion of tropical wet forest into an agricultural commodity maximizes short-term returns while at the same time it irrevocably precludes even the possibility of future sustained yield, since much of the resultant deterioration is irreversible. The decision to extirpate the forest, therefore, tacitly assumes that the harvests of short-lived monocultural substitutes are worth more than the forest itself. Not only are the monoculture harvests limited in duration, but their benefits accrue largely away from the people directly involved in the development process. Part of the solution lies in appreciating the value of the forest.

Present problems stem basically from a difference of opinion between scientists and governments concerning the value of the tropical wet forest: the same development ethic that eliminated nearly all wild lands from temperate zones. Scientists, aware of the value of biological diversity and of the irrevocable nature of tropical wet forest destruction, consider that the forest is more valuable in the long-term than are a few decades of agricultural or cattle production. Governments, on the other hand, consider the forest to be a resource of relatively negligible value until it is exploited. Resettlement of slum dwellers and the sale of forest in the form of palm oil, rubber, and beef exported in return for hard currency, is viewed by controllers as a worthwhile use of forest.

The value of tropical wet forest has been outlined as far as possible; monetary value now should be assigned to it. This will assist governments in evaluating their development plans. Should the scientific and international community value certain tracts of tropical wet forest more than the nation to which the forest belongs, that nation should be assisted by its scientists and by the international community in showing that no loss of earnings will be incurred by such forest preservation (cf. Nicholls, 1973). This process will start with small areas of especial significance, but as tropical wet forest becomes increasingly rare, more and more tracts should be incorporated into this process. Since nations rightly cherish their sovereignty over their natural resources, strenuous invigoration of the idea will be necessary.

The Case that Sustained Yield has not been Attained

The crucial difference between tropical wet forest and temperate ecosystems is that when removed on a wide scale, the tropical wet forest does not regenerate to its original condition. This ecosystem must therefore be used only on a sustained yield basis and not until then. The hazard of irreversible changes requires that present activities should be channeled elsewhere until sustained yield can be achieved. Rarely is it admitted or realized that sustained yield has not been achieved by exogenous societies. Dr. Gerardo Budowski (this volume and 1976) writes: (our translation) ". . . the mixed heterogeneous forest, especially in the humid zones, cannot, under existing conditions of knowledge and markets, be maintained under sustained yield management."

The case that sustained yield has not been achieved rests upon the distinction drawn between resource capital and interest derivable from investment of the resource capital. Since sustained yield consumes only interest while all other exploitative procedures spend capital, the latter systems must eventually run down. Methods of utilization span the gamut between rapid capital squandering — such as cattle export — to selective logging on long cycles which approach sustained yield; net export of products (timber) being balanced by net input (of nutrients from the atmosphere and rivers).

The main methods employed today for the utilization of tropical wet forest are ranked (Table I) into our approximate order of environmental preference. The methods at the head of the list approach sustained yield, while those lower down lead to rapid depletion of the capital resource base. The ranking is predicated on the fact that the forest ecosystem is a closed mineral cycle, and that anything removed must be replaced if the cycle is to be sustained. Moreover, Amazonian wet forest occurs almost entirely on soils of low fertility so that the most damaging export is that of mineral nutrients, whether from

Table I. Environmental ranking of forest utilization

Examples of sustained yield at the top are environmentally preferred over nonrenewable examples at the bottom.

INTACT FOREST	Scientific Repository, National Parks, Environmental Protection, Tourism, Recreation, Collecting, Gathering, Tapping, Game Harvest, Indigenous Reserves.
UTILIZATION OF NATURAL FOREST	Shelterwood Forestry, Selective Felling, Refining, Liberation, Clear-cutting, Regeneration.
TREE PLANTATIONS	Mixed-species Products, Monoculture Products (e.g. rubber), Mixed Timber plus Product, Mixed Species Timber, Monoculture Timber.
AGRI-SILVICULTURE	Intercropping (e.g. rubber with annuals), Taungya (annuals to trees), Treed Pasture (products, browse, graze), Subsistence Gardens (trees, perennials, annuals).
AGRICULTURE	Subsistence Annuals (e.g. manihot, rice, and beans), Cash Crop Annuals (tobacco, sugar export), Grass Pasture (cattle export).

increased leaching, oxidation, or erosion, or whether in the form of eutrophic products such as cattle.

Tropical wet forest is characterized by a superabundance of water and sunlight — essential ingredients severely restricted in many parts of the world. Products derived in the presence of sunlight, largely from water and air — such as carbohydrate, lignin and cellulose — are oligotrophic; they contain few mineral nutrients, so that their removal from the cycle is readily balanced. Exported eutrophic products rich in inorganic nutrients, such as cattle and annual grains, deplete the already-low stock of minerals. This distinction between eutrophic and oligotrophic products is not commonly drawn, but it is critical in attaining sustained yield. Similarly, any activity accelerating the loss of nutrients, even though it is not exported as a product, is harmful. Thus, the growing of annual and other herbaceous crops leads to greater nutrient loss than does the growing of woody perennial and tree crops. Although manihot tubers and cane sugar are oligotrophic, their production accelerates ecosystemic disruption.

Any cash or export crop can be more harmful than a subsistence crop which is consumed locally, with its nutrients, ideally, being returned to the site. At present, exports are exchanged largely for products not replenishing their loss, such as gasoline and manufactures. In general, agriculture is less sustainable in the wet tropics than silviculture or tree plantations. A few methods are practiced which combine both agriculture and silviculture (Roche, 1973; Douglas and Hart, 1976). Examples of agri-silviculture include intercropping, such as tobacco and ground nuts in rubber (Trenbath, 1974; 1975), and taungya — the planting of tree-crop seedlings along with a food crop (Baur, 1968; King, 1968).

Tree plantations protect the soils and reduce erosion — but only to the extent the canopy remains closed. Plantations producing a crop without removal of the tree are thus environmentally preferable to that of timber or pulp trees. Fruit trees, rubber, palm oil, and cocoa, for example, protect the soils while producing a relatively oligotrophic crop. Clear-cut plantations are less protective, depending on the age to maturity or the length of the harvest cycle. A ten-year-old *Gmelina* pulp harvest may thus be less desirable than a thirty-year teak plantation. Little is known of the relative protection afforded by canopies of different species, nor what nutrients are extracted, and in what proportions by different trees. Even so, most timber is oligotrophic, and if bark, leaves, and branches are allowed to remain in the forest, nutrient depletion will be slight. In both pulp and timber plantations and in forest logging, extraction mechanics disturb the ecosystem far more than does loss of nutrients contained in the timber itself.

The use of tropical wet forest for the production of timber is environmentally preferable to that of agriculture. Different logging techniques vary greatly in the amount of environmental damage they cause, but the degree of damage is not easy to rank. It appears that selective felling causes less environmental damage than clear-cutting. Even though many lesser trees are removed for access or extraction paths, or are damaged by the fall of the selected specimens, selective logging almost always causes less damage than clear-cutting. Removal of selected trees by balloon or blimp would minimize damage to both trees and to the environment (Stark, 1976, pers. comm.).

Clear-cutting, too, varies greatly in the degree of destruction to the environment. The use of a chainsaw wreaks much less damage than felling by bulldozer, for example. Similarly, the impact of clear-cutting can be mitigated by the rapid planting of tree seedlings together with a rapid and vigorous cover crop of vines and legumes (e.g. Watson et al., 1964). The ground is protected if stumps and unexploitable timber are left in place with seedlings planted between them. The bulldozing of stumps and unexploitable trunks into windrows which are allowed to dry and then burned increases environmental damage. Chipping of branches, trash and unexploitable trees in place is preferable to bulldozed and burned windrows. The ecological basis of tropical wet forest silviculture is outlined by Whitmore (1975), who is experienced in Malaysian forestry practices, which are among the most sophisticated in the tropics (Wyatt-Smith, 1963; Baur, 1968).

Gradual modification of the species composition of natural forest by 'refining' or 'liberation' is environmentally judicious (Dawkins, 1955, 1963; Wong, 1966; Troup, 1952; cf. Meijer, 1973). Such methods encourage desired species by reducing competition from less desirable species by means of poisoning, girdling, or removal, thus maintaining the protective canopy largely intact. Conventional so-called 'sustained yield forestry' is in essence a hopeful euphemism for selective logging on a cycle longer than has been customary. The Shelterwood system formerly practiced in Nigeria for 60 years may indeed be sustainable (Janzen, 1976, pers. comm.).

Least damaging of all forest exploitation is the use of forest products without felling — fruits, nuts, seeds, oils, latex, resins, loose bark, flowers, leaves, branches, and dead material. As practiced by the Amerindians today, such hunting and gathering is truly sustainable, but supports only a small human population (Gross, 1975). Part of the sustained yield approach may include harvesting without perforating the canopy for leaf-protein (Pirie, 1971) or green-crop fractionation (Bray, 1976), for example. Both firewood and charcoal can also be permanent and renewable products (Earl, 1975). The production of bio-gas (methane), methanol or ethanol, and possibly the hydrogenation of wood, can all be achieved on a small scale with little technology. The product — transportable, oligotrophic and fairly valuable — also leaves a nutrient-rich sludge suitable for fertilizing crops. Also of a permanently sustained yield nature are recreation and tourism, two activities which have become surprisingly lucrative in a number of countries (Budowski, 1976). As has been mentioned, intact forest as a repository of scientific material and in environmental protection has great, though at present largely unappreciated, value.

Although not yet widely appreciated, the preference of forest over silviculture over agriculture over cattle export is straightforward to environmentalists but not profitable quickly enough to controllers. This preference is reinforced by being intimately linked with human ecology. The environmental preference closely parallels an equally important sociobiological preference. The way in which agro-ecosystems are utilized directly influences both rural society and rural economy. For example, regions in which ecosystems are manipulated to produce bananas for distant consumption share certain characteristics. Because banana plantations are possibly the most widespread substitute for the tropical wet forest that this section of the book proposes to preserve, the situation merits examination.

A significant characteristic of world agricultural commodity trade is the poverty of tropical exporting nations, particularly those in the banana trade. Several countries rely on bananas for as high as 54 percent of their export earnings, as for example, Ecuador, 48 percent; Panama, 54 percent; Honduras, 44 percent; Costa Rica, 29 percent. Though the importers are the world's richest nations, consumers in those nations are paying less for bananas. Their price has fallen 30 percent in the last 20 years, while the price of the products bought by banana producers soars annually, e.g. fertilizer, pesticide, machinery, and fuel. The banana trade is lucrative — for the transporters, marketers and distributors who reap 80 percent of the price, while the producer receives only 10 percent or so. When bananas leave the country of origin (and often before), prices are out of their hands, just as with most other tropical crops — sugar, cocoa, coffee, tea, cotton, vegetable oils, and rubber, for example. Conversion of tropical wet forest into an export crop can become debasement: a tractor 'costing' three tons of bananas in 1960, 'cost' eleven tons in 1970. In 1960 Malaysia earned enough money by exporting 25 tons of natural rubber to buy six tractors. Today, the same 25 tons of rubber buy only two tractors (CIDA, 1976). To the extent that rubber and bananas represent cut and transmuted forest, such export is a powerful and, of necessity, increasing force accelerating the removal of forest.

The environmental implications of such imbalanced trade are severe. Tropical export monocultures (or, at least, oligocultures) can maintain high profits either with large and ever-increasing expenditures for fertilizers, pesticides, machinery, and fuel, or by extirpation of further areas of forest for plantations. The first choice shackles the region to exchanging a low-priced commodity for an expensive one, while the second choice essentially exports forest wealth for only a few years and culminates in enormous environmental problems and loss of the resource base. Tropical export prices are determined by the purchaser and are unrelated to their production cost and even less so to their environmental cost (Janzen, 1973).

Interim Alternatives to Extirpation

Agro-ecosystem methods can be used as a deliberate instrument of social policy. At one extreme, government-sponsored agricultural colonization schemes resettle the impoverished with the tacit but false assumption that farming is an occupation for which adequate skills and training are innate. Successful small-holding in tropical wet forest is at best difficult for a rich tropical agronomist or a person with centuries of village tradition to follow, but far more so for a relocated slum-dweller, urbanite, or drought victim. At the other extreme, agro-ecosystems can be manipulated to slow the flight from the land to the towns, to reduce vulnerability of the region to world trade vagaries and to domestic failures, to increase independence, and to promote self-reliance. It is ironic that in Brazil, for example, the third largest importer of US grain, the percentage of economically active population in agriculture slipped from 57.5 percent in 1950 to 44.3 percent in 1970 (Turner, 1976).

Research to pinpoint site-specific, sustained yield, agro-ecosystems is possibly the most effective means of achieving the goal of this book — the preservation of biotic diversity. Such research will take time, and until it yields results, the forest will continue to be destroyed unless suitable alternatives are available and effectively communicated. The alternatives here proposed (Table II) are admittedly inadequate palliatives for the long-term, but they are preferable to the present course and will at least grant a grace period for necessary research. The first five alternatives suggest utilization of tracts not bearing on the vulnerable forest, while the sixth is more specific to Brazil since it suggests using the enormous, robust, and contiguous cerrado. This proposal is amplified elsewhere (Goodland, 1975a; Goodland and Irwin, 1975).

Table II. Interim alternatives to clearing more virgin forest

Region	Utilization
Várzea	Seasonal Enrichment Flooding
Water Bodies	Aquatic Animals, Fish, Vegetation
Fertile Soil Pockets	Agriculture, Perennials
Non-Forested Tracts	Savanna, Campo, Grassland
Second-Growth Areas	Intensive Tree Plantations
Cerrado	Appropriate Technology

Várzea, the seasonally-flooded land adjacent to rivers, occupies as much as 2 percent of the total area of Amazonia. Seasonal floods enrich the várzea soil with fertile silt, while at the same time reducing the pest and weed menace. Such areas should be exploited preferentially for both subsistence and other annual crops. Water bodies such as the rivers themselves, as well as lakes, backwaters, oxbows, and swamps, are all amenable to increased exploitation for fish (Lowe-McConnell, 1975) and other aquatic, riparian (ducks, geese), and domestic wildlife. Large floating mats of vegetation, matupá (Goodland, 1975b), and vegetation floating at the water's edge all can be harvested without damaging the environment, and represent a potential resource of no mean value.

Little more than 10 percent of Amazonian soils have been surveyed even at a most cursory stage, according to Falesi (1974), but each survey reveals small, isolated pockets of more fertile soil within the widespread infertile latosols. Conventional agriculture can be practiced on these tracts of richer soil with greater chance of success. Similarly occurring as discontinuous islands within the forest are patches of non-forest vegetation — grassland, savanna, and open woodland. Although cattle are unlikely to thrive in such sites, they can be raised there with less environmental damage than would occur in pasture created from forest. Non-forest patches can be 'improved' by promoting nitrogen-fixing legumes and by encouraging browse, as detailed later on. Use of such areas reduces pressure on the forest itself.

Extensive areas of Amazonian forest have already been obliterated. It is more appropriate to manage these areas as tree plantations than to cut still more forest. Successful commercial tree plantations supplying raw material for various forest industries (pulp, paper, compressed woods, boards, charcoal, timber) will reduce the pressure to cut down more forest, while creating a livelihood for a sizeable population. "The fact that a hundred hectares of well-managed plantation can sometimes provide more utilizable wood products than thousands of hectares of natural forests, considering the necessities of the local populations, should not be overlooked" (Budowski, 1974). The world averages about one hectare of forest land per person and Latin America has more than five hectares of forest per person (King, 1975). Yet the region imports more timber products, in terms of value, than it exports. Amazonia has a powerful opportunity to export its abundant solar energy and water in the form of oligotrophic timber products.

ADVANTAGES OF CERRADO OVER AMAZONIAN DEVELOPMENT

Nearly all of central Brazil is clothed with cerrado, a savanna-like vegetation occupying 1.6 million km^2 — more than the total area of land now being cropped in Brazil — and larger than the total harvested cropland in the United States (Figures 2 and 3). Cerrado represents one of the world's last and largest vacant lands suitable for agricultural development. Existing knowledge already is adequate to establish successful sustained yield agriculture which could support more people with less cost than the exploitation of Amazonia. Judicious agricultural colonization of cerrado is suggested as a means to alleviate pressure to develop Amazonia, thus buying time to learn sustained yield practices there.

The most important reason that cerrado development is so greatly preferable to Amazonian development is that the cerrado environment is relatively robust and resilient, whereas damage to fragile Amazonian environments can be irreparable. The cerrado biota is depauperate when compared with that of Amazonia. Rare and endangered species

Fig. 2. Brazil: Cerrado Vegetation. Map showing main region of cerrado (after IPEA, 1973).

223

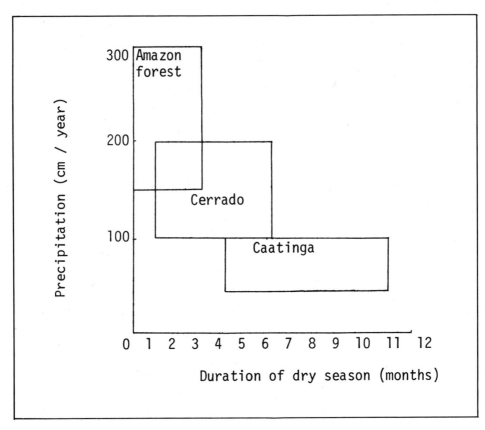

Fig. 3. Simplified diagram showing the effect of rainfall and seasonality on Brazilian vegetation.

certainly occur in the cerrado and new ones still are being discovered by thorough collectors (e.g. Irwin and Arroyo, 1972, 1973, 1974), but the pockets of endemism are relatively well-known. In any case, the extent of the unknown is considerably less than in Amazonia. Substitution of the natural vegetation of the cerrado does not lead to widespread and damaging erosion since the canopy is so sparse in any case and much of the topography is gently undulating or flat. While not generally fertile in the agricultural sense, cerrado soils are more fertile than Amazonian soils and more amenable to agriculture. The profile structure and drainage characteristics of cerrado soils is, in general, excellent. There are few extensive hardpans, indurated horizons, water-logged zones, or soils susceptible to laterization. As a substrate, therefore, cerrado soils are propitious for crops.

Table III. Distribution of the central continuous cerrado area

State	Cerrado Area (km² approx.)	Total Area of State (km²)
Mato Grosso	436,000	1,231,459
Goiás	381,750	611,302
Minas Gerais	217,500	589,725
Bahia	72,750	561,026
Maranhão	c. 15,000	328,663
(São Paulo)	(c. 11,000)	(247,898)

(Note: Cerrado in São Paulo is severely fragmented by agriculture and is not now continuous.)

The second major advantage of cerrado over Amazonia is the alternation of a rigorous dry season with a predictable and uniform wet season in Central Brazil. The diagram of Figure 3 shows the relationship of precipitation to the duration of the dry season in Brazil. Cerrado is endowed with three to six months of almost no rain, but with about one-and-a-half meters of annual rainfall, occurring mainly during the wet season. Both temperatures and humidity are almost ideal for plant growth and human enjoyment (Table IV). The advantages conferred by a dry season are not yet generally appreciated by agronomists. A dry season is an effective pest control feature which is especially valuable in the tropics, where insect pests can breed a tolerance to pesticides faster than the industrial chemist can invent and deploy new ones (Janzen, 1973). Furthermore, a dry season pulses agricultural activity, reduces vegetative growth and provides a period for ripening of grains, etc. The dry season must be appreciated as a boon and should not be irrigated into oblivion. Some irrigation may be appropriate to shorten the dry season in certain cases, but the drought must be maintained with sufficient stringency to prevent pest proliferation. Much of the cerrado is reasonably well-watered by small creeks. The water table is frequently deep in cerrado (about 10-30 m) but appears plentiful and most appropriate for wind-powered wells where necessary.

Table IV. Cerrado climate (annual means: 1931-1960, after IPEA, 1973)

	Units	Minas Gerais Cerrado Stations	Goiás Cerrado Stations	Mato Grosso Cerrado Stations
Precipitation	mm	1459.0	1569.0	1428.0
Humidity	percent	73.9	71.7	75.0
Temperature	°C	21.4	22.4	23.6

Assuming that the cerrado offers abundant light, equable temperatures, adequate water and a superior physical substrate, the only remaining factor for plant growth is mineral supply. Although cerrado soils are infertile, acidic, and contain toxic levels of aluminum (Goodland, 1971a, 1971b; Lopes, 1975), they are more successful for agricultural development than practically all Amazonian soils. Most fortunately, the cerrado contains large and accessible outcrops of limestone — sixteen sources are mapped in Figure 4. Some of this is dolomitic, so that with insignificant environmental impact and little capital expenditure, the acidity and most of the aluminum toxicity can be neutralized cheaply. Limestone outcrops bearing rare or endangered species, or those which are rich in archaeological artifacts, need not be exploited. At least three of these cerrado outcrops contain enormous tonnages of the even more useful phosphatic rock, (Abreu, 1973) which is appropriate for fertilizer. Araxá (MG) contains 90 million tons of 15 percent P_2O_5 rock, Catalão (GO) contains 80 million tons of 10 percent rock, while Patos de Minas (MG) contains 256 million tons of 13 percent P_2O_5 rock. Potassium is scarce in cerrado soils and the main sources of potash mineral are distant, e.g. Carmópolis in Sergipe, so that Brazil annually imports nearly one million tons of potash fertilizer.

The final major plant nutrient — nitrogen — also sorely deficient in cerrado soils, may not impede agricultural development because of two interesting features. Brazil leads the world in the fascinating discovery of nitrogen-fixation in grasses and on a scale large enough to be economically valuable. Dobereiner et al. (1973) and Day et al. (1975) have demonstrated rhizosphere nitrogen fixation principally by *Azotobacter* in eight genera of abundant tropical grasses. It appears likely that this extraordinary potential can successfully be selected for. The second feature concerning cerrado nitrogen is that cerrado biomass probably contains more legumes than any other family and the flora contains more genera of legumes than any other family and possibly more species. Among this enormous group of plants, many will be found to fix nitrogen. This group should be identified, improved by selection and widely planted — in the pasture sward, in rotational fallows, as shade, hedge, fence-post, browse, and fire-land trees. Already, the common indigenous cerrado legume Townsville lucerne *(Stylosanthes gracilis)* is mixed with pasture grasses. Such emphases will go far in redressing the low levels of nitrogen in cerrado soils.

The final advantage of cerrado is that at the moment it is underpopulated, so there is room for a much larger population in the 1.6 million km^2 area. In spite of being almost empty, it is more accessible than Amazonia, and boasts a more advantageous

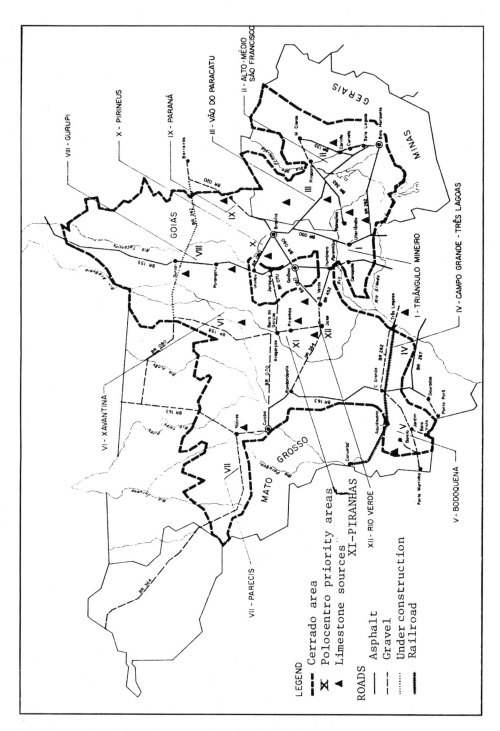

Fig. 4. Cerrado infrastructure: Map showing the chief towns and roads in the cerrado region (from Polocentro [Ministério de Agricultura, 1975]).

infrastructure of roads, communication networks, and market towns (Figure 4 and Table V). Cerrado is nearer to the regions containing people who are emigrating — principally from the drought-stricken Northeast. Although the cerrado can absorb more people, the optimal population size should be calculated so that demographic balances can function to maintain this optimal size. With no such planning, even the cerrado could rapidly deteriorate to supporting quantitatively an excessive population at qualitatively the lowest denominator of livelihood. The cerrado is one of the few remaining places in the world where the opportunity still exists to create a decent life for an optimally-sized population.

Table V. Main "cerrado towns" (c. 1974)

Town	State	Population
(Belo Horizonte)	Minas Gerais	(500,000)
Brasília	Distrito Federal	600,000
Goiânia	Goiás	500,000
Campo Grande	Mato Grosso	150,000
Uberlândia	Minas Gerais	125,000
Uberaba	Minas Gerais	125,000
Montes Claros	Minas Gerais	120,000
Anápolis	Goiás	105,000
(Cuiabá)	Mato Grosso	(101,000)
Patos de Minas	Minas Gerais	80,000
(Imperatriz)	Maranhão	(80,000)
Sete Lagoas	Minas Gerais	67,000
Ituiutaba	Minas Gerais	65,000
Itumbiara	Goiás	65,000
Araguari	Minas Gerais	64,000
Rio Verde	Goiás	60,000

Note: This list contains most of those towns over 50,000 in size that occur in or depend mainly on cerrado. Belo Horizonte is situated outside the cerrado, but about 50 percent of its sustenance depends on the cerrado. Cuiabá and Imperatriz depend largely on forested areas. The population figures given here are rough estimates for each town, together with its surrounding urban area. It should be remembered that a town of about 100,000 in interior Brazil is very much smaller in all respects except population when compared with a town of 100,000 in Central Europe or Central U.S.A.

THE ECOLOGICAL DEVELOPMENT OF CERRADO

Current trends and plans (Ministério de Agricultura, 1975; Conselho de Desenvolvimento Econômico, 1975; PBDCT, 1976; Goodland, 1975a) emphasize increased use of fertilizers, irrigation, and mechanization for the cerrado to intensify production in 1.8 million ha of priority annual crops and on 1.2 million ha for beef cattle. In other words, largely-imported fuel (or its derivatives) will be substituted for labor, land, and rain, which already are in plentiful supply. This will result in a capital-intensive, agro-industrial system employing far fewer people than small-scale 'appropriate-technology' systems. Considering that Brazil wishes to develop cerrado to accommodate people from drought, overpopulated, or disaster areas, and since Brazil imports 83 percent of its annual petroleum consumption, and with fertilizer, pesticide, and machinery costs strongly influenced by petroleum costs, it seems rational to prefer labor-intensive plans which reduce overdependence on petroleum. Low soil fertility will be worsened by emphasis on annual crops over perennial, and by the export of beef.

Since this essay advocates the development of cerrado in order to alleviate pressure for development of the Amazon forest, it is appropriate here to outline methods to ensure the success of cerrado development. Successful long-term development of the cerrado should be based upon five ecological tenets: carrying capacity, closed ecosystem, sustained yield, diversity, and small-scale. The wisest and most rational plans for tropical ecological development are those devised by Janzen (1973, 1975).

Carrying capacity is the size of the population sustainable indefinitely at a given mean standard of living on a finite area. The sustainable agricultural productivity of the finite area is a constant and should be determined. Once known, the social decision has to be taken: how shall this constant production be distributed? If this decision is not consciously taken, the constant production tends to be divided among ever-increasing numbers of people with concomitantly ever-decreasing standards of living. Eventually, productivity declines from its sustainable state, soils are mined, and the environment deteriorates. The other extreme is politically unlikely in the foreseeable future: sharing the constant productivity among the smallest necessary population at the highest possible mean standard of living. It is hoped that any compromise between these two extremes emphasizes quality over quantity.

Closed ecosystems obtain their raw materials locally and consume their production locally. Those few raw materials deemed essential and not substitutable can be imported only when balanced by the excess of production over consumption. Ideally, humanly-conceived closed ecosystems first arrange to avoid the need for big imports. Nitrogen-fixation by legumes avoids the need for importation of nitrogenous fertilizer. Maintenance of woodlots obviates the need to import domestic cooking and heating fuels, for example. The approach to closed ecosystems and its corollary, sustained yield, creates unexpectedly valuable benefits. Self-reliance is engendered, as is a desire for a small family.

Sustained yield (Janzen, 1973), closely related to and entirely dependent upon the closed ecosystem, is production set at the level at which it can best be maintained in perpetuity, i.e., that yield sustainable indefinitely without the system running down. Sustained yield (SY) is significantly lower than maximum production because SY takes into account drought years and other unfavorable climatic vagaries, fallow periods and rotations, and the unavoidable time lag for long-term harvests such as lumber, fruit trees, and slow-maturing animals. SY will provide the needs of the population first, with less emphasis on producing exports. Autonomy in food, fuel, shelter, and clothing creates independence in the society. A stringent distinction should be maintained between export of production remaining after local consumption has been satisfied and export created to the detriment of sustainability. It is all too easy to produce enormous exports (and profits) from a region for a decade perhaps, but not indefinitely. Witness the coffee cycle of Brazil, which left vast tracts of depleted and eroded soils abandoned in its wake.

A closed-cycle ecosystem and sustained yield can be achieved only by combining temporal and taxonomic diversity. Temporal diversity increases the number of harvest seasons in a given time period by staggered plantings, the initial introduction of plants at diverse ages, and the use of varieties of crop plants. The results are several ephemeral crops per year, and temporally diverse yields from annuals, such as beans and grains, to perennials, such as oil and fruit trees; and to longer-term silviculture. Even the tree crops should be temporally diverse; they should be as uneven-aged as possible. Taxonomic diversity is the ecological means of reducing storage costs and pest damage, maintaining soil fertility, and mitigating climatically-caused crop losses (e.g., the 1975 frost on Brazilian coffee) while improving diet. The widest possible array of crops (polyculture) is strongly preferred over the monoculture or even the oligoculture which is practiced currently. The stronger the perennial or tree component of the mix, the more resilient the system, with less erosion, increased fertility, and lowered susceptibility to drought. Related to taxonomic diversity is the fundamental guide, "adapt the crop to the environment, rather than the environment to the crop."

Taxonomic diversity within one crop creates insurance against severe losses. Some strains will resist this patch of different soil or that pest, or this season's quirks better than will other strains. Selection of strains should be practiced by the people all the time. In this way, all environmental variables will be accommodated. Too often, strains are selected away from the long-term local environment, hence becoming vulnerable to severe losses during environmental extremes. Current practice stresses the converse, namely adapting the environment by providing irrigation, fertilizer, mechanical weeding (or herbicides), pesticides, and other industrial products. This approach destroys sustained yield and breaks the closed-cycle ecosystem, while shackling the region to the thrall of alien industrialists.

Fire is widespread in cerrado and is an influential, fairly frequent component of the ecosystem. In our opinion, the few advantages of uncontrolled burning are vastly outweighed by the disadvantages, particularly the loss of volatilized nitrogen and sulfur, and of ash, organic matter, soil moisture, and protective soil cover. Fire can be greatly reduced in area and damage by living firebreaks such as double lines of fairly closely-grown mango trees. The canopy shades out the ground cover to the point at which fire will not cross. Living firebreaks, browse phreatophytes, living fenceposts or hedge trees and windbreaks should be planned as an important feature of the agro-ecosystem.

Taxonomic diversity includes physiognomic as well as specific diversity. Thus, pastures will be rich mixtures of grasses with many legumes and forbs. Browse trees (Douglas and Hart, 1976) — especially leguminous, phreatophytic, and oil or fruit producers — will be encouraged, as will shrubs and vines. Beef cattle will not be widespread since as much as ten hectares of pasture are needed to produce a million calories of beef, but only one-third of a hectare can produce a million calories from sugar, for example. No beef will be grain-fed since it takes a lavish eight kilos of grain to produce one kilo of beef. Not only is beef inefficient in calorie production, but it is also wasteful in protein production. Beef from forage produces 49 kg of protein per acre, whereas lucerne produces 675 kg of true protein, and soybeans 260 kg from the same area (Bray, 1976). Furthermore, since beef is eutrophic, its export unbalances the ecosystem cycle far more than export of oligotrophic sugar does.

Diversified communities will include animals as well as the plants just mentioned, but the current emphasis on large-scale cattle ranching will be redirected. "O boi não povoa" (cattle don't populate) but consume 2 1/2 times as much food as they produce. The raising of animals to thrive on unused materials or on materials which would otherwise be wasted, will close the ecosystem cycle still further. Rabbits, cavies and agoutis feeding on household wastes, pigeons and ducks gleaning from the fields and fishponds, and honeybees working the neighborhood, will all boost sustained yield without degrading the environment.

Cattle raising is far less damaging in cerrado than in pasture that replaces Amazonian forest, and the rare ability of ruminants to convert cellulose to protein is valuable in zones unsuitable for crops. Even so, since it supports very few people ("Onde o boi entra, o homen sai": "where cattle enter, man leaves") and for the reasons already detailed, cattle ranching is not appropriate in most sustained yield closed ecosystem regions. Ranching can be made less destructive by increasing the diversity of the sward, particularly with legumes, and by encouraging browsing as much as possible. Otherwise the cattle lose much weight during the dry season. The use of browsing mammals, such as the eland, is less destructive than conventional beef cattle. Nutrient loss can be usefully diminished by slaughtering and dressing the beef in the cerrado, exporting only the meat. The slaughterhouse should recycle the bones and all other residues in such forms as fertilizers, hides, and glue, for example.

The fifth and final ecological tenet for achieving long-term successful development of the cerrado is the promotion of small-scale over large, villages over cities, and gentle-, appropriate-, vernacular-, or soft-technology over alien and complex imported industry (Lovins, 1975; MacKillop, 1975; Williams, 1975; Village Technology Handbook, 1973; The Futurist, 1974). The closed-cycle ecosystem is successful where there is no waste. Normally-wasted crop residues are retained within the cycle as fertilizer or mulch from composting or as ashes from burning for fuel. Animal and crop residues should be digested to produce a fertilizing sludge and biogas for fuel (Penagos, 1967; Makhijani, 1975). Leachate, treated sewage effluent, and domestic wastes can be used to fertilize fishponds, for example. These methods are more suited for use on the village scale rather than to individual families. In petroleum-deficient Brazil and especially in nutrient-deficient cerrado, the abundant and free sunlight and water should also be taken advantage of in the form of biomass and its derivatives: alcohol, biogas, and firewood for fuel, with compost and sludge for fertilizer. Solar energy, particularly for domestic cooking and agricultural crop drying, is appropriate for central Brazil. Wind power harnessed for crop processing (e.g. dehusking, milling), water pumping, irrigation where appropriate, and possibly some electrical generation all will decrease the need for fuel imports and will close the ecosystem cycle.

Although in most ways the proposed development is simpler and cheaper than the current agro-industrial plants, it still will not be easy. Education and experimental plots directed toward sustained-yield, closed-cycles, and conservation, will be most important. There is no single panacea, for the carrying capacity of each region will be different and the best plans will be strictly site-specific, although the entire cerrado should be zoned with land-use regulations from the start. The initial period will be especially trying until yield, recycling, standard of living, and population are all determined, stabilized, and tuned. Successful achievement of sustained yield in the cerrado will deflect pressure from Amazonian exploitation and will decrease reliance on foreign and expensive imports, while at the same time it will increase the independence of the nation as a whole. This is a glorious opportunity for Brazil and one that all ecologists will encourage.

ACKNOWLEDGEMENTS

This paper has been greatly improved by the most constructive comments of Professor Daniel Janzen, University of Michigan, of Professor Nellie Stark, University of Montana, and Jan M. Fanto, New York. We are most grateful for this assistance.

DISCUSSION

Dr. C. Heusser: Please comment on what you think the carrying capacity of the cerrado will be in terms of population, and whether this will require population control.

Dr. R. J. Goodland: The cerrado, at the moment, does not have a population problem, and in fact, at present it supports very few inhabitants per square kilometer. Sustained yield is the prime goal to be aimed at and population stability is a necessary condition. However, I don't think that is the entire answer. World population has soared. We — the temperate exploiters of the chief tropical products — have removed the very key factors upon which population stability depends. When a country produces a product for a distant market and when that producer has no control over prices, when the product is not storable and when it takes some years to get into production, then normal feedback loops of population stability are broken. There is no means by which the populace can know that its own numerical limit is being exceeded or when it is growing too rapidly. But where people have a feel of how much land is available, then they know that a stable community improves qualitatively while a growing population reduces the quality for all. In other words, I submit that when people have some sort of internal feedback, they will want to reduce or maintain their *own* population, which is a more favorable situation than imposition of controls from outside.

Dr. G. Budowski: Yesterday the point was made by Alvim that there is a great future in harvesting the different phases of the successional process of the tropical rain forest. Unfortunately, there is another factor which bears some influence and must be taken into consideration. Products and industry are more easily worked in secondary succession rather than in the primary forest. For example, the Japanese chipping industry is offering great advantages for chipping rain forest. Recently they offered Peru 500 million dollars for the privilege of chipping the rain forest there. They argue that if you remove only the chip and leave other organic matter, you will get secondary forest with which you can work.

I would be interested to know how you would counter such an argument.

Dr. R. J. Goodland: Chipping itself is a more reasonable activity than is the export of beef cattle, and we agree to that extent. Chips themselves are relatively oligotrophic. Most of the nutrients are in the leaves and branches and bark, and they are left in situ. Chipping has the disadvantages of being a low-value product generating little employment, but I see no reason why chipping shouldn't be limited to plantations in secondary growth areas or even from the secondary growth itself. Debasing intact forest into chips is irrational. In relation to the product, I propose that chips, sawdust, and other relatively oligotrophic products should be made into more valuable commercial products such as methane for fuelling essential machinery and for alcohol. Alcohol is a very useful product which only contains carbon, hydrogen and oxygen and therefore it does not represent the export of nutrients. Moreover, it is lighter than water and could be floated down the Amazon in barges. It doesn't deteriorate rapidly, which also is a strong export feature. It may even be possible to separate undigested pulp for the Japanese, while at the same time retaining nutrients in the residue.

Professor P. Richards: Have you made any calculations about the amount of fertilizer needed between the cerrado and forest? I always thought the cerrado was on extremely poor soil. Is not the reason for the cerrado vegetation the poor soil?

Dr. R. J. Goodland: Yes, the cerrado soils are very poor. However, they are not quite as poor as those of Amazonia. The drainage and structure of cerrado soils are, by and large, excellent, and other agricultural restrictions are also less onerous. This suggests to me that it is preferable to use cerrado soils. The answer to your other question: yes, calculations of soil fertility do exist. There are a number of excellent soil fertility plans of the cerrado showing this type of comparison. This represents the amount of fertilizer to be put on in an industrial ecosystem, which would raise production up to approximately that of the temperate production level.

It may be preferable to aim at a somewhat lower production of crop but with an increase both in tolerance of infertile soil and in the tolerance of pests. That is the sort of breeding we should be aiming at, not one which insists that we apply vast quantities of imported pesticides and fertilizer. The fertilizer which is most critical in the cerrado is limestone, but this generally is available nearby. Limestone outcrops are dotted through-out the cerrado. This is cheaply converted into agricultural calcium and transported to areas where it is needed. The use of locally available limestone should be encouraged, rather than the use of increased amounts of modern, highly-industrialized fertilizers.

Professor P. Richards: In the region of Aragarças there are places where the richest soil is in the forest areas.

Dr. R. J. Goodland: Yes, because in the Aragarças region the forest is deciduous or semi-deciduous and it grows on richer soil than under the Amazonian wet evergreen forest.

Professor P. Richards: Is that not a better alternative for cultivation than the cerrado itself?

Dr. R. J. Goodland: Yes, it certainly is, but there is so little of it that it won't relieve pressure on Amazonian forest. Cerradão and other transitional forests cover a rather limited area in contrast to the enormous cerrado which covers well over a million square kilometers.

Dr. P. de T. Alvim: You emphasized the point about doing research on the cerrado, and I think you are right. We must do more research before we try to develop the cerrado. But I don't think we should emphasize too much our preference for the cerrado because both areas are relatively little known. The cerrado has poor soil which is often poorer than the Amazon forest and it also has a stronger dry season. Furthermore, all the cerrado without exception is on poor soil, whereas in the Amazon there are patches of good soils. I mentioned yesterday that there are at least two million hectares of good soil in Amazonia. There is calcareous soil in the upper Xingú, and there are also six million hectares of soil with very good potential in the Amazonian várzea.

Dr. R. J. Goodland: I think you may have missed the first part of my talk. I stressed that várzea should receive preferential treatment for exploitation and, in my opinion, the cerrado dry season can be an advantage rather than a disadvantage. According to EMBRAPA, there are areas of excellent soils in the cerrado region. Although more cerrado research is highly desirable, enough is already known to develop the cerrado. I don't think this holds true for the Amazonian wet forest.

Dr. P. de T. Alvim: We have to work in both areas.

Dr. R. J. Goodland: Yes, definitely. The main point of my talk is that if we develop the cerrado we buy time for a research program to work out what is a sustained yield for the exploitation of tropical rain forest. We don't know at present. Research should tell us in a few years.

LITERATURE CITED

Abreu, S. F., ed. 1973. Recursos minerais do Brasil. São Paulo, Ed. E. Blucher, Vol. I: 324 pp.; Vol. II: 321-754.

Baur, G. N. 1968. The ecological basis of rain forest management. New South Wales, VCN Blight, Govt. Printer. 499 pp.

Bray, W. J. 1976. Green-crop fractionation. New Scientist 70(995): 66-68.

Budowski, G. 1974. Forest plantations and nature conservation. Bull. IUCN 5(7): 25-26.

_____. 1976. Las perspectivas de la educación forestal en la America Latina Turrialba, Costa Rica, CATIE, Ciencias Forestales (stencil). 12 pp.

CIDA, 1976. Contact. Ottawa, Canadian International Development Agency (March) 50:6 pp.

Conselho de Desenvolvimento Económico (CDE). 1975. Programa de desenvolvimento dos cerrados — Polocentro. Brasília, Ministério de Interior, Decreto No. 75.320 de 29 janeiro de 1975. n.p.

Dawkins, H. C. 1955. The refining of mixed forest: a new objective for tropical silviculture. Emp. Forestry Rev. 34:188-191.

_____. 1963. The productivity of tropical high forests and their reaction to controllable environment. Oxford Univ. thesis (stencil).

Day, J. M., M. C. P. Neves and J. Dobereiner. 1975. Nitrogenase activity on the roots of tropical forage grasses. Soil. Biol. Biochem. 7:107-112.

Dobereiner, J., J. M. Day and P. J. Dart. 1973. Rhizosphere associations between grasses and nitrogen-fixing bacteria. Soil Biol. Biochem. 5:157-159.

Douglas, J. S. and R. A. de J. Hart. 1976. Forest farming. London, Watkins. 197 pp.

Earl, D. E. 1975. Forest energy and economics. Oxford, Clarendon. 128 pp.

Falesi, I. C. 1974. Amazônia: A terra é pobre. Rio de Janeiro. Opinião, 18 de março. p. 4.

Futurist, The. 1974. (whole issue). The Futurist 8(6) December: 263-298.

Goodland, R. 1971a. The cerrado oxisols of the Triângulo Mineiro, Central Brazil. Anais. Acad. Brasil Ciênc. 43:407-414.

_____. 1971b. Oligotrofismo e alumínio no cerrado. Pages 44-60 *in* M. G. Ferri, ed. III Simpósio sôbre o cerrado. São Paulo, Univ. São Paulo. 239 pp.

_____. 1975a. Cerrado of Brazil: state of knowledge report on tropical ecosystems. Paris, UNESCO, Div. Ecological Sciences. 103 pp.

_____. 1975b. Glossário de ecologia Brasileira. Manaus. Inst. Nacional Pesquisas Amazônia. (INPA) Imprensa Oficial do Estado, 96 pp.

_____. and H. S. Irwin. 1975. Amazon jungle: green hell to red desert? New York, Elsevier. 155 pp.

Gross, D. R. 1975. Protein capture and cultural development in the Amazon Basin. Am. Anthropol. 77(3): 526-549.

IPEA. 1973. Aproveitamento atual e potencial dos cerrados. Brasil, Inst. Planejamento Económico e Social, (IPEA). 197 pp.

Irwin, H. S. and M. T. K. Arroyo. 1972. A new species of *Periandra* (Leguminosae: Lotoideae) from the Planalto of Brazil. Brittonia 24(3): 327-330.

Irwin, H. S. and M. T. K. Arroyo. 1973. New endemic species of *Harpalyce* (Leguminosae: Brongniartieae) from south-central Brazil with a key to the Brazilian species. Brittonia 25(1): 21-25.

_____. 1974. Three new legume species from South America. Brittonia 26: 264-270.

Janzen, D. H. 1973. Tropical agroecosystems. Science 182 (4118):1212-1219.

_____. 1975. Ecology of plants in the tropics. London, Arnold. 68 pp.

King, K. F. S. 1968. Agri-silviculture — the taungya system. Bull. Dept. For., Univ. Ibadan.

_____. 1975. Putting the emphasis on tropical forestry. Unasylva 27(110): 30-35.

Lopes, A. S. 1975. A survey of the fertility status of soils under 'cerrado' vegetation in Brazil. Raleigh, North Carolina State Univ., Dept. Soil Science (MS thesis). 138 pp.

Lovins, A. B. 1975. World energy strategies. San Francisco, Friends of the Earth. 132 pp.

Lowe-McConnell, R. H. 1975. Fish communities in tropical freshwaters. New York, Longmans. 338 pp.

MacKillop, A. 1975. Why soft technology? London, Methuen. 25 pp.

Makhijani, A. 1975. Energy and agriculture in the third world. Cambridge, MA, Ballinger. 168 pp.

Meijer, W. 1973. Devastation and regeneration of lowland dipterocarp forests in Southeast Asia. BioScience 23: 528-533.

Ministério de Agricultura. 1975. Programa de desenvolvimento dos cerrados — Polocentro. Informações. Brasilia, Min. Agric. 26 pp.

Nicholls, Y. I. 1973. Sourcebook: Emergence of proposals for recompensing developing countries for maintaining environmental quality. Morges, Switzerland, International Union for the Conservation of Nature (IUCN), Environmental Policy and Law Papers no. 5: 142 pp.

PBDCT. 1976. II Plano basico de desenvolvimento científico e tecnológico. Brasília, Presidência da Republica. 217 pp.

Penagos-G., M. D. 1967. Plantas biológicas: solución practica e inmediata de graves problemas nationales. Guatemala, Congr. Nacional de Ingeniería. 48 pp.

Pirie, N. W. 1971. Leaf protein: its agronomy, preparation, quality and use. Oxford, Blackwell. 192 pp.

Roche, L. 1973. The practice of agri-silviculture in the tropics. Ibadan, Univ. Ibadan (FAO/SIDA Conf. Shifting Cultivation and Soil Conservation). 29 pp.

Terborgh, J. 1974. Preservation of natural diversity: the problem of extinction prone species. BioScience 24(12): 715-722.

_____. 1975. Faunal equilibrium and the design of wildlife preserves. Pages 369-380. *in* S. B. Golley and B. Medina, eds. Tropical ecological systems. New York, Springer Verlag.

Trenbath, R. B. 1974. Biomass productivity of mixtures. Adv. Agron. 26: 177-210.

_____. 1975. Diversify or be damned? Ecologist 5(3): 76-83.

Troup, R. S. 1952. Silvicultural systems. Oxford, Clarendon, 216 pp.

Turner, F. C. 1976. The rush to the cities in Latin America. Science 192(4243): 955-962.

Village Technology Handbook. 1973. Mt. Ranier, MD, Volunteers in Technical Assistance, Inc. 387 pp.

Watson, G. A., W. P. Weng and R. Narayana. 1964. Effect of cover plants on soil nutrient status and on growth of *Hevea*. Jour. Rubber Res. Inst. Malaya 18 (2): 80-95.

Whitmore, T. C. 1975. Tropical rainforests of the Far East. Oxford, Clarendon. 282 pp.

Williams, R. H., ed. 1975. The energy conservation papers. Cambridge, MA, Ballinger. 416 pp.

Wong, Y. K. 1966. Poison girdling under the Malayan uniform system. Malayan Forester 19: 193-201.

Wyatt-Smith, J. 1963. Manual of Malayan silviculture for inland forests. Malay. Forest Rec. 23 (v.p.)

PRESERVATION OF THE FLORA OF NORTHEASTERN BRAZIL

Dárdano de Andrade-Lima

Instituto de Pesquisas Agronómicas, Pernambuco, Brazil

The Brazilian Northeast includes seven states: Piauí, Ceará, Rio Grande do Norte, Paraíba, Pernambuco, Alagoas and Sergipe, in addition to the northern part of another one: Bahia. The entire region extends from about 2°54' to 17°21' South and from 35° to 46°30' West.

This area, by and large, can generally be described as being flat, especially the western part. The highest mountains there are no more than 1,100 m, while in the eastern section, not far from the coast, there are a series of low, undulating hills reaching no more than 300 m high.

Four main climate types[*] of Koeppen (from Andrade, 1972) occur in the area. On the coast between Rio Grande do Norte and Sergipe, there is Aw with its variant Aw[1]; in Piauí state and part of Ceará, BSh deep in the interior; and Csa on some of these higher mountains with large plateau areas, such as Garanhuns in Pernambuco.

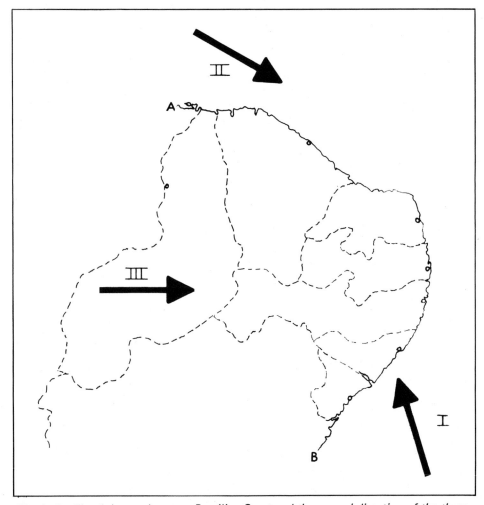

Fig. 1. Profile of the northeastern Brazilian Coast and the general direction of the three main atmospheric disturbances which bring rain to the area.

[*]*See key to climate types at end of article.*

The rainfall — or lack of it — in northeast Brazil is due to a peculiar system of atmospheric disturbances (Fig. 1). Considering Line A-B of Fig. 1 as the profile of the northeastern Brazilian coast, arrow I indicates the general direction of the cold Atlantic Polar Fronts. Arrow II shows the southeast movement of the intertropical convergence (ITC) which brings rain over the northern half of Piauí, most of Ceará, Rio Grande do Norte and parts of Paraíba and even Pernambuco. Arrow III is the incursion over the central western part of Brazil by the continental equatorial air mass (Ec).

None of these three atmospheric disturbances affects northeast Brazil with full force since each is weakened by the long journey from its origin. Most of the area's rainfall emanates from the cold Atlantic Polar Fronts.

These Fronts serve to refresh the warm winds from the Atlantic Ocean (trade winds) which become more humid during their sweep across the ocean. Actually, the Polar Fronts only reach Bahia as a result of unusual atmospheric intensity, at which time they frequently branch westwards from Sergipe. Eventually, the fronts with the warm winds result in heavy rains as they move in during February and April (fall in the southern hemisphere).

The heavy rains do not penetrate over much of the continent. In the interior, for example, rainfall varies between zero in Northeast Rio Grande do Norte to only about 120 mm in Northeastern Alagoas. In contrast, the fall-winter rainfall can measure up to 2,360 mm/year in Barreiras, Pernambuco. As a direct consequence of all this rainfall, an ombrophilous forest covers those hills with soils which originate from Pre-Cambrian rocks, and those Plio-Pleistocenous low tablelands of the Barreiras group which lie along the coast.

The width of such forests is not constant from north to south. In the northern part of the State of Rio Grande do Norte, these forests decrease to only a few km from east to west while they grow increasingly wider to the south, depending on relief variations. In northern Pernambuco, it is about 80 km wide; on the parallel of Recife, it is no more than 50 km wide; in the southern part of the same state, on the boundary with Alagoas, the forest's western limit is more than 120 km inland from the coastline.

One must not expect to find this area densely covered by humid forests with tall, beautiful, vigorous trees growing close together and full of wild animals strolling on the ground. Most of these forests have been cut down for sugarcane plantations, timber, and even for charcoal production. Only occasional remnants of the primitive cover are now to be seen. Nevertheless, the sight of even one bit of green forest produces a spark of hope, because it is an indication that all is not yet lost.

It is still possible to find some spots where the forest maintains the calm air of four centuries ago. Herbs, vines, and epiphytes are still to be found in their original, primitive places, and large trees with well-developed buttresses and with dark, soft bark full of mosses can occasionally be seen also.

Except for the southern part, coastal Alagoas still retains a sizeable area of good forest, though axes and saws are working hard for timber production and are actively opening new areas for more sugarcane plantations. The forests are increasingly narrow to the south. In Sergipe state, they are again reduced to irregularly distributed areas only a few kilometers from the coast.

The amount of annual rainfall gradually decreases as one moves inland, behind the forests.

The rainfall in the northeast *caatingas* (dry scrub forest) is concentrated during a short period: November-February (approximately - summer) in the west and south of the area, with the progression of the Continental equatorial Air Mass (Ec); March-April (beginning of fall) in the north of the area, under the autumn oscillations of the *doldrums* (adapted from Andrade, 1972). This does not mean that the caatingas receive water from rain only during these months, but these are the months with the greatest precipitation. During these short periods, the rain comes as very heavy concentrated showers with frequent thunderstorms.

In the caatingas during the dry season, the average monthly temperature is 26-27° C (Andrade, 1972), with an appreciable variation between day and night. During the rainy season, the average temperature usually is lower. Maximum and minimum temperatures vary greatly throughout the caatingas. The highest are registered in the State of Piauí and in western Paraíba. This is the BSh climate country.

Under such climatic conditions, the soils of the caatingas are shallow in most places due to superficial runoff and erosion. The mineral content of these soils is usually high. There are, of course, areas of sedimentary origin, where sandy soils that are poor in nutrient content and with very low water retention capacity, exacerbate the general situation even more.

Small mountains are scattered throughout the caatingas. These possess particular topoclimates which are expressed by higher rainfall, lower temperatures and higher and longer-lasting clouds because they are either exposed to the southeastern humid cooler winds or to the northwestern rainmaker winds (ITC). These factors all contribute to the development of a forest, the density of which depends on the productivity of the soil. In all of northeast Brazil, these humid mountain summits with their cloud forests bear the name of *brejos,* a word which literally means marshes but is used here with a quite different meaning.

In the central and southern parts of Piauí state, quite a large area is covered by cerrado vegetation which actually is a continuation of the vast central Brazilian Cerrados.

The tree and shrubby varieties of cerrado are present here with, occasionally, the appearance of even the open field extreme type containing the same general characteristics and problems.

In sum, coastal humid forests, caatingas, brejos and cerrados are the four main vegetation environments to be considered in this report.

COASTAL HUMID FORESTS

These forests used to cover 10-15 percent of northern Brazil. From a physiognomical point of view, most of this forest could be considered vastly different from the Amazonian high-ground forests (mata-de-terra-firme). One could assume that it represents the joining zone of two floras — the Amazonian and the southeastern floras. Species belonging to each of these two groups are represented here, as well as some purely local ones. Among the Amazonian species, *Parkia pendula, Coumarouna odorata, Tapirira guianensis, Pourouma guianensis* and *Didymopanax morototoni* could be cited as good examples.

Those of the southeastern flora are found on the higher hills, where the temperature is somewhat cooler; a typical representative is *Anchietea parvifolia.* The Amazonian species do not occur at these altitudes but are restricted to the warmer low grounds.

Among the species of the local regional flora, *Couepia rufa* is quite characteristic.

Some genera, as *Euterpe* and *Virola,* belong to both Amazonian and southeastern floras, each with a single species in the northeast.

THE CAATINGAS

Though these are well known in the botanical and phytogeographic literature, it is impossible to know exactly how they looked before the Europeans first saw them. The early records of Brazil were written by people who did not penetrate far inland.

Today, the caatingas are a complex variety of landscapes, ranging from thorn-forest to open, bushy land and even semi-desert areas with sparse, shrubby and herbaceous vegetation.

This complex ecosystem was called *caatinga* by the Indians of that region, and the name has been kept in common use and scientific literature. The term means "open forest" or "open vegetation." It is impossible to say whether the whole area was once entirely covered by "open forest" which, through human influence, has been transformed into the present landscape containing wide stretches of low or almost nonexistent vegetation, or if the present condition prevailed prior to human destructive action. Perhaps neither of these possibilities represents the truth since forest and shrubby vegetation might always have coexisted, depending on local physical conditions. The evidence, however, is that forests used to cover a larger area than they do today and that the present reduction in such forests is due to man's influence.

We still do not know when the northeastern part of Brazil was invaded by the xerophytic species which are seen there today. Neither do we know the migration routes which were taken during the speciation process which culminated in the present flora. There is geological evidence of past climatic fluctuations in the area. These would have forced the caatinga to recede and humid or semi-humid forest to have encroached to some extent and would have retreated later on. Today's remnants of mountain summit forests, surrounded by semi-arid caatingas, are a further indication of that movement. Exactly when these variations took place, and how many times they occurred, is one more problem to be solved.

Floristically, caatinga is not completely isolated from any other flora of the world. A few examples are useful to prove this: *Anadenanthera macrocarpa* and *Amburana cearensis* of the caatingas occur also in the Chaco region, in northeastern Argentina and in southern Paraguay; *Amburana cearensis* belongs also to the flora of Acre (western Amazonia); *Tillandsia tricholepis* occurs both in the caatingas and in Rio de Janeiro, as well as in many other places. Several other examples could be cited and even more at the generic level: for instance, *Cavanillesia,* which grows both in northeast Brazil and in Acre.

The present isolation of the area is possibly due to the slow climatic changes, but exactly when this isolation occurred is difficult to pinpoint.

The fact is that most of the Brazilian northeast is covered by a vegetation which, no matter how complex it is, cannot be described as a "forest vegetation." In the caatingas of today, arboreous communities (thorn-deciduous forest) are not common, but they do form a sparse community with some undergrowth during the rainy season. Open communities with scattered low trees and shrubs, where Cactaceae *(Pilosocereus gounellei* and others) are well-represented, occupy a much larger area today.

THE BREJOS (humid forest mountain summits).

Various mountains of more than 500 m high in the Brazilian northeast have summits which are covered by a dense, varied forest, a direct result of the high humidity and better soil conditions there. Trees of 30 to 35 m tall are common, and ferns, mosses, *Begonia* and all sorts of ombrophilous undergrowth species are also well-represented.

The surplus of water that is neither used by the plants nor evaporated, runs down the mountains and improves the caatingas below, which consequently grow taller and denser.

THE CERRADOS

In general, the cerrados of Piauí do not differ very much from any other cerrados in South America. However, they have a somewhat drier climate than most of the areas do which are covered by this vegetation type. The Piauí cerrado contains a higher percentage of species with small leaflets, and often the whole community becomes leafless during the dry period of the year.

PROBLEMS

The forests

From the very onset of Portuguese colonization, the dense, humid coastal forests of northeast Brazil were destroyed in order to permit sugarcane and timber production. Some relatively small forest areas were left standing but most of it was cleared. Although other agricultural products have been cultivated in the area from time to time, their disturbing influence was of minor importance to the vegetation itself.

The caatingas

There are two problems here: 1) the total and localized clearing of the vegetation for agricultural purposes; and 2) the gradual and widespread use of the vegetation by cattle and goats, which were introduced into the area during the seventeenth century. Major fires have never been a problem in this area.

The brejos

Most of the *brejos* were partially cleared many years ago for coffee plantations. However, the culture of coffee was abandoned several times because of the fluctuations of coffee prices, and other species were cultivated in its stead. The trees of the original forest had been cut down previously.

In some of these *brejos,* a few species of Orchidaceae *(Cattleya labiata, Cattleya granulosa* [also from the coastal forests] , *Cattleya elongata* and *Phragmipedium sargentianum)* used to grow wild. Not surprisingly, the well-known and ubiquitous "plant hunters" have gone after them so eagerly in recent years, that these species can actually be considered endangered today.

PRESERVATION

In the forest zone

Sugarcane itself has helped to protect the soils of the coastal forest zone, but most of the forests are gone.

In Pernambuco, the first official step towards preservation was a Forest State Department ruling which stipulated that only the forests on the slopes and valleys could be cut. The upper parts of the hills were to remain in their natural vegetation in order to protect the soils, animal life and the plant species themselves.

A second constructive step was taken by the Universidade Federal Rural de Pernambuco when the Tapacurá Ecological Research Station was created. This included an area of 100 ha of sub-deciduous forest, in addition to 230 ha of open ground in the São Lourenço da Mata county, about 40 km from Recife.

More recently, the Brazilian Government decided to buy from private owners a fairly large area still covered with primary forest, in order to save it from total destruction. Now it is possible to find such residual areas only in southern Bahia, northeastern Alagoas and southeastern Pernambuco.

In the caatingas

The Brazilian Government policy for improving the present situation of impoverishment of the caatingas vegetation is twofold:

1. the establishment of National Parks or Ecological Stations in floristically and physiognomically different areas in order to preserve the local floras and landscapes;
2. the development of special programs of proper caatingas management together with its utilization by cattle.

The first Caatinga Ecological Station has already been decided upon by the Special Secretary for the Environment of the Brazilian Government. It is located in northern Bahia, between the Paulo Afonso Falls and Geremoabo, and has officially been named the "Raso da Catarina Ecological Station." It covers an area of about 150,000 ha of well-preserved caatinga on fairly sandy soil and has an average rainfall of 350-400 mm.

The management program of the caatinga-cattle complex is being developed by the federal agency, EMBRAPA, on the grounds of the Research Center of the Semi-arid Brazilian Tropics in Petrolina, Pernambuco.

The primary goal of this program is to achieve the correct quantification of the cattle population which can feed there under natural conditions on a permanent basis. This is aimed at improving — or at least not reducing — the productive potential of that ecosystem.

As a second step, there is the overall improvement of the plant community to be achieved, hopefully, through the introduction of drought-resistant species which also provide food for cattle.

In the brejos

Attempts to preserve the vegetation and flora of the *brejos* have been the most successful so far, though it is difficult to attribute this success to any one specific factor.

The National Forest Park of Ubajara in the Ibiapaba Plateau, northwestern Ceará and the Araripe Forest Park, southern Ceará have been created by the Brazilian Government.

In Pernambuco, the Serra Negra *brejo* is now a Biological Reserve, and its vegetation is rapidly regenerating.

The forest of the county-owned *brejo* of Caruaru in Pernambuco, where a population of *Podocarpus sellowii* is still vigorously growing, is presently under the protection of the Brazilian Institute for Forest Development (IBDF).

The Morro do Chapéu State Park (up to 1,100 m) in northern Bahia also has been created. A few rare *Vellozia* and succulent *Euphorbia,* in addition to many other species, including *Podocarpus lambertii* from southern Brazil, will be preserved in it.

In the cerrados

In Piauí, where the Sete Cidades National Park was recently established, the vegetation and curious sandstone formations will be preserved.

In such a variety of landscapes, ranging from open fields fringed by forests and pure grasslands to dense arboreous caatinga, we hope to overcome all these preservation problems, so that the lives of future generations will be enriched by the flora of Northeast Brazil.

LITERATURE CITED

Andrade, G. O. 1972. Os climas. *In* Aroldo de Azevedo, Brasil, a terra e o homem, **1**: 397-462. Cia. Edit. Nac. São Paulo.

Key to climate types:

Aw = Tropical climate with a distinct dry season (6 cm or less in the driest month). Savanna vegetation.

Aw^1 = As in Aw, with the rainy season delayed until autumn.

BSh = Dry climate, too dry to support forest, but with a short wet season; mean annual temperature $> 18°C$; mean temperature of coldest month $< 18°C$.

Csa = Rainy season maximum monthly rainfall equal or larger than 3 times that of the hottest month; driest month with less than 30 mm rainfall; 4 months with mean monthly temperatures above $10°C$; mean temperature of hottest month above $22°C$.

THE PROBLEMS OF THREATENED AND ENDANGERED PLANT SPECIES AND PLANT COMMUNITIES IN ARGENTINA

Elías R. de la Sota[1]

La Plata National University, La Plata, Buenos Aires, Argentina

INTRODUCTION AND SOME EXAMPLES OF ENDANGERED SPECIES

Many plant species and vegetation types are seriously threatened today in Argentina, and in this paper, I propose to offer a positive and factual statement of this often ignored problem. It is always easy to criticize but I believe it would be more useful and constructive to suggest guidelines so that my country, with its many human and material resources, can act wisely, while there is still time to do so.

I will present the problem in terms of endangered species, genera or communities, and I also will discuss the reasons why they are threatened today.

1) The species of *Nothofagus* are very common along the Andes in Patagonia, where there are several national parks. These plants frequently have two natural enemies, *Cyttaria* and *Misodendrum*. The Government, in its eagerness to increase local and foreign tourism, introduced European deer to replace the native one, *huemul,* which almost had disappeared due to intensive uncontrolled hunting. The European deer reproduce more rapidly than the native species do and they threaten the *Nothofagus* forests because they eat the seedlings, young plants, and bark.

A similar but graver situation occurred with other trees in the Patagonian Andes. *Aristotelia machi* L'Herit. *(maqui)* and *Schinus patagonicus* (Phil.) Johnst. *(laura),* are especially endangered on the island of Victoria on Nahuel Huapi Lake.

2) *Schinopsis lorentzii* Hieron. *(quebracho colorado santiagueño)* is a characteristic tree of the plain forests in Chaco, Salta and Santiago del Estero provinces. However, at present it is only possible to find few specimens of it there. The loss is attributed to the construction of railroads extending from Buenos Aires to the interior of the country. Trees were felled to make railroad ties and also to use the wood as fuel, instead of coal. Wood burns well and its hardiness and durability make it a readily available and economical substitute for coal.

This situation is underscored by the rainfall analysis. The decrease of rainfall has now limited the area which can be used for the culture of sugarcane to the region near the Pampean and Subandine Mountains in Tucumán, Salta and Jujuy provinces, though it should be noted here that during the last century it was still possible to raise sugarcane in Santiago del Estero. The lowered rainfall also permits another type of vegetation to thrive — shrubby and drier types — with plenty of *Zygophyllaceae* and *Cactaceae,* which we call *Monte.* This also has contributed to the displacement of the original forest.

Jorge H. Morello and his disciples have been working for several years in the Chaco forests, studying not only floristic composition of these forests, but also their dynamics, natural repopulation, and germination processes. In addition, some interesting peri-domestic aspects are being studied, including that of ruminant animal influences, particularly goats, since these are usually the animals which cause most of the forest destruction and the formation of "*Opuntia* rings," which are coincident with the grazing habits and patterns of ruminants around their watering places. Sometime ago, Morello produced, under the University of Tucumán sponsorship, a documentary film in color and with folkloric music, entitled "Un bosque saqueado" (A plundered forest). During the showing of the film, this ecologist stated clearly that this urgent problem was not yet overcome.

---

[1]Guggenheim Fellow, 1975-1976: Professor of Morphology of Vascular Plants, Faculty of Natural Sciences and Museum, La Plata National University, La Plata, Argentina; Research Fellow, C.O.N.I.C.E.T., Buenos Aires, Argentina.

3) *Trichocereus pasacana* Britton & Rose, *T. terscheckii* Britton & Rose, the well known *cardone* trees, are frequent columnar giants on the slopes and in the deep passes of the *Prepuna,* in northwestern Argentina. These trees are often felled in order to use their dry vascular tissue to make walls and furniture, or such objects for tourists as jewel boxes, chests, etc. This has been a common practice for many years, as can be seen in the church of Los Molinos, a village on the Calchaquíi River, not far from Cachi, Salta.

4) *Azorella yareta* Hauman *(yareta)* and some species of *Adesmia (cuerno)* in the Calchaquíi Ridge of Tucumán province. Since there are no trees in the high mountains, the local people use these plants as a substitute for wood fuel for heating and cooking. Fortunately, very few families live at those altitudes.

5) *Podocarpus parlatorei* Pilger *(pino del cerro)* is the only conifer tree in northwestern Argentina. This species forms forests in several places in the mountains of Tucumán, Salta and Jujuy. At present, people are indiscriminately felling these trees in order to plant such faster-growing species as *Pinus* and *Eucalyptus,* which also produce more wood for fuelling the blast furnaces of Zapla (Jujuy) and for obtaining cellulose.

6) *Dicksonia sellowiana* Hook. *(xa-xim)* was a common tree fern in the swampy forests of Misiones. These are being cut excessively in order to use the external layer of roots as a support for some epiphytes *(Orchidaceae, Peperomia, Araceae,* ferns, etc.).

7) *Adiantum lorentzii* Hieron. *(culantrillo)* sometime ago became endangered in the Anconquija Park, only 12 km from Tucumán City. During the winter holidays, both the native people and the tourists take these pretty and delicate ferns for decorating their houses indoors, but the ferns soon die because the humidity in patios and rooms is too low for them.

8) *Equisetum giganteum* L. *(cola de caballo)* was quite common in the marginal forest of Punta Lara (Buenos Aires province), but now it has almost disappeared. Many people from Buenos Aires and from La Plata spend summer vacations and weekends there, picking the plants for use as folk medicines, thus destroying the native flora.

None of these examples is new, and they therefore are not to be considered as a complete checklist of endangered species in Argentina, though they are typical of the situation in my country at the present time. The examples were gathered from a variety of monographs and books, including Cabrera and Dawson, 1944; Dimitri, 1972; Hueck, 1966; Morello, 1958; Morello and Saravia Toledo, 1959; with some, like Hieronymus (1874), dating from the last century. In addition, the above information is also based on my own observations during field trips within the country and from my own teaching activities.

THE VEGETATION COMMUNITIES AND ADVENTITIOUS PLANTS

Some of the vegetation types in Argentina are endangered not only because people destroy their components, but also because foreign species, more aggressive than the native ones, are introduced without control and previous study.

The Punta Lara marginal forest, mentioned above, is a good example to illustrate this situation. This forest was considered a biological reserve about 35 years ago. It was the most austral one belonging to the South Brazilian forest complex. But in spite of that, little was done to preserve its unique quality. For example, the following species of Pteridophytes are endangered: *Equisetum giganteum, Selaginella marginata* (H. & B.) Spring, *S. muscosa* Spring, *Pleopeltis macrocarpa* (Bory ex Willd.) Kaulf., etc.

The problem in Punta Lara has become much worse not only because of the increased felling of trees and the uprooting of herbs for ornamental or medicinal purposes, but also by the introduction of such exotic elements as *Ligustrum lucidum* Ait. and *Rubus ulmifolius* Schott. Cabrera and Dawson (1944) mention more than 300 species of vascular plants for Punta Lara and of this total, nearly 10 percent are introduced. In the *Nothofagus* forests, where the European influence is very strong, the percentage of foreign

241

elements in the local flora is high (cf. Dimitri, 1972). In addition, Punta Lara suffers from water pollution caused by petroleum, oil, and industrial residues. Obviously all these factors seriously endanger the aquatic flora.

Returning to the northwestern part of Argentina, I wish to comment about two types of vegetation which are presently endangered or entirely destroyed there.

The deciduous forest at the foot of the Anconquija mountains in Tucumán has almost disappeared — a direct result of the human population growth. Our actual knowledge of the original type of vegetation rests on various historical reports based on the expeditions made to the area by Hieronymus and other botanists during the 19th century. Fortunately, the characteristic trees of the deciduous forest of the Anconquija region, *(Enterolobium, Piptadenia, Tabebuia, Tipuana)* are frequent along thoroughfares and in public squares and parks. Therefore, though these species still exist, they have disappeared from their natural habitat.

Something similar, but not as extreme, is now happening with the mountainous forest on Sierra de San Javier, near Tucumán City. Many years ago, tree-felling was carried out by removing several of the typical elements of the low altitude, such as *Cedrela* and others. In addition, people clear the trunks and branches of the trees, destroying the unique epiphytic communities, because they do not understand the difference between epiphytism and parasitism.

There are many more examples of species and vegetation types in Argentina which need effective and fast preventive action in order to avoid irreversible alteration.

GENERAL SUGGESTIONS FOR CONSERVING THE ENDANGERED SPECIES AND VEGETATION COMMUNITIES

The examples cited up to this point reveal the type and scope of the problem. Superficially it may be interpreted as a criticism that our government does not control the natural resources of the country as much as necessary. That is not really quite an accurate conclusion. The policy for conserving should be built upon other bases which, like strong pillars, can be more effective towards a wise course of action. In order that this objective may be carried out successfully, I wish to offer the following suggestions:

1) To educate all the social levels of the population, including workers and executives, children and old people — to explain to all these people why and how they can and ought to conserve nature. We have to know and understand plants in order to love and take care of them. The government has newspapers and both radio and television stations under its control. These media are a good way to communicate information about conservation, and to gradually instill a love for nature in the people. They have the resources with which to carry out this educational process with the maximum effect and result.

2) To improve and to orient the right concept of nationalism: to save our natural resources, to increase our own national health, and to show tourists that which is genuine in each region. The examples of "foreignism" are frequent in our popular customs: only exotic trees and herbs are planted in some metropolitan areas, ignorance of the local flora (even among students from the universities), false pride of comparison, for example, "Buenos Aires is like Paris." To aspire to be a copy, to ape other cultures, draws us strongly away from our authenticity which we really should be so proud of.

3) The problems are not only restricted to the endangered species, but they also endanger several natural ambients. It would be ideal to protect ecological niches within a minimum area which would permit the preservation of the soil, the climate and the living community without too many changes. That is a very difficult goal but I believe it is still possible. One of the convenient ways to protect various types of vegetation could be to enlarge the number — but not necessarily the area — of biological parks and reserves. Apparently, such areas do not seem attractive for tourists or useful for men.

Argentina has fourteen national parks; seven are along the Austral Andes, two are in Patagonia, two are in the Chacho woodlands, one is in Misiones, one is in Entre Rios and only one in the entire northwestern region. These are too few and wrongly distributed.

Concerning northwestern Argentina, the following types of vegetation need care:

a) Mountainous woods of *Alnus, Podocarpus, Polylepis.* The botanical importance of these forests is based on the presence of some boreal species of plants or complexes *(Athyrium dombei* Desv., *Dryopteris parallelogramma* ([Kze.] Alston) and genera *(Fragaria, Lathyrus, Prunus).* These taxa are useful and significant for helping us to resolve some of the most fascinating questions in Biology: e.g., to determine the pathways of migration along the ranges of mountains, in order to explain the geographical distribution of plants and their disjunction areas in the South Cone of America.

b) High Andean swampy grounds.

c) Angiosperm peat beds, with *Cyperaceae, Juncaceae,* etc., on the elevated valleys of the Calchaquíes peaks in Tucumán, from 3,750 m upwards. The pollen and spore analysis of these peats can yield the necessary information which would clarify some problems about the ecology, the climate, the altitudinal limits of certain species in the past, and also for comparison with the similar data from the *páramos* in tropical America (Colombia, Venezuela).

4) The choice of species and vegetation types which need care, and the order of the emergency priorities, are both the biologists' tasks. They know very well the diversity of plants, their ambient demands, their instability, and evolutionary potential. It sounds obvious, but it is necessary for our social method to emphasize this point of view, since agronomists are trained to handle only farms, and not wild communities.

5) To control very carefully the introduction of foreign species. *Ligustrum lucidum* and *Rubus ulmifolius,* in the biological reserve of Punta Lara, are good examples of this type of "dangerous invasion". The *ligustros,* which multiply from the gemmiparous roots, are competing and displacing the native woody plants of that interesting marginal forest. That situation could become very serious in the near future, and could prove as disastrous as the invasion of *Salvinia molesta* Mitchell on Lake Kariba (Rhodesia) and other places (Ceylon, Australia), where it reproduced so rapidly on the artificial lakes behind the dams that it became a threat to each of those projects.

6) When these previous steps have been taken, and when the government has the whole structure ready, it would then be opportune to legislate the process in order to protect it. Rules and laws do not have to be dogmatic, but everyone would like to know why and how restrictions are imposed. Obedience sprouts spontaneously when there is understanding. We can use and even consume the green world; but this must not be allowed to happen.

7) The universities must actively participate and offer their full cooperation. A university is not an isolated island, and research is not its own adventure. Universities are portions of the human community and as such, they must not be like factories for making diplomas and papers. Rather, they must serve as permanent reservoirs of knowledge and as efficient developers of culture. They must teach the people how to take a critical position in view of the known facts. To offer exhibitions, to give public lectures, to publish popular and inexpensive books on this problem and to establish academic recognition and degrees for the protection of wild vegetation, must be a fundamental concern of our universities and high schools.

LITERATURE CITED

Cabrera, A. L. and G. Dawson. 1944. La Selva Marginal de Punta Lara en la Rivera Argentina del Río de la Plata. Rev. Museo La Plata, n. ser., Bot. **5**: 267-382, 15 figs. 10 photos.

Dimitri, M. J. 1972. La región de los bosques andino-patagónicos. **381** pp., 37 figs., 109 photos., tabs. A-I. I.N.T.A., Buenos Aires.

Hieronymus, J. 1874. Observaciones sobre la vegetación de la Provincia de Tucumán. 161 pp., 1 map. Instituto de Estudios Geográficos de la Universidad de Tucumán, Tucumán.

Hüeck, K. 1966. Die Wälder Südamerikas. 422 pp., 253 figs. Gustav Fischer Verlag, Stuttgart.

Morello, J. 1958. La Provincia Fitogeográfica del Monte. Opera Lilloana **2**:155 pp., 59 figs., 58 photos.

_____. **and C. Saravia Toledo.** 1959. El bosque chaqueño. I. Paisaje primitivo, paisaje natural y paisaje cultural en el Oriente de Salta. Rev. Agron. Noroeste Argentino **3**(1-2): 5-81, 21 figs., 46 photos.

THREATENED AND ENDANGERED SPECIES IN ARGENTINA

Angel L. Cabrera

Instituto de Botanica Darwinion, San Isidro, Argentina

INTRODUCTION

Argentina is a country actively undergoing rapid development. Its varied landscape contains not only large plains which are ideally suited to extensive agriculture and cattle-raising, but also mountainous regions which are covered by tropical forests, and temperate woodlands where forestry activities are growing more intense daily. An inevitable result of these commercial pursuits is the rapid destruction of the original plant communities and an equally rapid decrease of the areas occupied by useful native plants.

On the other hand, many species tend to disappear for reasons other than direct exploitation or destruction by man, reasons such as changes in ecological conditions or intrinsic plant characteristics. Herewith are some examples of the problems faced by Argentina's endangered species and plant communities:

Plant extinction resulting from intrinsic causes.

Many plant species have been collected either a few times or only once, and then seem to have disappeared. Perhaps they were species which existed for a short time because of poor adaption, survival fitness, or poor reproductive capabilities. *Senecio sandwithii* was collected only once in the Neuquén mountains. *Aylacophora deserticola,* a monotypic genus of the Compositae, has been found only near Plaza Huincul, Neuquén, and every year it becomes increasingly scarcer. The same trend seems to be occurring with many other species of different families. But it should be emphasized that these plants grow in uninhabited regions, and therefore perhaps more careful exploration will rediscover them. *Mutisia tridens,* for instance, was collected only once by Poeppig nearly a hundred and fifty years ago in the region of Antuco, Chile, but it reappeared last year in the boundary zone of Neuquén. It is quite possible that many species of plants, heretofore considered rare or extinct, will be found again if the correct region is visited at the right time.

Disappearance of plants through ecological causes.

Some of the main reasons for the destruction of species or plant communities by environmental "nonhuman" causes are the following: volcanoes, earthquakes, landslides, and changes in water table levels, as well as the competition of other plant species. The proliferation of wild herbivores may also be a threat to plants. In addition, climatic fluctuations can prove fatal to plant communities by causing floods, draining of lakes, etc. The dead woodlands of *Prosopis,* for example, found in some regions of Catamarca and La Rioja, seem to be the direct result of the lowering of the water table. Many other species of plants, such as *Senecio grisebachii* and *Aristida pallens,* which are not eaten by the herbivores, enlarge their geographical areas at the expense of other plants.

Extinction of communities and species of plants due to human influence.

DIRECT HUMAN ACTION

(1) *Destruction of whole plant communities due to agriculture or forestation.* Fire is the fastest method of total destruction of original vegetation. In forest and woodlands, the pristine community is destroyed through the felling of trees or fire, in order to make these areas suitable for agricultural or commercial forestry activities. In the provinces of Salta, Jujuy, Tucumán and Misiones, the original tropical forest was destroyed in large areas and replaced by woodlands of *Eucalyptus* or *Pinus,* the former used as fuel and the latter for paper manufacture. At other times, the forest is supplanted by groves of *Citrus,* or by plantations of sugarcane or orchards. In the Chaco region, the original "quebracho" woodlands (woodland dominated by species of *Schinopsis,* Anacardiaceae), were destroyed in order to obtain timber and charcoal. If the zone has a good water supply,

the woodlands are replaced by trucking crops such as tomatoes, chilies or beans, or by large cultures of cotton. But if there is no available water for irrigation, the regions are abandoned, only to become deserts after a period of time, a direct destructive result of the timber and charcoal removal.

In the Pampa, the climax grassland has been almost totally destroyed or replaced by extensive agriculture and cattle-raising.

(2) *Selective exploitation and destruction.* In this case only certain species of economic value are destroyed by man. Some greatly exploited species which are therefore endangered in Argentina are the following:

Schinopsis balansae and *Schinopsis lorentzii,* the *quebrachos colorados,* originally dominant in the Chaco woodlands and now nearly extinct. The wood is being used in the manufacture of supports, pillars, railway ties and for the extraction of tannin. The growth of these species is very slow and the young plants are browsed by free-roaming goats.

Cedrela spp., the American cedars which furnish very valuable timber, are nearly destroyed today.

Araucaria araucana, the *pehuen* or *pino* of southern Chile and northwestern Patagonia, which is used in the manufacture of plywood. Its growing area is rapidly being reduced.

Podocarpus parlatorei, the *pino* of northwestern Argentina, is progressively being destroyed because it is used in the manufacture of paper-pulp.

Polylepis tomentella, the quenoa of the Puna, the only tree of this interesting region is being progessively destroyed by its intensive use as fuel.

Fitzroya cupressoides, the *alerce* of southern Argentina and Chile, has become very scarce in both countries.

Acrocomia chunta, the chonta palm of tropical Salta which has a small area of distribution, is being exploited for pillar manufacturing.

Euterpe edulis, the *palmito* of Misiones with edible terminal buds, also is being overused.

Many other species of trees are being subjected to uncontrolled exploitation and therefore are considered to be threatened with extinction.

Some species of plants are intensively sought by man for various other reasons. Many plants are considered aphrodisiac, such as *Haplopappus rigidus (bailabuena)* and *Lycopodium saururus (cola de quirquincho)* and are collected so intensively as to almost cause their extinction. Others are used as fuel in the arid region. Typical of these are the cushions of *Azorella caespitosa,* the *yareta,* and several species of *Adesmia,* commonly known by the name of *cuerno de cabra,* which also have the same cushion-like habit. Several other species with known medical properties are collected in great quantities and sent to the pharmacies of Buenos Aires. These species are also threatened by overcollection.

Other species of Cactaceae, especially those from the genera *Lobivia, Rebutia* and *Parodia,* are collected by thousands for export to Europe, where they are sold to cactus lovers. In the mountains of Tucumán, a beautiful Compositae, *Chuquiraga calchaquina,* is actively collected and sold in flower shops in the cities.

(3) *Destruction through engineering works.* The building of towns and the construction of roads, railways, or dams result in the destruction of all vegetation in the vicinity. The broad, 100 m wide roads that run through Argentina have destroyed almost all of the original vegetation. An extreme example of this is the area around the future dam of Salto Grande, on the river Uruguay. Nearly 250,000 hectares of land, including two towns, several villages, and extensive areas of marginal forests, will be permanently flooded as a result of the dam's construction.

INDIRECT HUMAN ACTION

(1) *Livestock.* Cattle-raising always results in altered plant communities because of the elimination of the species most palatable to the herbivores. Nearly the entire

246

Argentinian territory in the Chaco, the Pampa, and in Patagonia, has been altered by the raising of livestock in those vast areas.

(2) *Transport of seeds.* Man inadvertently transports many seeds of weeds along with those of cultivated plants, in his vehicles or on his clothes. Frequently these are annual species which are so aggressive that they disturb the original plant communities. In the agricultural zones of Argentina, it is not uncommon to observe large communities of European thistles, Cruciferae and Umbelliferae, for example, which invade the fields and take over from the indigenous species. Even in the national parks there is remarkable penetration of invader species. *Ligustrum lucidum,* for instance, a tree largely cultivated in streets and gardens, is a dangerous invader in the marginal forests of the Delta del Parana and the Rio de la Plata. In the national parks of Patagonia, the primitive herbaceous layer is being progressively replaced by such European species as *Erodium cicutarium, Crepis virens, Hypochoeris radicata* and *Hordeum murinum.* A beautiful invader is *Chrysanthemum leucanthemum.* Shrubby species of *Rosa* and *Rubus* invade the woodland. It is quite probable that in the not too distant future, man will only have cultivated plants and anthropophilous weeds around him, with the original vegetation subsisting only in zones of extreme environmental conditions where human activities are not possible.

PROTECTION OF THE NATURAL VEGETATION IN ARGENTINA

Argentina has had a National Parks Law for nearly fifty years. Today, eighteen national parks and biological reserves totalling more than two million hectares exist in different regions of the country. The best known of these are the parks of Iguazú, of Nahuel Huapí and of Lago Argentino. Notwithstanding the numerous laws and regulations, however, the attraction of tourists to our national parks is held to be more important than is the conservation of the original ecosystems. In many areas, it is possible to see the introduction of exotic animals and plants, the presence of domestic herbivores such as cows and goats, and the exploitation of the forests for lumber. There are very few trained naturalists and ecologists employed by the National Parks System and of course those who are so employed have hardly any influence on the Parks Administration itself.

As far as I know, there is no official list of endangered plant species requiring protection. The destruction of valuable plants on private land is not at all restricted. All in all, a great deal still needs to be done on this subject.

CARLOS MUÑOZ PIZARRO (1913 – 1976)

An Appreciation

by Calvin J. Heusser

Editor's Note: During the Symposium, Carlos Muñoz Pizarro was scheduled to speak on Threatened and Endangered Species of Plants in Chile. The paper was never completely delivered since Prof. Muñoz died in action while presenting it. In place of his paper, therefore, we are publishing a biography of Prof. Muñoz, including his list of Threatened and Endangered Species of Plants in Chile.

The following message was sent to Chile by the participants in the Symposium:

"The participants in the Bicentennial Symposium on the Endangered and Threatened Plants of the Americas at The New York Botanical Garden wish to express to the family and the Government of Chile their heartfelt sympathy at the death of their world-renowned colleague, Prof. Carlos Muñoz Pizarro, while delivering his contribution on Threatened Plant Species of Chile. His loss is a serious blow to world plant conservation, playing as he did a leading role in Chilean plant conservation. While his loss is irreplaceable, we are confident that his many students will ensure the continuance of his pioneer work in the conservation of a unique natural heritage for Chile and the world."

Carlos Muñoz Pizarro, photographed at the Symposium (photo by Paulo de T. Alvim).

Carlos Muñoz was born in 1913 in the north of Chile in the port city of Coquimbo, "my province," as he fondly called it. His father was a port official there, and through him he came to know the ships from many nations that anchored in the harbor. It was during these early days of his youth that the first awakenings of a curiosity for foreign lands began. In his later life, Carlos Muñoz traveled widely to lecture, attend scientific meetings, and to visit herbaria and botanical gardens throughout the world. He visited such distant places as Honolulu, Tokyo, Christchurch in New Zealand, Florence, Geneva, Madrid, Paris, London, Copenhagen, New York, Washington, and Mexico City. But his emotional "roots" always remained in his beloved Coquimbo, his homeland, and he returned to it whenever the demands of his busy life permitted him this luxury.

His early education began at the Liceo Gregorio Cordovéz in La Serena, adjacent to the port of Coquimbo, and continued in Santiago at the Internado Nacional Barros Arana. He entered the Escuela de Agronomía at the Universidad de Chile in 1932 and graduated with distinction in 1937. During these years, he was assistant to the taxonomists, Francisco Fuentes M. and Victor M. Baeza. Under their guidance, he wrote his thesis on Chilean weeds (Contribución al Estudio de las Malezas Chilenas).

These formative years were shaped by a variety of influences, but mostly through work in agricultural botany, particularly with forage grasses. During the summers of 1935 and 1936, in the course of his university training, he returned to Coquimbo to study the plants of the *Aextoxicon-Drimys* cloud forests of Talinay and Fray Jorge. These forests are the northernmost in Chile and are located in the semiarid coastal ranges

(30°30' S. Lat.).   Almost daily collecting and study trips were made on horseback from nearby farms up onto the Altos de Talinay, and ultimately, a collection of 459 species and 48 genera of plants was assembled. Following two additional field seasons, the study was published in 1947 *(Estudio de la Vegetación y Flora de las Parques Nacionales de Fray Jorge y Talinay.* Agricultura Técnica 7, No. 2, Ministerio de Agricultura, Santiago). It remains to this day, thirty years later, the definitive reference for these distinctive forest communities.

Carlos Muñoz traveled to the United States for the first time in 1938. As a Fellow of the John Simon Guggenheim Memorial Foundation, he was to spend a year working in the Gray Herbarium at Harvard University. The occasion of his arrival in Boston was one of great expectation. But feelings of anxiety, so typical in someone about to take up foreign residence, were with him when he was arriving by train from New York. Arrangements had been made to meet a resident Chilean, who would accompany him during his early days in Cambridge and help get him settled. But how were they to meet in the busy train station in Boston? It turned out to be no problem. His fellow countryman was standing atop a bench singing the Chilean national anthem!

The year spent at Harvard brought him in contact with leading figures in botany. He studied with Oakes Ames, M. L. Fernald, Ralph Wetmore, Kenneth Thimann and Ivan Johnston. Dr. Johnston, his mentor at Harvard, had long been a student of the flora of northern Chile. Carlos Muñoz later acknowledged his debt to him by naming *Puccinellia johnstonii* in his honor in 1948 *(Cinco Especies Nuevas de Plantas para Chile.* Agricultura Técnica 8, No. 3) and by dedicating to him his *Sinopsis de la Flora Chilena* (Ediciones de las Universidad de Chile, 1959, 2ª edición 1966). Appropriately, the inscription reads: "A Ivan Murray Johnston quien me enseñó a observar la flor" (To Ivan Murray Johnston who taught me to observe the flower).

In 1939, his Guggenheim Fellowship was renewed. This second year was spent in Washington, D. C., working under the direction of Agnes Chase and Jason Swallen in the Grass Herbarium at the Smithsonian Institution. Concerned with the problem of forage in the drier parts of Chile, he reviewed the collections, systematics, and nomenclature of the Chilean grasses housed at the Smithsonian. The result was the publication in 1941 of the comprehensive *Indice Bibliográfico de las Gramíneas Chilenas* (Boletín Técnico No. 2, Ministerio de Agricultura, Santiago).

When Carlos Muñoz returned to Chile in 1940, he joined the Ministerio de Agricultura, holding various posts there until 1957. He was the Director, Departamento de Genética y Fitotécnia (1946 — 1948) and also served as Director, Departamento de Investigaciones Agrícolas (1948 — 1957). A noteworthy outcome of these years spent at the Ministerio was the publication in 1950 of *Siete Años de Investigación Agrícola* (Dirección General de Agricultura, Santiago). He both edited and co-authored this work, which stressed the importance of agricultural science to crop production. During this same interval, he rose through the ranks to become Profesor de Botánica at the Universidad de Chile and subsequently, was in charge of the herbarium (Sección Fanerógamas) at the Museo Nacional de Historia Natural.

When he took over supervision of the herbarium in 1942, he found that the collections sorely suffered from neglect. Painstakingly, he completely reorganized and modernized the material, mounting and labeling accumulated collections, remounting damaged and antiquated sheets, and housing specimens in newly-constructed cabinets. At the same time, he instituted a collection program in order to increase and update the holdings of the herbarium. He encouraged botanists and others to collect specimens, especially in the botanically little-known parts of Chile. His own collections, as a matter of record, are notably from the cloud forests of Fray Jorge and Talinay (1935, 1936, 1941, 1947), from the high cordillera of Bío-Bío (1941), from Atacama (1943, 1944), from Laguna de San Rafael in Aisén (1959), from the Archipiélago de Juan Fernández (1965), from Parque Nacional Puyehue in Osorno (1972, 1973), and lastly, from Isla de Chiloé (1974, 1976).

Perhaps his most important contribution to the herbarium at the Museo was the particular attention he paid to assembling the type material of Chilean plants. In 1961, the Rockefeller Foundation awarded him a grant, and with the help of his wife, Sra. Ruth Schick de Muñoz, he studied and photographed type specimens in herbaria in Kew, Paris, Turin, and Glasgow. Today, as a result of this labor, the herbarium contains over 80,000 specimens, some 10,000 of which are of grasses. In addition, the collections contain 95 percent of the types of Chilean grasses preserved as fragments and 6,100 photographs of types held elsewhere in world herbaria. The herbarium is now under the supervision of his daughter, Sra. Mélica Muñoz Schick, who has followed in her father's professional footsteps.

In 1948, Carlos Muñoz returned to the Gray Herbarium at Harvard as a Fellow of the Rockefeller Foundation. During this stay, he pursued still further his studies of Chilean plants in addition to adding to his knowledge by visiting herbaria in various parts of the United States. Two years later, on a grant from the United Nations, he traveled extensively throughout Europe, visiting over thirty agricultural institutions for the purpose of learning about the production of forage plants and cereals. In 1955 and 1969, he was the Chilean delegate to the Arid Lands Conferences in New Mexico and Arizona. He represented Chile at the IX International Botanical Congress held in Montreal in 1959, and also at the Tenth Pacific Science Congress in Honolulu in 1961. At both the X and XI International Botanical Congresses, convened in Edinburgh and Seattle in 1964 and 1969, he was named Honorary Vice-President. He was accorded this honor again in 1972 at the I Congreso Latinoamericano Botánica celebrated in Mexico City.

His long career in botany was a most productive one. His outstanding works not previously cited include: *Las Especies de Plantas Descritas por R. A. Philippi en el Siglo XIX* (Ediciones de la Universidad de Chile, 1960); *Flores Silvestres de Chile* (Ediciones de la Universidad de Chile, 1966); and *Chile: Plantas en Extinción* (Editorial Universitaria, Santiago, 1973). Works in press include: *Eucrifiáceas, Familia de Plantas que une Continentes* (Editorial Jurídica, Santiago) and *El Atlas de Claudio Gay. Reproducción Facsimilar de sus Principales Láminas* (Editorial Jurídica, Santiago). Among works in preparation are: *Los Arboles Chilenos* and *Los Géneros de las Plantas Chilenas*. This latter undertaking, planned in five volumes, is supported by a grant from the National Science Foundation.

His labors in botany, agricultural science, and conservation brought him recognition both at home and abroad during his lifetime. At home, he was honored by the Sociedad Agronómica de Chile with the *Premio al Mérito* (1961), by the Colegio de Ingenieros-agrónomos with the award *Mejor Investigador* (1966), and by the Academia de Ciencias of the Instituto de Chile, which made him *Académico de Número* and presented him with the *Premio de Ciencias Juan Ignacio Molina* (1967). Outside Chile, the Organization of American States (OAS) recognized his contribution to agriculture by awarding him the *Medalla Agrícola Interamericana* (1973), and the Prince of the Netherlands bestowed the honor of *Grade de Caballero de la Orden del Arca de Oro* (1974), this last deservedly given him as a scientist dedicated to the protection of natural resources.

The lasting heritage that Carlos Muñoz leaves for botany is a real and viable one. His ideals, his dedication to his profession, his love for mankind and for the natural environment, and the conduct of his life evoke the deepest feelings of admiration in all who knew him. He has earned a place of distinction with Claudio Gay, R. A. Philippi, Karl Reiche, Carl Skottsberg, and Ivan Johnston, botanists before him, who, like Carlos Muñoz, all loved the land and collected, studied, and described the Chilean flora.

THREATENED AND ENDANGERED SPECIES OF PLANTS IN CHILE

Carlos Muñoz Pizarro

Museo Nacional de Historia Natural, Santiago, Chile

FILICES
Arthropteris altescandens (Colla) J. Sm.
Dicksonia berteriana (Colla) Hook.
Dicksonia externa Skottsb.
Trismeria trifoliata (L.) Diels
Dryopteris spectabilis Macloskie & Dusen

PODOCARPACEAE
Dacrydium fonkii Ball
Saxegothaea conspicua Lindl.
Podocarpus andinus Poepp.
Podocarpus nubigenus Lindl.
Podocarpus salignus D. Don.

ARAUCARIACEAE
Araucaria araucana C. Koch.

CUPRESSACEAE
Austrocedrus chilensis (D. Don) Florin & Boutelje
Pilgerodendron uviferum Florin
Fitzroya cupressoides I. M. Johnston

LACTORIDACEAE
Lactoris fernandeziana Phil.

LAURACEAE
Persea lingue Nees
Beilschmiedia berteroana (Gay) Kosterm.
Beilschmiedia miersii (Gay) Kosterm.

GOMORTEGACEAE
Gomortega keule I. M. Johnston

BERBERIDACEAE
Berberis littoralis Phil.

KRAMERIACEAE
Krameria cistoides Hook. & Arn.
Krameria illuca Phil.

PROTEACEAE
Orites myrtoides Benth. & Hook.

FLACOURTIACEAE
Berberidopsis corallina Hook.

CARICACEAE
Carica chilensis (Planch. ex A.D.C.) Solms.

CACTACEAE
Trichocereus atacamensis (Phil.) Marshall & Bock
Browningia candelaris Britton & Rose

EUCRYPHIACEAE
Eucryphia glutinosa Focke

ROSACEAE
Polylepis tarapacana Phil.
Quillaja saponaria Poir.

LEGUMINOSAE
Balsamocarpon brevifolium Clos.
Caesalpinia spinosa (Milina) O. Kuntze
Prosopis chilensis Stuntz
Prosopis atacamensis Phil.
Prosopis tamarugo Phil.
+ *Sophora toromiro* Skottsb.
Sophora masafuerana Skottsb.
Sophora fernandeziana Skottsb.

MYRICACEAE
Myrica pavonis C. DC.

FAGACEAE
Nothofagus glauca (Phil.) Krasser
Nothofagus alessandrii Espinosa

SANTALACEAE
+ *Santalum fernandezianum* F. Phil.

BALANOPHORACEAE
Juelia subterranea Asplund

ANACARDIACEAE
Haplorhus peruviana Engl.

UMBELLIFERAE
Laretia compacta Reiche
Diposis tuberosa

OLEACEAE
Mendora linoides Phil.

SOLANACEAE
Nicotiana cordifolis Phil.

SCROPHULARIACEAE
Calceolaria picta Phil.

+Extinct

COMPOSITAE
Yunquea tenzii Skottsb.
Dendroseris litoralis Skottsb.

IRIDACEAE
Alophia lahue (Molina) Espinosa
Calydorea xiphioides (Poepp.) Espinosa
Tigridia philippiana I. M. Johnston

TECOPHILLACEAE
+*Tecophilaea cyanocrocus* Leyb.

ALSTROEMERIACEAE
Leonotochir ovallei Phil.
Alstroemeria gayana Phil. var. humilis

AMARYLLIDACEAE
Gethyum altropurpureum Phil.
Ancrumia cuspidata Harv. ex Baker
Garaventia graminifolia (Phil. f. ex Phil.) Looser.
Erinna gilliesioides Phil.

PALMAE
Jubaea chilensis Baill.
Juania australis Drude ex Hook.

ARACHNITACEAE
Arachnitis uniflora Phil.

CYPERACEAE
Chillania pusilla Roivainen

GRAMINEAE
Orthachne breviseta Hitchcock
Podophorus bromoides Phil.
Gramerium convolutum Desv.
Cinna valdiviana Phil.
Bromus mango Desv.

+Extinct

Section 4

Plant Groups Especially Prone to Endangerment

NEOTROPICAL SPECIES THREATENED AND ENDANGERED BY HUMAN ACTIVITY IN THE IRIDACEAE, AMARYLLIDACEAE AND ALLIED BULBOUS FAMILIES

Pierfelice Ravenna

Universidad de Chile, Santiago, Chile

During my travels throughout Latin America (1959-1976) in search of Iridaceae and other families of bulbous plants, I have observed the unfortunate destruction of woods, forests, cerrados, and even the vegetation high up on mountaintops, intentionally burnt, in most cases, by man, or even ruthlessly knocked down by bulldozers in order to establish pine or eucalyptus plantations. I have frequently visited the same area several times over a few years and have noted the rapid and utter disappearance of a plant population, due to such indiscriminate practices.

Some of the genera I am most concerned with cannot adapt themselves to live outside the forest, while other species I study have had their populations seriously reduced. Only a few groups can withstand even moderate changes in the ecosystem. *Cipura paludosa,* some *Trimezia* species, most of the *Trifurcia* species in the Iridaceae, *Zephyranthes* and *Habranthus* in the Amaryllidaceae, and *Nothoscordum* in the Alliaceae, are among these.

Below I list species that are in danger of extinction mainly because of the limited number of their individuals. Others that were saved by being cultivated, or for other reasons, are also included in the list.

IRIDACEAE

1. Trimezia sabinii (Lindl.) Ravenna, comb. nov.[*]

    Marica sabinii Lindley, Trans. Hort. Soc. **6**:75.tab. 1. 1826.

This species belongs to the group of plants included in *Neomarica.* Both *Trimezia* and *Neomarica* overlap in such a manner that it is not possible to maintain them as different genera. The recent discovery of several intermediate species has demonstrated and underlined this fact.

This handsome species, supposedly discovered in Africa, was seemingly native to the very reduced forests near Salvador, Brazil. Fortunately, it is now commonly cultivated in that town, as well as in Rio de Janeiro.

2. Trimezia northiana (Schneev.) Ravenna, comb. nov.

    Moraea northiana Schneevogt, Icones Plant : tabulae 41, 42. 1795.

Trimezia northiana inhabits the rocky cliffs above the sea near Rio de Janeiro, the neighboring islands, and also some small areas to the north. It was first collected on Rasa Island by Sir Joseph Banks. In 1964 I visited this island and noted the disappearance of the plant, apparently because of the people settled there.

A different form of the same species, which grows in the low woods of the *restinga* near Yacarepagua, Rio de Janeiro, is also endangered. The development of the area, and the increasing activity centering around the beach, have affected the permanence of this plant population.

3. Trimezia silvestris (Vell.) Ravenna, comb. nov.

    Iris silvestris Vellozo, Fl. Flum. 1:34, tab.82. 1825.

    Neomarica heloisamariae Occhioni, Rodriguesia **10**(20):80. 1946.

This delicate species is still to be found in very reduced populations in a few scattered places of the Brazilian States of Rio de Janeiro and São Paulo.

---

[*]Editor's note: The author has asked for our permission to make a number of nomenclatural new combinations in this paper in order to use the same names that will appear in his monograph to be published later.

The State of Espírito Santo has several *Trimezia* species which are allied to this, but they will soon be extinct unless they can be saved for horticulture. The main problem here is extermination of forests by fire, since these plants grow on flat ground where they have no protection.

4. Trimezia variegata (Mart. & Gal.) Ravenna, comb. nov.

Marica variegata Martens & Galeotti, Bull. Acad. Brux. **10**(2):112. 1843.

This species is often wrongly described in herbaria as "Neomarica gracilis," appearing so in Standley and Steyermark "Flora of Guatemala." Its range extends from Costa Rica to Mexico. In the latter country it is endangered by the increasing development of forest land for colonization.

5. Trimezia humilis (Lodd.) Ravenna, comb. nov.

Marica humilis Loddiges, Bot. Cab. 11:tab.1081. 1825.

This species still survives, growing on the rocks in the Floresta de Tijuca, above Rio de Janeiro.

6. Trimezia caerulea (Ker-Gawl.) Ravenna, comb. nov.

Marica caerulea Ker-Gawler, Edwards' Bot. Reg.: tab.713. 1823.

Several forms of this magnificent species are distributed from Rio de Janeiro to Rio Grande do Sul, in Brazil. It is related to *T. rupestris,* a species from Minas Gerais. Although seriously endangered in its wild state, it is commonly cultivated in several countries, including Brazil.

Fig. 1. Threatened and Endangered species of bulbous plants: A, Trimezia candida *(Hassl.) Rav.*

7. Trimezia glauca (Bak.) Ravenna, comb. nov.

Marica glauca Baker, Jour. Linn. Soc. London **16**:149. 1877.

This variable species inhabits the States of Minas Gerais, Rio de Janeiro, and São Paulo. It also is affected by the continuing destruction of the forest, its natural environment.

8. Trimezia candida (Hassl.) Ravenna, comb. nov.

Marica candida Hassler, Physis **6**:359. 1923.

As *T. caerulea,* this is endangered in its wild condition. Fortunately, it is widely cultivated in Argentina, Brazil, Paraguay, and the United States.

9. Trimezia brachypus (Bak.) Ravenna, comb. nov.

Marica brachypus Baker, Jour. Linn. Soc. London **16**:149. 1877.

This is apparently a rare species from Trinidad, and it may be affected by the same problems as its relatives.

10. Trimezia steyermarkii Fost.

Due to the rapid development of Guatemala's Department of Alta Verapaz, the forests of that region are being destroyed. *Trimezia steyermarkii* was at one time rather frequent in the vicinity of Cobán, growing under the shade of *Liquidambar styraciflua, Quercus* sp. and *Pinus* sp. The species is now extremely rare. Living plants were recently distributed to institutions or horticulturists of Guatemala City, Caracas, Rio de Janeiro and Santiago.

11. Trimezia lutea (Klatt) Fost.

This small cormous herb inhabits the hilly fields of Minas Gerais, Goiás, Maranhão, and the "llanos" of Central Colombia. The customary use of fire for land-clearing by peasants has endangered it considerably. At present, this herb is very rare, especially in Minas Gerais. Other species such as *T. violacea* and *T. spectabilis,* are in the same jeopardized situation.

12. Tigridia minuta Ravenna

High altitude agriculture has apparently reduced the range of this Peruvian *Tigridia* to a few small areas in the Department of Apurimac and Ayacucho. In 1969 it was described as being a very rare species.

13. Tigridia lutea Link, Klotzsch & Otto.

Tigridia lutea is confined to a few coastal *lomas,* coastal hills (described by Dr. Ferreyra) in this symposium, ranging from Lima to the northern village of Pativilca, in Peru. *T. lobata* (Herb.) Macbr., is another synonym for this species. The practice of grazing goats on the *lomas* is affecting the existence of this pretty plant.

14. Tigridia tepoxtlana Ravenna

This Mexican species inhabits a very restricted area near the ruins of a small Aztec temple in the State of Morelos, at the top of the Sierra de Tepoztlan. Although the latter is the correct name of the town, I have followed the label data of the type-specimen in naming the species. Tourism is modifying the environment, with the inevitable result that the number of individuals in this species is noticeably being reduced.

15. Tigridia flammea (Lindl.) Ravenna, comb. nov.

Rigidella flammea Lindley, Edwards' Bot. Reg.: tab.16. 1840.

The group of species hitherto placed in *Rigidella,* is closely related to *Tigridia pavonia,* the type species of *Tigridia;* much closer, in fact, than any of the other species of *Tigridia.*

Tigridia flammea, a native of the State of Michoacan in Mexico, is a rare species, probably already reduced in number of individuals as a result of man's activities in the area.

259

Fig. 2. Threatened and Endangered species of bulbous plants: B, Tigridia lutea *Link, Klotzsch & Otto.*

16. Tigridia immaculata (Lindl.) Ravenna, comb. nov.

Rigidella immaculata Lindley, Edwards' Bot. Reg. Misc. **62**, tab.68. 1841.

Since Guatemala is a comparatively sparsely populated country, this species is at present not yet affected or endangered. In view of this, prompt action should be taken in order to protect the plant for the future.

17. Tigridia orthantha (Lemaire) Ravenna, comb. nov.

Rigidella orthantha Lemaire, Fl. des Serres, Ser. II, **1**:107. 1845.

Growing at high altitudes in the State of Oaxaca, Mexico, where it is not seriously affected by man's activities, this species stands a better chance of surviving than do other members of its family.

18. Tigridia inusitata (Crud.) Ravenna, comb. nov.

Rigidella inusitata Cruden, Brittonia **23**:222. 1971.

A rather infrequent species from the State of Guerrero, Mexico.

19. Cypella pusilla (Link, Klotzsch & Otto) Benth. & Hook.f.

Though this insignificant species was recently common on the hills and dry fields near Porto Alegre, Rio Grande do Sul, Brazil, it has already disappeared from this area.

Fig. 3. Threatened and Endangered species of bulbous plants: C, Cypella osteniana Beauv.

Fig. 4. Threatened and Endangered species of bulbous plants: D, Nemastylis convoluta Rav.

20. Cypella osteniana Beauv.

This is another rare species, whose already few individuals are diminishing. It is to be found in only a few places in the Department of Mina, in Uruguay.

21. Nemastylis convoluta Ravenna

Described as recently as 1968, this species inhabits a very small area near the highway that joins Colima and Manzanillo, in the State of Colima, Mexico. Man's nearby activities will undoubtedly endanger this species in the future.

Nemastylis mcVaughii Crud. is a synonym for this species.

22. Eleutherine bulbosa (Mill.) Urb.

This species inhabits the tropical forests in Guatemala, Honduras, the West Indies, Brazil, and Bolivia. It is believed to bring good luck to people when it is planted in front of their houses. Because of this superstition it is widely cultivated, and therefore is not threatened at present.

23. Mastigostyla brevicaulis (Bak.) Fost.

This *Mastigostyla* species inhabits the mountain ridges above La Paz, Bolivia, and also the northern Bolivian valley of Sorata. Due to agriculture and erosion, it is becoming endangered.

24. Mastigostyla mirabilis Ravenna

This very pretty species is found only in a few places of the Calchaquíes mountains, in north Tucumán, Argentina. Each plant produces only a single, terminal, one-flowered spathe. Cattle seriously affect the already small populations.

Fig. 5. Threatened and Endangered species of bulbous plants: E, Mastigostyla mirabilis *Rav.*

25. **Libertia tricocca** Phil.

This Chilean member of the family is severely endangered by both fire and the modification in the ecosystem produced by man. It always forms small populations.

ALSTROEMERIACEAE

1. **Alstroemeria pelegrina** L.

This beautiful plant was first collected and illustrated in the 18th century by Feuilliée. At that time, and at least until the end of the last century, the species was rather common along the coast, in the provinces of Aconcagua, and in Valparaíso, especially near the town of Valparaíso. Now, unfortunately, it is confined to a few scattered places.

The species is found in stony or sandy sites near the seacoast but the increasing development of beaches for tourism has endangered it. Moreover, peasants and bathers alike find the flowers so attractive that they cut them for decorative purposes.

Fig. 6. Threatened and Endangered species of bulbous plants: F, Amaryllis aviflora *Rav.*

AMARYLLIDACEAE

1. Amaryllis aviflora Ravenna
A very small population of this handsome species was discovered at Rosario de la Frontera, Province of Salta, Argentina. Some years ago I attempted to find it in its native area, without success. Possibly, the spot was the remainder of a larger population. This is unknown.

2. Amaryllis arboricola Ravenna
This is another recently described species which was found to be growing, at almost 25 meters, on tall trees near El Dorado, Province of Misiones, Argentina. The forest now is being destroyed as a result of human settlement.

3. Amaryllis santacatarina Traub
This species inhabits the bogs of the States of Paraná and Santa Catarina in Brazil. I recently discovered it also in northern Rio Grande do Sul. Because cattle eat the leaves and flowers, the plants are unable to produce fruits and seeds and the weakened bulb does not produce offsets. Since the species multiplies only by means of seeds, it is becoming very rare. Only three to five individuals are found in an average population.

4. Amaryllis scopulorum (Bak.) Traub & Uphof
This fine species from Bolivia grows in small restricted areas near Sorata. On three successive opportunities when I was in the region I noted the gradual disappearance of the species. It is possible that the natives may cut the flowers or even dig the bulbs.

5. Amaryllis reginae L.
This beautiful species was one of the stocks from which the cultivated forms originated. The plants grow on high rocks and usually are out of reach, and therefore the species is not endangered at present.

6. Griffinia hyacinthina Ker-Gawl.
Most of the species of this genus are threatened because of the destruction of their natural forest environment. These plants cannot exist in open areas. *Griffinia hyacinthina* inhabits the rain forest near the coast of São Paulo.

7. Griffinia liboniana Lem.
This was discovered by Mr. Libon in the forests near Serra do Cipó, which are all but destroyed today. It has already disappeared from that area, but fortunately it was recently rediscovered in another reasonably undisturbed place. (see Ravenna, *in* Pl. Life 30:67. 1974).

8. Griffinia concinna (Mart. ex Roem. & Schult.) Ravenna
This originally was discovered by Martius in woods near Ouro Preto, Minas Gerais, Brazil. It was described as *Crinum concinnum,* and recently transferred by me to *Griffinia,* its proper genus. In 1963, I explored the region of Ouro Preto, but since the woods there had virtually disappeared, no plants of this species were to be found. Several other *Griffinia* species are similarly threatened in other regions.

9. Phaedranassa tunguraguae Ravenna
I found this species on the rugged slopes above the Pastaza River, near Banos, Province of Ambato, Ecuador. Only a few plants, possibly the remainder of a larger population, were to be seen. The same trail was walked by Richard Spruce in 1857.

10. The Urceolina species
All the *Urceolina* species, some of which were previously included in *Eucharis,* are now threatened because of the destruction of the rain forest. The genus ranges from Bolivia and Brazil to Panama.

LILIACEAE

1. Paradisea stenantha (Ravenna) Ravenna, comb. nov.

Anthericum stenanthum Ravenna, Bol. Soc. Arg. Bot. **11**:147. 1967.

I previously described this species as *Anthericum stenanthum* Rav. Recently, however, after reexamining the entire collection of the type-material (at TRP), I reached the conclusion that it rather belongs in *Paradisea* Mazzuc. The genus, therefore, contains three species: *P. liliastrum* (the type species) in western Europe, *P. bulbifera* in Tibet, and *P. stenantha* in Peru. The genus was hitherto unknown in the New World.

Paradisea stenantha occurs on a single low hill called Cerro de las Cabras, in the coastal formation of northern Peru. During the 'green' season (August-September) goats graze on this hill, where they eat the tender leaves of this and other species. The plants, whose roots extend rather superficially, grow on rocks and among mosses. Since it is quite easy to eradicate the plants, either accidentally or otherwise, the species is seriously endangered by herds of goats.

2. Chlorophytum brasiliense (Nees & Mart.) Ravenna, comb. nov.

Hagenbachia brasiliensis Nees & Martius, Nov. Act. Nat. Cur. **11**:19, tab. 2. 1823.

Hagenbachia brasiliensis was placed by the authors in the Haemodoraceae.

Thanks to Dr. P.J. Maas, of Utrecht, the Netherlands, who kindly lent me the type-material, it was possible to ascertain that this species belongs to *Chlorophytum*, in the Liliaceae.

Prof. E.P. Heringer recently found another species of the genus, growing in dark soil among calcareous rocks, at Fercal, near Brasília City. Apparently, this is the restricted habitat of the species. It was very unfortunate that shortly after it was found, a cement factory was constructed in the area, with the result that the entire calcareous hill where the species lived, was utterly destroyed. I visited the area recently but was unable to find the species.

ALLIACEAE

1. Speea humilis (Phil.) Loes.

Speea humilis is a Chilean member of the tribe Gilliesieae, of the Alliaceae. It is found in very small groups wedged among rocks. Both fire and wandering goats endanger them. As a matter of fact, all the genera in the tribe are becoming rare. Most of the flowers are zygomorphic except in this genus, and like some species of orchids, they resemble insects to attract their pollinators.

2. The Gilliesia species

Gilliesia, formed by four species, is also a Chilean genus. With the exception of *G. graminea,* the type species, the rest of the species are threatened, especially by fire and domestic cattle.

The above list includes only a small part of the endangered or threatened species in these families. In addition, many others are declining, and these also will inevitably disappear unless preventive action is taken soon. But what can be done to save these threatened species from extinction? The ideal solution, of course, would be to have a concerted, nationwide, mass educational program, one which would instill in the people themselves the sound, basic principles of ecology and wildlife protection. But though such extensive programs might be very effective, they unfortunately are very impractical in many countries because of the huge expenses involved. A good beginning in this direction can be made, however, by including basic courses on these subjects at the primary and high school levels. This certainly would be a positive step that would reap dividends for many years to come.

It also would be highly desirable if governments could evolve more rational use of their newly-developed regions, which would preserve portions of the wild areas as biological reserves or national parks. Advisory committees should be formed to work closely with local governments towards this goal. These committees should include specialists such as ecologists, taxonomists, economists, agronomists, and various other experts or technicians. Such committees could emphasize that wild areas, with their resources of raw material and energy, really are part of a nation's irreplaceable heritage, and therefore they must be preserved. This concept, admittedly, is difficult to instill in many of the new 'third-world' countries that are in the process of emerging and developing today, because the prevalent attitude in these 'young' countries so frequently is that "conservation of nature goes against progress." But some of the above-named specialists could be very effective if they worked with advisory committees to help overcome this erroneous attitude on the part of so many of these rapidly growing smaller countries.

Perhaps the most direct method of preserving these species would be to reproduce them horticulturally. The representatives of the families enumerated above show some noteworthy conditions. Most of them have bulbs, rhizomes, or tubercles, and they reach their reproductive size a few years after the seed germination. A reference to their ornamental value seems unnecessary here. However, domestication is assured only after several generations have been reproduced by seed.

It seems, therefore, that a relatively certain, albeit artificial, method for preserving the species from extinction could be their introduction to horticulture. Several successful past examples give strength to this alternative, the most striking being the case of *Tecophilaea cyanocrocus*. Though already extinct in its natural habitat, this Chilean species is intensively cultivated for its deep-blue flowers in certain European establishments.

Plant collectors also are beginning to gather living bulbs, offsets, or seeds, for distribution to specialists, botanic gardens, or plant nurseries. This is encouraging, and leads us to hope that these beautiful neotropical species, though endangered in their native habitat, will be saved from extinction.

ENDANGERMENT AT THE SPECIFIC
AND GENERIC LEVELS IN PALMS

Harold E. Moore, Jr.

L. H. Bailey Hortorium, Cornell University, Ithaca, New York 14853, U.S.A.

Palms are characteristic components of many tropical ecosystems (Moore, 1973a). They occur in a diversity of habitats, ranging from seacoasts, mangroves, desert oases, and open savannas to swamp forest, lowland and montane rain forests, and even to deciduous forests of warm-temperate parts of the world. Sometimes they form nearly pure stands of one species, as *Mauritia flexuosa* L.f. in the basins of the Amazon and Orinoco Rivers. At other times they may be abundant both in kinds and in numbers, as they are in the lowland rain forests of America and Indomalaysia, but they often are represented by very limited populations. Palms frequently serve as indicators of soil types, drainage patterns, or vegetation types (e.g., Eiten, 1974; Pérez Jiménez, 1974; Read, 1974; Romney, 1959), and they may also be very precise markers, such as the species of *Geonoma* in certain montane forest types of Venezuela (Otto Huber, pers. comm.). They occasionally are known to influence the formation of soil, as suggested by Furley (1975) for *Orbignya cohune* (Mart.) Dahlgr. ex Standl.

Though interrelationships between palms and animals are poorly documented, they certainly are important. Palm/insect relationships are perhaps best known, having been considered in general by Lepesme and Paulian (1947). The sometimes elegant methods of pollination, however, were not treated by the two authors and this subject has only recently begun to receive the detailed study it deserves (Brown, 1976a; Essig, 1971, 1973; Read, 1967, 1975; Schmid, 1970), as have the adaptive morphology and anatomy of the plants (Uhl and Moore, 1977). Palm flowers, inflorescences, and fruits are utilized by insects other than those that actually pollinate, as sources of food and sites for oviposition (Essig, 1973; Schmid, 1970). Bruchid beetles, for example, may feed on nectar of *Sabal palmetto* (Walt.) Lodd. ex Schult. & Schult. f. as adults and pass their larval instars in the endosperm of the fruit (Brown, 1976b).

The fruits of many palms are fleshy and colored, sometimes against bright red or orange inflorescence axes, and obviously adapted for dispersal by animals (Corner, 1966; Van der Pijl, 1969). Birds are probably the palm's chief feeders and disseminators (see Brown, 1976b; Keppler, 1970; Leck, 1969; Read, 1960), though mammals, from rodents to primates, including man, also feed on their fruits and disperse their seeds (Burtt, 1929; Enders, 1935; Janzen, 1971), even those such as *Caryota,* which are filled with irritant crystals (Dransfield, 1974).

Man benefits enormously from palms throughout the tropics, as accounts of many explorers amply demonstrate (for example Wallace, 1853). They serve him in almost every aspect of life, ranging from shelter to clothing, food, drink, stimulants, medicine, arms, and religion (Braun, 1968; Burkill, 1935; Dransfield, 1976a; Gowda, 1951; Hodge, 1963; Miller, 1964; Schultes, 1974). Palm products also play an important role in our industrial society (Hodge, 1975; Kitzke and Johnson, 1975).

Many other examples can be cited to show the importance of palms in the ecosystem. The family also fulfills another important scientific, though less immediately obvious, role, as a subject for the study of evolution. Distinctive pollen of the Asiatic mangrove palm, *Nypa,* is one of the earliest fossils identified as to family and genus (Muller, 1970), dating from the Senonian (Upper Cretaceous), about 70 or more million years ago. Although contemporary *Nypa fruticans* Wurmb. seems to differ little from its fossil antecedents and retains some characteristics considered primitive (Moore and Uhl, 1973), it is advanced in other characteristics and must represent a significant span of evolutionary time beyond the origin of the even more primitive palm stock from which it evolved. This long history probably accounts for the great diversity we find at the subfamilial, tribal, and generic levels within the family today (Moore, 1973b, 1975).

Diversity at the specific level is less well understood, owing to the fact that in the past, species were frequently described from fragmentary specimens that are difficult to compare with the often more complete specimens of modern collectors. Moreover, the great size of many palms deters most botanists from including them in their collections at all. As a consequence, it is frequently difficult to assess endangerment at the specific level, especially in the Americas. Because palms also provide us with exceptional material for the study of evolution at the generic level, and because more than one-third are monotypic and nearly one-half have only one or two species, I have chosen, with permission of the organizers of this Symposium, to broaden my approach in order to call attention to endangerment in both hemispheres since some of the more clearly documented examples come from the Old World.

Despite their versatility in the ecosystem, palms as a group have a great disadvantage. A few are notable colonizers of disturbed habitats, examples being *Pigafetta filaris* (Giseke) Becc. after clearing in the Celebes (Dransfield, 1976b), *Prestoea montana* (Grah.) Nichols. *("Euterpe globosa")* after hurricanes in the Lesser Antilles (Beard, 1945, 1976), an unidentified palm on volcanic flows in Costa Rica (Gary Hartshorn, pers. comm.), and *Acrocomia* in Costa Rican pastures (Janzen, 1971). Most palms, however, appear to require precise conditions for germination and establishment, although few adequate studies have been made in this regard, those of Bannister (1970) and of Vandermeer et al. (1974) being exceptions. Palms are often commanding presences left standing when the forest is cleared, but they do not regenerate until their requirements for shade and moisture are met by regrowth of forest following shifting agriculture. When cleared land is retained in pasture, as in the Sarapiquí Valley of Costa Rica or on the slopes of the Andes in Colombia and to an increasing degree elsewhere, regeneration is severely limited or fails to occur at all.

Palms have another disadvantage. They are often overutilized by man. Each stem has a single growing point, and when this is cut for the tender "heart" or terminal bud, the stem or the plant, when the stem is solitary, is destroyed. Such destruction appears to have been a major factor in the virtual elimination of palms as wild plants on Mauritius and a similar elimination of *Euterpe macrospadix* Oerst. is at present taking place in Costa Rica (Balick, 1976). A less immediate threat, but one which in time is expected to become more serious, is the constant collection of fruit or seed in the wild for sale, or the continued cutting of young leaves to be used for hats, baskets, and other items. Another constant threat to palms is the excess cutting of mature leaves for thatch or for sale as greenery (Vosters, 1975).

Palms in cultivation (and potentially in the wild state) also are jeopardized by the increasing incidence of lethal yellowing, a disease attributed to the presence of mycoplasmalike organisms in the phloem of palms, transmitted by an as yet unknown vector (Fisher et al, 1973; Parthasarathy and Fisher, 1973; Romney et al, 1976).

The Threatened Plants Committee of the International Union for Conservation of Nature and Natural Resources has recently set up a Threatened Palms Subcommittee. Six species of particular interest are already listed as vulnerable — *Caryota no* Becc., *Johannesteijs-mannia altifrons* (Rchb. f. & Zoll.) H. E. Moore, *Juania australis* (Mart.) Drude ex Hook. f., *Lodoicea maldivica* (J. F. Gmel.) Pers., *Maxburretia rupicola* (Ridl.) Furtado, and *Phoenix theophrasti* Greuter — and two — *Medemia argun* (Mart.) Wuerttemb. and *Neoveitchia storckii* (H. Wendl.) Becc. — as endangered. Five more, which are probably endangered or even extinct, have been documented for consideration by the Threatened Plants Committee, but our work has only begun, as the following comments on palms of America, Africa, Asia, and oceanic islands will suggest.

ENDANGERED PALMS IN THE AMERICAS

There are so few proper monographic treatments of palms in the American tropics that it is difficult to assess endangerment in larger genera, many species of which can only be listed as insufficiently known; that is, they are suspected of being rare, vulnerable, or endangered, but current information is insufficient to categorize them. Too many species are still known only from a single collection and too much of the area still needs to be explored for palms.

The monotypic *Itaya amicorum* H. E. Moore from Peru may serve as an example. Discovered originally in 1960 while crossing from the Itaya River to the Amazon, and described twelve years later (Moore, 1972) after several attempts to obtain more complete material, the species is still known from fewer than 100 individuals in what constitutes, essentially, a single population adjacent to a clearing that is being extended into the forest and in the vicinity of some dwellings. If one judges from current evidence, *Itaya* must be considered endangered, yet there has been no effort to determine the extent of its range and any attempt to do so will be severely handicapped by the difficulty of travel in the region. The single introduction of this palm, truly one with potential as an ornamental, failed in 1974 when the entire shipment of fruit was "cooked" because some seeds were found to be infested with larvae.

Similarly, on the basis of current knowledge, at least three other species of Peru — *Chrysallidosperma smithii* H. E. Moore, *Iriartella ferreyrae* H. E. Moore, and *Socratea salazarii* H. E. Moore — may be endangered. Each is known from only two small areas in Peru, one near Aguaytía, the other near Yurimaguas. The region near Aguaytía where these palms grew has been much modified since 1960, and although a few individuals of *Socratea salazarii* were seen in 1974 in a ravine much disturbed by debris from a road cut, the natural vegetation of the region where *Chrysallidosperma* had been found appears to have been cut down. It is likely that other populations exist, but the determination of the range and the size of populations is very difficult to ascertain.

Since I have just returned from field work in Colombia, let me introduce some situations that are clearer. Populations of wax palms *(Ceroxylon* spp.) in the Andes from Venezuela to Peru and Bolivia occur mostly at high altitudes, where forest has been or is being cleared and kept in grass, or more rarely at elevations as low as 1,500-1,900 m in the region where coffee is grown. Only recently has the identity of the original species, *C. alpinum* Bonpl. ex DC. from the Quindio Pass in Colombia, become clear (Moore and Anderson, 1976). Because of the forest, the 80 km journey from Ibague to Cartago over the pass, required 10-25 days in the early 1800's (Bomhard, 1937), whereas a paved road now carries one there in hours. On the eastern side of the pass, the road winds through pastures where *Ceroxylon quindiuense* (Karst.) H. Wendl. (Colombia's national tree, once cut by the thousands for wax, [André, 1878]) is still extant though with little evident regeneration. On the western side, at lower elevations among coffee plantations, *Ceroxylon alpinum* is occasionally to be seen, though also with little evidence of regeneration. On the other side of the Cauca Valley, populations of *C. alpinum* share the same fate. One can still find specimens, but low population levels and lack of regeneration suggest that the species should be considered endangered.

Ceroxylon quindiuense, growing as it does at higher elevations, may still be seen in small patches of forest, as at Tenerife in the Departamento del Valle, but it is vulnerable. A similar situation prevails in Peru where only very limited populations of the complex centered on *Ceroxylon crispum* Burret are known (personal observation, 1960, 1974). Because of their habitat and the difficulty of finding appropriate regions in which to cultivate them, all species of *Ceroxylon* appear increasingly endangered unless steps are taken to protect wild populations and to plant and protect young trees.

Slightly lower, on slopes bounding the Cauca Valley at elevations from 900 to 1,200 m, *Syagrus sancona* Karst. once was abundant. Today the last remnants of forest where it occurs are being cut (Fig. 1a). The species is frequently left in pastures (Fig. 1b), where it does not reproduce. Though it is cultivated as an ornamental throughout the valley and even on the western slopes of the Cordillera Occidental, it is clearly endangered as a wild species. A comparable species is *Aiphanes caryotifolia* (HBK) H. Wendl., which is cultivated for the edible fruits, though it also is infrequent as a wild plant.

Elsewhere in South America, clearing of land, especially in montane areas, is reducing or eliminating palm populations. *Jubaea chilensis* (Mol.) Baill., the Chilean wine palm, once had a more extensive range along the west coast of Chile, but because it has been cut to extract the sap for honey and wine, the populations today are reduced to five from Cuesta Las Palmas in the north to El Almacigo in the south, and it is considered endangered by Chilean botanists. *Juania australis,* also monotypic, occurs only on the island of Masatierra in the Juan Fernandez Islands off the coast of Chile. Here, probably 500-1,000 or perhaps even more individuals still inhabit forests on the relatively undisturbed and often inaccessible upper slopes of one part of the island (Moore, 1969b) where it regenerates well. The islands were declared a national park in 1935, and the

269

Fig. 1. Syagrus sancona *once was widespread in the Cauca Valley of Colombia in forest like the remnants (a) now being felled near Alcalá, Valle. Although still remaining in fields and pastures as at La Virginia, Risaralda (b), it does not reproduce effectively without the intervention of man.*

Fig. 2. Gaussia princeps *is restricted to limestone haystack hills (mogotes) in Pinar del Río, Cuba, where it would be vulnerable if the hills were mined for lime, as they are in Malaya.*

palm therefore is now protected by law. So long as the sale of tourist items made from the hard black fibers of the trunk is controlled, *Juania* is probably only vulnerable, though possibly endangered.

In the West Indies, *Gaussia princeps* H. Wendl. is restricted to the *mogotes* (haystack hills) of dogtooth limestone (Fig. 2) in Pinar del Río province of Cuba (Leon, 1946) and must be considered vulnerable. If these hills were to be mined for limestone, as similar hills are in Malaya, the species would clearly be endangered. The related *Gaussia attenuata* (O. F. Cook) Becc. of Puerto Rico, occupies a more extensive limestone area in the north as well as limestone hills in the southwest (Little and Wadsworth, 1964) but it is considered endangered (Roy O. Woodbury, pers. comm.). *Calyptronoma rivalis* (O. F. Cook) L. H. Bailey, also from Puerto Rico, is known from only one locality three miles east of San Sebastián in the northwestern portion of the island (Little et al., 1974) plus two recently discovered small Puerto Rican populations (Roy O. Woodbury, pers. comm.) or, if Wessels Boer (1968) is followed, from two additional localities in Hispaniola. It must surely be listed as endangered. Efforts should be made to introduce this species into cultivation in Puerto Rico as well as elsewhere.

Some other palms in the West Indies also occur in very limited populations and should be considered for conservation. On Hispaniola, *Zombia antillarum* (Descourt.) L. H. Bailey is known from a few localities (Bailey, 1939; Jiménez, 1960), *Coccothrinax ekmanii* Burret, sometimes separated as *Haitiella ekmanii* (Burret) L. H. Bailey, is reported from two or perhaps three localities, and *Pseudophoenix ekmanii* Burret and *P. lediniana* Read are known from very limited populations (Read, 1968). The genus *Coccothrinax,* now being studied by Read, may provide additional examples of species that require action on the part of our subcommittee.

The genus *Chamaedorea,* abundantly represented in Mexico and Central America, contains several species that are apparently very rare and presumably endangered — *C. stolonifera* H. Wendl. ex Hook f., from Chiapas, *C. tuerckheimii* (Damm.) Burret from Guatemala being two examples. *Colpothrinax cookii* Read, is reported from two widely separated localities in Guatemala and Panama (Read, 1969) and is probably endangered in both localities. Two species of *Brahea, B. decumbens* Rzedowski and *B. moorei* L. H. Bailey ex H. E. Moore, occur in limited populations, the latter with *Chamaedorea radicalis* Mart. on limestone outcrops of eastern Mexico, where they may be considered vulnerable. As in South America, tropical North America still requires exploration and study of its palms before all the possibly endangered species can be listed.

At least two species in the United States are endangered. *Roystonea elata* (Bartr.) F. Harper, which was at one time quite abundant in Florida's Fakahatchie Swamp, is now much depleted and is protected only in a small area at Royal Palm Hammock in Collier County in that state. Further taxonomic study is desirable to determine whether this species is truly distinct from the more abundant Cuban (and probably Mexican) populations of *Roystonea regia* (HBK) O. F. Cook. The second species, *Pseudophoenix sargentii* H. Wendl. ex Sarg. subsp. *sargentii* is represented only by a few individuals remaining on Elliott Key, Long Key, and Sands Key, Florida, and probably at its two localities in Mexico and one in Belize as well. Other populations in the Bahamas, Cuba, and Hispaniola represent a different subspecies (Read, 1968).

AFRICA AND THE MEDITERRANEAN

Africa is a continent with a limited complement of palms today, though the continent is considered to have been part of the original center of palms (Moore, 1973a). In the south, *Jubaeopsis caffra* Becc., a monotypic genus, is limited to the northern banks of two rivers in South Africa — the Msikaba and the Mtentu — where its status in terms of endangerment is presently being investigated. The genus is of particular interest both because of its apparently unspecialized nature among the cocosoid palms and also its disjunct distribution. All other genera except *Cocos* itself and one species of *Elaeis* are now restricted to the Western Hemisphere.

In the north, the monotypic genus *Medemia argun* is listed as endangered. It occurs as single or few individuals in three localities in Egypt and one in the Sudan (Ahti et al., 1973; Boulos, 1968) and although fruits have been found in Egyptian tombs, the nature of staminate and pistillate flowers is insufficiently known. Thus the relationship between this species and *Bismarckia nobilis* Hildebrandt & H. Wendl. from Madagascar, which is sometimes considered a species of *Medemia,* cannot yet be fully understood and may never be understood if conservation is not effected.

Wissmannia carinensis (Chiov.) Burret (Fig. 3) is a third monotypic genus known from two limited regions in Africa — the Oasis of Uncad or Uncud in the Somali Republic, and a series of springs in the Monts Goda, in the Territoire Français des Afars et des Issas —

Fig. 3. A small group of Wissmannia carinensis *is framed against the forbidding desert hills among which it grows near Ronda, in the vicinity of the main population at Bankoualé, Territoire Français des Afars et des Issas.*

and one in Arabia (Monod, 1955; Moore, 1971), although the species has recently been introduced into cultivation. The population at Uncad is reported to be much reduced while that in the Hadhramaut (Burret, 1943) is little known. If the principal population at Bankoualé is at all typical of the other six at Monts Goda, then the species is clearly endangered. At Bankoualé, about 90 individuals cluster by a trickle of water. All are mature and of substantial height. Reproduction is apparently absent because the ubiquitous goats eat every shred of green, including the fallen leaves. If a proposal to fence this population has been acted upon, then there may be some hope of limited regeneration. Since the palm has many of the ornamental attributes of *Washingtonia robusta* H. Wendl., it may ultimately be reproduced in cultivation.

The island of Crete is the home of *Phoenix theophrasti,* known from five localities along the coast, where use of the largest grove by tourists and campers prevents regeneration. Since other groves consist of only a few individuals (Barclay, 1974; Greuter, 1967), its status is vulnerable. The present extent of the sole European species, *Chamaerops humilis* L., has not been studied but its distribution is certainly more limited than formerly and a survey of localities and numbers is in order.

ASIA AND INDONESIA

Several species of continental Asia and the larger islands of Indonesia are vulnerable, endangered, or perhaps even extinct. The two species of the genus *Calospatha* from Perak, West Malaysia — *C. confusa* Furtado and *C. scortechinii* Becc. — are known only from the fragmentary type collections, each from a different mountain where some forest remains but where a recent effort to locate *C. confusa,* at least, met with failure. If these species are extinct, we will never have a complete understanding of the morphology or biology of the genus, which is a problematic one in terms of position among the scaly-fruited lepidocaryoid palms of the rattan group. If they are not extinct, these species must surely be considered endangered.

Ceratolobus glaucescens Bl., another rattan of more than usual interest is, so far as is known, now restricted to a single population at Sukawayana, West Java. It is estimated at 30 plants or fewer, and is about equally divided between staminate and pistillate individuals. The genus is a small one, characterized by a greatly reduced inflorescence enclosed within a prophyll that opens only at the tip. Both reproductive cycle and morphology are therefore of exceptional interest to the student of palm evolution.

One of the most primitive of palms, the monotypic coryphoid genus *Maxburretia rupicola,* is confined to three limestone hilltops within 40 kms of Kuala Lumpur in West Malaysia (Whitmore, 1971), where probably fewer than 1,000 individuals now remain. These hills are relicts of an ancient calcareous mantle and have a rich endemic flora that is threatened by quarrying at Batu Caves and by fire at one or both of the other localities despite their protected status. Again, a genus now listed as vulnerable is threatened, and two of its closest relatives, also monotypic (and one as yet undescribed), are very rare indeed.

All four species of another coryphoid genus, *Johannesteijs mannia* (Dransfield, 1972), appear to be threatened despite their occurrence in forest preserves and *J. altifrons* is listed as vulnerable. It does not survive clear-felling of trees, though it can survive with some damage when logging is selective. It does not appear in secondary regrowth.

The status of other species of continental Asia and Indonesia is presently being studied and a longer, perhaps much longer, list of vulnerable and endangered species is to be expected.

OCEANIC ISLANDS

Some of the most unusual palms are endemic on both large and small islands, especially those of the Indian and Pacific Oceans. Two relatively large islands, Madagascar and New Caledonia, offer an interesting comparison. Madagascar, in the Indian Ocean, is about 1,000 miles long. It is of interest to the student of palms because of the large number of species there (about 115) that are described in relatively few genera. Of the 18 genera listed by Jumelle and Perrier de la Bathie (1945), five are also African

(*Hyphaene, Borassus, Raphia, Elaeis, Phoenix*) and it is doubtful that even the species are distinct. One genus, the monotypic *Bismarckia,* is closely related to or sometimes even considered generically identical to the endangered *Medemia* of Africa. Twelve genera are endemic to the island or to its outliers as far north as Pemba. An added genus, *Marojejya,* was described by Humbert in 1955. Generic limits are not yet completely worked out and it is likely that there will be further consolidation among *Chrysalidocarpus, Neophloga, Neodypsis,* and *Antongilia.* Doubtless, many species of the larger genera are threatened by the unceasing destruction of forests, especially along the east coast, which in many areas has become a jungle of *Ravenala.* One of the classic localities for palms, the Forêt d'Analamazoatra near the railway station at Perinet, halfway between Tananarive and Tamatave, is now much degraded and reduced in size, with only remnants of its former palm flora remaining (Moore, 1965). *Vonitra utilis* Jumelle, known only from this forest, persists as fewer than half a dozen trees, according to my count in 1963: *V. fibrosa* (C. H. Wright) Becc. [*V. thouarsiana* (Baill) Becc.] is not to be found there, though it occurs elsewhere. *Louvelia lakatra* Jumelle and *L. madagascariensis* Jumelle & Perr. have been searched for in vain on several occasions. *Ravenea robustior* Jumelle & Perr. is reduced to a few individuals, while *R. latisecta* Jumelle has not been seen, and the monotypic *Beccariophoenix madagascariensis* Jumelle & Perr. has apparently become extinct there, though it is reported on the Masoala peninsula.

It is the decimation of the generic representation that is particularly distressing. *Beccariophoenix* is apparently being exploited to near extinction in its remaining habitat on the Masoala peninsula, according to reports, even though it is not yet fully understood botanically. The monotypic *Masoala madagascariensis* Jumelle is known to me only from three individuals in the Forêt de Mahavinitra near Ambohitralanana, along with a limited population of the similarly monotypic *Sindroa longisquama* Jumelle, an exceptional palm that occurs elsewhere in limited numbers on the peninsula (Bernardi, 1974) and is related to *Orania* of Indomalaysia. The monotypic *Marojejya insignis* Humbert is known only from three localities — eastern slopes of Marojejy, rivière Anove on the east coast (Humbert, 1955), and in the Forêt d'Ankiririryra, where I visited the very small population in March, 1971, following directions given me by the late René Capuron. All of these monotypic genera, and probably *Louvelia* and *Vonitra* in their entirety as well, are apparently endangered.

New Caledonia in the Pacific is, in contrast, about one-fourth as long as Madagascar and much narrower. It has an extraordinary assemblage of 18 endemic genera, five of them as yet undescribed (Moore, in manuscript), and about 30 species. Low population density, selective felling in the forested areas, poor agricultural quality of many soils, and the preservation of the Panié Massif in a botanical reserve, have thus far all served to relieve pressures on most New Caledonian palms. Even so, the only known population of *Kentiopsis oliviformis* Brongn. near Bourail, the small population of a new species of *Cyphophoenix* on Lifou, the populations of *Cyphophoenix elegans* (Brongn. & Gris.) H. Wendl., and an undescribed genus from the Haute Mayavetch, as well as the two very limited populations of *Burretiokentia hapala* H. E. Moore, may be considered candidates for a list of protected species, all being somewhat vulnerable.

The exception to the above is the monotypic genus *Pritchardiopsis* that once grew near Prony. It was apparently cut by convicts for the cabbage and has been searched for on the ground and by helicopter without success. It must be considered extinct unless some very limited population elsewhere has escaped the intensive search.

Among smaller islands, the four endemic palms of Lord Howe Island off the coast of Australia are examples of palms that are carefully husbanded by the local inhabitants as a source of commercial seed, although it is necessary to protect, with wire, the inflorescences of the small population of *Lepidorrhachis mooreana* (F. Muell.) O. F. Cook, in order to prevent loss of the seeds to rats. And in the Ryukyu Island, the very small population of *Satakentia liukiuensis* (Hatus.) H. E. Moore (Fig. 4) on Iriomote is remote from habitations (Moore, 1969a), the population on Ishigaki (Fig. 5) has been set apart as a reserve, and plants appear to be doing well in cultivation. Although all are probably vulnerable, the indigenous palms of the Seychelles, including the extraordinary coco-de-mer, *Lodoicea maldivica,* are now protected in reserves at the Vallée de Mai and Fond Ferdinand on Praslin Island or occur *(Deckenia, Roscheria)* in proposed nature

Fig. 4. Satakentia liukiuensis *in the wild on Iriomote, Ryukyu Islands.*

Fig. 5. The population of Satakentia liukiuensis *at Yonehara Village on Ishigaki Island consists of perhaps a thousand trees and is protected as a national monument.*

reserves on Mahé (Swabey, 1970). And it is to be hoped that *Neoveitchia storckii* will be protected in Fiji (Gorman and Siwatibau, 1975).

Not so comforting is the situation in the Mascarenes, three islands set in the Indian Ocean, which once contained an apparently prominent complement of palms that were decimated by man many years ago for construction purposes and for food. Five genera are native to the islands — *Latania* with three dioecious species (which unfortunately hybridize in cultivation), *Hyophorbe* with five species, *Dictyosperma* and *Acanthophoenix*, each with one variable species, and an undescribed monotypic genus (Moore, in manuscript). On Réunion, *Latania lontaroides* (Gaertn.) H. E. Moore persists as individuals remaining in

Fig. 6. The last individual of true Hyophorbe amaricaulis *on Mauritius.*

tilled land, *Dictyosperma album* (Bory) H. Wendl. & Drude ex Scheff. is planted and an exceptional specimen or two has been seen on nearly inaccessible cliffs. Both are endangered. Since *Hyophorbe indica* Gaertn. has a bitter cabbage and also grows in almost inaccessible places or on land unfit for agriculture, there is a series of small populations totalling more than 100 but probably fewer than 500 individuals. *Acanthophoenix* has natural populations in the Forêt de Bebour and in forested patches near the sea. Both are vulnerable, though *Acanthophoenix* is cultivated as an ornamental and is being investigated as a potential commercial source of palm heart.

Mauritius, the largest and oldest of the islands in the Mascarenes, had its own species of *Latania*, *L. loddigesii* Mart., which apparently was once abundant in coastal savannas. Today, only a few individuals persist on tiny islands off the coast — Coin de Mire and Round Island (where most regeneration is precluded by the activity of goats and rabbits) — or in cultivation: its status is presently endangered. A few individuals of *Dictyosperma album* also persist on Round Island and in cultivation. Occasional young plants, in addition to an individual or two of *Acanthophoenix*, are rarely encountered in the low scrub forest at the center of Mauritius. Though these individuals were probably disseminated by birds, they are usually cut as soon as the bud is large enough to eat, and their status is therefore endangered. Three species of *Hyophorbe* are also endangered: *H. amaricaulis* Mart. is represented by a single individual (Fig. 6), which was recognized only recently; *H. lagenicaulis* (L. H. Bailey) H. E. Moore barely survives on Round Island, but is cultivated both on the main island and elsewhere; and four individuals of *H. vaughanii* L. H. Bailey are known, all in cultivation. In the center of the island, 22 or 23 individuals of an undescribed genus related to *Acanthophoenix* are endangered, although four of these have already been protected and many of the remainder occur in a reserve. The palm produces little fruit and has not yet been successfully brought into cultivation.

Worst of all is the island of Rodrigues, now a botanical "poorhouse." Perhaps only four or five individuals of *Hyophorbe verschaffeltii* H. Wendl. persist outside cultivation, where it is far from abundant; a few individuals of *Dictyosperma* remain in tilled land or are deliberately planted; and the few *Latania verschaffeltii* Lem. that remained in a tiny patch of forest appeared to be in distinct danger in 1972.

The Mascarenes provide a splendid example of man's destruction of genetic potential in both palms and other plants, and regrettably this example is being followed elsewhere in the world. In late 1974, I had as a field companion a recent graduate of a small college who, travelling on a fellowship, teamed up with me for two months. At one point in our travels, he asked me why I collected palms and I replied to the effect that I was trying to make a record for future generations of what had existed. There was no immediate reply, but nearly a year later, following the destruction for road materials of a *campina* area in Amazonian Brazil upon which the young man had worked, he wrote to tell me that he had thought my reply pompous at the time, but that now he understood exactly what I meant. I can only hope that enough additional humans also will understand this important message while there still is time to save some palms other than coconuts and African oil palms.

ACKNOWLEDGEMENTS

I am indebted to the Threatened Plants Committee Secretariat for information on those palms that have been documented thus far and to John Dransfield and Victor Manuel Patiño for discussions while in Colombia. Howard H. Lyon and Donald C. Steinkraus assisted with illustrations and Lucille S. Herbert with the manuscript. Much of the background for this paper was obtained through field work undertaken in association with National Science Foundation grants GA-239, GB-1354, CB-3528, GB-7758, and GB-20348X.

LITERATURE CITED

Ahti, T., L. Hämet-Ahti, and B. Petterson. 1973. Flora of the inundated Wadi Halfa reach of the Nile, Sudanese Nubia, with notes on adjacent areas. Ann. Bot. Fenn. **10**:155.

André, E. 1878. L'Amerique Équinoxiale. Le Tour du Monde **35**:224.

Bailey, L. H. 1939. New Haitian genus of palms. Gentes Herb. **4**:237-246.

Balick, M. J. 1976. The palm heart as a new commercial crop from tropical America. Principes **20**:24-28.

Bannister, B. A. 1970. Ecological life cycle of *Euterpe globosa* Gaertn. Pages B299-B314 *in* H. T. Odum and R. F. Pigeon, eds. A tropical rain forest. Div. Tech. Info., U.S. Atomic Energy Comm., Oak Ridge, Tenn.

Barclay, C. 1974. A new locality of wild *Phoenix* in Crete. Ann. Mus. Goulandris **2**: 23-29.

Beard, J. S. 1945. The progress of plant succession on the Soufrière of St. Vincent. J. Ecol. **33**:1-9.

_____. 1976. The progress of plant succession on the Soufrière of St. Vincent: observations in 1972. Vegetatio **31**:69-77.

Bernardi, L. 1974. Notulae ad *Sindroa,* genus endemicum Palmarum peninsulae Masoala, Madagascariae; cum digressione circum Humbertiam Madagascariensim. Candollea **29**:163-171.

Bomhard, M. L. 1937. The wax palms. Smithsonian Report for 1936:303-324. (Publ. 3429).

Boulos, L. 1968. The discovery of *Medemia* palm in the Nubian Desert of Egypt. Bot. Notiser **121**:117-120.

Braun, A. 1968. Cultivated palms of Venezuela. Principes **12**: 39-103, 111-136.

Brown, K. E. 1976a. Ecological studies of the cabbage palm, *Sabal palmetto.* I. Floral biology. Principes **20**:3-10.

_____. 1976b. *Idem.* II. Dispersal, predation, and escape of seeds. Principes. **20**:49-56.

Burkill, I. H. 1935. A dictionary of the economic products of the Malay Peninsula. 2 vols. Crown Agent for the Colonies, London.

Burret, M. 1943. Die Palmen Arabiens. Bot. Jahrb. Syst. **73**:175-190.

Burtt, B. D. 1929. A record of fruits and seeds dispersed by mammals and birds from the Singida District of Tanganyika Territory. J. Ecol. **17**:351-355.

Corner, E. J. H. 1966. The natural history of palms. Weidenfeld and Nicolson, London, 393 pp.

Dransfield, J. 1972. The genus *Johannesteijsmannia* H. E. Moore, Jr., Gard. Bull. Straits Settlem. **25**:63-83.

_____. 1974. Notes on *Caryota no* Becc. and other Malesian *Caryota* species. Principes **18**:87-93.

_____. 1976a. Palms in the everyday life of West Indonesia. Principes **20**: 39-47.

_____. 1976b. A note on the habitat of *Pigafetta filaris* in North Celebes. Principes **20**:48.

Eiten, G. 1974. An outline of the vegetation of South America. Proc. Symposium 5th Cong. Internat. Primatological Soc. 529-545.

Enders, R. K. 1935. Mammalian life histories from Barro Colorado Island, Panama. Bull. Mus. Comp. Zool., Harvard Univ. **78**:385-502.

Essig, F. B. 1971. Observations on pollination in *Bactris.* Principes **15**:20-24, 35.

_____. 1973. Pollination in some New Guinea palms. Principes **17**:75-83.

Fisher, J. B. 1973. Report of the lethal yellowing symposium at Fairchild Tropical Garden, Miami. Principes **17**:151-159.

Furley, P. A. 1975. The significance of the cohune palm, *Orbignya cohune* (Mart.) Dahlgren, on the nature and in the development of the soil profile. Biotropica **7**:32-36.

Gorman, M. L. and S. Siwatibau. 1975. The status of *Neoveitchia storckii* (Wendl.): a species of palm trees endemic to the Fijian island of Viti Levu. Biol. Conserv. **8**:73-76.

Gowda, M. 1951. The story of pan chewing in India. Bot. Mus. Leafl. **14**: 181-214.

Greuter, W. 1967. Beiträge zur Flora der Südägäis 8. *Phoenix theophrasti,* die wilde Dattelpalme Kretas. Bauhinia **3**: 243-250.

Hodge, W. H. 1963. Toddy collection in Ceylon. Principes **7**: 70-79.

_____. 1975. Oil-producing palms of the world — a review. Principes **19**: 119-136.

Humbert, H. 1955. Une merveille de la nature à Madagascar. Première exploration botanique du Massif du Marojejy et de ses satellites. Mem. Inst. Sci. Madagascar, ser. B, Biol. Veg. **6**: 1-210.

Janzen, D. H. 1971. The fate of *Scheelea rostrata* fruits beneath the parent tree: predispersal attack by bruchids. Principes **15**: 89-101.

Jiménez, J. de J. 1960. Novelties in the Dominican flora. Rhodora **62**: 235-238.

Jumelle, H. and H. Perrier de la Bathie. 1945. 30e Famille — Palmiers. Pages 1-186 *in* Humbert, H., Flore de Madagascar. Tananarive, Imprimerie Officielle.

Keppler, C. B. 1970. Appendix A: The Puerto Rican parrot. Pages E186-E188 *in* H. T. Odum and R. F. Pigeon, eds. A tropical rain forest. Div. Tech. Info., U.S. Atomic Energy Comm., Oak Ridge, Tenn.

Kitzke, E. D. and D. Johnson. 1975. Commercial palm products other than oils. Principes **19**: 3-26.

Leck, C. F. 1969. Palmae:hic et ubique. Principes **13**: 80.

Leon, H. 1946. Flora de Cuba **1**: 241.

Lepesme, P. and R. Paulian. 1947. Analyse biologique et synécologique du complexe palmier/insect. Pages 13-134 *in* P. Lepesme, ed. Les insectes des palmiers. Paul Lechevalier, Paris.

Little, E. L., Jr. and F. H. Wadsworth. 1964. Common trees of Puerto Rico and the Virgin Islands **1**: 42 (Agriculture Handbook 249).

_____ R. O. Woodbury, and F. H. Wadsworth. 1974. Common trees of Puerto Rico and the Virgin Islands **2**: 70 (Agriculture Handbook 449).

Miller, R. H. 1964. The versatile sugar palm. Principes **8**: 115-147.

Monod, T. 1955. Remarques sur un palmier peu connu: *Wissmannia carinensis* (Chiov. 1929) Burret 1943. Bull. Inst. Franç. Afrique Noire, Ser. A, **17**: 338-358.

Moore, H. E. Jr. 1965. Palm hunting around the world. I. Madagascar to Malaya. Principes **9**: 13-29.

_____. 1969a. *Satakentia* — a new genus of Palmae-Arecoideae. Principes **13**: 3-12.

_____. 1969b. The genus *Juania* (Palmae-Arecoideae). Gentes Herb. **10**: 385-393.

_____. 1971. Wednesdays in Africa. Principes **15**: 111-119.

_____. 1972. *Chelyocarpus* and its allies *Cryosophila and Itaya* (Palmae). Principes **16**: 67-88.

_____. 1973a. Palms in the tropical forest ecosystems of Africa and South America. Pages 63-68 *in* B. J. Meggers, E. S. Ayensu, and W. D. Duckworth, eds. Tropical Forest Ecosystems in Africa and South America: A Comparative Review. Smithsonian Institution Press, Washington, D. C.

_____. 1973b. The major groups of palms and their distribution. Gentes Herb. **11**: 27-141.

_____. 1975. The origin of and main trends of evolution within the Palmae. Abstracts, XII International Botanical Congress, Leningrad, p. 98.

_____. and A. B. Anderson. 1976. *Ceroxylon alpinum* and *Ceroxylon quindiuense.* Gentes Herb. **11**: 168-185.

_____ and N. W. Uhl. 1973. The monocotyledons: their evolution and comparative biology. VI. Palms and the origin and evolution of monocotyledons. Quart. Rev. Biol. **48**: 414-436.

Muller, J. 1970. Palynological evidence on early differentiation of angiosperms. Biol. Rev. Cambridge Phil. Soc. **45**: 417-450.

Parthasarathy, M. V. and J. B. Fisher. 1973. The menace of lethal yellowing to Florida palms. Principes **17**: 39-45.

Pérez Jiménez, L. A. 1974. Some ecological notes on *Sabal yucatanica* in Mexico. Principes **18**: 94-98

281

Read, R. W. 1960. Palms as bird food. Principes 4: 31-32.

_____. 1967. A study of *Thrinax* in Jamaica. Ph.D. Thesis. University of the West Indies, Mona, Jamaica. 228 pp.

_____. 1968. A study of *Pseudophoenix* (Palmae). Gentes Herb. **10**: 169-213.

_____. 1969. *Colpothrinax cookii* — a new species from Central America. Principes **13**:13-22.

_____. 1974. The ecology of palms. Principes **18**: 39-50.

_____. 1975. The genus *Thrinax* (Palmae: Coryphoideae). Smithsonian Contr. Bot. **19**: i-iv, 1-98.

Romney, D. H., ed. 1959. Land in British Honduras. Colonial Research Publications 24: 286-302.

_____. 1976. Second meeting of the International Council on lethal yellowing. Principes **20**: 57-69.

Schmid, R. 1970. Notes on the reproductive biology of *Asterogyne martiana* (Palmae). II. Pollination by syrphid flies. Principes **14**: 39-49.

Schultes, R. E. 1974. Palms and religion in the northwest Amazon. Principes **18**:3-21.

Swabey, C. 1970. The endemic flora of the Seychelle Islands and its conservation. Biol. Conservation **2**:171-177.

Uhl, N. W. and H. E. Moore, Jr. 1977. Correlations of inflorescence, flower structure, and floral anatomy with pollination in some palms. Biotropica **9**: (in press).

Vandermeer, J. H., J. Stout, and G. Miller. 1974. Growth rates of *Welfia georgii, Socratea durissima, and Iriartea gigantea* under various conditions in a natural rainforest in Costa Rica. Principes **18**:148-154.

Van der Pijl, L. 1969. Principles of dispersal in higher plants. Springer-Verlag, Berlin. 154 pp.

Vosters, J. 1975. Commercial use of *Chamaedorea elegans.* Principes **19**:149-150.

Wallace, A. R. 1853. Palm trees of the Amazon and their uses. John Van Voorst, London. 129 pp.

Wessels Boer, J. G. 1968. The geonomoid palms. Verh. Kon. Ned. Akad. Wetensch. Afd. Natuurk., Tweede Sect. **58**: 67.

Whitmore, T. C. 1971. *Maxburretia rupicola.* Principes **15**: 3-9.

PRESERVATION OF CACTI
AND MANAGEMENT OF THE ECOSYSTEM

Lyman Benson

*Department of Botany, Pomona College,
Claremont, California 91711, U.S.A.*

I am not old — it's just that I have been around for a long time. I remember New York when the Brooklyn Bridge could be bought for a very low retail price. This city was the great metropolis of the country, and there were few other large centers, such as Chicago and the San Francisco Bay area. An upstart called Los Angeles was just beginning to extend its city limits so far out that volunteers had erected large signs in British Columbia and China. The country as a whole was rural, with vast areas of peaceful countryside and wilderness.

Contrary to the laws of genetics, just before the middle of the Twentieth Century, a generation of country bumpkins gave rise to a generation of city slickers, who soon transformed the fields and forests into pavements and factories. It was during this time that a textbook editor wrote in blue pencil on the margin of my manuscript, "Why use this plant for an example? Why not use a well-known one like corn?" This prompted me to ask a class of 140 college freshmen how many of them had ever seen corn growing. The result of this poll was yes, 39, no, 101. After some discussion, we chose for the example, not the supposedly common, everyday corn, but Saturday-night orchids instead.

As the population continued to increase and to spread, the generation of city slickers begat a generation of suburbanites, who appreciate nature. A suburbanite is a city man who drives around the surrounding countryside until he finds a beautiful large oak tree. Then he buys the property, rents a bulldozer, pushes out the oak tree, builds a house, and plants an acorn!

This world is becoming crowded, and in the United States, the current areas of most rapid growth are in San Diego County, California; in the vicinities of Phoenix and Tucson, Arizona; near the cities of Texas; and in Florida. All these areas also happen to be cactus country, and not even my beloved deserts are sacred any more. When we had a population explosion down on the farm, we'd just put some of the kittens in a sack and take them down to the river. Well, of course, we can't do *that* with *people!* We don't have enough water. The population explosion is the reason we are here today — dedicated to the proposition that this shall *not* be the century in which the wilderness got lost!

Now that this population boom is engulfing the warm areas, including the dry ones of the Southwest, the threat to the cacti is particularly dangerous. These plants are desirable for cultivation because they are novel and beautiful. The leafless succulent stems covered with spines of various shapes and colors catch the human fancy, and so do the flowers, which in some species are larger than the plant itself. Beyond this, many of the species are rare, with about one-fourth of those which are native to the United States and Canada falling into this category. Rarity catches the imagination of collectors and especially so when a new species has just been discovered and a name for it has recently been published. There then is a rush to get the plant into one's garden before other collectors have it. Some of the plants I have named have appeared in advertisements within a month or two after publication, despite my caginess about revealing the exact locality of known occurrence. The initial asking prices for these plants have ranged up to $25, $35, or even $50 each. In these cases, obviously, the plants for sale were field-collected and not grown from seeds. During the course of time, these prices dropped to $5 or $10 a plant, a direct reflection of several factors: the passing of the initial rush to be the only one with the new plant, competition among suppliers, and, hopefully, successful propagation of larger numbers from seeds. However, there is no guarantee of the last point unless numerous relatively young plants of about the same size are offered for sale.

The cacti are more vulnerable than most other plants because whole individuals are transplanted easily from the field to pots. Except in a few cases, they quickly sprout roots and grow well in ordinary potting mixtures. Removal of entire plants rapidly depletes the populations of rare species, which usually are quite sparse, though there may be local concentrations.

Removal of seeds has relatively little effect on a natural population, which tends to produce not only all the individuals the area can support, but most frequently, far more, though many young plants often die during the seedling stage. The vast overproduction of seeds is a guarantee that if the population is depleted, some new individuals will replace those which are lost. Thus, propagation from seeds is desirable. It not only creates public interest in plants, but also provides an enjoyable natural history hobby for many people throughout the world. More than twenty major cactus and succulent societies have been founded in various countries, and there are many local ones also. The only difficulty arises from the scramble by commercial collectors to obtain as many species as possible for quick sale, with some unscrupulous dealers even going into the field for plants instead of seeds.

Danger of extinction is usually associated with rarity, and to some extent a rare plant is, *per se,* a threatened one, while a *very* rare one lives precariously. Even a slight change in the environment may eliminate the taxon, either by its direct effect or more often through favoring competitors. Nevertheless, some rare taxa have special features which enable them to survive. Among the cacti of the United States, special modes of survival may be illustrated by the following examples:

A beautiful and very rare barrel cactus occurring in Arizona had been collected by only three or four people until it was described in 1969. This is *Ferocactus acanthodes* var. *eastwoodiae* (Fig. 1), known from only two widely-separated desert localities. The larger and more accessible area of occurrence would be much exploited except that the plant most frequently is a cliff-hanger. Individuals about 2 m high and 5-6 dm in diameter must weigh about 150 kg. Yet frequently they grow in cracks or on ledges of cliffs far out of reach of collectors. More vulnerable plants similarly are protected by great weight and by a dense covering of beautiful golden-yellow spines, mostly 5-7.5 cm long.

A small hedgehog cactus, *Echinocereus reichenbachii* var. *albertii,* is known from a single locality in Texas, but it is protected by dense thickets of mesquites and their tangle of spines. The mesquites are most effective barriers and prevent us from learning even the complete distribution of the taxon. They are a deterrent to those who would make off with the plants, but the variety is vulnerable to land-clearing operations and, in the end, it may lose out completely.

The Mesa Verde cactus, *Sclerocactus mesae-verdae,* of the southwestern corner of Colorado and the northwestern corner of New Mexico, is protected by edaphic tolerance, because it grows only on a special soil on which no one would expect to find a cactus. Furthermore, since it will not grow in potting mixtures, it has not become popular in cultivation.

The golf ball-sized *Pediocactus bradyi* of Arizona can be seen easily only during the two or three days of the year when it flowers. For a short time before and after flowering, it can be located by prolonged searching of the rocky desert floor on hands and knees or stooping. During the rest of the year, the stem, which is about two-thirds underground, shrivels, and the plant body retracts to ground level or below. The wind covers it with dust, and I have been able to locate the plants only when I sat down to rest.

Pediocactus knowltonii (Fig. 2) of the San Juan River of northern New Mexico is the size of a dime or a quarter, but the pink flowers are the diameter of a fifty-cent piece. This very rare species also goes underground for much of the year. Unfortunately, most of its range has been flooded by the San Juan Dam.

Pediocactus peeblesianus var. *peeblesianus* has resisted human search since at least 1938. Innumerable man-hours have been spent on attempts to locate the marble-sized stems, mostly without success. The known range is about 5 to 7 miles (8-11 km) in gravelly hills north and northwest of Holbrook, Arizona. The special adaptations include, as with most of the other rare and small species, going underground and adaptation to specialized soils requiring drastic physiological adaptation not possible for competitors.

Figs. 1-2. Rare Species of Cactaceae.

Fig. 1. Arizona Golden Barrel Cactus, Ferocactus acanthodes *var.* eastwoodiae. *Plants up to more than 2 m tall, 1.5 m in diameter; weighing about 150 kg. Rare, localized, and in demand, but protected by growing on steep mountainsides and cliffs. From Lyman Benson,* The Cacti of Arizona, *ed. 3, Univ. Ariz. Press, Tucson.*

Fig. 2. Pediocactus knowltonii. *A very rare local species; northern edge of New Mexico; endangered by commercial collectors and flooding from the San Juan Dam. Plant emerging from underground, where it has spent the dry season; dirt remains between the spine-bearing stem tubercles, each smaller than a match head. The whole stem is about 2 cm in diameter. Three flower buds are emerging. (Robert L. Benson.)*

285

The most promising methods for protection of the cacti, as well as of other plants, are by *legislation* and by *habitat preservation*. But to be effective, laws must have public support. If they lack such support, they are either forgotten or flouted, as were the Eighteenth Amendment and the accompanying Volstead Act. To gain the backing of the public, the laws must be worded so carefully that they are obviously fair and equitable. Many people must be consulted in order to avoid unexpected side effects which might cause unintentional injury to the interests of various individuals or groups. Somehow, the good will of most segments of the public and sometimes even corporations must be obtained. Such support often can be achieved by consultation with concerned individuals or groups. Dr. A. J. Haagen-Smit of the California Institute of Technology, who formulated the relationship between automobile exhaust fumes, ozone, and smog, and who served on many local, state, and national environmental committees and as Chairman of the California Air Resources Board, told me, for example, how he secured the cooperation of an oil company. He quietly analyzed the smoke from a refinery, then successfully demonstrated to the company president that he was losing half a million dollars a year up the chimney. Devices not only to curb pollution but also to recover the contents of the smoke were installed at once. Discussion of problems *before* a law is passed often will produce solutions benefitting both sides or minimizing difficulties for one or both protagonists. Through consideration of the problems arising from plant protection laws, we can either achieve full cooperation or at least a degree of promise which could not only serve to reduce opposition but possibly even gain support.

The most effective plant protection law I have ever known is one that probably doesn't exist. For at least fifty years and probably even much longer, residents of southern California have been sure there were special laws protecting both the yuccas and the California poppies. They considered these laws to be sound, just, and highly-important. When I tried to obtain special permits for specimen collection for scientific purposes, I was never able to find evidence that these laws existed either as county ordinances or state rulings. Admittedly, I have not made a thorough search, but the existence of such laws has not yet come to light through inquiry among various agencies. Nevertheless the thought that there were and should be such statutes on the books has provided long-term protection for the plants, and this situation continues to this day. Eloquent proof that if public opinion favors a law, little enforcement is necessary.

Preservation of some of the natural habitat is the most promising method of saving rare and endangered species, not only of the cacti but also of other plants and animals as well, which usually are saved by conserving the habitats of the plants. Presumably, when an area including rare species is set aside, it needs only to be shut off from outside influences, especially from those not present before the coming of European man. In many areas, this is all that is necessary for preservation of the desired species, but sometimes even this does not prove adequate. The question involves the nature of the original climax, the possibility of its restoration, and whether the rare and endangered species were really favored by the climax or were barely able to hold on in spots where their competitors were eliminated by natural disturbance.

But even if the original climax is known, this occasionally is insufficient to restore it because some changes such as the introduction of diseases or of alien species may be irreversible. Furthermore, the nearest attainable subclimax may not favor the rare species. The rarity of these plants or animals is due either to tolerance to a most narrow range of environmental conditions or to vulnerability to competition with other species. The one certain point is that the rare species exist under the conditions of the present. Presumably, they are better adapted to the original climax vegetation, but this may not necessarily be so, and we will not know until the climax is restored.

In such regions as the glaciated areas in Canada and slightly in the adjacent states, some rare species are early invaders living in the still-unsettled conditions following the Pleistocene ice sheets. As the northern forests approach an undetermined climax, a few species are squeezed out by slower-travelling invaders now gradually taking their places. The present species of the spruce-fir forests are mostly the mobile ones that got there just after the ice retreated. As might be expected of early invaders, most of these are wide-ranging taxa, though some have now become rare. At first, these were interpreted as relics on unglaciated areas, but current evidence points to them as early invaders capable of survival only in the early subclimaxes. Disturbance may do them more good than harm, and it is unlikely that attempted protection will favor them.

An extreme example of a floristic or vegetation type that has had little disturbance is the Arctic tundra, growing north of timberline. In the high Arctic, there is no agriculture and little human disturbance, and there may even be no weeds. In 1950-52, I found no evidence of weeds at Point Barrow, Alaska, and in 1959, I saw none either on Cornwallis Island or at points farther south in the Canadian Arctic Archipelago. Perhaps some weeds of high Arctic origin may invade such areas as those disturbed in the new oil fields, but any extensive development of a weed flora is unlikely. The climate is too severe, and major disturbance has not been sufficiently widespread over a long period. Except for such groups as the Eskimos, who are hunters, man has not tended to penetrate very far beyond timberline, and he has attempted no agriculture in the far north. He has not stimulated the evolution of weeds, as he did along the Mediterranean Sea.

In many temperate regions, the original ecosystem has been sporadically preserved in small patches, and a return to it may be possible. However, this is not true everywhere, and in some areas, the nature of the original forest is not easily established. Introduction of the chestnut blight in the Southeast and of the Dutch elm disease throughout large Eastern areas has not only made a return to the original climax impossible, but introduced trees and shrubs also have confused the succession.

Figs. 3-4. Preserved Desert Habitats.

Fig. 3. Organ Pipe Cactus National Monument. Pima County, Arizona. Protected cacti include many smaller species hidden in the desert shrubbery. Visible are young saguaros, Cereus giganteus; *the organ pipe cactus,* Cereus thurberi; *and a cholla,* Opuntia acanthocarpa *var.* major *(lower right).*

Fig. 4. Anza-Borrego State Park. San Diego County, California. Protected cacti include several small species among the rocks, several barrel cacti, Ferocactus acanthodes *var.* acanthodes, *and a rare cholla,* Opuntia echinocarpa *var.* wolfii *(center). A young, leafless (during drought) ocotillo,* Fouquieria splendens, *is just left of center, and there are several desert century plants,* Agave desertii. *During the wet season a host of wild flowers (many of them rare) appears.*

In many forests of tropical lowlands, as for example on Oahu in Hawaii, the original plants are gone, and hardly a single native species persists at low elevations there. Natural vegetation still occurs at higher levels, but it is different, and on the higher Hawaiian Islands, this vegetation is even temperate instead of tropical. The possibility of restoring the original lowland ecosystem is gone.

In the deserts, where cacti are abundant, the ecosystems are particularly fragile, and once they are destroyed, the loss is irrevocable. Some have even claimed that there is no succession, and in any event, succession is very slow and restoration of the climax is difficult. Habitat preservation is almost the only practical means of preserving endangered desert species, and commonly it is effective (Figs. 3, 4). Fortunately, on the alluvial fans, hills, and mountains, relatively few invaders can replace the native plants. The rare species occur mostly in these areas, while the weeds are along roadsides and in disturbed valleys or plains.

Where water is somewhat more plentiful in the dry woodlands or chaparrals, or brushlands, preservation of the habitat is the only feasible method of species conservation. However, the problems in such areas are different from those in the deserts. For example, both woodlands and chaparrals are not only subject to fire but also they are correlated with its occurrence. Development of the oak woodland, which is prevalent in northern California (Fig. 5) but occurs only in droplets in the chaparral of southern California, is dependent upon flash grass fires that kill seedlings of shrubs and the young trees, including the oaks.

If fire is prevented, as it has been for thirty years or more, shrubs and young trees grow up under and between the well-spaced original oak trees (Fig. 6). Throughout most of northern California, it has become impossible to find a good representation of the formerly common open, parklike woodland that forty years ago was dominant within a 500 mile (800 km) area. The older trees are now engulfed in thickets of young saplings and shrubs. When a fire does break out, it is a hot one and the trees are killed outright, instead of being merely singed as they formerly were by grass fires. They do not sprout from the root crown, but the shrubs among them do. Consequently, after a hot fire, oak woodland tends to be replaced by chaparral (Figs. 7, 8). Along with the oak trees, a host of herbaceous endemics is lost. Thus, fencing an area of oak woodland and keeping fire out does not reestablish the climax but, instead, destroys both it and the rare species it contains. (For another type of dry woodland, see Fig. 9.)

Only droplets of chaparral occur in the oak woodland of northern California, but chaparral or brushland is the prevailing vegetation of the lowlands of southern California and of the hills and mountainsides up to 1,200 or 1,700 m elevation. This is a fire type of vegetation dependent upon burning once in ten, twenty, or thirty years. If the period between fires is too long, the brush becomes overgrown, and when a fire eventually does come, it is very hot and tends to kill the shrubs. This is true even though the crown-sprouting of the shrubs after a normal fire restores the climax almost immediately.

Cacti do not grow in the climax chaparral of southern California, because they cannot

Figs. 5-8. Relationship of Fire to the Ecosystem.

Fig. 5. California Oak Woodland — Normal. As it occurred through 500 miles before advanced methods of fire suppression: open and parklike with numerous endemic species of herbs and intermingling introduced grasses. Flash grass fires did not harm the large trees, but they killed the seedlings of trees and shrubs.

Fig. 6. California Oak Woodland — After 30 Years of Complete Fire Suppression. The old trees (see spreading branches at the right and left margins) and a host of young ones, normally killed as seedlings by flash grass fires. The stage is set for a hot fire, which will kill all the trees.

Fig. 7. *California Oak Woodland and California Chaparral. Oak woodland on sandstone of the north-facing slope at the upper center; chaparral on the volcanic soil of the south-facing slope diagonally across the middle of the picture. A flash grass fire in the woodland may become a very hot brush fire in the chaparral near by. The tall Digger pines,* Pinus sabiniana *, occur in either woodland or chaparral.*

Fig. 8. *California Oak Woodland and California Chaparral Following a Hot Fire. The scrub oak in the foreground,* Quercus dumosa, *is a typical chaparral plant, crown sprouting soon after the fire. The blue oaks, typical of the oak woodland, cannot sprout from the bases, and they have been killed by the heat of the fire in the chaparral adjoining them. The chaparral shrubs will take over the edge of the oak woodland and any areas where young trees and some shrubs had grown up during grass fire suppression. The oak woodland ecosystem is destroyed if fire is not present, and the endemic herbs are unable to grow among the shrubs of the succeeding chaparral.*

endure the brush fires. However, several rare and endemic taxa grow in disturbed areas where the brush has been cleared away by man, on rock outcrops where the shrubs are scattered, or along the washes where light winter floods wash away the grasses or cover them with sand and gravel. The weedy Mediterranean grasses burn every few years, but some cacti escape burning where grasses are sparse or where thicket-forming cacti exclude the grasses, thus being singed only along the margins. Restoration of the climax chaparral would restrict or eliminate the cacti.

Grasslands are especially vulnerable to the activities of man. Such grasslands may be plowed at once without clearing or they are soon overgrazed if the grasses remain. Originally some, like the Desert Grassland (Fig. 10) or those of the Great Plains, existed in part because of annual prairie fires set by the Indians. The fires killed the young woody plants and favored the grasses and grassland instead of forest. The grasses were perennials not killed by the fires. It is unlikely that fencing and exclusion of outside influences will restore the prairies to the condition which existed before white settlement took place. It probably will favor something wholly different, such as a relatively dry

Figs. 9-10. Dry Areas Bordering the Deserts.

Fig. 9. Juniper-Pinyon Woodland. Zion National Park, Utah. Numerous species of cacti occur in this floristic association. Fortunately, several areas are preserved in National Parks and Monuments and on other public lands. However, not all the rare species occur in these preserves, and more protection of habitats is needed. In this area fire is unimportant, because the woody plants are spaced widely and the grass cover is sparse.

Fig. 10. Desert Grassland. East of Sedona, Arizona. Both the Desert Grassland and the Great Plains Grassland were burned annually by the Indians. Because of a combination of overgrazing and lack of fires, the seedlings of trees and shrubs are taking over both the upper and lower portions of the Desert Grassland. At 1,500-1,800 m elevation the South-western Oak Woodland is moving down from the mountains; at 750-1,150 m the Arizona Desert is moving upward. In many places the zone of grassland has become narrow. This presents management problems for preserves for endangered endemic species of the grassland.

forest or woodland, and the rare species may be lost.

The most beautiful flower fields in the world are frequently composed of native species occurring on disturbed soils or in altered vegetation types. Examples of these are: the great fields of bluebonnets or lupines in Texas, and the enormous areas of "African daisies" or "Namaqualand daisies" *(Arctotis* and its relatives) in Cape Province of South Africa just south of the Orange River. These vivid South African flower fields are fallow grain fields cleared from brushland resembling the chaparral of southern California. The most brilliantly colored flower fields of North America are at the southern tip of the San Joaquin Valley near Bakersfield, California (Fig. 11). They are developed in grassland that has been grazed since soon after the Spanish missions were first founded in 1769. All three flower areas include many endemic or nearly endemic and some rare species, and it is questionable whether restoration of the climax would help them.

For many years I worked for a wild flower preserve in Kern County, California, which includes Bakersfield. From 1931 to 1938, I taught at Bakersfield Junior College. During a three-year period, the rainfall there was above the average of 13.75 cm or 5.5 inches and evenly distributed throughout the winter and spring, from October until March. It probably is needless to add that springtime there was absolutely glorious. Since then, however, there have been only four such years — 1941, 1952, 1958, and 1973. When an area, the Sand Ridge Reserve, southeast of Bakersfield, was set aside by the Nature Conservancy several years ago, I was invited to speak at the outdoor dedication. My theme was essentially, "Now we've got it, what are we going to do with it?" We did not really know what factors favored the native and especially the rare species. We might guess that exclusion of cattle and sheep would benefit the native plants, and, obviously, overgrazing would spoil everything. But was it desirable to have no grazing there at all? Would light grazing by horses or even cattle be beneficial? An amazing suggestion was made at the time by Ernest Twisselman, who was both a botanist (author of a flora of Kern County) and a practical rancher. He recommended light sheep-grazing during the early spring in order to reduce the population of annual introduced Mediterranean grasses, thus favoring the native plants. Such a possibility would not have occurred to me, and had the idea come from anyone else but Mr. Twisselman, I would have considered it as the next step above grazing by goats. Need I add that the feral goat, to my mind, is a work of the Devil — created in his own image!

Obviously, the management of such an ecosystem presents problems. The goal cannot be restoration of the original climax because we do not even know what the original climax was. Early explorers preceded the science of ecology, and they left us no accurate or detailed descriptions of the natural vegetation. Early exploration by American and European botanists produced plant specimens but no idea of the ecosystems from which they came. Even the localities of specimen collection were vague. The upper San Joaquin Valley had been overgrazed by Spanish cattle for sixty or eighty years before California became a state in 1850, and when John C. Fremont explored the area in the 1840's, the ecosystem was already altered both by disturbance and by introduction of a large number of Mediterranean weeds, especially the weedy grasses whose seeds had been brought over in the wool of sheep and on the hooves of range animals. Until about seventy years after Fremont's explorations, nothing was known of the ecosystem, and when scientists first became really aware of it, their chief interest and concern was an attempt to describe its original composition, and perennial bunch grasses were assumed to dominate every climax for grasslands. In absence of replacement of the climax as an objective, other goals must now be formulated. Presumably, these include preservation of the gorgeous fields of wild flowers, even though they may develop only periodically, and with them the maintenance of many native species, including the rare and local ones.

The brilliant flower fields of the 1930's stretched for many miles along and across the arc of the upper San Joaquin Valley, especially south of Bakersfield. Both the floor of the valley and the surrounding alluvial fans and foothills looked as if a veritable giant had painted the area with a brush dipped into each of several paint buckets. I described this in glowing terms to the late Dr. W. L. Jepson of the University of California. Jepson smiled and observed, "You should have seen it in the '90's!" Thus, I presume much had been lost already. During the Second World War, the valley floor was plowed up for planting of wheat, cotton, and potatoes. Until the glorious spring of 1973, when heavy and well-distributed rainfall blanketed California and Arizona, I worried about the flower fields, wondering if overgrazing during the many dry years had destroyed them. However, in 1973 a gigantic horseshoe curving along the fans and foothills for 25 or 30 miles (40-48 km) and mostly about two to three miles (roughly 3-5 km) wide turned to color as brilliant as that of the earlier times I had known. In view of this, perhaps the enlightened grazing practices of the Tejon Ranch on which most of the flower fields occur may represent the best management of the floral displays, as well as the best method for beef production. It is difficult to imagine better results.

The present flora of the Valley is composed of a large number of native species and a smaller number of Mediterranean weeds with great numbers of individuals. These elements have reached a dynamic equilibrium that shifts from place to place, according to local conditions and management. The problem is proper control of the equilibrium, and this must be learned from studies on the ranches. The great need now is for more information

Figs. 11-13. Preservation of Flower Fields and of a Cactus Growing in Them.

Fig. 11. Wild Flower Field in the Upper San Joaquin Valley near Bakersfield, California. The most conspicuous plants in the picture are the California poppy, Eschscholtzia californica var. crocea; the taller ones farther back are thistle sage, Salvia carduacea. The area of the picture includes about 30 native species, lost in black-on-white. Various species dominate other areas of the Pacific Grassland. (Frasher Photograph; courtesy of the Pomona Public Library.)

Fig. 12. The Prickly Pear of the Flower Fields. Opuntia basilaris *var.* treleasei *essentially endemic in the Pacific Grassland near Bakersfield; now restricted to areas grazed by sheep early in the spring and not known from the vast Tejon Ranch, grazed by cattle for about 200 years and perhaps exterminated by them. The cactus photosynthesizes through the long, hot, dry summer (Fig. 13) while the neighboring annual plants live over only as seeds. Management of the area in one way may favor the wild flowers, in another the cactus. Will any method favor both?*

Fig. 13. The Wild Flower Fields in Summer. The Great Valley (Sacramento and San Joaquin) of California. During the summer there is no rain (from May to October), and the hills are dry. Only a few perennial plants live over. These include the cactus of Fig. 12. California Oak Woodland appears just above the grassland, at lower levels only on the north-facing slopes (to the right here). From Lyman Benson, Plant Taxonomy, Methods and Principles. © Ronald Press Co., New York, 1962. Used with permission.

that can only be obtained through experimentation with the ecosystem. In my own studies, I have found the cooperation of ranchers easy to gain. Naturally, one cannot expect to make major alterations in a ranch for the sake of experiment, but often the elements of an experiment already exist. For example, a fence dividing similar areas may confine a small number of breeding bulls to one side and the more abundant steers to the other, thus producing a difference in the intensity of grazing. The fence-line between two ranches may mark the limits of different grazing practices, and an analysis of the vegetation on the two sides of the fence may indicate the survival of some species under one set of conditions and of others under another. So long as one is careful not to give the impression that he is trying to determine who is the better stockman but rather only the answer to an academic problem, opposition usually is not aroused and often the ranchers even become interested in the scientific work conducted on their property.

Preservation of an endemic cactus, *Opuntia basilaris* var. *treleasei* (Fig. 12), confined essentially to the flower field area near Bakersfield, is dependent upon solution of the problem of conservation of the flower fields and their annual species. This plant was abundant as recently as the 1930's, but expansion of Bakersfield has reduced it to a small fraction of the number of individuals present at that time. Some plants are preserved in the Sand Ridge Reserve, but under the new management, other taxa may thrive better than they do. According to old photographs taken by David Griffiths about 1910, the cactus once densely covered areas as large as forty acres, with plants bearing cerise flowers about 1 dm in diameter. Such patches were already gone by the 1930's, and in all likelihood they never can be restored. The question to be answered is what practices favor the cactus. Information is scanty.

The rare cactus in the flower fields may have largely different ecological requirements from its annual neighbors, and it may be favored by a different management of the ecosystem. The annuals require no rain during the long summer drought of the typical Mediterranean climate of California — from about the first week of April until October or early November. They live through the summer as seeds ready to sprout in the fall (Fig. 13). If early rains start the seedlings but there is drought during parts of December or January, many flower seedlings will die, leaving the more resistant young plants of the introduced weedy grasses to benefit from later rains which may occur during February and March.

Relationships of the cacti to the seasons are quite different. They are adapted to survival through dry months or even years by storage of water in the succulent stems. The shallow, widespreading root system of a cactus can live and remain ready to absorb water even if the small roots are exposed to dry air. Whenever there is even only a light rain, the roots soak up the water from just below the surface of the soil. For example, *Opuntia basilaris* var. *treleasei* does not die down during the summer, though its metabolism is slowed. It remains alive, living on stored water and photosynthesizing food according to the common succulent plan. The stomata are open at night when water-loss to the atmosphere is low and they are closed during the heat of the day. During the night, carbon dioxide is stored in organic acids, and it is released for food manufacture in the daytime. With little water, the cactus is able to produce food all summer, even when day temperatures reach 100° to 115° F (38°–46° C). At this time, there is no competition from other plants. During the spring season of a moist year, shading by competitors probably becomes a problem, but the food deficit does not last long and it may be made up during the summer. Thus, the cactus is able to survive under a number of possible versions of management of the flower fields through controlled grazing. However, there is at least one potential hazard. Cattle in the deserts often become addicted to eating cacti, especially prickly pears and chollas *(Opuntia),* and the spines are neutralized by great balls of mucus formed around them in the intestines. The plants are sources of water and some food for the cattle. The cactus of the flower fields east and southeast of Bakersfield is not known to occur on the Tejon Ranch but on foothills and fans not grazed by cattle. During the past 45 years — and probably much longer — these hills have been grazed heavily each February and March by sheep, with the result that the competing grasses were kept low. So far as has been observed, it is unlikely that the sheep eat the cactus, but they probably do reduce the wild flower population, except on the steep slopes. The cactus disappears near the boundary of the cattle ranch, and there are no specimens indicating past occurrence on the Tejon property. However, the first plant collectors visited the area only about 130 years ago, and cattle have grazed it during the past 200 years, at first very heavily for production of hides and tallow, which were the only products exportable by sailing ship. Whether the prickly pear occurred on the Tejon Ranch site before Spanish settlers arrived there is a matter of conjecture; so is the question of whether cattle eat this particular cactus. In all likelihood they do, because the plant has only small spines presenting no great barrier to animals. A system of management involving cattle probably would prove fatal to the prickly pear. As in all problems of ecosystem management, this raises questions for which no answers are yet available. The answers are vital to choice of management for a preserve, according to an objective of either favoring the flower fields or the cactus growing in them or, hopefully, both.

REFERENCES

More detailed information concerning species of Cactaceae may be obtained from the following publications by the author:

1940, 1950, 1969. The Cacti of Arizona. eds. 1-3. Univ. of Ariz. Press, Tucson.

1961-1962. A Revision and Amplification of *Pediocactus* I-IV. Cactus and Succ. Jour. **33:** 49-54; **34:**17-19, 57-61, 163-168.

1965. The Southern Californian Prickly Pears — Invasion, Adulteration, and Trial-by-Fire. Ann. Missouri Bot. Gard. **52:** 262-273. (With David L. Walkington.)

1966. A Revision of *Sclerocactus* — I-II. Cactus and Succ. Jour. **38:** 50-57, 100-106.

1969. The Native Cacti of California. Stanford Univ. Press.

1970. Cactaceae in C. L. Lundell, Flora of Texas **2:** 221-317.

1970. Cactaceae in Donovan S. Correll and Marshall C. Johnston. Manual of the Vascular Plants of Texas. 1087-1113.

1975. Cacti, Bizarre, Beautiful, But in Danger. National Parks and Conservation Magazine **49:** 17-21.

In press. The Cacti of the United States and Canada. Stanford Univ. Press.

ENDANGERED AND THREATENED CARNIVOROUS PLANTS OF NORTH AMERICA

George W. Folkerts

*Department of Zoology-Entomology, Auburn University
Auburn, Alabama 36830, U.S.A.*

INTRODUCTION

The North American carnivorous plants are included in the families Sarraceniaceae *(Sarracenia, Darlingtonia)*, Droseraceae *(Drosera, Dionaea)*, and Lentibulariaceae *(Pinguicula, Utricularia)*.

Most carnivorous species are either characteristic of or adapted to sites with acid soils that are often low in available nutrients. Consequently, the ability of carnivorous species to obtain nitrogenous compounds and other substances from the bodies of their prey enables them to occupy a habitat where competition is less severe than in many other areas.

Because of their novelty, carnivorous plants have for many years attracted the interest of horticulturists and plant fanciers. Most species can be cultured without difficulty.

Darwin (1899) and Lloyd (1942) discussed carnivorous plants from a scientific viewpoint. More recent popular publications are by Dean (1975), Pietropaolo and Pietropaolo (1974) and Schwartz (1974).

The taxa in the following discussion are arranged in a traditional taxonomic order. Where taxonomic considerations may be important to an understanding of the status of a species, comments are included under that species.

A number of taxa are discussed that are not, in my opinion, endangered or threatened. These are included because opinions concerning their status may vary and some of the information may assist the reader in making an independent decision about the status of these plants. If the taxa are considered endangered or threatened, this is noted after the scientific name at the beginning of the discussion of the species.

The definitions of "endangered" and "threatened" used in this paper are those included in the Endangered Species Act of 1973. Though these definitions are subject to considerable latitude of interpretation, I have attempted to take a conservative viewpoint. Efforts to protect vanishing species have been seriously hurt in the past when biologists "cried wolf" a little too often. Further loss of public confidence will seriously hamper our efforts.

The term "species" as defined in the Endangered Species Act of 1973 also includes infra-specific taxa, and has been loosely interpreted to apply to any "population-segment" of a species. This interpretation allows populations of evolutionary or phytogeographic significance to be protected under the Act.

Judgements as to the status of each of the taxa included here were based on my opinion of the future of the species throughout its range. It is realized that some states, as a protective measure, may want to list as endangered or threatened, a species which might be peripheral or rare in that state but not jeopardized elsewhere. Such considerations, while laudable and beneficial in result, have not influenced my decisions on the status of any of the taxa involved.

Many species not listed as endangered or threatened are nevertheless deserving of immediate protection and monitoring. Only by stopping degradation in its early stages can we prevent more species from reaching the point where they must be placed in one of these categories.

The following material constitutes a rather superficial review of the topic and should not be considered an exhaustive treatment. Much of the more detailed literature is not cited and many available pieces of information have had to be omitted to keep within the page limit of this paper.

FAMILY SARRACENIACEAE

Included in the North American Sarraceniaceae are two genera of pitcher plants,

The pitchers of Sarracenia oreophila, *a threatened species largely confined to northeastern Alabama.*

Sarracenia, with ten species and *Darlingtonia,* with one. The genus *Heliamphora* is extra-limital, the six species being confined to northern South America. Aspects of the systematics and ecology of the family have been treated by Bell (1949), DeBuhr (1975), Harper (1918), Macfarlane (1907), Thanikaimoni and Vasanthy (1972) and Uphof (1936), among others.

All of the North American species are discussed, albeit briefly, below:

Genus *Sarracenia* L.

McDaniel (1971) has recently mono-graphed the genus. Bell (1952) and Bell and Case (1956) have discussed natural hybridization among the species. Although there is general agreement as to the status and level of most of the taxa, specialists disagree on some points, especially in the *Sarracenia rubra* complex. These differences of opinion are not deemed important here since our major concern is with the conservation of unique populations regardless of their taxonomic rank.

SARRACENIA OREOPHILA (Kearny) Wherry — THREATENED

This species is largely confined to a relatively small area in northeastern Alabama. One population also survives in Elmore County, in central Alabama (T. C. Gibson, pers. comm.). Although

The flower of Sarracenia oreophila.

possibly occurring in Tennessee and Georgia (Bell, 1949, McDaniel, 1971), Mr. Randall Troup, who has intensively investigated the distribution of this species, informed me that he knows of no extant populations in these states.

The ecological requirements of *S. oreophila* are unusual among the species of the genus. Populations occur in shaded sites along sandy stream edges and in moist sites in mesic woodlands. Troup (pers. comm.) informed me that it also occurs in open or shaded depressions where acid conditions have developed and is often associated with mild ecological disturbance.

This species is more common than many botanists realize (Troup, pers. comm.), and several sizeable stands still exist. It is not in imminent danger of extinction throughout its range and therefore cannot qualify for endangered status. However, impending changes within its range may damage or destroy certain populations and the species, therefore, should be considered threatened.

Some populations of *S. oreophila* were probably inundated by the construction of Weiss Reservoir on the Coosa River in Alabama. More recently, commerical collectors and construction associated with urban expansion have reduced populations. Increased pressures to strip-mine coal within the range of this plant may also create additional habitat degradation. Additionally, road construction planned within the range may disturb some sites.

Although some populations are within Alabama's DeSoto State Park, authorities there have no effective way to monitor these populations in order to prevent removal of the specimens by collectors.

SARRACENIA FLAVA L.

This species ranges from southwestern Alabama and north-central Florida north to southeastern Virginia. Populations in upper South Carolina and adjacent central North Carolina may be disjunct and worthy of immediate protection.

Sarracenia flava is primarily an inhabitant of open moist savannas although it occasionally occurs in bogs. It often also occurs in ditches and other disturbed sites and may persist for long periods of time. Spectacular stands which should be considered for protection still occur in the Carolinas (Radford, Ahles, and Bell, 1968).

The extensive range and relative abundance of this species precludes considering it endangered or threatened.

SARRACENIA ALATA (Wood) Wood

This species exists in two disjunct populations. The eastern population ranges from southwestern Alabama to the eastern Florida parishes of Louisiana. West of the Mississippi delta, the species occurs sporadically in scattered acid bogs in western Louisiana and eastern Texas (McDaniel, 1971). The latter populations should be studied in order to determine if the species is endangered in the western portion of its range.

Sarracenia alata is still quite abundant in the eastern portion of its range. In the past, populations extended almost continuously from the coastal savannas of Mobile County, Alabama, to the Louisiana border. Over a thousand bogs containing this species are still present in Hancock and Harrison counties, Mississippi. Plans should be made immediately to protect some of these extensive stands. Past experience indicates that the populations in DeSoto National Forest are not safe. Unfortunately, many Forest Service administrators in the South are unaware of or unconcerned about unique plant species. The species, as a whole, is neither endangered nor threatened.

SARRACENIA MINOR Walt.

This species ranges from the eastern panhandle and central peninsula of Florida along the Coastal Plain to southeastern North Carolina. Its ecological tolerances seem wider than do those of many of its congeners. It inhabits savannas, wet flatwoods, and bogs, seemingly doing well in both wet and dry sites. It is frequent and sometimes abundant on moist road rights-of-way.

This species cannot presently be considered endangered or threatened. Many hardy stands can still be seen in the Okefenokee National Wildlife Refuge in Georgia, where they will presumably be protected for generations to come.

SARRACENIA LEUCOPHYLLA Raf.

This white-topped pitcher plant is confined to the Gulf Coastal Plain, and ranges from extreme southeastern Mississippi to southwestern Georgia. It is typically an inhabitant of wet hillside bogs and seldom forms conspicuous stands in savannas. It is less tolerant than many other species and seldom persists long after major environmental disruption within its immediate habitat.

Because the pitchered leaves are so conspicuous, it is often used by florists for decorative purposes. It also is frequently picked or pulled up by tourists and curiosity seekers.

Sarracenia leucophylla still survives in substantial numbers in many hillside bogs in Alabama and Florida. Some of the strikingly beautiful stands should be considered for immediate protection but the species cannot be considered endangered or threatened.

SARRACENIA RUBRA Walt.

Case and Case (1974, 1976) have divided *Sarracenia rubra (sensu* McDaniel) into three species, one of which has two subspecies. These taxa are treated separately under the appropriate headings.

Sarracenia rubra (sensu Case and Case) ranges from the central panhandle of Florida along the Coastal Plain to southeastern North Carolina. The population in the Florida panhandle is apparently disjunct (Frederick W. Case, pers. comm.).

Because the leaves and flowers of this species are smaller than those of *Sarracenia flava* and *S. leucophylla,* its common associates, it is often overlooked in cursory surveys. However, it is present at many sites in relative abundance. It occurs in roadside ditches and seems to tolerate disturbance. Although it is listed *(sensu* McDaniel) as threatened in the Family Lists of Candidate Endangered and Threatened Plant Species in the Continental United States (Smithsonian Institution, 1975), I cannot agree that this species conforms to the accepted definition of "threatened" and I therefore suggest that it be removed from the list.

SARRACENIA ALABAMENSIS subsp. ALABAMENSIS Case and Case — ENDANGERED

This subspecies occurs only in Autauga, Chilton, and Elmore counties in central Alabama. The habitat consists of sandy-gravelly bogs and damp spring heads. It is more tolerant of shade than are most other members of the genus (Case and Case, 1974). Harper (1922) reported on the floristics of these sites.

Less than a dozen populations of this subspecies remain and only two of these are of significant size (Gibson, pers. comm.). None can presently be considered safe from degrading factors. Plant collectors have been avidly seeking specimens from these populations. In 1975, a well-known collector ran an advertisement in the Chilton County News offering a $20.00 reward for verified localities and additional fees for collecting and mailing specimens.

When I recently visited the habitat, I found many areas overgrown with honeysuckle. Case and Case (1974) mentioned the advent of honeysuckle as being the greatest change that has taken place in the habitat of this species. Herbicide spraying along the railroad right-of-way has also degraded the habitat.

Immediate steps must be taken to protect this subspecies. Without doubt it qualifies as endangered.

SARRACENIA ALABAMENSIS subsp. WHERRYI Case and Case — THREATENED

This subspecies ranges in extreme southwestern Alabama, adjacent portions of Mississippi and the western panhandle of Florida. It inhabits bogs and boggy flatwoods, and almost always is associated with *Sarracenia leucophylla* or *S. alata*.

Although this subspecies is still abundant at a number of sites in Baldwin and Mobile counties, Alabama, in my opinion it should be classified as threatened. For the last seven years, I have been visiting sites where this taxon occurs, and during this time a number of populations have been drastically reduced by both road construction and the creation of pine plantations. Additional road construction through certain sites is now pending, and forest practices which degrade pitcher plant habitats are likely to accelerate.

SARRACENIA JONESII Wherry — ENDANGERED

This species occurs in a limited area of the Blue Ridge Mountains in eastern North Carolina and adjacent upper South Carolina. According to Wherry (1972), the species is characteristic of "mountain-front cataracts." It also occurs in boggy seepage areas and moist stream-edge sites. According to Case (pers. comm.), many of the previously known populations have been damaged by overcollecting on the part of commercial dealers, amateur enthusiasts and even professional botanists. One population was severely decimated by the construction of a golf course. There seems to be only one sizeable population remaining in the vicinity (Case, pers. comm.).

The extant populations are located in an area that is rapidly developing as a center of tourism. The associated changes stimulated by tourism are likely to cause the extinction of this species, which is undoubtedly already endangered. More than any other member of the genus, its future seems bleak and it needs immediate attention.

SARRACENIA PURPUREA L.

This is the most widely distributed species of the genus, occurring from the southwestern states north to Labrador and Hudson Bay and west to Minnesota and Manitoba.

It occupies sphagnum bogs in the northern part of its range and is quite common. In the Southeast, it occupies a variety of moist habitats and occurs in both open and shaded areas.

This species is far from being endangered or threatened at the present time. The populations in the Blue Ridge Mountains of North and South Carolina may deserve attention because of their phytogeographic significance. However, there are nearby populations in the adjacent Piedmont.

SARRACENIA PSITTACINA Michx.

This is the smallest species of the genus and ranges from eastern Louisiana along the Coastal Plain to east-central Georgia. It typically occupies drier sites in bogs and savannas, and also occurs in ditches and along the edges of canals.

Although it is included as threatened in the Family Lists of Candidate Endangered and Threatened Plant Species in the Continental United States (Smithsonian Institution, 1975), it should immediately be removed from the list. The impression some people have of its rarity probably results from the fact that it cannot be detected at a distance. Its small size and usually procumbent leaves make it inconspicuous. Actually, it is abundant in approximately sixty percent of the bogs along the Gulf Coast and occurs in moderate to low numbers in an additional thirty percent. It would certainly be unwise to accord this species threatened status.

DARLINGTONIA CALIFORNICA Torr.

The cobra plant or western pitcher plant of Oregon and California differs from members of *Sarracenia* in leaf organography and floral morphology, but it occupies a similar ecological niche. The species ranges in the coastal areas of Oregon north to approximately fifty miles south of the Columbia River (Larry E. DeBuhr, pers. comm.). Trappe and Gerdemann (1974) reported an extant population at Sand Lake, Tillamook County, Oregon. In California, the plant ranges on the coast to Del Norte County and south, in the Sierra Nevadas to Plumas and Nevada counties. The populations in the Sierras seem to be disjunct (DeBuhr, pers. comm.).

Darlingtonia inhabits bogs, seepage areas and trickling streams, often being associated with sand or serpentine rock. Partial shade frequently characterizes the habitat.

Darlingtonia is included as threatened in the Family Lists of Candidate Endangered and Threatened Plant Species in the Continental United States (Smithsonian Institution, 1975). However, its status differs little from a number of other species (e.g. *Sarracenia leucophylla*) that were not included.

I have seen a few populations in Oregon which were damaged or obliterated by siltation resulting from a nearby logging operation. In spite of this, DeBuhr (pers. comm.) informed me that forest practices have not significantly damaged this species.

Darlingtonia may seem rarer than it is because many populations occur in areas lacking easy access by good roads. In addition, neither the flowers nor the leaves are conspicuous at a distance and populations may be overlooked unless an intensive search is made for them.

DeBuhr (pers. comm.) thinks that the most significant factor currently affecting *Darlingtonia* is that of commercial collecting. Although this activity should be controlled perhaps by state laws in California and Oregon, such collecting has not yet decimated the species to the extent that it can be considered endangered or threatened.

Many of the known *Darlingtonia* populations exist on federal land (National Park Service, Forest Service, Bureau of Land Management) and can be protected by appropriate administrative action. This has already been accomplished at two sites. One fairly healthy population also is being protected at Darlingtonia Wayside near Florence, Oregon.

FAMILY DROSERACEAE

Two genera, *Drosera* L., the sundews, and *Dionaea* Ellis, the Venus flytrap, comprise this family in North America.

Of the eight Nearctic species of *Drosera,* many are widespread and a number of these are essentially circumboreal. The species are characteristic of bogs and other moist habitats, and populations do not seem to be seriously affected by environmental modifications unless the changes are drastic. Some of the species could be considered "weedy." In the Southeast, thick stands of *Drosera capillaris* often turn roadside ditches red. *Drosera filiformis* frequently occurs by the thousands on disturbed roadsides. For these reasons, none of the species of *Drosera* are endangered or threatened and no further mention of them will be made.

DIONAEA MUSCIPULA Ellis — THREATENED

The Venus flytrap is one of the most remarkable plants in the world. Its strange appearance, carnivorous habit, and the rapidity with which the leaves close on prey have fascinated botanists since its discovery in 1770. Many of the general works mentioned previously provide information on the biology of *Dionaea.*

The species has a restricted range, occurring from Charleston County, South Carolina north to Beaufort County, North Carolina and inland as far as Moore County, North Carolina. Coker (1928) clarified the distribution of *Dionaea,* and Roberts and Oosting (1958) studied the distribution and ecology in considerable detail.

Dionaea is characteristic of ecotonal areas which lie between grassy savannas and moister, more heavily vegetated pocosins. Soils in its habitat are usually poor. Specific occupied sites are usually unshaded and possess only a sparse vegetative cover. Although shading does not kill the plants, heavily-shaded specimens seldom flower, and populations that are shaded for a number of years usually disappear. *Dionaea* not only is tolerant of fire but also requires periodic fires to reduce competition and stimulate growth (Roberts and Oosting, 1958).

Coker (1928) expressed pessimism about the survival of some populations but, contrary to the findings of Roberts and Oosting, did not imply that the species "was doomed to extinction." Roberts and Oosting (1958) were optimistic about the continued survival of the plant. Writing when they did, these botanists could not foresee the cataclysmic changes that have occurred within the plant's range in the past decade. Unfortunately, these changes are expected to continue at an accelerated rate during the coming years.

For *Dionaea,* the most significant changes are those relating to current forest practices. Many of the former savannas and pocosins are now covered by even-aged pine stands. Site preparation before planting eliminates *Dionaea.* Even without site preparation, however, shading by the pines and fire suppression destroys flytrap populations. Ditched fire lanes further traumatize the habitat by reducing soil moisture.

At least three *Dionaea* populations which I visited in the late 1960's have now disappeared as a result of the areas becoming pine monocultures. Jeffries *et al.* (1971), also could not locate the plant at a number of previously recorded stations and contended that the range was diminished.

It is rumored that *Dionaea* has become "weedy" in some areas where it occurs "by the millions." I have not been able to establish the existence of such populations, which obviously would have to occupy extensive ecotones where other vegetation was essentially absent. In some cases, it seems that such rumors originated with unscrupulous plant collectors wishing to justify their continued collection of large numbers of plants from native populations. The plant is easily propagated in culture and little justification exists for continued collection from the wild.

Herbicide spraying, road construction, and urban expansion further darken the future for *Dionaea* and the plant must be considered threatened. Endangered status is not warranted since at least 100 stations still exist.

Although *Dionaea* is protected by law in North Carolina, the law has not been effective in preventing its decline.

FAMILY LENTIBULARIACEAE

This family includes two North American carnivorous genera, *Utricularia* L., the bladderworts, and *Pinguicula* L., the butterworts. Komiya (1973) recently studied the systematics of supraspecific groups in the family.

Some of the species of *Utricularia* are not carnivorous because they lack bladders or occur in habitats where the bladders seldom function. Of the remaining dozen or so carnivorous species, none are known to be endangered or threatened. A number are widespread, some being circumboreal. The aquatic and semi-aquatic sites where many species grow have not suffered the degradation that characterizes so many bogs and other moist terrestrial sites. However, except when in flower, species of *Utricularia* are quite inconspicuous. Accurate knowledge as to the status of a number of species is at present incomplete or lacking.

Pinguicula includes nine Nearctic species. Ernst (1961) revised the genus. Wood and Godfrey (1957) and Godfrey and Stripling (1961) treated the southeastern species.

Two species of *Pinguicula* were included in the Family Lists of Candidate Endangered and Threatened Plant Species in the Continental United States (Smithsonian Institution, 1975). In addition to these, which are discussed below, *Pinguicula primuliflora* Wood and Godfrey needs further investigation to determine its status.

PINGUICULA IONANTHA Godfrey — THREATENED

This species is restricted to the area of Pleistocene and Holocene terraces in the lower Apalachicola River valley of Florida. Records are from Bay, Franklin, Gulf, and Liberty counties.

Pinguicula ionantha grows in bogs and flatwoods depressions, occupying sites where shallow standing water is often present. The rosettes are frequently submersed (Godfrey and Stripling, 1961). It also occurs in ditches and shallow drainage canals.

Although listed as endangered in the Family Lists of Candidate Endangered and Threatened Plant Species in the Continental United States (Smithsonian Institution, 1975), Godfrey (pers. comm.) is of the opinion that the species is not in danger of extinction throughout its entire range. However, habitats within the range of the species are suffering some degradation due to increasingly intensive forest practices. In addition, succession in the ditches where some populations occur will destroy these demes if mowing operations cease (R. K. Godfrey, pers. comm.). *Pinguicula ionantha* should therefore be considered threatened and should be closely monitored. The presence of a number of populations in Apalachicola National Forest provides potential for protection through awareness and positive administrative action by Forest Service personnel.

PINGUICULA PLANIFOLIA Chapman

This species is listed as threatened in the Family Lists of Candidate Endangered and Threatened Species in the Continental United States (Smithsonian Institution, 1975). It has been reported from Bay, Franklin, Gulf, Leon, Liberty, and Walton counties in Florida and from George, Harrison, and Jackson counties in Mississippi (Godfrey and Stripling, 1961; Jones, 1975). I have found the species in Santa Rosa County, Florida, and further investigation may show that its range is more extensive than previously assumed.

Pinguicula planifolia occupies habitats similar to those which support *P. ionantha*, typically occurring in shallow water in flatwoods depressions, ditches, and canals (Godfrey and Stripling, 1961). It also occupies areas of shallow standing water in bogs.

Godfrey (pers. comm.) is of the opinion that the species should not be considered threatened at the present time. Its range is considerably more extensive than that of *P. ionantha* and in my experience, its habitat requirements are less stringent.

FACTORS AFFECTING THE SURVIVAL OF THE SOUTHEASTERN CARNIVOROUS PLANTS

The southeastern United States is an important area of carnivorous plant diversity and endemism. All of the species considered endangered or threatened in the foregoing discussion are restricted to this area. The following comments are made in order to draw attention to a number of factors which are currently affecting the ecosystems in which carnivorous plants occur.

Forest Practices

The past decade has seen a tremendous increase in lumber and pulp production in the southeastern part of the United States. Officials in the forest industry have expressed an intent to make the South the "wood-basket" of the Nation.

The most dramatic change correlated with this increased emphasis is the conversion of vast acreages to pine monocultures. The preferred method of harvest has also changed from the selection method to that of clear-cutting large tracts. Herbicides are sometimes used to kill hardwoods. Both the construction of canals and ditched fire lanes serve to dry out bogs, pocosins, and other moist sites, making them more suitable for pine plantations. With the encroachment of competing plants, shading, and needle cover caused by the pines, and with the reduction of fires, few carnivorous plants can survive.

Harvesting in these pine plantations usually takes place every thirty years. Replanting is often preceded by mechanical site preparation which breaks up the soil. Extensive siltation often follows.

These practices are not limited to private land but are also carried out on Forest Service land throughout the Southeast. If alterations of this type continue into the next century, the future of many carnivorous plants and their associated biota is bleak.

Agricultural Drainage

During the past few years, increasing economic pressures have caused rural landowners and corporation farms to seek ways of using wetlands that were formerly left fallow or pastured only occasionally. Under Public Law 566, the Soil Conservation Service made plans to channelize over 20,000 miles of streams in the United States. Over 70 percent of this mileage is planned for the Southeast.

Stream channelization involves straightening, deepening, and widening streams in order to increase channel capacity. During the construction phase, many wetland habitats adjacent to the stream are destroyed. Deepening the channel often causes a lowering of the water table. More intensive use of the former wetlands usually follows.

Associated with drainage is the conversion of wetlands to pastures. Although in my experience, and in that of others (Case, Gibson, Troup, pers. comm.), cattle seldom feed on carnivorous plants and grazing may be helpful in eliminating competitors, intensive pasturing often involves discing, and the planting of forage crops such as bahia grass. Eleutarius and Jones (1969) reported that the addition of 6-12-12 fertilizer and ammonium nitrate caused a decrease in a *Sarracenia alata* population in southern Mississippi. Pullen and Plummer (1964) discussed floristic changes in the Georgia pitcher plant habitats that were previously studied by Harper (1906). Some of the changes were attributed to agricultural drainage since the sites had not received fertilizer.

In summary, it seems that use of carnivorous plant habitats as unimproved pastures is not detrimental and may even be of benefit to the plants. However, drainage and the resultant intensive use of the land will tend to eliminate the plants.

Farm Pond Construction

The so-called "hillside bogs" of the Gulf Coastal Plain where *Sarracenia alata, S. leucophylla* and other carnivorous species often occur, result from the presence of perched zones of saturation caused by a restrictive soil layer which prevents rapid downward percolation of water. Such areas make ideal sites for the construction of farm ponds. Typically, an earthen dam is constructed near the base of the bog, the area above it being partially excavated to form a pond. Construction, flooding, and alterations in the soil moisture level often destroy the bog. The presence of the pond often stimulates landowners to develop adjacent areas for pasturage by discing, fertilizing, and planting forage grasses.

I have examined over a dozen hillside bogs that have been damaged or destroyed by pond construction. An employee of the Soil Conservation Service informed me that "hundreds" of ponds have been constructed on such sites.

Since employees of the Soil Conservation Service are usually involved in giving landowners advice on the technical aspects of farm pond construction, increased awareness of the habitats may enable these public servants to advise landowners to choose alternate sites.

Reduction of Fire and Ecological Succession

It is generally recognized that periodic moderate fires are necessary to reduce the encroachment of competing plants and stimulate the growth of *Sarracenia, Dionaea, Pinguicula,* and other bog and savanna inhabitants. A number of researchers have discussed the effects of fire in such situations (Eleutarius, 1968; Eleutarius and Jones, 1969; Garren, 1943; Harper, 1943; Komarek, 1965; McDaniel, 1971). Moderate fires have probably occurred naturally for as long as these habitats have existed. However, fire control during the last half-century has reduced the frequency of such fires and allowed the accumulation of heavy undergrowth. Less frequent, more severe fires then occur and may damage species normally considered to be fire tolerant.

The encroachment of competing vegetation occurs rapidly after fire. If drainage has also occurred, the changes take place more rapidly and some carnivorous species may disappear entirely within a decade. Eleutarius (1968) found larger rhizomes and more leaves per rhizome in *Sarracenia alata* on burned sites. This was partially attributed to the release of nitrogen from litter and soil resulting from fire.

It is clear that fire is a natural event in the carnivorous plant habitat. This fact must be taken into account if populations are to be managed to insure their continued survival.

Collecting by Commercial and Amateur Interests

Because of their unique beauty and unusual mode of life, carnivorous plants have inevitably attracted the attention of many non-biologists. The recent houseplant craze has further stimulated this interest. Almost all species of North American carnivorous plants can now be obtained from commercial sources, while a number of dealers specialize almost exclusively in this group. *Dionaea, Darlingtonia,* and members of the genus *Sarracenia* receive the most collecting pressure. Even where some species are protected by law, plants continue to disappear.

Since all species now endangered or threatened can successfully be grown in culture, it is not necessary to collect from natural populations in order to obtain specimens. However, for commerical dealers and some amateurs, collecting from wild populations is cheaper and demands less effort and little horticultural expertise.

The fact that known localities for rare taxa receive heavy pressure from collectors has made many of the professionals who work with carnivorous plants reluctant to report their localities or has caused them to present vague or obscure locality data. Unfortunately, an attitude of mutual distrust exists even among some of the botanists.

Collectors sometimes obtain locality data by writing to herbaria, requesting copies of the data on herbarium sheets. Randall L. Troup, who informed me of this practice, suggested that curators begin to monitor these requests carefully in order to determine whether or not they stem from legitimate scientific needs. Though this will be possible in some cases, herbaria which are supported by public funds may run into difficulty on this score if they refuse such requests from the public.

Florists put additional pressure on some species of carnivorous plants. On the Atlantic and Gulf coasts, the leaves of *Sarracenia leucophylla* and *S. flava* frequently appear in floral arrangements sold by local florists. Removal of the leaves for this purpose usually does no permanent damage to the plant, but repeatedly picking leaves from the same rhizome may eventually kill it.

Professional botanists are not entirely innocent. Overcollecting is known to have damaged populations of a least one species of *Sarracenia* (F. M. Case, pers. comm.). The compulsion to have a few sheets of a rare species is difficult to resist by individuals in whom the desire to collect is inherently strong.

Federal listing of the endangered and threatened species will result in their protection by law, but enforcement will be difficult. The pressures that will eliminate rapacious collecting must come from within the groups of plant dealers and amateur enthusiasts, as well as professional botanists themselves.

Introgressive Hybridization

Hybridization and the possibility of extensive introgression are only important in certain of the southeastern pitcher plants. In the genus *Sarracenia,* thirteen naturally occurring hybrids are known and additional hybrid combinations have been successful in the greenhouse (Bell, 1949, 1952; Bell and Case, 1956; Harper, 1918; McDaniel, 1971). Individuals resulting from natural backcrossing have been reported by the aforementioned authors and by Wherry (1934).

Except for seasonal isolation, which sometimes fails, prezygotic isolating mechanisms are poorly developed in *Sarracenia.* Postzygotic mechanisms, such as hybrid inviability, seem to be ineffective, especially when the habitat has been disturbed. The potential then exists for genetic exchange among syntopic forms and introgression may result.

Bell and Case (1956) spoke of introgression when referring to populations of *Sarracenia alata x purpurea* in Baldwin County, Alabama. Some years ago, after a lengthy period of highway construction and other disturbances near Pensacola, Florida, nearly all the plants I examined in a savanna appeared to represent various combinations between *S. flava* and *S. leucophylla.*

Surprisingly, environmental disturbance seems to be a factor which promotes hybridization, or at least allows hybrids to survive (McDaniel, 1971), though the reasons for this are still unclear. However, the possibility exists that continued and accelerated environmental disturbance in *Sarracenia* habitats may cause loss of genetic uniqueness of the populations in some areas.

RECOMMENDATIONS

In the final analysis, the only way to preserve species is to preserve the habitats in which they occur. Species-by-species discussions of the vanishing biota are helpful, but in the future, we will probably have to concentrate on endangered and threatened ecosystems more extensively.

The habitats which support carnivorous plants also contain a variety of other interesting components. Commonly-associated plants are noted by Eleutarius and Jones (1969), Harper (1906), and Pullen and Plummer (1964). Some of the animals associated with carnivorous plants in obligatory or more casual ways are noted by Lloyd (1942). More recent studies on animal associates are by Brower and Brower (1970), Judd (1959) and Swales (1969, 1972). Ecosystem protection will benefit these species also.

The following brief comments merely touch on some possible methods for insuring the survival of carnivorous plant ecosystems. Considerable study and much trial and error will be necessary to determine which methods are most effective. Extensive detailed surveys are first needed to accurately assess the status of many of the species.

Some researchers have suggested that as long as viable cultivated stock is present, concern about wild populations should be minimal, since restocking could presumably be done. I totally disagree with this viewpoint for several reasons: first, when the habitats are destroyed there will be no appropriate sites left to restock; second, accidents could easily eliminate rare genetic types in cultivation unless they were extensively grown and distributed; and third, plants from every known deme would have to be retained in order to insure preservation of the natural genetic variability. Finally, the greatest worth of these plants lies in their ecological, scientific, esthetic, and educational value as components of the natural ecosystems in which they occur. All species should certainly be maintained in cultivation, but neither this, nor transplanting specimens to sites where they are not native, can be considered solutions to conservation problems.

Purchase of land by the federal or state governments or by private conservation groups may provide a partial answer. Healthy populations of many species are presently available for purchase. Such sites should not merely be protected. They must also be carefully managed in order to prevent degradation and to insure that natural processes, such as fire, continue to occur.

Many known populations are located on land owned or controlled by federal, state, or municipal agencies. Officials within these agencies should be properly informed so that appropriate administrative steps can be taken. If, for example, all populations occurring on Forest Service land could be protected, it would be a significant step.

The species previously noted as endangered or threatened should be protected without delay under the Endangered Species Act of 1973. Enforcing the regulations prescribed in this Act and in others promulgated by the United States Department of the Interior will be difficult. It will be especially hard to control collecting by commercial and amateur interests. Nursery inspectors might be trained to monitor commercial stocks and quarantine officials could be partially effective in controlling interstate and international shipments.

Controlling environmental degradation also will be difficult or even impossible in many cases, especially where private lands are involved, and if the alterations are not related to the use of federal funds.

Ultimately, the survival of carnivorous plants and many other components of natural ecosystems will depend on the degree of success which biologists and conservationists have in educating the public to the importance of natural systems in the long-term welfare of mankind. During the past decade, a small but significant start has been made in this direction, but much work still remains to be done.

ACKNOWLEDGMENTS

I owe much to several specialists currently working with carnivorous plants. The following individuals graciously gave me information resulting from their studies and observations and I have accepted their comments as authoritative: Frederick W. Case, Jr., Larry E. DeBuhr, Thomas C. Gibson, Robert K. Godfrey, and Randall L. Troup. For a variety of favors, I would also like to thank Daniel A. Botts, Dan W. Brooks, Ronald S. Caldwell, Robert A. Defilipps, Lois A. Donavan, James F. Duke, Audrey E. Goins, Ben F. Hajek, James D. Harper, R. Harold Jones, Thomas R. Jones, William Kirkpatrick, Alfred T. Lamb, Robert H. Mount, Thomas M. Pullen, Jr., William H. Redmond, Teresa Rodriguez, and David H. Word. Mrs. Sharon Harper kindly typed the manuscript on short notice.

LITERATURE CITED

Bell, C. R. 1949. A cytotaxonomic study of the Sarraceniaceae of North America. Jour. Elisha Mitchell Sci. Soc. **65**: 137-166.

_____. 1952. Natural hybrids in the genus *Sarracenia.* 1. History, distribution, and taxonomy. Jour. Elisha Mitchell Sci. Soc. **68**: 55-80.

_____. and **F. W. Case.** 1956. Natural hybrids in the genus *Sarracenia.* 2. Current notes on distribution. Jour. Elisha Mitchell Sci. Soc. **72**: 142-152.

Brower, J. H. and A. E. Brower. 1970. Notes on the biology and distribution of moths associated with the pitcher plant in Maine. Proc. Entomol. Soc. Ont. **101**: 79-83.

Case, F. W. and R. B. Case. 1974. *Sarracenia alabamensis,* a newly recognized species from central Alabama. Rhodora **76**: 650-665.

Coker, W. C. 1928. The distribution of *Dionaea muscipula.* Jour. Elisha Mitchell Sci. Soc. **43**: 221-228.

_____. 1976. The *Sarracenia rubra* complex. Rhodora **78**: 270-325.

Darwin, C. 1899. Insectivorous Plants (2nd ed.). D. Appleton, New York.

Dean, A. 1975. Carnivorous plants. Lerner Publications, Minneapolis.

DeBuhr, L. E. 1975. Phylogenetic relationships of the Sarraceniaceae. Taxon **24**: 297-306.

Eleutarius, L. N. 1968. Floristics and ecology of coastal bogs in Mississippi. M.S. Thesis. Univ. Southern Mississippi, Hattiesburg.

_____. and **S. B. Jones, Jr.** 1969. A floristic and ecological study of pitcher plant bogs in south Mississippi. Rhodora **71**: 29-34.

Ernst, A. 1961. Revision der Gattung *Pinguicula.* Bot. Jahrb. **80**: 145-194.

Garren, K. H. 1943. Effects of fire on vegetation of southeastern United States. Bot. Rev. **9**: 617-654.

Godfrey, R. K. and H. L. Stripling. 1961. A synopsis of *Pinguicula* (Lentibulariaceae) in the southeastern United States. Amer. Midland Natur. **66**: 395-409.

Harper, R. M. 1918. The American pitcher-plants. Jour. Elisha Mitchell Sci. Soc. **34**: 110-125.

_____. 1922. Some pine-barren bogs in central Alabama. Torreya **22**: 57-60.

_____. 1943. Forests of Alabama. Geol. Surv. Ala., Monogr. **10**, 230 pp.

Harper, R. W. 1906. A phytogeographical sketch of the Altamaha Grit Region of the Coastal Plain of Georgia. Ann. N.Y. Acad. Sci. **17**: 1-415.

Jeffries, D. B., T. Minton and D. Vodopich. 1971. Notes on the distribution of *Dionaea muscipula.* Jour. Elisha Mitchell Sci. Soc. **87**: 155 (abstr.).

Jones, S. B., Jr. 1975. Mississippi Flora IV. Dicotyledon Families with aquatic or wetland species. Gulf Research Reports **5**: 7-22.

Judd, W. W. 1959. Studies of the Byron bog in southwestern Ontario. X. Inquilines and victims of the pitcher plant, *Sarracenia purpurea.* Can. Ent. **91**: 171-180.

Komarek, E. V., Sr. 1965. Fire-ecology — Grasslands and man. Proc. 4th Annual Tall Timbers Fire Ecology Conference, 169-220. Tallahassee.

Komiya, S. 1973. New subdivision of the Lentibulariaceae. Jour. Jpn. Bot. **48**: 147-153.

Lloyd, F. E. 1942. The carnivorous plants. Chronica Botanica, Waltham, Mass.

McDaniel, S. 1971. The genus *Sarracenia* (Sarraceniaceae). Bull. Tall Timbers Res. Sta. **9**: 1-36.

Macfarlane, J. M. 1907. Observations on *Sarracenia.* Jour. Bot. **45**: 1-7.

Pietropaolo, J. and A. Pietropaolo. 1974. Carnivorous plants of the U.S., Stoneridge, Bramford, New York.

Pullen, T. M., Jr. and G. L. Plummer. 1964. Floristic changes within pitcher plant habitats in Georgia. Rhodora **66**: 375-381.

Radford, A. E., H. E. Ahles and C. R. Bell. 1968. Manual of the vascular flora of the Carolinas. Univ. North Carolina Press.

Roberts, P. R. and H. L. Oosting. 1958. Responses of the Venus fly trap *(Dionaea muscipula)* to factors involved in its endemism. Ecol. Monog. **28**: 193-218.

Schwartz, R. 1974. Carnivorous plants. Praeger, New York.

Smithsonian Institution. 1975. Report on the endangered and threatened plant species of the United States. U.S. Congress, Committee on Merchant Marine and Fisheries, Serial No. 94-A, U.S. Government Printing Office, Washington. 200 pp.

Swales, D. E. 1969. *Sarracenia purpurea* L. as host and carnivore at Lac Carré, Terrebonne Co., Quebec. Naturaliste Can. **96:** 759-763.

_____. 1972. *Sarracenia purpurea* L. as host and carnivore at Lac Carré, Terrebonne Co., Quebec. Naturaliste Can. **99:** 41-47.

Thanikaimoni, G. and G. Vasanthy. 1972. Sarraceniaceae: Palynology and systematics. Pollen et Spores **14:** 143-155.

Trappe, J. M. and J. W. Gerdemann. 1974. A northern extension of the range of *Darlingtonia californica.* Madrono **22:** 279.

Uphof, J. C. T. 1936. Sarraceniaceae. In A. Engler and H. Harms. Die natürlichen Pflanzenfamilien. (2nd ed.) Bd. **17:** 1-21.

Wherry, E. T. 1934. Exploring for plants in the southeastern states. Scientific Monthly **38:** 80-85.

_____. 1972. Notes on *Sarracenia* subspecies. Castanea **37:** 146-147.

Wood, C. E., Jr. and R. K. Godfrey. 1957. *Pinguicula* (Lentibulariaceae) in the southeastern United States. Rhodora **59:** 217-230.

THREATENED AND ENDANGERED SPECIES OF ORCHIDS

Carl L. Withner

Brooklyn College, New York 11210, U.S.A.

During the past few days, we have all seen and heard of the widespread destruction of forest areas, particularly in tropical America. Orchids, among the many victims of forest devastation, are particularly vulnerable to disturbance.

Orchids are members of a vast family, so vast that it is almost beyond imagination, and wherever orchids are found, they bring forth a positive response. Depending upon who is counting, there are between 12,000 and 30,000 species, but no one really knows. The whole New England flora only consists of about 5,000 species, and I wonder if anyone but orchidologists will ever appreciate the scope of this complex family. Many orchids are so showy and ornate that the very mention of the word conjures up, to many people, a vision of something large, floppy and purple that is generally worn upside down on some ample lady's shoulder! The highly decorative and commercially familiar species are often in the genera *Cattleya, Cymbidium, Paphiopedilum, Phalaenopsis,* or *Vanda,* or their many hybrids produced in cultivation. Thousands of hybrid combinations have been registered with the Royal Horticultural Society, where such records are diligently and lovingly kept.

Orchids enjoy a worldwide distribution in both terrestrial and epiphytic habitats. In addition to the few showy species with greatest commercial value, there are literally hundreds of more insignificant types, many of which are only a few centimeters in height, with flowers just a few millimeters in diameter. With these tiny "botanicals," as they are often referred to by orchid growers, it frequently is only the very dedicated specialist who gets to know one genus, perhaps, in sufficient detail to really understand it. Such genera as *Pleurothallis* or *Stelis* may have hundreds of species in the single genus, and there are many examples of unusually large genera within the family.

Orchid plants in general, and epiphytic ones in particular, respond to the same favorable environmental factors as do other plants. By and large, they are light-loving, appreciate high humidity, and enjoy what is euphemistically referred to as "buoyant" air when one tries to reproduce it in the greenhouse. Orchids grow among a great variety of other epiphytic vascular and non-vascular plants, and sometimes it is hard to find them amidst all the other vegetation on the branches, though at other times they may be the dominant plants on a given tree. They are florally and vegetatively highly specialized in both their epiphytic and terrestrial environments, and they grow in harmony with the seasons. This vast family of plants may propagate vegetatively by tubers, corms, creeping rhizomes, offsets, sideshoots, or divisions, but for sexual methods a much more complicated balance of factors is necessary. For instance, orchids require for their seed germination certain mycorrhizal associations. Of the thousands of seeds produced in a single capsule, only a minute fraction will reach the correct microniche where all the proper conditions for nutrition and growth are present. In certain terrestrial species, the seedlings may remain underground for as many as ten to fifteen years, nourished all the while by the fungal symbionts, before even a first tentative green leaf is put above ground. And, of course, before the seed can be produced, the proper conditions to promote flowering must prevail, and the right pollinator must be present.

The interdependence of many factors over a period of years for the success and propagation of a natural population points up the fact that all of these factors, in effect whole ecosystems, must be preserved in order to perpetuate the life cycle. It is unquestionably *such interdependences which make orchids more vulnerable than other plants to habitat destruction.* Sanford (1974a, 1974b), who has made a particular study of these relationships, points out that "conservation should not be accepted as meaning saving a particular plant or bird or scenic site. Rather, conservation is best thought of as the management of natural ecosystems in such a way that they benefit man not only now but in the future."

We are only just beginning to understand the complex external requirements and internal metabolism of these epiphytes. Recent research with my graduate students and other colleagues (Rubenstein et al, 1976) who were studying the carbon dioxide fixation patterns, has shown that there are at least three patterns for orchid photosynthesis which have evolved over the years. We find, for example, that orchids such as *Coelogyne,* which grow in shady humid places, have regular Calvin-Benson C_3 photosynthesis patterns, as does *Cymbidium* also. For plants such as *Schombocattleya* and *Paphiopedilum,* we confirmed previous reports — by work with chromatography, stomatal behavior and carbon dioxide fixation patterns — that many such orchids show crassulacean acid metabolism, i.e., the de Saussure effect. This process takes place at night, with the plants absorbing carbon dioxide in the dark. A particularly interesting finding during this research was that the *Paphiopedilum* combined the C_3 pattern during the day with the CAM pattern at night. Furthermore, we found that the *Schombocattleya* combines the CAM at night with Hatch-Slack C_4 carbon dioxide fixation during the day. In other words, orchids are remarkably well adapted to their specific habitats in terms of photosynthetic efficiency.

These critical microniches have many parameters and are easily disturbed. When one considers how we attempt to grow orchids in cultivation with plants from different locations and different environments often growing side by side, one wonders how it is possible to duplicate natural factors reasonably well so the plants will exist at all. In any attempt to conserve orchids, either by cultivation or in various natural preserves, these factors affecting photosynthetic efficiency need to be better understood and more carefully worked out for a great variety of species. Indeed, other physiological factors also must be considered. For example, nutrition, light intensity, pH, etc., all affect the productivity of orchids and their role in a particular ecosystem.

How long can an orchid plant survive? How well does one species compete with another? Dunsterville (1961) found that in Venezuela as many as 48 species of orchids can grow on a single tree, while in Nigeria, Sanford (1974a) counted up to 14 species per tree. This clearly reveals that there are at least some kinds of orchids which will grow and reproduce under the same general set of circumstances. Occasionally, a few hundred plants of the same species can be found on a single tree — more than enough to provide stock for cultivation around the world for a long time to come; and there may be many such trees until the forests are removed. Since other species, however, are found with only an occasional plant here and there, it may be difficult to find a population to study. The distribution patterns vary with the species and range from local to widespread occurrence.

Some plants or their divisions reach venerable ages. In cultivation, various famous clones of certain species have been known for over a hundred years. Smith (1966) found an *Epidendrum gracile* plant growing in the Bahamas which, by counting the backbulbs, he was able to estimate at 128 years! An *Epidendrum concolor* plant that I collected from an oak in southern Mexico had 27 pseudobulbs, each progressively larger, so that we know it was surviving for at least that number of years, not to mention how many before the pseudobulbs were large enough to show up. And yet, at the other extreme, we have seen *Oncidium pusillum* plants flowering as epiphytes on mango leaves in Ecuador where they grew from seed to flower before the leaf dropped from the tree, a matter of two or three years at most. *Zeuxine strateumatica* even manages to be an introduced weed in Florida, not only growing but seemingly thriving in lawns, on roadsides, in sidewalk cracks, and in flower pots along with other orchids. It can grow from seed to flowering in only ten months (Luer, 1972).

Endemism of orchid species, which is extremely developed in the family, is another prime factor that makes them susceptible to destruction. Orchidologists generally accept this as evidence of a family in an explosive state of evolution (Garay, 1960). In Papua, New Guinea, there are over 2,600 species in 128 genera. Hundreds of species are endemic, as are 19 of the genera. In Java, there are about 1,000 species in 139 genera, and in Madagascar, a comparable array of hundreds of species can be found with the same story of endemic locations. These statistics, however, are at best preliminary counts and should only be looked upon in that light. There are experiences such as Dunsterville's in Venezuela, when, recuperating from an operation a few years ago, he decided to make some drawings of the orchids of that country. He enlisted the coauthorship of Harvard's Dr. Leslie Garay for taxonomic expertise, and thought that the flora could be completed in two volumes. Based on previous records, he estimated that about 250 species would complete the project. Their sixth volume is now published, and the number of species known from their field research in Venezuela has already reached the 1,000 mark. They are still finding species of orchids in that country which are completely new to science. This experience is not unique to Venezuela, but can be repeated elsewhere. Donald Dod and his friends in the Dominican Republic, for example, found 65 genera and about 100 species recorded in the literature. Their field work has shown that only 51 of the genera were valid records for native orchids. Since then, they have found 17 new records, making a total of 75 genera and over 300 species. Dod and his friends have found twenty species of *Lepanthes* alone, which have not yet been named. There is still no way to determine accurately exactly how many orchids there are in certain areas. The continued destruction of forests at the present time, combined as it often is with localized distribution patterns, means that many species will, unfortunately, die unknown. No world flora of orchids has yet been compiled, and there is not even a local flora available for critical orchid distribution areas involving specific countries. In fact, many important genera have not yet been critically monographed either. The current available data would indicate that such projects should be undertaken with all due haste, because another fifty years of the so-called "progress" which our civilization practices today might very well see the end of many major habitats and intact ecosystems.

Orchids, because of their showiness, have been horticulturally important for many years. It is difficult to find a group of plants which offer greater challenges to growers. They have refinement, elegance, grace, fascinating variety, and scent — all qualities that make them proper subjects for horticulturists and botanists alike. These very characteristics are implied in the Chinese word "lan", meaning love and beauty, or the Japanese word "ran", both of which are symbols for the orchid. The Chinese were cultivating these plants for esthetic pleasures before the time of Christ, and they have been appreciated in the Orient since then. Incidentally, the plants were valued as much for their leaves as for their flowers, and also for the grace with which they grew.

Most of our present horticultural knowledge about orchids originated in England. The early history of orchid-growing in that country actually parallels both the evolution of greenhouse technology and England's expansion and colonization around the world. The English people were always interested in plants and gardening, and they carried orchids back to their homeland as prized specimens from their worldwide travels. Records show that tropical orchids first flowered in England in 1732, and by 1794 Kew was cultivating 15 species, mostly brought by Admiral Bligh from the West Indies. Indeed, the "close, glazed glass cases" of Dr. Ward were, by the 1840's, helpful in transporting such plants on the return voyage to England. By the middle to the end of the last century, there was a lively but nonetheless serious competition among various landed gentry to obtain the finest orchid imports for their estate collections. Auction records reveal that thousands of plants were sold during those years, with prices as high as a thousand pounds being paid for a single plant. Collectors were sent out, but they often gave the wrong locations for their finds, thus causing record problems that are still being sorted out today. In spite of this, these past records continue to provide a good historical literature on orchid collecting that goes back for over a century — an unusually complete record, when one considers most plant groups.

Many fine orchid books with hand-colored illustrations offer vivid views of what species were once imported and sold, and then grown and flowered. These books produce a fantastic record of all the variations that were present in the more abundant and popular showy species. Orchidologists are, indeed, fortunate in having such a permanent record, even if world floras are not yet prepared. One of the major weaknesses of such records, however, is that they do not include the thousands of lesser, though equally interesting, species.

Through study and actual measurement of these old plates, and by comparing them with living flowers of the same types, we have found that these old drawings are extremely accurate, even down to depicting deformities and leaf spot afflictions. They are so accurate, in fact, that when these flowers are studied today for numerical data, for instance, the measurements are quite useful. We have published (Withner and Adams, 1960) one such study, based partly upon such data, that concerns the hybrid

Lc. Elegans from Santa Catarina Island off southern Brazil, where three separate species produced a fascinating introgressive hybrid swarm.

Vanilla, which is sometimes called the orchid of commerce, was discovered by Cortez in Mexico. It is the only orchid in which a sizeable trade exists today, except for plants or flowers which are grown for sale. Cortez found the Aztecs making a drink, which was supposed to fortify the hearts and clear the minds of government officials and others of high rank. This special chocolate drink was flavored with chili and vanilla. After Cortez took vanilla back to Spain, the Spanish king introduced it to the rest of the world. In Mexico, the vanilla-growing center was the town of Papantla, famous also for its "flying eagles." But, sad to say, when I visited Papantla about 15 years ago, the only vanilla I could find was in a tile painting on the back seat in the zócolo. Oil had been discovered in the immediate vicinity of Papantla, with the result that vanilla-growing and vanilla habitats are hardly to be found there today. Most vanilla now comes from the island of Madagascar.

Orchids present many fascinating scientific problems which often require study of the plants in their natural habitats. Orchid ecology is in its infancy. Speciation problems abound, with many concomitant problems concerning geographic distribution and evolutionary relationships. There is, for example, the matter of understanding and then saving certain species and clones because of the gene pools they represent, which may or may not already be conserved to some extent by growing the plants in cultivation. Of the thousands of species, only six or seven hundred have had chromosome counts made of them, and there are still many genera without a single count known. During the last ten years, orchid alkaloids have been "rediscovered", particularly in Asiatic orchids. Not too much is yet known about their value or importance, or even the technical structure of the compounds. Pheromones and fragrance studies in pollination of orchids, a field of botany relating insect and plant interaction and evolution, has only recently begun to be sufficiently appreciated. Many pollinators for specific species are still unknown. Phytoalexins are now known in orchids. There are myriad ways in which these plants are of interest scientifically for biology at large (see Withner, 1974) — as well as horticulturally. It is obvious that there is still much field research and data collecting to be done while large habitat areas are still more or less intact.

Certain species have, undoubtedly, been overcollected in past years, and the old auction records show how plentiful some of these plants must have been in times past. But, so far as I can determine, no showy and desirable orchid species has been made extinct as a result of the collector alone during the more than 100 years of orchid explorations. In spite of many years of active export laws in such countries as Colombia (Ospina, 1969), Venezuela, and Mexico (Hagsater, 1976b), most of these plants are still available, in fact, from the country of origin. Though the populations of showy taxa, undoubtedly, have been diminished, there has to date been comparatively little concern shown for the lesser species of no commercial value, though these have just as much botanical importance. There are now undoubted examples of species most likely extinct in the wild. Dr. E. A. Schelpe has written to me about *Disa* from South Africa, and Dr. Guido Pabst has written to me of *Scuticaria* from Brazil, as being presently extinct in the wild. Locations of orchids, particularly those of new species, should never be published or given widespread, or even local, publicity, because such information is frequently misused.

A rare or endangered species, particularly if it is on an official list or is a national flower, presents a challenge to most growers. Dunsterville (1975a) has wisely written about the sheep whose eyes develop a "goatish gleam" when confronted with the opportunity of adding a rare plant to his bag. In an effort to alleviate this situation, since specific clones of exceptional quality or color are highly desirable horticulturally, growers in both the tropics and elsewhere are now "selfing" these species and raising quantities in cultivation. There is little danger of extinction *in cultivation* for such taxa, and the growing has resulted in many vastly improved techniques for managing these particular types of plants (See Young, 1974; Rands, 1975; Steele, 1975; Duveen, 1975; and Channon, 1974).

Putting these commercial types of species or any others on a list automatically produces problems by stimulating the "trophy syndrome". But since lists are quite arbitrary in the first place, what about the many species which are also being destroyed,

317

not so much by collectors as by deforestation? A true orchid list would already contain hundreds of such species. I maintain that the existing lists demonstrate, primarily, a frustration, an acknowledgment that a problem exists, and that little has been done to curb effectively the destruction of natural habitats. It should be obvious by now that I am not in favor of these arbitrary lists. I feel that in the long run, they may do more harm than good.

It is also possible that certain species will be kept alive *only* by being collected and kept in cultivation. Two examples, with whatever accompanying species grow with them, come immediately to mind. The habitat of *Laelia milleri,* which was discovered only about fifteen years ago, is, according to botanists in Brazil, already completely destroyed by mining operations, and the species is no longer to be found in the wild. Another example is *Oncidium macranthum* from Ecuador. This species lived in a forest area that has been reduced mostly to charcoal. Kennedy (1975) has written that a situation such as this occurs only where roads traverse an orchid-rich area, and that there are plenty of acres and mountains in between such roads where plants are still to be found for the exploration. But the destroyed habitats are often the only known ones for the species involved. As mentioned above, endemism can be a disadvantage from this point of view. Many take exception to Kennedy's views (Dunsterville, 1975b; Tarlow and Tarlow, 1976; Hagsater, 1976a) and feel that all import and export of orchids — except those certified as grown in cultivation — should be stopped. In addition, collecting permits would also be required, and no one should encourage local collectors by purchase of wild plants.

One of the interesting developments in the orchid world has been mericloning with the possibilities of cultivating thousands of plants of a given clone, either a rare species or a fine hybrid with particularly desirable qualities. Added to this are the new refinements of seed growing, such as green-podding, but still there are certain orchids that have not yet been successfully grown in cultivation. This applies particularly to temperate zone terrestrial species, rather than to the tropical epiphytic types. Noel Gauntlett in Jamaica and a few interested individuals in other areas are taking these greenhouse-grown plants and restocking them in the wild, thus reestablishing native populations. This seems to be a largely unexplored possibility that can be applied to many, if not most, orchid species when proper preserves or botanic gardens are set up for orchid conservation.

With over 10,000 orchid hybrids now registered by the Royal Horticultural Society, another focus has been found for many orchid growers. Interest in hybrids can take the pressures of orchid depletion away from the species. Many growers prefer hybrids to species for ease of growth, floriferousness, a certain season of bloom, and so on. There are now over 200 hybrid genera (Garay and Sweet, 1974), some with four to six different genera involved, so that an index is a necessity in dealing with these vast new populations.

The orchid family is obviously ripe for computerization. Increasing numbers of people (Hoffman, 1976; Sanford, 1974b) are calling for such recording, not only for the registration of hybrids, but also to list what kinds of orchids are present in collections throughout the world and what sort of a reservoir in cultivation actually exists. No one now knows. Also, computers can constructively be used in the identification process, for data storage, for population and ecological studies, as well as for other types of research. Although there is almost universal agreement on how helpful such a step might be, the questions of where to place such facilities, who might do the work, and how it could be financed, remain unanswered. Yet, with the continued destruction of habitats and population pressures steadily rising, orchidologists throughout the world probably have only about thirty or forty years left for them to gather the data into a permanent record.

There is no question that intense concern and awareness of the orchid problem exists, both locally and throughout the world. The American Orchid Society, for example, has had a Conservation Committee for more than ten years. There has been much publicity about the problem and many articles have been written about it (Alphonso, 1966; Anderson, 1968; Ayensu, 1975; Dickinson, 1968; Hunt, 1968; Mathisen, 1974; Melville, 1971; Moir, 1967, 1970; Peterson, 1974; Pradhan M.G., 1974; Pradhan U., 1971, 1975a, 1975b, 1975c). There are 16,000 members in the American Orchid Society, an international organization, and there are many more people all over the world who are interested in growing orchids but who are not members of any organized group. All told, there exists a force of between 20,000-25,000 people who could be used for constructive purposes in the pursuit of orchid conservation, if there were some one individual or some way to organize them. There are more people interested in orchids than there are professional botanists in the world.

Pollard in Mexico used to say that the Spaniard's worst gift to the people of this hemisphere was the machete. Dod (1975) in the Dominican Republic has said the worst threat was fire. But the effects of native agriculture — the slash and burn variety — is miniscule these days compared to the effects of power machinery providing an armament of power saws, bulldozers, trucks, and tractors. The old *incendio,* the campesino with his machete, is no match for modern methods at all. We have heard various statistics, but destruction of up to or over 5,000 hectares of forest is actually taking place every day.

Two examples to emphasize my point will be sufficient: Costa Rica has 5.1 million hectares of territory. In 1950, 3.9 million hectares, or about 76 percent of the land, were in forests. In 1973, only 52 percent of the country was forested, a decrease of a quarter of the forests in 23 years! This produced about 1.5 million feet of lumber a year, the balance of the wood either being converted to charcoal or just burned to dispose of the debris which included tons of epiphytes. Eric Hagsater from Mexico writes (1976b) that there were 70,000 kilometers of roads there in 1970, but by 1975, 150,000 kilometers had been constructed. With roads and so-called "progress," there is always an influx of people and the further destruction of forests follows in its wake. Most unlikely places are then used for agriculture, with subsequent erosion and destruction of the soil and depletion and impoverishment of the land.

In Haiti, where population pressures are so intense and the land so limited, natural vegetation survives only in the most inaccessible spots. In the Andes, where agriculture has continued for many centuries, even the tops of large ridges have been leveled off through cultivation and erosion. Of course, the orchids and forests have accordingly disappeared.

CONCLUSIONS

On the positive side, there seems to be one program, at least, which is working well — a plan to stabilize agriculture by eliminating slash and burn techniques, a plan that proposes to ease population pressures and still leave forests and ecosystems reasonably intact. That is the "Plan Chiapas" now operating in southern Mexico. The plan was conceived, prepared and activated by Walter L. Hartmann (1974), Technical Coordinator of the Council for the Protection of Nature of the State of Chiapas, Mexico. Though it specifically concerns orchids, the Plan has many more far-reaching implications and can readily be adapted for the conservation of natural areas in other regions.

With the support and cooperation of local government agencies, and as a result of a concerted program of education in the schools, the destruction of orchid and forest habitats in Chiapas has now definitely been abated. The Plan has included 42 lectures and other events for teachers to use in high schools and agricultural institutes as part of a basic educational program. It was widely publicized on radio and TV, and in newspapers. Through the Plan, bulletins have been published concerning alternative agricultural methods in order to prevent burning and destruction of natural vegetation, and newly-built fire towers have decreased burning in that area by 75 percent. Under the Plan, 82 Councils for the Protection of Nature in Chiapas were formed, with permits now required for collecting and exporting. The people there have begun to cultivate orchids and other plants for sale to tourists, scientific institutions, botanic gardens or nurseries. The local people thus derive some benefit from the disposal of their assets besides their conservation for the future. They have also trained rangers, and they have set up a tourist park, an educational park, and a botanic garden (Hagsater, 1976b). In an effort to reforest depleted hillsides, trees are actively being replanted, but not eucalyptus or pine, which are detrimental to other species and to the growth of epiphytes. The people in Chiapas are justifiably optimistic, and the Plan should serve as a model for other groups to follow.

In summary, we now can see how orchids are vulnerable to destruction because of complex life cycles and the variety of factors necessary for continued reproduction in nature. We also know of their past records, their potentials for growth and propagation in cultivation, and how they may be of use scientifically. We are aware of population pressures, endemism, and how and why forest ecosystems are being destroyed today. The many thousands of orchidists — botanists and horticulturists alike — form a potent, informed source that, with proper direction, could contribute much toward the preservation of these plants and their natural habitats for future generations. The Plan Chiapas is one workable scheme that has already shown positive results.

ACKNOWLEDGEMENTS

My sincere thanks go to the following correspondents who replied to my letters: Dr. Edward S. Ayensu, Dr. Anton Ghillány, Walter L. Hartmann, Clarence Kl. Horich, Prof. George Kennedy, W. W. G. Moir, Rebecca Northen, Ganesh Mani Pradhan, Udai C. Pradhan, Guido Pabst, Dr. E. A. Schelpe, Jorge Verboonen, and Prof. Kenneth Wilson. They have provided many of the important points in the talk I presented, and, indeed, have provided me with even more specific details of species and localities than I could include in this article. I am particularly appreciative of the fine letters, news clippings, and "before and after" photographs of Costa Rican forests which were so generously sent to me.

LITERATURE CITED

Alphonso, A. G. 1966. The need for conservation of Malaysian orchid species. Proceedings of the Fifth World Orchid Conference.

Anderson, L. 1968. Methods of orchid conservation. Amer. Orchid Soc. Bull. **37:** 293-294.

Ayensu, E. S. 1975. Endangered and threatened orchids of the United States. Amer. Orchid. Soc. Bull. **44:** 384-394.

Channon, G. 1974. Save the species? Amer. Orchid Soc. Bull. **43:**123-124.

Dickinson, S. 1968. Further thoughts on orchid conservation. Orchidata **8:**88-89.

Dod, D. D. 1975. Los incendios. Boletin "Jardin Botanico Dr. Rafael M. Moscoso" **2:** 1, 17.

Dunsterville, G. C. K. 1961. How many orchids on a tree? Amer. Orchid Soc. Bull. **30:** 362-363.

_____. 1975a. Love us, but please do not raze us! Orquidea (Méx.) **5:**55-58.

_____. 1975b. A letter to orchid conservationists. Amer. Orchid Soc. Bull. **44:** 883-885.

Duveen, D. 1975. From Emperor's favorite to practical conservationist, or a century of orchidology. Amer. Orchid Soc. Bull. **44:** 314-317.

Garay, L. 1960. On the origin of the Orchidaceae. Proceedings of the Third World Orchid Conference.

_____ and H. Sweet. 1974. Natural and artificial hybrid generic names of orchids. In The Orchids, Scientific Studies, C. L. Withner, ed. John Wiley and Sons.

Hagsater, E. 1976a. Can there be a different view of orchids and conservation? Amer. Orchid Soc. Bull. **45:** 18-21

_____. 1976b. Orchids and conservation in Mexico. Orchid Review **84:** 39-42

Hartmann, W. 1974. The "Plan Chiapas" and its fulfillment. Orquidea (Méx.) **4:**124-127.

Hofmann, M. 1976. Index systems for special groups as illustrated by a collection of Orchidaceae. Orchid Review **84:** 62-63.

Hunt, F. 1968. Conservation of orchids. Orchid Review **76:** 320-327.

Kennedy, G. 1975. Orchids and conservation — a different view. Amer. Orchid Soc. Bull. **44:** 401-405.

Luer, C. 1972. The Native Orchids of Florida. New York Botanical Garden.

Mathisen, J. A. 1974. Dilemma of the orchids — orchid conservation. Amer. Orchid Soc. Bull. **43:**1043-1048.

Melville, R. 1971. Conservation of orchids. Orchid Review **79:** 21-22.

Moir, W. W. G. 1967. Final report of the Conservation Committee to the American Orchid Society. Orchidata **7:** 106-110.

_____. 1970. Conservation of native orchids. Amer. Orchid Soc. Bull. **39:**425-426.

Ospina, M. 1969. Colombian orchids and their conservation. Proceedings of the Sixth World Orchid Conference.

Peterson, R. 1974. Conservation conversation. Amer. Orchid Soc. Bull. **43:** 99.

Pradhan, M. G. 1974. Orchid conservation in India. Amer. Orchid Soc. Bull. **43:** 135-139.

Pradhan, U. 1971. Orchid conservation attempts in Sikkim and E. India. Amer. Orchid Soc. Bull. **40:** 307-308.

_____ 1975a. Conservation of Eastern Himalayan orchids. Problems and prospects. Part I. Orchid Review **83:** 314-317.

_____. 1975b. Conservation of Eastern Himalayan orchids. Problems and prospects. Part II. Orchid Review **83:** 345-347.

_____. 1975c. Conservation of Eastern Himalayan orchids. Problems and prospects. Part III. Orchid Review **83:** 374.

Rands, R. J. 1975. Phragmipediums — and their future. Amer. Orchid Soc. Bull. **44:** 235-238.

Rubenstein, R., D. Hunter, R. McGowan and C. Withner. 1976. Carbon dioxide metabolism in various orchid leaves. Northeastern regional meetings Amer. Soc. Plant Physiologists. Rutgers University. Abstracts of papers, p. 10.

Sanford, W. W. 1974a. The ecology of orchids: In The Orchids, Scientific Studies, C. L. Withner, ed. John Wiley and Sons.

_____. 1974b. Some urgent problems of orchid ecology. First Symposium on the Scientific Aspects of Orchids. University of Detroit. Harry Szmant and J. Wemple, eds. pp. 63-74.

Smith, J. E. 1966. Desirable orchids of the Bahamas. Amer. Orchid Soc. Bull. **35:** 970-975.

Steele, A. 1975. Species orchid seed — conservation and distribution. Amer. Orchid Soc. Bull. **44:** 514-515.

Tarlow, C. and A. Tarlow. 1976. Editorial and Letters to the Editor about conservation. Orchidata **15:** 5, 7-8, 11-13.

Young, J. L. 1974. Seed banks. Amer. Orchid Soc. Bull **43:** 124-125.

Withner, C. L., ed. 1974. The Orchids, Scientific Studies. John Wiley and Sons.

_____ and H. Adams. 1960. Generic relationships and evolution among the cattleyas and their relatives. Proceedings of the Third World Orchid Conference.

RARE AND ENDANGERED PTERIDOPHYTES
IN THE NEW WORLD AND THEIR PROSPECTS FOR THE FUTURE

John T. Mickel

The New York Botanical Garden, Bronx, New York 10458, U.S.A.

The discussions in this Symposium have centered around the flowering plants. In most floristic considerations all vascular plants are included, since indeed the ferns and fern allies do share most of the same collection techniques, broad life history outline, and taxonomic problems with the seed plants.

The ferns and fern allies exhibit certain features, however, that make them somewhat more sensitive perhaps to the pressures of our times than are many of the higher plants. First, their method of sexual reproduction is much less efficient — by spores rather than seeds. It is not uncommon for a single fern plant to produce literally millions of spores in a season, yet the world is not overrun by ferns. Actually, relatively few new plants are produced by this method. Probably more individual plants are produced by vegetative means — by buds produced on the roots, rachis, pinnae, or frond apex, or even on the gametophytes in the form of gemmae. Ferns are perennials and usually slow-growing, often taking a year or two to reach spore-producing size. They are generally not aggressive in competition, either in the gametophyte stage or as sporophytes.

CAUSES OF ENDANGERMENT

The causes of rarity and endangerment in the ferns are largely the same as for the other plants discussed in this volume. Some are naturally rare because of special habitat such as páramo, serpentine rock, aluminum-bearing soil, or because of reproductive ineffectiveness, but mostly because of reasons we do not yet understand. At times, well-established taxa may become significantly diminished in numbers through natural disasters. Recent droughts in Costa Rica have greatly reduced the number of epiphytic ferns in some areas, and rare species in these areas might be truly endangered.

Commercial harvesting of pteridophytes has taken place in certain groups. For example, large numbers of the resurrection plant *(Selaginella lepidophylla)* have been collected and sold as novelties over the years, but in spite of this, it is neither uncommon nor threatened at present. A greater market has been found for osmunda fiber, the root systems of *Osmunda regalis* and *O. cinnamomea*, which is used for epiphytic soil mix or for orchid support, but neither of these species can be considered rare or threatened either. Osmunda fiber breaks down within a few years and for this reason tree fern fiber is favored and consequently causes more of a problem from the standpoint of conservation. Tree ferns have a vertical trunk that reaches 4-20 m in height. The stems may be 6-13 cm in diameter and the fronds 2-3 m in length. They are among the most impressive ferns in their size and beauty. Nearly all belong to the family Cyatheaceae. In some species the vertical stems are clothed with a mantle of adventitious roots which serve in the wild as a habitat for many epiphytic plants. In horticulture, tree fern roots have been found to be extremely effective as support for epiphytes, especially orchids. Tree ferns are harvested in large quantities as whole trunks, then cut and sold as slabs, baskets, posts, troughs, and boxes, or chopped and used as part of a potting medium. One side effect, nature's revenge perhaps, is that those engaged in extensive cutting of the tree fern fiber are often afflicted with "black lung" disease from breathing the dust. Tree fern fiber varies in coarseness and durability of the roots, tightness of the root mass, and thickness of the root mantle, some species being favored for commercial use over others. Cutting for this purpose seems to be carried out wherever the tree ferns occur and produce the root mantle.

The most extensive root mantles, and consequently the most widely-cut, are found in species of *Dicksonia: Dicksonia karsteniana* in Venezuela and *D. sellowiana* in southern South America. Several species of cyatheoid tree ferns also produce great root masses. In Brazil, *Trichopteris atrovirens* is extensively used; in Venezuela, *Trichopteris tryonorum* and others. Most of the tree fern fiber used in horticulture in the United

States comes from Mexico, Guatemala, and Costa Rica, threatening especially *Nephelea mexicana* and *Alsophila salvinii*, both of which were cited in the Convention on International Trade in Endangered Species of Wild Fauna and Flora (1973). Other species of tree ferns in Mexico and Guatemala are more widespread and consequently are not threatened; many are smooth-trunked. In Costa Rica and further south there are richer tree fern floras and a higher degree of endemism, so extensive cutting and harvesting is of even greater significance in these more southern areas from the standpoint of conservation. Diminishing supplies and greater legal restrictions are making tree fern fiber increasingly difficult to obtain now. One saving grace for the tree ferns is that many of them produce spores at a relatively small size before a significant root mantle is formed, thus lessening the chances of total extinction.

Horticultural collecting of ferns apparently is not a serious problem in the tropics. Both commercial and amateur fern collecting is minimal there. Within the United States, however, some danger exists from individual private gardeners who collect rarities for their gardens. For example, the hart's-tongue fern *(Phyllitis scolopendrium)*, though common in Europe, is greatly restricted in its North American variety to serpentine soils in Michigan, Ontario, New York, and a sink hole in Tennessee. There is a close relative *(P. lindenii)* in Haiti and southern Mexico. This peculiar disjunction needs more study and deserves preservation. The hand fern, *Ophioglossum palmatum,* is rare throughout its range and is difficult to grow, but because of its curious form it is frequently taken by collectors. Scott's spleenwort, *Asplenosorus ebenoides,* is a rare hybrid between the walking fern *(Camptosorus rhizophyllus)* and the ebony spleenwort *(Asplenium platyneuron)* and is one of the most sought-after ferns in North America. Commercial fern suppliers generally deal in the common fern species, and because of ease of collecting and growing, they do not pose a serious threat to rare fern species.

By far the greatest danger to pteridophytes, as to other plant groups, is the wholesale habitat destruction throughout the New World. This comes from many causes — farming, lumbering, grazing, and settlement. The lowland wet forests are nearly gone in many areas throughout tropical America. Fortunately, most of the species of these areas are very widespread. Recent agricultural development in the Amazon basin has destroyed vast areas, but most of the pteridophyte species are so wide-ranging that there is little threat of immediate endangerment. There are some exceptions, of course. The bizarre *Dictyoxiphium panamense,* for example, is rare, but it also hybridizes with a species of *Tectaria* to form *Pleuroderris michleriana,* an extremely rare but morphologically significant plant. In Oaxaca, Mexico, *Bolbitis umbrosa* had been collected once in 1842 and four times in the last eight years. Of its four known localities, two are now destroyed.

Of greater importance are the middle and higher elevations where the pteridophyte flora is more diverse and the taxa are often of more limited range. The cut-and-burn agriculture of some of the Indian groups has not been widespread enough to seriously threaten the overall forest, but improving technology and increasing population pressure has wiped out great areas. In the middle elevations, the forest is being cut wherever it is not too steep. In some areas forest remnants can be found along streams, but often even these remnants are eliminated. There is great danger to such pteridophytes as *Schaffneria nigripes,* which is very rare on limestone on the Atlantic slopes of Mexico and Guatemala, *Solanopteris brunei* in Costa Rica with its ant colonies, *Hyalotricha anetioides* in Costa Rica with its unknown affinities, *Onocleopsis hintonii* of Mexico and Guatemala growing in streams, *Lycopodium pithyoides,* rare in treetops of Middle America, and *Elaphoglossum,* the largest and most difficult genus of ferns in America with some widespread species complexes and much endemism.

The Guayana Highlands are unique and it is encouraging to know they are not in immediate danger, but this situation may not last as Man becomes more pressured by population increases to utilize all land surfaces. In this region there are two fern genera — *Hymenophyllopsis* and *Pterozonium* — and many species not known from elsewhere.

In the cloud forests of Central America and Mexico which are also being threatened, we have some of the most interesting species of ferns and a high degree of endemism. The highest peak in southern Mexico, Cerro Zempoaltepetl, which has over 250 species, with cloud forest pockets harboring undescribed species of *Lycopodium* and *Elaphoglossum,* is being lumbered and cultivated at higher elevations each year.

In Costa Rica, lumbering and charcoal manufacture on the Cerro de la Muerte has significantly reduced the cloud forest of great oaks. Species of *Polystichum, Plecosorus, Asplenium,* and *Grammitis* are often of limited range and are seriously threatened. In Oaxaca, Mexico, there is a famous locality, the Llano Verde, that was visited by Galeotti in 1839. Martens and Galeotti (1842) cited 28 species of ferns from the trail to the Llano Verde, including several new species. The forest is of great oaks, pines, *Cupressus,* and *Taxus,* intermixed with small meadows or llanitos. In 1969, with the help of Boone Hallberg of Oaxaca, I visited this locality and collected nearly 130 species on the same trail, rediscovering Galeotti's species and finding several additional ones, including new or rare taxa in *Asplenium, Grammitis, Phanerophlebia,* and *Phyllitis.* In 1972, logging operations were approaching the area.

The paramos of the Andes and Central America are the home of much of the genus *Jamesonia,* several *Lycopodium* species, and a diversity of other pteridophytes. Gradual encroachment on these lands is threatening these plants in several regions.

Another habitat that is at present being greatly reduced are the wetlands — swamps, ditches, ponds — either for drainage for crops or through lowering of the water table from increased water use. In Mexico this has virtually wiped out the water ferns, such as *Marsilea, Salvinia, Azolla, Ceratopteris,* and the fern ally, *Isoetes.* In North America, *Schizaea pusilla* is found in Nova Scotia, Newfoundland, the New Jersey pine barrens, and on the tip of Long Island, though greater housing developments on Long Island have lowered the water table there, reducing the bogs and threatening the curly-grass fern.

Ceratopteris is threatened or endangered from Mexico to the Guianas. Each population is somewhat different, so that loss over a great area, though the taxa may be common elsewhere in the world, destroys a segment of the genotype.

A boggy area in the Cerro de la Muerte of Costa Rica has recently been drained; the once conspicuous tree *Blechnum, B. buchtienii,* is now uncommon and a new species of *Isoetes* has apparently become extinct before it was described.

Another specialized habitat is acid, sandy soils. One group particularly suited for this habitat is the genus *Schizaea. Schizaea pennula* in Trinidad is reduced to a single sandy ridge, too small an area to remain acid for long; it is gradually becoming less acid. *Schizaea incurvata* is found on the white sand campinas of the Amazon basin and these areas are becoming fewer.

The aluminum soils of the Brazilian Planalto cerrados are the home for most of the genus *Anemia.* Sixty-five of the 95 species in the genus are found in Brazil. Some are very rare, and as the areas are gradually becoming more actively used for recreation, undisturbed habitats are becoming less common.

What is endangered? As Dr. Chambers pointed out, many lists are not of truly endangered species, but rather of rare, or morphologically unique or otherwise interesting plants. We should not restrict our concern to keeping the species from total extinction but instead we should be concerned with maintaining remnants throughout a taxon's range. Frequently, a wide-ranging species will display clinal variation in its morphology, or more obscurely in its physiology. Extremes of its range and especially disjuncts should deserve special consideration. We just do not know enough about the plants to determine whether these extremes or variations or disjuncts are the same or not. Broad geographical ranges and wide disjuncts can teach us a great deal about the basic biology of the plants — aspects of their dispersal, reproduction, ecology, competition, etc. Certain common species of the southwestern United States are known from single localities in the eastern United States. *Asplenium septentrionale, A. monanthes, Pellaea wrightiana, Cheilanthes castanea,* are examples. Though *Grammitis nimbata* is frequent in the West Indies, it is known only from a single locality in North America — behind a waterfall in North Carolina. *Asplenium exiguum* and *Ceterach dalhousiae* of the Himalayan Mountains are found in southwestern United States and Mexico. *Dryopteris filix-mas* and *Phyllitis scolopendrium* are common in Europe but rare in North America and must be saved. *Anemia hirta* and *Microgramma vacciniifolia* of the West Indies are also found in southeastern Brazil. What causes disjunctions? Are the disjunct elements really the same taxon? We cannot afford to waste them and find out too late that they were different. Apparent disjunct elements of *Microgramma tecta* of the Andes and southeastern Brazil have quite different spore sizes, strongly suggesting that one is a polyploid. *Trichomanes alatum* and *T. trigonum,* common in the West Indies, are found to bear gemmae on the leaf margins in Trinidad, Tobago and Grenada. Supposed disjuncts between Africa and Brazil in *Elaphoglossum* are in most cases different taxa.

Certain groups of pteridophytes are particularly threatened. We have already mentioned the tree ferns in regard to habitat destruction and harvesting. The quillworts *(Isoetes)* and water ferns are being lost through drainage of standing water. Another group especially sensitive to habitat disturbance is the club-mosses *(Lycopodium).* It is generally believed that their spores are slow to germinate. Many epiphytic species require well-established trees of a mature forest, but little is known of most of the species. Observations have suggested that at least a few species can develop on a diversity of tree surfaces in secondary forests; for example *L. linifolium* has been observed on cacao in both Trinidad and Costa Rica.

Human disturbance is not always a negative influence for pteridophytes. The omnipresent bracken *(Pteridium)* thrives in disturbed areas, becoming an agricultural nuisance in many places. Even some plants thought to be rare, such as *Ophioglossum* species, are frequently found in cemeteries in southern United States. Certain genera are actively expanding their ranges. Species of *Thelypteris* from the Old World, such as *T. dentata* and *T. torresiana*, and more recently *T. opulenta,* are becoming abundant in much of tropical America.

SUGGESTED MEANS OF CONSERVATION OF SPECIES

How can we best direct our efforts to preserving the diversity of ferns and fern allies? One important step in this direction would be to coordinate the experience of pteridologists and other botanists in preparing a list of the rare, endangered, threatened or unique species of pteridophytes. The list should be left broad due to the difficulties of pinpointing degrees of endangerment.

Ideally, the best methods of preservation of plant species is to preserve habitat samples of significant size and diversity throughout the Americas. Whole ecosystems must be preserved, not just tiny fragments. There is little hope of preserving all the individual taxa individually — we lack the space, expertise, time, and manpower to do so. Besides, we must learn how the pieces of the puzzle fit together. Since we still also lack the knowledge to reproduce the ecosystems, the preserved areas could provide us with permanent laboratories for study as well as demonstrations of mankind's natural heritage. The preserved areas of Durango in Mexico, with concentric rings of varying security and purposes, should be the model for other areas.

Parks and preserves of many types have been described for different areas — of varying sizes, controls, regions and sponsorships (government and private), both temperate and tropical. There is an extensive array of parks in North America, and fewer in Latin America, but the preserves in all regions are pitifully small in number and size relative to the need.

Others at this Symposium have emphasized the need for education as the most effective route to conservation. Students in Mexican secondary schools are receiving instruction in better means of agriculture as well as in the ways to develop local botanical gardens for useful plants and also in the appreciation for the native vegetation. Mr. Boone Hallberg has been involved in developing local projects and programs in over 250 communities, ranging from Baja California to the Yucatan.

But these Mexican preserves are tenuous and often violated, and an educated people is a long way off. Additional means of preserving what we can of the pteridophytes should be explored.

Botanical gardens hold a very small sampling of miscellaneous fern plants, a few of which are rare or endangered. In the tropics, the only botanical garden with a fern collection of any size is Las Cruces Tropical Botanical Garden at San Vito de Java, Costa Rica. Similarly, few temperate botanical gardens have appreciable fern collections, and these are largely those institutions where there is an active professional pteridologist, such as the New York Botanical Garden and the University of California at Berkeley, but there are also good numbers of ferns at Longwood Gardens, Montreal Botanic Garden, and the Fairchild Tropical Garden. Some fern collections are found in Europe, such as at Kew, Munich and Berlin. A concerted effort should be made to obtain, propagate and disseminate specimens of rare and endangered pteridophytes before they are lost.

In addition to dissemination of plants between botanical institutions, efforts should be made to introduce more species into general horticulture. Some species are clearly becoming more widespread in cultivation than in the wild, such as *Didymochlaena truncatula,* which is currently enjoying new popularity. Work at the New York Botanical Garden has shown that two rarities of Mexico are ideally suited for popular cultivation. *Schaffneria nigripes* is small, attractive in form, color and soriation, and propagates well from spores. *Asplenium exiguum* has, under high humidity of a cloud chamber, developed three methods of vegetative propagation — extended rachis, buds on the rachis, and, most interesting, plantlets at the tips of the pinnae, forming a necklace of plantlets around the plant. Both these species will do well in terraria or bottle gardens, which are quite popular now. We expect to have these in sufficient quantities by next year to distribute them more widely.

We must develop cultural techniques to preserve more pteridophyte groups than are now possible. Some genera are difficult or impossible to grow in cultivation. These include *Schizaea, Lindsaea, Jamesonia, Grammitis, Loxsomopsis,* the Gleicheniaceae, and the filmy ferns (Hymenophyllaceae). We must conduct research on the plants' life cycles and ecology. This will help explain the causes of rarity and provide assistance in developing cultural techniques.

There is also need for more exchange and experimentation with temperate hardy plants. Hardy ferns in temperate areas have not been grown widely. Those from Japan, northern Europe, and both northwestern and northeastern United States have been cultivated locally but not exchanged extensively. We should test their hardiness in other regions to see their range of tolerance. High elevation plants of tropical America have not been tested to any degree. *Dryopteris parallelogramma* from Mexico, for example, is found to be hardy in New York. Plants of New Zealand, the Himalayas, and Africa also should be tried.

Difficult tropical ferns should be grown on varying media and under other environmental conditions. This can be done using different kinds of personnel. At the New York Botanical Garden we have an excellent fern horticulturist in Dr. Bruce McAlpin, who is addressing himself to many of these problems. We also utilize a great deal of volunteer help at the Garden. Furthermore, members of the American Fern Society assist by sowing spores of many ferns at home and sharing the plants with the Garden. They also test tender ferns in their homes, greenhouses, and other modified environments for possible use in general horticulture. These trials sometimes result in unexpected scientific rewards. Mrs. Virginia Otto of Stamford, CT, recently grew *Asplenium (Phyllitis) delavayi* from China from spores. It had recently been placed in the otherwise monotypic Mexican genus *Schaffneria.* Having living material with which to compare it soon made it clear that they were not closely related after all.

Greater and more systematic use should be made of spore exchanges. Several fern societies have spore banks and many botanical gardens regularly exchange spores. Individual collectors and growers can be utilized more fully. In extreme cases, herbarium specimens also can provide a source of viable spores. Many fern spores have even proved viable after more than 50 years.

Finally, substitute materials for tree fern fiber should be found for use in epiphyte cultivation. Though the tree fern supply is diminishing, it is hoped that regulations will bring tree fern harvest under control. Sassafras branches have been used for small orchids with some success. Shredded wood or artificial substances might be tried.

In summary, there are many species of lower vascular plants in the Americas that are endangered or threatened, largely from habitat destruction. By launching a multifaceted attack on this problem now — through the preservation of natural systems samples, backed up with horticultural and basic botanical studies — the prospects for long-term survival are very good indeed.

DISCUSSION

Dr. B. MacBryde: Do you see any problem or any pressure on our native ostrich fern as a result of its use as a food? In certain areas of North America, such as the St. Lawrence River in New Brunswick, Ontario, upstate New York, etc., the fiddleheads are harvested quite extensively.

Dr. J. T. Mickel: I do not see any danger from the standpoint of species survival. Certainly for local use this can cause a problem if you harvest all the fiddleheads at one time, but harvesting of these usually is done either selectively or at a particular time, and the plant can still send up enough fronds for the year to build up a food supply for the coming year.

Dr. B. MacBryde: You mentioned having your list broad because of partitioning in the future. I urge you to categorize the list so that reluctant governments can take it in pieces.

Mr. J. Malter: In regard to the food question, several workers have shown that some ferns produce intestinal cancer in rats. It has been suggested that ferns be taken off the edible food list in the U.S.A. This has yet to be done.

Dr. J. T. Mickel: The only cancer tests that I know of have been done on bracken fern. The one species that is commonly eaten, of course, is the ostrich fern, and I don't think any studies have been done on that.

LITERATURE CITED

Convention on International Trade in Endangered Species of Wild Fauna and Flora, Appendix II. 1973. Washington, D.C.
Martens, M. and H. Galeotti. 1842. Mémoir sur les Fougères du Mexique. Brussels.

Section 5

Special Topics

THE PROBLEM OF RARE AND OF FREQUENT SPECIES: THE MONOGRAPHER'S VIEW

K. Kubitzki

Institut für Allgemeine Botanik, Universität Hamburg
Hamburg, Federal Republic of Germany

In our present-day plant world, it is an accepted fact that though not all species are equally threatened or endangered, the rare ones are more especially threatened. If we ask, however, what is a rare species, and why it is considered rare, it quickly becomes apparent that biologists have heretofore failed to pay attention to the theoretical and practical aspects involved in the phenomenon of the commonness and rarity of the thus far different evolutionary end-products that have resulted in present-day species.

The question, "What is a rare species?" can be answered meaningfully only in reference to a specific concept. For example, the total number of individuals of any one species living at any given time in any given area can be compared with the total number of individuals of another species living in the same area, or of the same species existing in still another area. This means that it is not the absolute frequency, but rather the relative frequency of a species which is a meaningful parameter. As to the frequency of neo-tropical plant species to which I want to restrict myself, for too little is known about their relative frequency. It is only too well known to the taxonomist that a large fraction, if not the majority of tropical tree species, is known only from quite a limited number of localities, though this does not entitle us to call these species really rare (since, often, e.g., enormous amounts of wood of such putatively rare species are going to be exported!).

The major factor contributing to the false impression of rarity of many tropical tree species is the overwhelming richness of the ever-wet tropical lowland forest. For a long time, there has been a difference of opinion as to whether the sclerophyllous shrub of the South African Cape Province, or the wet tropical lowland forest is the community richest in species on earth. Projects to ascertain species density of Amazonian rain forests have encountered between 60 and 87 tree species on one hectare (Black et al., 1950), and 179 species on three hectares (Pires et al., 1953), though it was suggested that the true number of tree species in the plots studied might be even larger. Recently, however, Klinge (1973) has shown that all these estimates must be far too low, since he found at least 502 tree and shrub species growing on an area of 2,000 m^2 in a forest of Central Amazonia. There is little doubt, therefore, that as far as green land plants are concerned, species diversity on earth is greatest in the wet tropical lowland forest, and that this diversity has the consequence of an extremely low population density for many species. Whereas Black et al (1950) estimated that the population density for many species of trees in Amazonia is less than one individual per hectare, it now seems that this number may be even lower for many species. However, since the tropical lowland forests often encompass huge areas, a low population density does not necessarily imply rarity in the sense of a low global abundancy. (It may be added, parenthetically, that one of the few examples of really rare taxa known to me from the Neotropics is the endemics of Juan Fernandez, which, due to the limited extension of appropriate habitats, comprise populations of extremely low numbers of individuals.)

Another factor contributing to the general impression of rarity is the inaccessibility of the often widely scattered habitats and frequent inadequacy of exploration. This is exemplified by the palm *Barcella odora* (Trait) Drude which was believed to be extremely rare, but which subsequently was shown, via helicopter exploration, to be widely distributed in certain habitats (J. M. Pires, pers. comm.). With regard to Amazonia, the statement of Ducke and Black (1953, p. 16) that we ignore almost completely the flora of the uplands between the navigable rivers is virtually still valid.

A third reason for putative rarity is endemism, but endemic species are sometimes quite frequent within their limited areas of distribution. For example, *Mahurea exstipulata* Benth., a shrub or small tree inhabiting a small area of distribution on and around the Guyana Highland (Kubitzki, in press, 2) has been reported to be locally the most frequent, if not the only tree species of special habitats surviving repeated fires. These are just a few of the many examples which could be cited to emphasize the fact that the concept of rarity is a rather vacillating one: inadequacy of knowledge and/or exploration, the high amount of species richness, and endemism are some of the varied factors which contribute to the unwarranted impression of a taxon being a rare plant.

If we look for some reliable criterion of frequency, it is clear that the total number of living individuals of any taxon (its "global abundancy") can never be established with accuracy. Similarly, the data brought together by ecologists are mostly too heterogeneous, — if not too incomplete — to serve as a firm basis for considerations of relative frequency.

There is, however, a rich source of relevant information readily available in taxonomic studies. The numbers of specimens deposited in museums and frequently cited by monographers, if used with caution, form an immensely valuable reflection of the global abundancy of related species. This would not be true, of course, with regard to the flora of floristically well-known areas like parts of Europe or North America, where the rarer species undoubtedly will turn out to be extremely over-represented in collections. This may also apply to a lesser degree to tropical regions, because the competent collector clearly avoids the common species. However, in regions with an extremely diversified flora, it is often impossible to draw such a distinction between rare and trivial species in the field. In other words, the true differences between the rare and the frequent species still will be much more prominent than those indicated by the data extracted from monographic studies.

Table I. *Collection data for 5 recently monographed woody Neotropical genera.*

Taxon and Source	Number of Known Species	Total Number of Known Collections	Average Number of Specimens per Species	Date of Collecting of Specimens		
				Up to 1900	1901 - 1950	1951 - present
Licania (Chrysobalanac.) (Prance, 1972)	152[1]	2,961	19.5	14.8%	50.6%	34.5%
Caraipa (Bonnetiac.) (Kubitzki (1))	21	502	23 9	9.6%	46.4%	44.0%
Doliocarpus (Dilleniac.) (Kubitzki, 1971, 1973)	31	517	16.6	24.7%	46.2%	29.2%
Davilla (Dilleniac.) (Kubitzki, 1971, 1973)	18	548	30.4	26.6%	48.4%	25.0%
Eperua (Legum./Caes.) (Cowan, 1975)	14	551	39.3	12.3%	33.6%	54.1%

[1] plus 1 in Asia. For 6 very frequent species, Prance cites only "representative specimens." These numbers have been doubled.

In Table I, I have collated data which has been extracted from some recent revisions of neotropical ligneous groups, including one larger genus, in addition to four small ones. The ratio between the total number of known collections and the number of species, reveals that each species is covered by an average of 16 to 39 collections (a figure which, as we will see, is rather misleading). I also have drawn a distinction between the dates of collecting, which reveals, as can be expected, that only the minor fraction of specimens, viz. between 10 and 27 percent, were collected up to the year 1900. Nearly the same percentage of specimens became available since 1951 as in the first half of our century. This very clearly points up the fact that the material basis virtually enforces botanical progress!

Let us now look somewhat closer into some figures. In the genus *Licania,* which comprises 152 American species plus one Asiatic, 25 species are covered by only one collection, whereas 19 species are covered by a mere 2 collections each. 104 out of 152 species are known from one to ten collections, and these are documented by a total of 373 from 2,961 specimens. This means that in *Licania,* 12.6 percent of the known material refers to 68.4 percent of all described species. On the other hand, there are only a few frequent species, e.g., 12 species with more than 50 known collections each, but these are covered by 1,886 collections, which means that 63.7 percent of all known specimens refer to only 7.9 percent of the species.

The frequency distribution of the other genera depicted in Fig. 1 also underlines the fact that there are, in practically any genus of green land plants, a few frequent (and at the same time most widely distributed) species plus many rare ones which often are of limited distribution. If compared with the other genera in Fig. 1, *Eperua* is somewhat anomalous for its lack of the "tail" of rare species (i.e., there are not a lot of rare species as in the other genera of Fig. 1). However, the monographer, Dr. Richard Cowan, fully expects a number of rare species still to be discovered. (Another reason for the unusual situation, if it is *really* unusual, may be seen in the extreme restriction of the distribution area of this genus.)

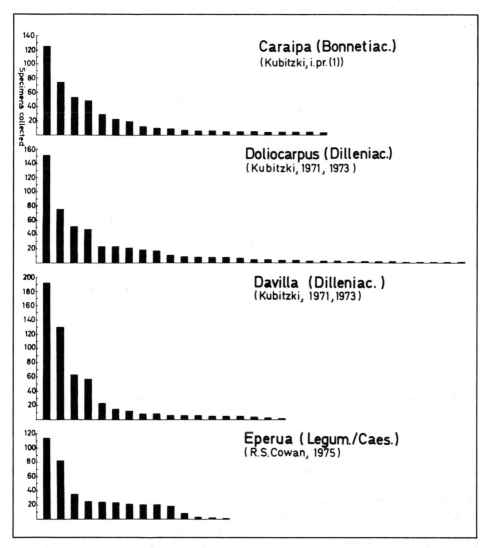

Fig. 1. Relative frequency of individual species of four genera, as indicated by the numbers of cited specimens per species. Each bar refers to one species. Subspecific taxa not considered.

The comparison of the relative frequency of different species of one higher taxon seems to be especially meaningful from the evolutionary point of view, since this allows for direct measurements of the evolutionary success of different taxa by their individual representation. As Stebbins has pointed out (see, e.g. Stebbins, 1974), one of the main trends in flowering plant evolution is improvement in the ability to disperse. This would mean that the most successful, (i.e., the most frequent and/or the most widely distributed) species should be expected to be derived phyletically. As I have shown elsewhere (Kubitzki, 1975), this is really the case in some tropical ligneous groups which have enjoyed a rather undisturbed evolutionary history. A good example is that of the genus *Davilla,* for which I have tried to assess the evolutionary status of each species by allocating marks for primitive character states to individual species. The score arrived at by considering the character states has been used for delineating the areas of the individual species, thus leading to an interesting result (shown in Fig. 2): The species which are

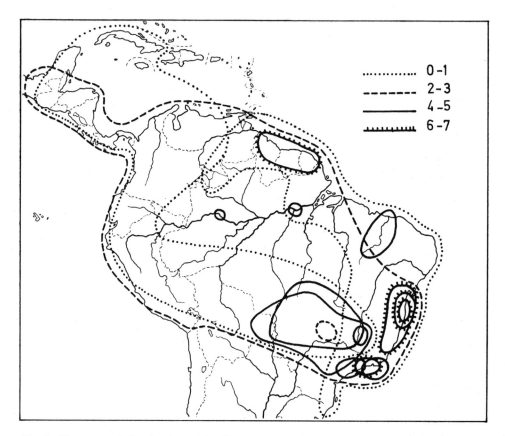

Fig. 2. *The areas of distribution of* Davilla, *with an indication of the number of primitive characters for individual species (after Kubitzki, 1975).*

carriers of primitive characters are not only limited in distribution but are also rather rare at the same time. Only three species have acquired a large area of distribution in tropical America, and it is not surprising that these are the genus species which are most variable and most difficult to delimit. While one of them *(D. nitida)* is a savanna species, the others are frequent in secondary growth; one, *D. kunthii,* appears after cutting of the primary forest, and the other, *D. rugosa,* appears after burning. The remaining species are confined to undisturbed types of vegetation. It is striking to note that several of the more primitive species are crowded together at the eastern fringe of the Brazilian Shield. This is a region which, together with the Guyana Shield, belongs to the most ancient continental areas of South America which, at least since Mesozoic times, have never been inundated but have been exposed to the buffering effect of the oceanic climate. That this region has also served as a refugium for other primitive groups has now also been demonstrated for some bambusoid grasses (Soderstrom and Calderon, 1974).

So we see that it is now possible to distinguish between rare and frequent species, and that the rare ones, in certain cases, are characterized by the possession of phyletically primitive features which are absent from the frequent species. Moreover, there may be certain spatial patterns with regard to the relationship between rare and frequent species. It is a general experience of biogeography that in today's world, the primitive is rare and the advanced is frequent, as Corner has pointed out on several occasions. Presumably, in many cases the rare species have previously had a wider distribution, so that they now represent what has been called a paleoendemic. In the absence of a well-documented geological history, however, this can never be ascertained for extant groups. It must also be mentioned that another evolutionary pattern may prevail, e.g., in herbaceous groups of temperate regions where often widely distributed older species have given rise to a number of endemics of more recent origin (Valentine, 1972).

If we now ask for the reasons that living species are either rare or frequent it would certainly be rather misleading to assume that the rare species are rare only because they are primitive with regard to some character states. Characters classified as primitive or advanced may have an adaptive value or not, but the distinction between them is not made in order to assess the plant's evolutionary success but rather in order to follow up the phylogenetic pathway of a plant group (Kubitzki, 1975, p. 226). The relationship between organism and environment, as expressed by the adaptive value of different features, is quite a different thing, and in principle at least, this can be verified by bio-systematic studies. It should be mentioned here, however, that in most cases it is impossible to define the adaptive significance of an isolated trait since characters are adaptive mostly as parts of an adaptive complex (Stebbins, 1974).

Perhaps the most important peculiarity which governs the different frequency of living species is the genetic constitution of their populations, i.e., the richness of ecotypes, a fact pointed out many years ago by Stebbins (1942). It is commonly accepted that it will be extremely difficult, if not impossible, to find out something about ecotypic variation in tropical tree species, because this would necessitate the collateral cultivation of different populations under systematically varied conditions. Therefore, this field will remain largely a matter of conjecture.

What does all this mean within the framework of our symposium, as set out by its organizer? Elton (1975) has presented evidence that species diversity of invertebrates of the wet tropical lowland forest is very high, and therefore numbers per species are proportionately very low, as is the case in many tree species. Such low population density inevitably makes these animal and plant populations vulnerable to local extinction, and the maintenance of these highly diversified systems depends upon the continued existence of extensive habitat areas from which recolonization can occur. Therefore, Elton (1975) claims that the conserved areas of wet lowland forest in the tropics will need to be large, in order to ensure long-term survival for many species. But an even more serious threat for tropical forest communities, I feel, is a factor which no longer discriminates between the rare and the frequent. This is tropical silviculture, with its liking for monocultures of an extremely reduced number of species of *Pinus, Eucalyptus,* and a few other genera, which inevitably will lead to the displacement and subsequent disappearance of whole gene pools. All attempts to introduce local tree populations in tropical forestry should therefore receive the widest possible support.

LITERATURE CITED

Black, G. A., Th. Dobzhansky and C. Pavan. 1950. Some attempts to estimate species diversity and population density of trees in Amazonian forests. Bot. Gazette **111**: 413-425.

Cowan, R. S. 1975. A Monograph of the Genus *Eperua* (Legum.: Caes.). Smithsonian Contr. Bot. No. **28**. 45 pp.

Ducke, A. and G. A. Black. 1953. Phytogeographical notes on the Brazilian Amazon. An. Acad. Brasil. Ciênc. **25**: 1-46.

Elton, C. S. 1975. Conservation and the low population density of Invertebrates inside Neotropical rain forest. Biol. Conserv. **7**: 3-15.

Klinge, H. 1973. Struktur and Artenreichtum des zentral-amazonischen Regenwaldes. Amazoniana (Kiel) **4**: 283-292.

Kubitzki, K. 1971. *Doliocarpus, Davilla* und verwandte Gattungen. (Dilleniaceae). Mitt. Bot. Staatssaml. München **9**: 1-105.

_____. 1973. Neue und bemerkenswerte neotropische Dilleniaceen. Mitt. Bot. Staatssaml. München **9**: 707-720.

_____. 1975. Relationships between distribution and evolution in some heterobathmic tropical groups. Bot. Jahrb. **96**: 212-230.

_____. (in press, 1). A monograph of the genus *Caraipa* (Bonnetiaceae). Mem. New York Bot. Garden.

_____. (in press, 2). The genus *Mahurea* (Bonnetiaceae). Ibid.

Pires, J. M., Th. Dobzhansky and G. A. Black. 1953. An estimate of the number of species of trees in an Amazonian forest community. Bot. Gazette **114**: 467-477.

Prance, G. T. 1972. Chrysobalanaceae. Flora Neotropica. Monogr. No. **9**. 410 pp. Hafner, New York.

Soderstrom, T. R. and C. E. Calderon. 1974. Primitive forest grasses and evolution of the Bambusoideae. Biotropica **6**: 141-153.

Stebbins, G. L. 1942. The genetic approach to problems of rare and endemic species. Madroño **6**: 240-258.

_____. 1974. Flowering Plants. Evolution above the species level. 400 pp. Edward Arnold, London.

Valentine, D. H. 1972. Patterns of geographical distribution in species of the European flora. Symp. Hung. **12**: 15-21.

COMPUTERS AS AN AID
TO SOLVING ENDANGERED SPECIES PROBLEMS

Theodore J. Crovello

Department of Biology, University of Notre Dame,
Notre Dame, Indiana 46556, U.S.A.

Through much of this Bicentennial Symposium we have heard and seen examples of the rapidly degenerating conditions of endangered species. Now let us turn our attention to action. What steps can we take? How can computers help? I firmly believe that the *wise* use of computers can help us to solve many of the problems associated with endangered species throughout the Western Hemisphere. "Wise" use implies that computers are not a "cure-all" for endangered species problems. I also believe that computers cannot be used effectively without considerable forethought by botanists like us.

MAJOR STAGES IN SOLVING ENDANGERED SPECIES PROBLEMS

To appreciate fully the role that computers can play in helping to solve endangered species problems, we should consider the procedure of studying, recognizing, and legally recommending endangered species to be a multistage, decision-making process. This concept was introduced by Crovello (1970) for use in any type of study in systematics. We can recognize three major stages in the process of solving problems of endangered species:

1. Determine exactly what species are endangered.
2. Communicate the status and importance of endangered species and relevant legislation to biologists, politicans, and all other concerned citizens.
3. Continue monitoring and communication activities that will assure the continued preservation of endangered species.

The fact that this is a multistage process implies that it is a complex task, rather than an easy, quick procedure. And because it is a multistage decision process suggests that the choices we make along the way may affect the outcome. For example, some of our earlier decisions may affect what is, and what is not, recommended to governments as an endangered species.

Upon reflection, many of the substages of previously listed *Stage One* are seen to be similar to the stages of a regular taxonomic monograph. This is a reasonable assumption, because determination of threatened or endangered status of any species or group of species requires that their taxonomic relationships be reasonably known and based upon sound systematic study. Whether it be a purely academic monograph or one whose particular purpose is to determine endangered species status, botanical researchers still must inventory herbaria, determine the correct taxonomy, summarize geographic and chronological distributions as completely and as accurately as possible, and provide the means to allow other workers to identify the species to which unknown specimens belong. The positive role of computers in the taxonomic process has been discussed in detail by Blackith and Reyment (1971), by Crovello (1970), by Pankhurst (1975), and by Sneath and Sokal (1973), among others.

Particular ways in which computers can contribute to satisfying the goals of *Stage One* and its substages, include the following: Computers can help us to determine what species are endangered by displaying, as either a printout or on a cathode ray tube display screen, the changing geographic distributions over time of putatively endangered species. The raw data can be based on either herbarium specimens that have been correctly identified, or on direct field observations that also have been incorporated into a computerized Endangered Species Data Bank. Computers also can recall very quickly for us detailed information on phenology, and provide specific information on collection sites. This, combined with other information, can serve to increase field survey efficiency. Finally, by using a digitizer or similar piece of equipment that is connected to a computer, much data can be quickly input into the expanding Endangered Species Data Bank, e.g., outlines of large or small political boundaries, or actual morphological measurements.

Stage Two: Communication of the status and importance of endangered species and of relevant legislation to biologists, politicians, and all other concerned citizens. This category also can benefit greatly from the utilization of computers. It is not enough for botanists to just determine candidates for endangered species status. Their suggestions and the supporting evidence they offer must be communicated to many people in all walks of life. And they must act promptly — before it is too late — to bring about the legal action required to prevent the species from becoming more severely endangered, or even extinct. This need for rapid communication is made still more difficult because those who must receive this information are in diverse parts of government or business, often are geographically scattered, and frequently they are not even botanists. Computers can help us here by permitting fast production of letters, addressed envelopes and other clerical tasks. The Western Union Telegraph Company already has a service which transmits fixed information to people on a mailing list of any size. The list is provided by the sender on a magnetic computer tape. This service easily can be applied to information concerning endangered species. For example, if a species were thought to be threatened by several development projects, a staff member at an Office of Endangered Species need only submit one copy of a form letter (describing the problem and what data need to be collected, e.g., on local abundance) to Western Union. Western Union's computer then electronically merges the letter with each name on the mailing list already in its data bank. The letter is then sent *electronically* to the Western Union branch nearest each person on the mailing list, printed out there, and delivered by local mail. All this can take place in a matter of hours, regardless of the weather, with almost no clerical help needed, and with a minimum of clerical errors.

There also is a need to communicate reliable information on the status of relevant legislation. A computerized law data bank exists in which court decisions involving different legal issues have been cataloged, and from which information can be retrieved rapidly. We need such a legal data bank for endangered species, one which easily can retrieve not only court decisions, but also laws for each country, state, and perhaps even county and township. This last category might appear extremely specific, but as Dr. Countryman reminded us, certain townships have laws wherein a citizen may not pick or harm plants, but a township can spray herbicides at will! We also need legal models for other counties and states to consider for adoption.

Stage Three: Continued monitoring and communication. This activity, perhaps, can utilize computers even more effectively than can the preceding two stages, because once a species is formally recognized as threatened — or if it just missed being recognized — it is essential that its condition be monitored to determine if the situation is either improving or getting worse. An Endangered Species Data Bank that could do this is best compared to the ongoing computer monitoring of a hospital patient who is seriously ill. Planned and established floristic projects that use computers, e.g., Flora Veracruz, Projeto Flora Brazil, Flora North America, etc., all could perform this task efficiently and accurately. In a related activity, computers also can help assure the preservation of endangered species still further, by monitoring changes in land development. They could indicate the effect that such development has on the remaining populations of threatened plant species. They also can provide an efficient way to continuously update checklists of each species with respect to their occurrence and abundance in different geographic localities. In addition, computers can be used to make more accurate recommendations about natural ecological areas that need preservation and protection.

Finally, it should be emphasized that throughout this multistage decision process, we must not consider endangered species problems at the global level only. We must consider the problem at any political level which could produce effective support. For example, though species in Indiana are globally threatened, they also are locally threatened within the state. We must therefore carry out activities at both levels. A state representative, the newspapers, teachers, and citizens in general are much more concerned about the endangered flora of Indiana than they are about the total endangered flora of the United

States or of the hemisphere. This is human behavior, and we must respect it and take it into consideration.

COMPUTERS AND THEIR USES

Before we look more closely at how computers can solve endangered species problems, we must understand exactly what computers are and how they are used.

Computers can be considered *programmable* hand or desk calculators. But though they carry out the comparatively simple tasks of a desk calculator, they also can perform some of the intermediate calculations which do not have to be entered into the machine by the user. These intermediate calculations are stored in the memory of the computer. It is with good reason that computers also are called **"number crunchers."** This is because they can carry out many operations on numbers very rapidly and with good accuracy. Computers are also *information processors,* because they can handle alphabetical as well as numerical information with ease. This is one of the major functions that computers perform in helping to solve endangered species problems. They quickly can store, summarize, retrieve, and check much information in computerized data banks related to endangered plant species and their distribution.

Computers are also *simulators* of biological systems. We often read how computers are being used to model the world (for example, the project of the Club of Rome), or to model various ecological systems. Computer simulation can be useful to endangered plant species studies in two major ways. The first is to model the population ecology of endangered species as one component of an ecosystem. This is a rather restricted ecological model, but it can provide information on the effects of environmental change. In the second way, computers can be used for educational purposes to model all three stages of the endangered species process mentioned previously. For this, a simulation model would be constructed that takes into account not only the ecology of the species, but also the social, political, and economic aspects of endangered species problems. This includes decisions made by government officials and other citizens in order to recognize and protect endangered plant species. This is a most promising use of the computer. Today at the University of Notre Dame, and at many other campuses, even undergraduate students are introduced to world problems by means of this simulation approach with computers. In one of the courses I teach, "Plants and Human Affairs," ninety students from business, engineering, political science, etc., use the agricultural part of the World Computer Model at the University of Notre Dame to comprehend more completely the interrelationships of the different parts of the world agricultural system with the rest of society. In this way, students can immediately see the effects of any policy changes on society that they might consider.

In much the same way, the endangered species multistage decision process could be simulated on a computer. Biologists, politicians, planners, and others could then study the many effects of either recognizing a certain species as endangered (protecting a gene pool, depressing the economy of an area, etc.), or of not recognizing it. An important point to keep in mind always is that it is *not* necessary for one to be a computer expert to operate the machine. Rather, one should be an expert in his own specialized field, with a desire to use whatever techniques are available to enable him to make decisions on the firmest possible foundation. We also should remember that this simulation "game" can be played by everyone, not just by environmentalists. When planners, politicians, ecologists, etc. all use the computer together in the same room, they will develop greater respect and appreciation for each other's problems and viewpoints. Such mutual under- standing inevitably will result in better solutions to endangered species problems.

Finally, computers can be considered an *extension* **of the human mind.** This is logical, when the amount of data of value to studies and decisions about endangered species increases to the point where, as individuals or as a group of workers, we lose efficiency and accuracy in the recall of such information and in what meaning this information has. In the final analysis, it is up to us to determine whether computers will ultimately be a replacement, or an extension, of the human mind.

THE VALUE OF USING COMPUTERS IN ENDANGERED SPECIES PROBLEMS

There are five major reasons why we should use computers in endangered species problems:

1. Computers offer us the means to improve on what we already are doing. For example, they can easily provide state and county checklists of any desired area. They can both provide species distribution maps, and update them when necessary. Moreover, such results can be communicated rapidly via computer networks across a country or hemisphere, regardless of where the planner, politician, biologist, or concerned citizen is who desires such information.

2. Computers enable us to easily do what may otherwise be difficult and time-consuming. They can provide details of phenology and of geographic variation analysis. They can correlate geographic distributions with climatic and other environmental factors in order to determine which of them, or which combinations, may be limiting species distributions. This is helpful in determining the actual and potential distributions of endangered species. By allowing the computer to do the work, these data can be rearranged to produce better collection locality information. This geographic thesaurus of collection sites then can even be improved by subsequent collecting at the same sites. Computers also can help to reduce the problems of changing political boundaries due to treaties, or simple name-changing of counties in the United States.

3. Computers can increase interest in endangered species problems. In the state of Indiana, for example, every public library has available a computer terminal, usually a teletype. It is connected with the Indiana University School of Business via regular telephone lines. Any citizen, free of charge, can ascertain the latest information on economics, unemployment, etc., for any county in the State. In a similar manner, one could also ascertain the latest information about endangered species in a county, or even obtain a species checklist for a county. From such an information network it is easy to produce a series of adult education tutorials like, "Did you know . . . ." The content could be directed towards endangered species status, legislation, and critical information needs that concerned citizens can help to satisfy, etc.

4. Computers can decrease the many tedious tasks so often associated with endangered species activities. Taxonomists, particularly, are well aware of the large amount of time that must be spent in such routine clerical chores. Computers can help us to reduce this in many obvious ways.

5. Computers can permit laboriously collected taxonomic data to be more valuable over a longer period of time. Taxonomic data are extremely useful, but unfortunately they often are extremely difficult to retrieve. This is true whether the information is stored with specimens in museums or in the published literature, etc. By increasing the length of time over which these data can serve the purposes of humanity, a computer also increases the value of the information it stores.

TYPES OF SYSTEMATIC DATA OF USE IN ENDANGERED SPECIES STUDIES

Data that may be particularly useful in an endangered species data bank may come from several sources. These include the *literature,* including published floras, monographs, or environmental impact statements. *Experimental data* are of value to determine genetic diversity over the natural range of a species, as well as to determine how quickly we might expect transplanted individuals of one species to survive and to flourish in other areas of a species' range. *Museum collections* provide information from labels of specimens, as well as from actual measurements on the specimens themselves. *Living collections* provide similar information. *Nonbiological data* also are extremely important. These include geographic coordinates, weather data, etc. Additional *ecological data,* such as plant community type in which individuals are growing, pollution levels, etc., also have a place in the endangered species data bank. Finally, this systematic data should be complemented by relevant economic, sociological, legal, and political information.

We also must consider the level of data organization. For example, will the data bank contain information to be organized according to specimen (from museums and from publications, etc.), or according to taxon (obtained from floras or from populations), or at the community or ecological associates level ("critical habitat")? Care must be taken that the chosen level of organization will be able to provide information with detail and resolution sufficient to help solve the endangered species problem.

COMPUTERIZED DATA ACCUMULATION

Computers can automatically accumulate data that are of value in endangered species studies, thus making the process more efficient, while at the same time also reducing errors involved with the capture of data. In endangered species work, automated data accumulation can most readily take the form of capturing information from published distribution maps of a species (e.g., county distribution maps in the United States). Any map outlines that are needed for presentation of the results (e.g., county outlines in a state, or state outlines in a country), can be captured very quickly with a *digitizer*. A digitizer is a semiautomatic, electro-mechanical device that can provide the X, Y geometric coordinates of any point on a map, a leaf, a graph of interest, etc., directly to a computer. In one model, an office worker or undergraduate student merely moves the stylus on the end of the arm of the digitizer to the point on the map of interest, and presses a foot pedal. This causes the relative location of the point to be stored in the computer, without worry of further clerical errors. *Optical scanners* also are available. These are completely automatic devices that can scan plant specimens, distribution maps, etc., and automatically record the desired characteristics. Finally, much data are being accumulated on chemical constituents of plants. These data now are automatically provided in "computer compatible" form by many of the chemical pieces of equipment themselves. Such data are useful both to determine species and to delimit geographical races. The use of computerized data accumulation in systematics and in endangered species work is only just beginning, but we can expect it to play an increasingly important role in helping to overcome the large task of data accumulation.

TYPES OF RESULTS TO BE EXPECTED

Results can be of four types: 1) the actual data; 2) summary statistics; 3) summary graphics; and 4) decision aids.

Actual data may be either numerical or alphanumeric. Numerical data may simply be a display of the change in values of characteristics over geography. Alphanumeric data may be the result of retrieval of label information from museum specimens which possess certain characteristics, e.g., all specimens of *Cardamine longii* from New York (see Crovello, 1972).

Summary statistics may be of many types. One type in regular use at the University of Notre Dame is called CRUMATH. It is an early warning system of information about species that may be threatened or endangered. Of the species that I am presently monographing, ten to twenty are in this category. Having borrowed 45,000 specimens of several genera of the Cruciferae from more than 75 museums, I identified each one to species as best as I could. Label data and the "State Of The Specimen" were then captured for later computer processing. After carrying out computerized error-checking routines to reduce the number of errors in the data bank, programs that provide summary statistics were run. In particular, the program called CRUMATH provides, for one species at a time, detailed information on the distribution of that species in both space and time (see Table I). Such a table, one for each species, lists in one direction all the counties of every state in which specimens have been collected, and along the other dimension it provides information on the time periods during which the collections were made. By referring to such tables, one for each species, endangered species workers now have available an early warning system to indicate which species might be in danger of becoming extinct, *and* in what areas. For example, it indicates very clearly and efficiently those counties in which collections have not been made for many years. This does not automatically imply that the species is no longer growing there, but it does provide the worker with information on where, throughout a species' range, he might check to see whether or not the species still exists there. This type of table, combined with a detailed printout of the collection records of each specimen for that species, provides endangered species workers with an efficient means of locating previous critical collection locality sites.

Table I. Part of the geographic distribution of Cardamine bulbosa, *by county and by time period. Information based on collections from over 75 herbaria. Numbers in the table indicate total number of specimens from all herbaria combined.*

TIME PERIOD

State, Counties	1700 to 1800	1801 to 1900	1901 to 1920	1921 to 1930	1931 to 1940	1941 to 1950	1951 to 1960	1961 to 1970	1971 to 1975	No Date	Totals
Delaware											
Unknown	0	6	0	1	1	0	0	0	0	3	11
New Castle	0	1	0	1	1	1	0	0	0	0	4
Illinois											
Unknown	0	51	28	8	3	1	4	0	0	6	101
Adams	0	1	0	0	0	1	1	1	0	0	4
Alexander	0	0	0	0	0	2	2	0	1	0	5
Bond	0	0	0	0	0	0	0	1	0	0	1
Boone	0	0	0	0	0	2	1	0	0	0	3
Cass	0	0	0	0	0	0	4	0	0	0	4
Champaign	0	1	4	0	0	4	1	0	0	1	11
Christian	0	0	0	0	0	1	0	0	0	0	1
•	•	•	•	•	•	•	•	•	•	•	•
•	•	•	•	•	•	•	•	•	•	•	•
•	•	•	•	•	•	•	•	•	•	•	•
Woodford	0	4	0	0	2	7	0	1	0	0	14

Summary graphics are most useful for computerized mapping, either via the production of maps at high speeds as regular computer printouts, or through the use of graphics terminals. These can provide high resolution maps on remote terminals in one's own office. Crovello (1970) presents a detailed classification of Summarization Graphics, including the different types of pictorial results that use geographic information. The graphic types of results of numerical taxonomy studies are also useful to orient people who are just beginning to study a group of species. Sneath and Sokal (1973) give many examples.

Decision aids are helpful in several ways. They can be used to identify unknown species, as described in Pankhurst (1975). In addition, they can help us to decide where and when to collect, depending on the results of analyzing the computer printouts. At Notre Dame we also use another program, which is somewhat similar to CRUMATH. It summarizes the phenology, or flowering time of specimens throughout a species' geographic range.

Computers also can indicate what herbaria must be checked for county or other distribution records. This is most important for determining endangered status. While my work is still in a preliminary stage, the analysis of more than 35,000 computerized specimens from about 75 museums indicates that it is not adequate to merely examine specimens only at the largest herbaria, for example, of the top ten in the United States, if one is working in that country. Though these museums may contain a large number of specimens, many are not from the United States, and of those that are, some specimens are very old, and others have incomplete geographic information. Thus, specimens in these collections are most likely to bear the description of only, "Indian Territory," for example, instead of details about the state and county in which the specimen was found.

These older specimens are important in helping to determine the past distribution of species, but they are of less importance in determining the present distribution.

Finally, computers can help us to determine specifically what either should or should not be included on endangered species lists. For example, I have cooperated with Dr. Tom Cooperrider in creating a list of endangered species for the state of Ohio. In the taxa with which I work, I have been able to provide him with extensive summary information. This resulted both in the addition of species to the Ohio endangered list, as well as the removal of others which previously were thought to be rare and endangered in Ohio.

CURRENT COMPUTER PROJECTS OF VALUE TO ENDANGERED SPECIES STUDIES

To describe all of the current computer projects of value to an endangered species study would require a separate paper. Though many are under way at present, many more ought to be. Also, unfortunately, through no fault of their own, many researchers who work with endangered species frequently are unaware of the existing computer projects that are available. Finally, many of us are not sufficiently familiar with what a computer actually does to adequately use the wealth of computerized information that is already available. Obviously, we must educate ourselves before we can educate others!

Perhaps it is best to classify current computer projects according to which of the three major stages of the Endangered Species Decision Process it is most relevant. Thus, systematic data are most useful for the first stage, while human population census data, etc. (available on computer tape from the Federal Government) would be most useful for Stages 2 and 3. Because of space limitations, I shall only consider a few of the many projects in plant systematics. The use of computers in plant systematics is just beginning to escalate. Some projects have been under way for a number of years and summaries of them already have been published. Crovello and MacDonald (1970) published an index to computer projects in systematics. Currently, I am Chairman of The International Register of Computer Projects In Systematics. The Register is sponsored jointly by the International Association For Plant Taxonomy and the Society For Systematic Zoology. This computerized Register of Computer Projects contains information on computerized data banks of value to systematists, plus programs of use. Requests for questionnaires needed to enter a project into the Register, or for Search Request Forms to use the Register, should be directed to the author. We hope that by informing people of what currently is being done, the Register will reduce both duplication of effort and compatibility problems, and will permit workers to make systematic, as well as other decisions which are based on the greatest amount of available information.

Let us now consider several current computer projects in plant systematics, bearing in mind that these are cited only as examples. Crovello (1972) described the computerization of 65,000 specimens in the Edward Lee Greene Herbarium at Notre Dame. This is of special value, because it is a collection predominately of the last century, and contains many type specimens. His current work, using computers to study data from many herbaria, was referred to earlier in this paper. Adams et al. (1975) are overseeing a project designed to capture plant distribution information for the state of Colorado, and recently for several other states of the western United States as well. This project will provide information on plant distribution overlaid with information on the soil, elevation, etc. of each locality. Gómez-Pompa and Nevling's (1973) Flora Veracruz project promises to be the most detailed project in systematics. It includes not only information on academic systematics, but also data on the use of plants by local residents for horticulture, medicine, etc. The recently-inaugurated Projeto Flora of Brazil also is to rely heavily on computer processing. Forero and Pereira (1976) are beginning to computerize information about plant specimens in Colombia. At Kew, England, a novel approach is being undertaken. Researchers there are integrating the microfilming of specimens with retrieval by computer of information about the specimens. The American Horticultural Society's Plant Records Center, under the direction of Richard Brown, gathers and disseminates data on plants under cultivation in botanical gardens throughout the world. Finally, at the Hunt Institute For Botanical Information in Pittsburgh, Pennsylvania, under the auspices of the International Association for Plant Taxonomy, Index Nominum Genericorum is now being computerized. This will serve as the nomenclatural backbone

343

on which information about endangered and other plant species can be added.

NECESSARY RESOURCES

Effective and wise use of computers to help solve endangered species problems requires the proper combination of both people and computers. Though each situation may be somewhat unique, it will have much in common with other endangered species projects.

Human resources are always the most important ingredient in any project. No one person or one type of person can be expected to fully understand all aspects of the use of computers in helping to solve endangered species problems. The important point here is to remember that a chain of communication among people must exist, a chain which permits people in different parts of the project to communicate their needs and potential to each other. For example, the field systematist need not have to be able to talk directly with the project's computer programmers. But intermediate people who can converse meaningfully with both types, must be present. Similarly, in most situations, it would be a waste of human and other resources to expect a field botanist to learn how to program a computer. But he or she should be knowledgeable enough to know what to realistically expect a computer system to do.

The type of computer most useful for endangered species work would probably not be one purchased specifically for this project alone. Rather, it would be more economical and efficient to rent computer time, terminals, and even the programs from an established university, government, or commercial computer center. The only exception to this arrangement might be in large endangered species projects, where small "feeder" computers may prove useful in remote locations or as part of a mobile unit, in order to enhance the information accumulation process. Data captured at these field or museum sources could regularly be sent to the main computer center for further processing.

Finances required to computerize data for endangered plant species studies are not trivial, but neither should they be considered astronomical. This is true whether we consider only systematic data, or also economic, political, legal, and sociological data. We must remember that for some complex questions, the only accurate means of obtaining the answers are through computer data banks. Also, as mentioned previously, computers promise to enhance the usefulness and length of life of systematic data. The decision to computerize need not demand that large amounts of resources be assigned to it immediately. But it does require a firm, *long-term* commitment by the organization.

FACTORS TO CONSIDER WHEN CONTEMPLATING THE USE OF COMPUTERS

It is not a simple task to determine whether an organization concerned with endangered species should computerize at least some of its activities. Again, one must consider the entire range of activities involved in the three major stages in solving endangered species problems. Table II contains a series of questions which endangered species groups contemplating computerization should answer completely prior to making a commitment. Even if the decision is *not* to computerize, much can be learned about your project by forcing yourself to answer the questions in Table II.

PROBLEMS WITH COMPUTERIZATION

Many of the problems encountered in the use of computers to help resolve endangered species problems are not unique to electronic data processing. We must consider the magnitude of the task, particularly if it is to be done at the individual species level. Also, computer capture of specimen information is made difficult and inefficient by the past and current practice of distribution of individuals of the same field collection among many herbaria. Inaccuracies, both in terms of misidentifications and mistyping, are extremely problematic, just as they are in taxonomic and ecological studies that do not use the computer. There exists the problem of compatibility of recording label and other data, both among neighboring countries and among herbaria within the same country. Even when agreement is reached on what to capture, problems remain. For example, in the United States, most herbarium specimens include the county of the State in which they were collected. But unlike certain uniform grid systems, counties vary greatly in area. The largest county in the United States (San Bernardino County in California), covers more territory than the States of Vermont and New Hampshire combined!

Table II. How to determine if an endangered species organization should computerize some of its activities.

1. What are the purposes of the project (general; specific)?
2. What types of people and organizations does it serve (now, and with computerization)?
3. What services does it provide (now, and with computerization)?
4. Will it merge its data with that of other projects?
5. Do its administrators see the project as an information repository and information source?
6. What are the general and specific purposes of the proposed *computerization?*
7. Can several purposes be integrated into the same system?
   a. Research, Administration, Education, etc.
   b. Within Research, within Administration, etc.
8. What is the detailed procedure of the computerization?
   a. Purposes.
   b. Preliminary Data Preparation (e.g., specimen identification).
   c. Data Accumulation.
   d. Data Verification.
   e. Data Processing.
   f. Useful Results.
   g. File Maintenance.
   h. Marketing Research and Advertising.
9. Are the required types of personnel and computing facilities available?
10. Have all "hidden" costs been included in the budget?
11. Will funding be continuous?
12. Do the project leaders really want to improve the quality and value of its information, or only increase its quantity?
13. Can a pilot study be run whose results will be of value in themselves, even if the decision is made not to computerize further?
14. After considering the foregoing questions, administrators, contributors, *and* users of the computerization project should jointly determine whether the long-term benefits and goals are worth the major investment of time, money, and effort which the project involves.

To combat problems such as these, the American Society Of Plant Taxonomists has established a Committee On Data Processing Standards to help determine guidelines for data capture. In addition, the previously mentioned International Register Of Computer Projects In Systematics also can be helpful in reducing compatibility problems.

CONCLUSIONS

Our conclusions are of two types — passive conclusions, and those that lead us to action. Among the major passive conclusions is the realization that computers are being used in all areas of botany. Also, the limits to their valuable use in endangered species projects are money, know-how, and imagination. Computers are exciting tools because they permit previously impossible analyses, and they also can free us from drudgery. They are steadily decreasing in cost and are becoming easier to manipulate. Systematists and other botanists ought to be involved in all phases of computer planning and design, not just as blind users. For each endangered species project and its individual divisions, there exists some combination and level of computer usage which can help researchers reach their goals. Though computers can never replace good biological thinking, they certainly can enhance it. Finally, computers can help us realize greater human potential, because by freeing us from tedious clerical and other repetitive tasks, they provide us more time for intelligent, constructive thought.

Consideration of the use of computers in endangered species problems also should lead us to positive action. This might include the following:

1. Given the financial means and sufficient manpower, perhaps the question of how computers can be used should be considered from the *top* of one's organization down, instead of from the bottom up. For example, computers can enhance communication among workers in a country or on a continent by using computers to generate mailing labels. In Venezuela, computer workers are capturing and retrieving information on the distribution and production of botanical work on the flora of that country. Computers also can be used to study ecological associations. All of the above projects, however, may be brought about even more efficiently if, as in the Projeto Flora Brazil, the decision is made at the *top* level of Government.

2. Establish specific goals for each of the three stages in the endangered species process.

3. As data are collected, even if it is not now computerized, consider that it may some-day be computerized, and act accordingly.

4. Collect accurate and compatible data. Be aware of The Committee On Data Processing Standards and of The Register Of Computer Projects In Systematics.

5. Inventory established natural preserves as soon as possible. This will not only produce less friction with the public, but will also serve to broaden the base for funding.

6. Work hard to integrate efforts of various groups and agencies. Such coordination among diverse groups will increase efficiency and compatibility, and also contribute to the success of a long-term funding campaign.

LITERATURE CITED

Adams, R. P., D. H. Wilken, W. M. Klein, G. Bryant and R. G. Walter. 1975. RAPIC, The missing link? BioScience **25**:433-437.

Blackith, R. E. and R. A. Reyment. 1971. Multivariate morphometrics. Academic Press. New York. 412 pp.

Crovello, T. J. 1970. Analysis of character variation in ecology and systematics. Annual Review Of Ecology And Systematics **1**: 55-98.

_____. 1972. Computerization of the Edward Lee Greene Herbarium (NDG) at Notre Dame. Brittonia **24**: 131-141.

_____ and R. D. MacDonald. 1970. Index of EDP-IR projects in systematics. Taxon **19**: 63-79.

Forero, E. and F. J. Pereira. 1976. EDP-IR in The National Herbarium of Colombia. Taxon **25**:85-94.

Gómez-Pompa, A. and L. I. Nevling, Jr. 1973. Ordenación de datos para la descripción de especies para la Flora de Veracruz. pp. 34-41 *In* A. Gómez-Pompa and A. Butanda C., eds. El uso de computadores en la Flora de Veracruz. U.N.A.M. México.

Pankhurst, R. J., ed. 1975. Biological identification with computers. Academic Press. New York. 333 pp.

Sneath, P. H. A. and R. P. Sokal. 1973. Numerical taxonomy. W. H. Freeman Co. San Francisco. 573 pp.

THE BALANCE BETWEEN CONSERVATION
AND UTILIZATION IN THE HUMID TROPICS
WITH SPECIAL REFERENCE TO AMAZONIAN BRAZIL

Paulo de T. Alvim

Instituto Interamericano de Ciencias Agrícolas
and
Centro de Pesquisas do Cacau, Itabuna, Bahia, Brazil

It has been difficult for me to decide exactly what should be emphasized in a paper concerning the equilibrium between conservation and utilization in the humid tropics. But since most speakers in this Symposium will be emphasizing problems of conservation, I have chosen to devote this paper to problems of utilization. I hope this approach provides some balance, at least, to the subject matter of our proceedings.

I would like to stress at the outset that I do not think a balanced use of natural resources is important simply because we all "love nature" and want to protect endangered or threatened plants and animals. Though these are reasons often given by biologists or ecologists, I believe that the really important reason is the protection of man himself. We here are concerned not only about the future of birds and flowers, but the future of humanity is our prime concern. While other speakers discuss endangered plants and animals, I believe my assignment here today relates more to an endangered member of the animal kingdom: the human being.

The word *conservation* implies more than simply preserving the environment and saving wild plants and animals. I prefer the broader definition of the word, as proposed by the International Union for Conservation of Nature, which defines conservation as *"the management of resources of the environment with the purpose of achieving the highest sustainable quality of human life."* This definition implies that no conservation movement has any meaning or significance unless its ultimate goal is also to help human life. It would be either childish or cynical if we failed to admit that protection of humans is more important than protection of plants and wild animals. Only a mentally unbalanced person would say that plants and animals are more important, more useful, and in more need of care and love than the millions of people throughout the world — particularly those in the so-called "third world" — who are dying before their time because of poverty, disease and hunger.

Some conservationists find it difficult to accept the idea that management is an essential part of conservation. It should be pointed out, however, that even in areas set aside as true reserves or sanctuaries, some degree of management is always needed, at least in terms of vigilance and protection. On the other hand, experience has shown that the most effective way to protect an ecosystem is not by enclosing it with a fence, but by educating man on how best to utilize its resources on a sustained basis. This principle is applicable to any type of ecosystem, including the more fragile ones such as the tropical rain forest. The crucial point we face is to decide how far we can go in utilizing our ecosystem before irreversibly hampering its future utilization. This is admittedly a difficult decision to make, and the final answer cannot be provided solely by botanists, zoologists, ecologists, or other members of the scientific community. Let us, therefore, recognize our limitations and humbly admit that economists, sociologists, and even politicians share with us the responsibility for defining what is good for the ecosystem and what is equally good — or preferably even better — for the human being.

The essence of my presentation is the relationship between man and the rain forest ecosystem. This is, indeed, a very broad subject, from both an ecological and a geographical point of view. The humid tropics cover about one-sixth of all continents, or approximately twenty-five million square kilometers. In South and Central America alone, there are over eight million square kilometers of humid tropical land. This is equivalent to about half the total land surface cultivated by man throughout the world (FAO, 1969), but at present only a very small fraction of this humid tropical land is utilized for agriculture. There is little doubt that the need for cropland will expand to the humid tropics in the foreseeable future, particularly in the Amazon Basin of South America and in Sub-Saharan Africa, which are the only regions where sizeable portions of well-watered, potentially arable land still are available (Brown and Eckholm, 1975). According to an FAO report (1969), the rain forest region in six South American countries — Brazil, Bolivia, Colombia, Ecuador, Peru and Venezuela — has a combined area of approximately three to four million square kilometers suitable for man's utilization. This represents five times the total area that is at present being cultivated in these six countries. It is important to stress that this gigantic land reserve is situated precisely within the world's ecological belt in which biological activities are most intense, and where the primary productivity of the ecosystems reaches its highest values (Alvim, 1973; Lieth and Whittaker, 1975). We are, therefore, confronted with some of Nature's paradoxes: Why does this environment, tremendously productive from a biological point of view, produce so little from an economic point of view? Would biological productivity in this area be a handicap for economic development? What factors in the wet tropics have thus far prevented man from achieving the same standard of living that he has attained in subtropical or in temperate regions? Why has agricultural development been so slow in coming to the region? Can we hope to develop a suitable production system which will promote economic growth without irreversibly damaging the environment?

I cannot pretend to know all the answers to these questions, but I believe that some of the factors which have contributed in the past and still continue to hinder progress in the wet tropics are known.

From an ecological point of view, the two main limiting factors for commercial farming in the humid tropics appear to be *low soil fertility* and *excessive rainfall* (Alvim, 1973). From a cultural point of view, the chief obstacles are either failure of man to develop production systems suitable for this type of environment, or the lack of an efficient program to introduce such systems into the region. The question we must ask ourselves now is whether we can hope to find technical and economical solutions to these problems while there still is time to do so. In other words, can we hope to promote the development of the humid tropics by means of commercial farming? If so, what type of farming or production system should be recommended for the region?

I am fully aware of the fact that this is a very controversial matter which inevitably produces emotional reactions rather than scientific reasoning. Let us use the Amazon region of Brazil as an example. As a Brazilian, I know that my country has had some very bad publicity among conservationists during the last few years for having decided to build a network of roads through the dense forest of the Amazon region. I am sure all of you have read some of the articles criticizing this decision published mainly in newspapers and popular magazines. If you are interested in a good review of this bad publicity, I would recommend the recently-published book entitled *Amazon Jungle: Green Hell to Red Desert* by Goodland and Irwin (1975). According to the authors, the Amazon region should be considered, from an ecological point of view, as a "desert covered with trees" (sic): if the trees are removed, it becomes a desert without trees. One would have to conclude from these pessimistic statements that there seems to be no solution to the problem of developing the Amazon region: either way we go, we are stuck with the "desert."

As an agronomist with some experience in tropical agriculture, I do not believe that the situation is as bad as portrayed by Goodland and Irwin. Obviously, much basic research in the Amazon area is still needed before we can propose an ambitious program for agricultural development. Until we have the answers from this research, care must be taken to promote agricultural programs only in limited, well-selected areas, leaving most of the region untouched as reserves for future use. It is, indeed, regrettable that most institutions responsible for nature's conservation and environmental protection in the Amazon region are still very inefficient and need to be greatly strengthened. There are, however, promising signs that important changes in the right direction are now taking place. In Brazil, we are anticipating that efficient measures will soon to be taken to avoid indiscriminate expansion of land-clearing for cattle ranching of dubious success. On the other hand, some constructive research programs are in progress, involving several tropical crops, which are already showing very promising results. Successful agricultural programs in the wet tropics are not, after all, as mysterious as some people seem to think. In this connection, since I have already referred to a pessimistic publication about land use in the Amazon region, I would now like to mention an optimistic one, which concerns crops suitable for commercial farming in the wet tropics. This forthcoming book, *Ecophysiology of Tropical Crops* (Alvim and Kozlowsky, in press), is a collection of papers presented in a symposium held last year in Manaus. Contributors are leading research scientists from several continents with many years of experience in tropical agriculture. This book deals with a wide range of economic plants, including tree crops such as cacao, rubber, oil palm, and coconut; food crops such as rice, cassava, yams, and sweet potatoes; fruit crops such as pineapple, mango, cashew, and citrus, as well as sugarcane, and pastures. Obviously, these crops have different requirements with regard to soil, climate, and field management. A decision to grow them will depend also on social, economical, and cultural considerations. It was recognized at the Manaus meeting that the Amazon region has great potential for growing such crops, but, as it happens with agriculture anywhere, the cost/benefit ratio is always the critical factor for deciding what to grow and where to grow it. It is well-known that high dosages of fertilizers and intensive management frequently result in good economical returns in highly-valued cash crops, as has been amply demonstrated in Brazil, for example, with pepper *(Piper nigrum)*. However, not many other tropical crops will respond economically to such intensive fertilizer applications. On the other hand, there are many other important crops that do not demand the soil fertility that pepper does, and which can be commercially grown with little or no fertilizer applications at all. This is the case of most forest trees, several important industrial crops such as rubber and palm, a large number of nitrogen-fixing grasses and legumes, and oil.

Though I have, up to this point, emphasized only the utilitarian aspects of the rain forest, I do not underestimate the importance of preserving the gigantic pool of germ plasm there. It is quite obvious that scientists living in the humid tropics have not been as successful as they would like to be in convincing their governments of the importance of this problem. Brazil is no exception in this respect. Because conservation practices produce no immediate and obvious impact on economic development, it is understandable that they seldom attract the attention of politicians and decision-makers. A well-organized educational campaign is obviously needed to overcome this problem, but care should be taken to avoid the use of fallacious arguments or controversial issues which cause confusion and often bring more discredit than support to a conservation movement. There is no need, for example, of exaggerating the effects of deforestation on the ecology of the environment in predicting changes which actually do not occur. Dire predictions, such as a decrease in the oxygen concentration of the atmosphere, a decrease in total rainfall, the formation of true deserts, the drying up of rivers, and other imaginary consequences, all lack scientific support. There are plenty of valid scientific reasons to protect the environment, and I am sure that politicians and decision-makers, as well as many other concerned individuals, will be more receptive to scientific *facts* rather than to mere prognostications.

Unfortunately, man thinks of "conservation" only when some alarming destruction has already taken place. This is natural, for the problems of conservation are always the results of man's demand on natural resources. Primitive man lived by hunting and gathering, and in this way he maintained the same perfect equilibrium with the forest as animals did. Indian tribes of the Amazon, as well as the peasant farmers in Amazonia of the present century, reached a new form of equilibrium through the development of shifting cultivation. This type of subsistence agriculture is practiced in all rain forest regions of the world, and wherever it exists, it impedes agricultural progress. From an economic point of view, agriculture can no longer be looked upon as a simple art of planting and harvesting. Obviously, it is a profitable business today. It seems clear that development cannot take place where farmers produce only enough to meet the needs of their families. Most economists agree that agricultural development in the tropics can only become viable when the present type of subsistence farming becomes more commercial and produces increasingly for the market (Mosher, 1970).

The greatest challenge confronting agricultural scientists working in the humid tropics is to find a production system which is ecologically suitable for the area. Such a system eventually will bring man to a new state of equilibrium with the environment, and, undoubtedly, will require the replacement of the natural forest in selected areas. Unfortunately, we will probably have to wait many years before new production systems are found which can be extensively used in the humid tropics. This does not mean, however, that we are completely ignorant about what to do *now* in some selected areas of the tropics without threatening the environment. In the Amazon region of Brazil, some promising projects are already underway, and I would like to cite a few examples as illustrations.

Cacao

Cacao is an ideal cash crop for the wet tropics. It is usually grown under the shade of larger trees and affords excellent protection against soil erosion and leaching. Native cacao is found almost everywhere in the Amazon valley, but very few commerical plantings were established in the past. The Amazon region still produces only 4,000 tons of cacao per year, representing less than 2 percent of the Brazilian production, which comes mainly from the State of Bahia. Soil survey and experimental work carried out during the last few years, however, have demonstrated the existence of extensive areas in the Amazon where cacao can very successfully be grown. The best regions so far surveyed are located in the Territory of Rondônia and along the Transamazonian Highway near Altamira, where calcareous soils (Alfisols) of high fertility are commonly found. Yield figures of 1,500 to 2,000 kg/ha/yr have been obtained from hybrid cacao, 4 to 5 years old, planted in these areas without the use of fertilizers. It is estimated that there are at least one million ha of this type of fertile soil in the Amazon (Silva, 1976). Based on these results, the Brazilian government is now undertaking an ambitious program to plant about 200,000 ha of cacao in some selected areas of the Amazon within the next fifteen years.

Rice

Brazil produces about seven million tons of rice per year; this represents more than 50 percent of the total production of all Latin American countries. The main producing areas are located in Southern and Central Brazil, and the Amazon region contributes only a very small fraction of the total production of this crop. Irrigation is not yet extensively used in rice growing areas of Brazil, and this accounts for the relatively low productivity obtained, around 1.5 ton/ha/yr, as against 4 to 5 tons in other Latin American countries, where irrigation is used.

Irrigated (paddy) rice has great possibilities in the flooded *(várzea)* plains of the Amazon valley. This has been demonstrated both experimentally and by a few commercial plantings established in Pará and Amapá. In the Guamá region, near Belém, some farmers are obtaining yields of about 18 ton/ha from three crops per year. The Jari Agro-florestal Inc., in the Territory of Amapá, has at present about 600 ha of rice growing in "poulders," producing an average of 10 to 12 ton/ha from two crops per year. In view of these excellent results, plans are now under way to plant several thousand hectares within the next few years.

It is estimated that the total area of flooded plains suitable for paddy rice in the Amazon valley is well over one million ha. If one assumes a mean yield of 6 to 7 ton/ha/yr with only two crops per year, the várzea plains of the Amazon could produce as much rice as is now produced in the entire country.

Oil Palm

This is another important crop with great potential for commercial farming in the Amazon region. From a climatic point of view, the best areas are located in the Western region and in the vicinity of the Amazon delta, where rainfall shows a better annual distribution. The poor oxisols which predominate in these areas are not much different from the oxisols where successful oil palm plantations have been established in the Ivory Coast or in Malaysia. Eight years after planting, a 1,500 ha oil palm plantation near Belém is now yielding an average of 20 tons of fruit (about four tons of oil) per hectare. Small farmers in the same region are at present planting an additional area of 3,000 ha under a program supported by the State of Pará. A plan is also under way to plant 100,000 ha within the next 10-15 years, with financial assistance from the Superintendency for the Development of Amazonia (SUDAM), a Brazilian Government Agency.

Silviculture

In Amapá, the Jari Agro-florestal Inc. planted, during the past eight years, about 55,000 ha of the African tree *Gmelina arborea* and 25,000 ha of *Pinus caribea* var. *hondurensis*. These plantings were established without fertilizers in ordinary oxisols of "terra firme", which is the predominant type of soil in the Amazon region. The first plantings are now yielding an average 38 m^3 (about 28 tons) of *Gmelina* lumber and 27 m^3 (about 20 tons) of *Pinus* per year. These are among the highest figures so far reported for commercial timber yield in the entire world, and they clearly demonstrate the tremendous potential for commercial silviculture in the Amazon. Near Manaus, experimental plots established in oxisols by local research stations are also producing promising results with a wide range of native and introduced trees.

Brazil Nut

This is one of the most valuable trees of the Amazon region, but it has not yet been the subject of much research. There is still much to learn about selection, breeding, and the best cultural methods for this crop. A few plantations have been established, but production has not yet been very satisfactory. As indicated by an experiment near Manaus, the use of Brazil nut as a shade tree for cacao seems to be an extremely promising combination, both from an ecological and an economical point of view.

Rubber

"South American Leaf Blight", a disease caused by the fungus *Microcyclus ulei*, has made it practically impossible in past years to establish successful rubber plantations in the Amazon region. However, recent research in Brazil has shown that this disease can probably be controlled by aerial spraying fungicides (Medeiros, 1976), a finding, of course, which opens vast possibilities for the successful development of rubber plantations in the Amazon. A new Rubber Research Center is now functioning in Manaus, and the Brazilian government is expected to subsidize the planting of 200,000 ha of rubber within the next fifteen years.

Pastures

Pioneer research already carried out in Brazil has revealed that some species of tropical grasses are almost as efficient as leguminous plants in fixing nitrogen (Dobereiner and Day, 1975). Though these pastures are now being tried in the Amazon region and show promising results, ecologists are categorically against pasture formation in wet tropics. It has been shown that pastures not only provide good protection against erosion, but also appear to improve soil fertility in some regions (Falesi, 1974). I believe that the critical problem with tropical pasture is the need to develop good management techniques. According to most experts, wherever this problem has been solved, cattle raising in these areas of the Amazon valley — particularly in Southern Pará and Northern Mato Grosso — has been quite successful.

The above examples obviously reveal that Brazil does at present have some guidelines on how the rain forest ecosystem can best be utilized for improving the quality of human life. Of course, all these production systems modify the local fauna and flora but with proper management practices, they afford good protection for the soil by preventing erosion and leaching. The soil is obviously the main resource with which we should be concerned. For long-range protection of the flora and fauna, the only efficient and effective solution is the establishment of parks and biological reserves.

Some ecologists may object to the examples that I have given since all of them are based on the substitution of the natural forest. Some of the participants of this Symposium might have hoped that I would also recommend the utilization of the natural forest by selective harvesting, without the need of clear-cutting for the establishment of man-made communities. Selective harvesting is used very extensively in the Amazon, particularly near urban areas. However, the examples I cited were selected because of their potential use in commercial farming. Selective thinning of the forest appears to be in the same category as subsistance agriculture. We have no indication yet that this

351

method will ever be sufficiently profitable to raise the standard of living for the people in the tropics. We cannot, therefore, recommend this method, in spite of its obvious advantage from an ecological point of view.

I hope that I have proven my basic thesis that commercial agriculture in the wet tropics is not as discouraging or frustrating as some people think it is. An important point to keep in mind in this connection is that in the tropics barriers still exist which are not yet understood, particularly by those with little training in tropical agriculture. Some of these barriers are ecological, while others are cultural. Good scientific knowledge about the plant and its environment, particularly with reference to plant-soil relationship, is an essential ingredient for planning successful agricultural programs. If natural vegetation grows luxuriantly in the wet tropics in spite of the poor soils, why should we not be able to devise man-made plant communities which will imitate the natural ecosystems and grow equally as well as the rain forest, while simultaneously producing something valuable to man? I am convinced that the solution of this problem is not as difficult as, for instance, the astonishing scientific achievement of sending a man to the moon. This makes me wonder if the problem of developing the wet tropics would not have already been solved if tropical rain forest also existed on the moon. There is no doubt that ecological and agricultural research for the purpose of achieving better sustainable quality of human life in the tropics would be less expensive than space research, and would certainly serve a more noble cause.

LITERATURE CITED

Alvim, P. de T. 1973. Los tropicos bajos de America Latina: recursos y ambiente para el desarrollo agricola. *In* Simposio sobre el potencial del tropico bajo. Centro Internacional de Agricultura Tropical, Cali, Colombia. pp. 43-61.

Alvim, P. de T. and T. T. Kozlowsky. eds. Ecophysiology of tropical crops. Academic Press, Inc. (In press).

Brown, L. R. and E. P. Eckholm. 1975. By bread alone. Pergamon Press Ltd. 272 pp.

Dobereiner, J. and J. M. Day. 1974. Importancia potencial de la fijación simbiótica de nitrógeno en la rizosfera de gramineas tropicales. *In* Bornemiza, E. and Alvarado, A. eds. Manejo de suelos tropicales en la American Tropical. University Consortium on Soils of the Tropics. North Carolina State University. Raleigh, N. C. pp. 203-216.

Falesi, I. C. 1974. O solo da Amazônia e sua relação com a definição de sistemas de produção agrícola. *In* Reunião do grupo interdisciplinar de trabalho sobre diretrizes de pesquisa agrícola para a Amazônia. Embrapa, Brasília. Doc. 2, 17 pp.

FAO (Food and Agriculture Organization of the United Nations). 1969. Indicative world plans for agriculture development to 1957 and 1958. Rome. 266 pp. (Regional Study no. 2, South America, V. I.).

Goodland, R. J. A. and H. S. Irwin. 1975. Amazon Jungle: green hell to red desert. Elsevier Scientific Publishing Co. Amsterdam. 155 pp.

Lieth, H. and R. H. Whittaker. eds. 1975. Primary productivity of the biosphere. Springer-Verlag New York Inc. 339 pp.

Medeiros, A. G. 1976. Novos conceitos técnicos sobre controle químico do "mal-das-folhas" da seringueira. CEPLAC (Brasil) Boletim Técnico no. 35, 20 pp.

Mosher, A. T. 1970. The development problems of subsistence farmers: a preliminary review. *In* Wharton, C. R., Jr., ed. Aldine Publishing Co. 481 pp.

Silva, L. F. 1976. Disponibilidade de suelos para cacao en la Amazonia brasileña. Revista Theobroma (Brasil) 6(1):31-39.

THE BRAZILIAN HUMID TROPICS PROGRAM

Antonio Dantas Machado

Conselho Nacional de Desenvolvimento Científico e Tecnológico
Brasília, D.F., Brazil

The Brazilian Humid Tropics Program of the National Council for Scientific and Technological Development was established with one paramount goal in mind: to preserve the fragile ecological balance of Amazonia, and to coordinate the latest scientific and technological findings in order to achieve optimum human adaptation to that area's challenging characteristics.

In order to accomplish such a task, several specific objectives were established, and those that I judge to be of interest for this meeting are the following:

1. To integrate and consolidate the research programs of institutions active in the Brazilian humid tropics, aiming at full utilization of the material and human resources available in the area;

2. To orient research institutions from other regions of the country, including universities, in their search for solutions to the problems in the Program's area of activity;

3. To work in close collaboration with other regional scientific and technological programs in order to exchange information acquired in the handling of similar problems; and

4. To work together with international and foreign organizations, by means of previously established governmental mechanisms, so as to secure the knowledge and information required to accelerate the development of the Program.

The area under study is the so-called 'legal Amazon,' which includes the states and territories of Acre, Amazonas, Pará, Amapá, Rondônia, Roraima, and also parts of Maranhão and Mato Grosso. These areas together comprise 57 percent, or approximately 4.5 million km^2, of the entire Brazilian territory.

The population of this immense expanse is about 3.5 million, which is equivalent to a demographic density of only 0.97 inhb/km^2. It should be emphasized here that population growth in Amazonia is more the result of migratory flow rather than one of natural increase. It reveals irregular distribution, with population primarily concentrated in the more important urban centers and in areas of agricultural and mining activities.

As is well known, the Amazon region is characterized by a vast and unique expanse of tropical rain forest. Contrary to popular opinion, however, even in this singular region such vegetation is not always continuous. It is interrupted by non-forested areas covering approximately 300,000 km^2, and also by transitional areas *(caatinga alta* and *mata de várzea)* of about 500,000 km^2. In addition, there are constant changes of botanical species, due to the different environments throughout the region, i.e., fertility and depth of soils, drainage, soil aeration, and availability of water.

Though the Amazon forests basically are flourishing, they normally are rooted in low-fertility soils. This means, of course, that the enormous amount of organic matter is constituted almost wholly of vegetation alone. Yet in spite of this, a perfectly regulated biological cycle not only maintains this huge mass, but even more important, it conserves the nutrients in the soil. This latter fact is of major relevance when one is considering the introduction of modern techniques for the area's agricultural development.

The heretofore widely accepted concept that a lateritic cap covered the entire Amazon region has now been totally negated, as a result of recent pedological research. Studies based on radar images have shown that only 70 percent of all Amazonian soils are formed by latosol units and red/yellow podzolic soils, which are of poor fertility but with good physical properties. On the other hand, some good naturally fertile units have also been identified. These include the 'terra roxas' (purple soils), and the red/yellow eutrophic podzols, as well as the lowland flood plains soils along the riverbanks which cover an area of 60,000 km^2.

Present-day information on morphological characteristics of Amazonian soils, and results of recent experiments involving pastures and crop plantation in the region, clearly point to the fact that there is a major need for additional field research on soil physics and soil chemistry, before any additional development can take place there. Such research, combined with simultaneous experiments on improved management systems, will help immeasurably towards establishing efficient guidelines for the agricultural techniques most suitable for future Amazonian development.

In addition to our Program's ongoing work with satellite images, ichthyology, and tropical diseases, we also are deeply involved with the problems of human adaptation in the tropics, and of course, within this framework, agricultural development is near the top of our list of priorities. In terms of food crops, our activities are directed towards increasing productivity, genetic improvement, and better handling and treatment of crops, soils, and disease controls — all in an attempt to discover the most efficient and economically feasible production systems for the area.

Research also continues to be aimed at improving methods for introducing, cultivating, and managing crops on the flood plains and *terra firme* of Amazonia. In this connection, the main crops now under study are cocoa, rubber, oil palm, guaraná, (used for a popular soft drink in Amazonia), and black pepper. Efforts also are being directed towards establishing production systems which will resemble the forest ecosystem itself.

ECOLOGICAL STUDIES

Amazonia has been beset with many problems since development there was accelerated in recent years. In view of the different options known to be available for tropical soil management, those involving bio-geo-chemical cycles and the impoverishment of edaphic conditions, have been considered to be of prime importance. We therefore have given them top priority.

Considering the area's capacity for natural vegetation recovery — if utilization is not strained beyond critical limits — we believe that caution in regard to non-edaphic ecological changes is really of secondary importance in the Amazon region. Germ plasm reserves will be assured by strict observance of provisions in the official program for biological reserves.

The Humid Tropics Program also stipulates that priority be given to preventive research studies on the integrated control of biological pests, on soil conservation, and on the maintenance of nutrients derived from forest burning, rather than on descriptive studies of eutrophication of water systems and the sedimentation of hydrographic basins.

Research on ecological agriculture will be emphasized until such time that adequate production systems for the humid tropics are devised. Today our research projects basically are concerned with detailed surveys of environmental changes which result from forest clearing, in order to make way for different production systems (staple food crops, cash export crops, pastures, or joint exploitations). Of course, we also will incorporate into our research any new ecological concepts and principles which can be readily integrated into the existing programs.

In relation to the Amazon flora, a project is already under way: "Projeto Flora Amazônica." The main objectives of this project are as follows:

1. To carry out basic inventory and assessment of the area's native plants, with the ultimate goal of utilizing Amazonia's vast botanical resources for the betterment of man. This inventory will concentrate on plants of known qualities, or on those having potential economic importance. The list will include the geographic distribution, quantity, possible uses, and such other information as toxic and medicinal plants, lactiferous plants, tannin-producing or textile plants, oleaginous plants, forest trees for timber, wild fruits, forage plants, weeds, and those plants indicative of mineral deposits or pollution indicators, etc.

2. To identify areas which should be set aside for (a) conservation; (b) rational development; and (c) controlled extraction of some specific natural resources.

3. To identify areas which should be preserved intact as ecological reserves to be used specifically for scientific and environmental purposes; and

4. To establish regional centers, staffed by professionals who are capable of carrying out research on inventory and assessment of natural resources, ecological problems, and

conservation of natural resources and the environment.

Information gleaned from the above comprehensive program is fundamental for the rational management of Amazonia's native vegetation, to exploit its resources without squandering them, to clear the ground for agriculture or for industrial purposes, and for the establishment of human settlements. It also will serve as a basic guide for projects which will preserve natural areas and the priceless genetic material such areas may contain.

The Project encourages international cooperation, of course. Though many Brazilian scientists will be involved, there is enough work for all. Already, special mention should be made of the contributions by Dr. Bassett Maguire and his group from the New York Botanical Garden, including Dr. Ghillean Prance, who has helped so much to establish a postgraduate course in botany at INPA in Manaus. But in all of Amazonia, there are only a few senior botanists presently available to work on this Project. We also have at present only about fifteen young scientists who are actively engaged in various facets of the program. Our need for well-trained botanists is great. We need them not only to carry out the basic research, but also to help train more young Brazilians. We could say that in Amazonia right now, the most endangered species is the *Homo sapiens* subspecies botanist!

As can be seen, we have many serious problems to tackle in Amazonia. The world already is aware that the Brazilian government, in the process of developing and improving this region for the betterment of its people, also wants to keep the Green Amazon as green as possible for all the world. It wants to save areas that are covered with endemic species. It wants to perpetuate its vegetation and its irreplaceable genetic material.

We will never lose sight of the fact that proper utilization of the natural resources of a country will perpetuate the good quality of the environment for generations to come. We will never forget that goal.

DISCUSSION

Anonymous: Do you have any clue as to the agricultural productivity of the cerrado?

Dr. A. D. Machado: Though I may be looked upon as an expert in this field, I do not feel that I can offer conclusive advice on a situation that has not yet been thoroughly explored. We are very interested at present in the strategy of wheat cultivation in the cerrado. At the National Research Council we feel that the cerrado option is a very good one, because it avoids too much pressure on the Amazon area. In this way, we can test its limits, but as of now, we cannot definitely encourage development in the cerrado because we have not yet done sufficient research on its economic feasibility. The soils in the cerrado are poor, and large amounts of fertilizer have to be used. Also, in some parts of the cerrado, irrigation is involved. So you see, we do not yet feel that we can or should advise on a situation that we have not yet extensively studied.

CONSERVATION: RECENT DEVELOPMENTS
IN INTERNATIONAL COOPERATION

Grenville Ll. Lucas

Royal Botanic Gardens, Kew, Richmond, Surrey, Great Britain

One of the primary responsibilities of the Threatened Plants Committee (TPC) of the International Union for Conservation of Nature and Natural Resources (IUCN) is the task of gathering and disseminating information on the world's threatened plants. It is an exacting challenge, but with the wholehearted worldwide support of the botanical fraternity, this information is now beginning to be assembled and correlated by the TPC Secretariat at the Royal Botanic Gardens, Kew.

In his early work on the IUCN Red Data Book for Plants, Dr. Melville, my colleague at Kew, has estimated that the number of threatened species would be between 20,000-25,000, or about 10 percent of the flowering plant kingdom. The percentage is close to the figures in the Smithsonian Institution list (1975), which states that 1,999 species are endangered or threatened in the U.S.A., while the flora contains approximately 20,000 species. Thus the size and seriousness of the problem can be seen.

Other related figures are equally disturbing. According to the latest FAO estimates, 11,000,000 hectares of tropical rain forest are being removed each year. When converted into more understandable terms, this huge figure means that tropical rain forest is being lost at the rate of 49.2 acres a minute, or, putting it another way, an area equivalent to all of the New York Botanical Garden is being cleared every five minutes!

Therefore we have no time. The TPC survey could be a most interesting and leisurely lifetime occupation for many of us — but time is the one thing that is just not available.

The TPC needs your help now in gathering the necessary information before it is too late and here I ask anyone who can provide us with data on threatened plants to contact, as soon as possible, the appropriate regional group or the Secretariat direct.

Although the Secretariat works at the species level, it is in their original habitats that plant species should be conserved, both from a genetic and management point of view, where in the long run, of course, it will be much more economical. The structure of the TPC is best summarized as follows:

1. *Regional Groups:* Throughout the world, botanists, horticulturists, ecologists and others with special interest in and knowledge of their regional floras are being asked to become associated with the work of the Committee, and these correspondents are being asked to identify threats to floras, plant groups and individual species in their own areas. Part of their task will be to document adverse changes affecting floras and species, and to prepare recommendations for minimizing their effects through governmental or other action. They may be expected also to be involved in protection and rescue operations.

2. *Specialist Groups:* To complement the work of the regional TPC organization, panels are being set up to handle the problems of special plant groups such as the palms, orchids, and cycads. In part, these groups will have a taxonomic basis, and specialists will be invited to help accordingly; but it is anticipated that panels will be formed to deal with more comprehensive categories, based upon ecology or life form. The succulents form such a category.

3. *Institutional:* The third component in the TPC is made up of Botanic Gardens, University Departments, Research Institutes and other organizations which have both the facilities and expertise to maintain plants in cultivation or in seed banks. Already, the Food and Agriculture Organization of the UN is organizing a worldwide network of plant genetic resource centers for the husbanding of cultivated species and some of their wild relatives. The present aim now is to set up a complementary network to handle natural-source plant material not already covered by the activities of the FAO system.

The TPC Secretariat's role is to coordinate these efforts and to provide the best possible information service, not only about the threatened species — but also where further information can be obtained; for instance, on legislation with regard to plants. The simple and yet the most valuable aid that the Secretariat aims to provide is carrying out routine preparatory tasks in order to relieve the pressure on the overworked experts.

Without the support of taxonomists and other interested and concerned professionals and amateurs throughout the world, the TPC will not succeed in its first aim of determining exactly what *is* threatened. These experts, with either an interest in a specific group or region, given someone to write down their knowledge or provided with initial lists of species, can provide much valuable conservation information so that future action plans can be based on the actual status of threatened species rather than on broad theories.

In some articles, one has the impression that there is a controversial decision necessary between conservation of the habitat or species. I do not see any such problem. They are not alternatives but different approaches to provide information for the same groups of decision-makers. Those using the "habitat approach" can cite instances of threatened species to be found within the habitat they wish to conserve. This is yet another reason why we need the cooperation of all field workers, amateur and professional alike.

Experience with the animal kingdom indicates that providing the decision-makers with accurate information is of the utmost importance. Specific advice such as "these species are unique to this state (or country)," or "this is the only site that these creatures are to be found in the world," will clearly evoke a response far more favorable and protective than if a general plea is made "to save an area of forest." To the decision-maker it looks like any other piece of forest, and so if the unique and tangible qualities are stressed, the response will be improved. If we can identify centers of endemism where more than one unique species exists, a much stronger case can be made for conservation measures to be undertaken. This is only one approach of many and it is not the purpose of this paper to cover them here.

A few basic guidelines are needed before one can progress further and one such guideline is to define the categories by which the degree of threat can be calibrated. A system has been developed by IUCN for use on a world basis for animals, and with a few modifications, the system also serves for plants.

Extinct (Ex)

Endangered (E)

Taxa in danger of extinction and whose survival is unlikely if the causal factors continue operating. Included are taxa whose numbers have been reduced to a critical level or whose habitats have been so drastically reduced that they are deemed to be in immediate danger of extinction.

Vulnerable (V)

Taxa believed likely to move into the endangered category in the near future if the causal factors continue operating. Included are taxa of which most or all the populations are *decreasing* because of over-exploitation, extensive destruction of habitat or other environmental disturbance; taxa with populations that have been seriously *depleted* and whose ultimate security is not yet assured; and taxa with populations that are still abundant but are *under threat* from serious adverse factors throughout their range.

Rare (R)

Taxa with small world populations that are not at present endangered or vulnerable, but are at risk. These taxa are usually localized within restricted geographical areas or habitats or are thinly scattered over a more extensive range.

Indeterminate (I)

Taxa *known* to be Endangered, Vulnerable or Rare but where there is not enough information to say which of the three categories is appropriate.

Out of danger (O)

Taxa formerly included in one of the above categories, but which are now considered relatively secure because effective conservation measures have been taken or the previous threat to their survival has been removed.

Insufficiently known (K)

Taxa that are suspected but not definitely known to belong to any of the above categories because of the lack of information.

In practice, Endangered and Vulnerable categories may include, temporarily, taxa whose populations are beginning to recover as a result of remedial action, but whose recovery is insufficient to justify their transfer to another category.

However, almost every individual and certainly many countries have their own system of classification. It would be preferable to have only one — the IUCN system, of course — but in effect it is the information that is of vital importance, and so long as the scheme used fully defines the degrees of threat in any particular case, it is of no consequence which system is used.

The TPC Secretariat has also compiled a list of the major threats to the plant kingdom (Table 1). This list represents only a first attempt and no doubt it contains omissions. It was produced initially for use with work being carried out for the Council of Europe Threatened Species List but we feel it can be expanded and refined to be used on a world scale. Suggestions for additions, deletions, etc., are of course welcome, and should be sent to the TPC Secretariat.

Table I. TPC Secretariat list of major threats to the plant kingdom.

1. Grazing (goats, sheep or cattle)
2. Regeneration of scrub (and lack of grazing)
3. Changes in arable farming (e.g. spraying, mechanical compaction)
4. Ploughing of old grassland
5. Forestry (e.g. changes in practice, drainage and monocultures)
6. Traditional rural practices
7. Flooding
8. Drainage
9. Water pollution
10. Air pollution
11. Industrialization and urbanization
12. Road building
13. Tourism and tourist developments — coastal
14. Tourism and tourist developments — inland
15. Dam construction
16. Mining and quarrying
17. Pressure from introduced plants
18. Collectors for horticulture
19. Collectors for botany
20. Critically low population — for example, below 100 individuals (hence IUCN category Endangered)
21. Natural causes
22. Lack of pollinators
23. Disease (catastrophic introductions)

With these two aids, the next step is the input from specialists with information about their group or region. Many areas such as the USSR, Europe, and North America, for example, are already being studied, but a great deal of the world is not yet covered, and so your help is most earnestly requested.

Lists must be prepared at local level, and national red data books also must be produced. It is IUCN's task to prepare and publicize the problem at all levels, but it is at the *world level* that it must take the final responsibility, indicating those species threatened on a world scale, and recommending appropriate protective action.

It is pointless to blame "them" in conservation. We must all provide the correct information. We have to become bio-politicians, and those who do not want to enter the arena should support those who do — with accurate data.

It is the decision-makers we must talk to and convince of the need to conserve the world's future plant resources. It is foolish to attempt to stop development; conservation and development need not be mutually exclusive. *Development* is about utilizing resources. *Conservation* is about the wise use of those resources.

What we need is the information on which to recommend balanced development. We must show exactly how the species listed as threatened are important economically, scientifically, aesthetically, and educationally.

Many plant species will be irrevocably lost whatever we do, but nevertheless it is *our* task to ensure that the vast majority do not disappear so as to be able to maintain the world's plant genetic resources for the future (see Heslop-Harrison, 1973, 1974).

It is the very diversity of species and diversity within species that has ensured the covering of the earth in its life-giving and life-supporting plant mantle. Above all, it gave rise to the species which Man has so carefully nurtured into his crop plants, medicines, and stimulants, to give but a few examples.

It is usually this very genetic diversity within a species which provides them with the potential and capacity for successful self-renewal in a gradually changing environment. Recent massive losses in some pure crop cultures due to disease show only too clearly what happens when genetic diversity is lost.

It is not only our task in the TPC to *know* what is threatened, but to communicate this information in a form easily understood by decision-makers. The decision-makers usually work on a five- and at most a ten-year plan; however, we have to think for the long-term future. It is necessary, therefore, to maintain a continuing monitoring system. It is no good blaming "them." Rather, it is up to US to ensure that the planner has the information on which to make the best decisions for plant conservation, and, consequently, for Man himself.

DISCUSSION

Dr. A. Fernández: Because many of the endangered plants are from Latin America, your lists should be in Spanish, not in English. In fact the tropics have the highest species density per cubic area. Perhaps the lists should be in Latin.

Mr. G. Lucas: I agree entirely. These first lists are the trials. When they have been in circulation for perhaps a year or so, I hope we will then develop both a French and a Spanish edition. As you said, I am sure that, per unit area, Hawaii will probably show the greatest threat so far, although we have just received a list from the Philippines which contains 5,000 endangered or threatened species. This is the sort of figure that makes one shudder. We certainly will try to produce lists in different languages.

LITERATURE CITED

Heslop-Harrison, J. 1973. The plant kingdom: an exhaustible resource? Trans. Bot. Soc. Edinb. **42**: 1-15.
_____. 1974. Genetic resource conservation: the end and the means, Jour. Roy. Soc. Arts. 1974: 157-169.
Smithsonian Institution. 1975. Report on endangered and threatened plant species of the United States. Washington, D.C.: U.S. Government Printing Office, 200 pp.

PANEL DISCUSSION

Chairman, Dr. Robert Goodland

The last plenary session of the Symposium on Rare and Endangered Plant Species of the Western Hemisphere consisted of a panel discussion on a variety of topics related to the theme of the conference. Dr. Robert Goodland chaired the discussion. The opening speaker was Dr. Timothy Whitmore, author of a notable book on "Tropical Rain Forests of the Far East" (Clarendon Press, Oxford, 1975). He was asked to summarize important differences between conservation in Southeast Asia and Latin America.

Dr. T. Whitmore: My own experience of the tropical rain forest is in Asia and I have lived there for quite a number of years. I know virtually nothing about the tropics of the New World. For that reason, it is interesting to me to be fortunate enough to attend this meeting and to meet so many distinguished Latin American visitors whom I knew only by name and reputation before I came here. I can't say very much about conservation in America. It is possibly interesting to point to some contrasts with Southeast Asia, the eastern tropical rain forests, where in some ways things have probably gone a little bit further ahead. I have been rather depressed with the various pictures we saw of Central American countries, where every hillside appears to be deforested up to the skyline. This really doesn't happen in much of Southeast Asia, probably because there aren't the same population pressures there, with the main exception of the Phillippines, where the situation is very much like the one for Central America, and in the southern peninsula of Thailand where similar scenes occur. This is, in a way, a general lesson to anyone interested in conservation of the forests of the humid tropics. Namely, wherever there are huge population pressures, the problems of conservation are of a different order of magnitude from elsewhere, because it is really a question of whether one conserves plants or whether humans die of starvation. This is a difficult problem which arises in those countries where there really is an enormous human pressure on the land.

The only other thing I want to say and especially since we have heard a lot about pessimism or optimism, is that it may be interesting to relate what has happened on the conservation scene in Malaya over the last five or six years, since I think it is a tremendous ray of hope as to what can be achieved. Probably the whole Malay Peninsula is like one of the small endemic centres of Amazonia. It has a fantastically rich flora abounding in local endemics of plants and animals, and you can't pick out any phytogeographic divisions within it.

In about 1930, a big National Park was created in the middle of the country covering about four percent of the land area of Malaya. In about 1970, there were rumors that part of this park was to be licensed for logging. At that time the whole country, including the National Park, was under a Land Capability Survey, and teams of soil and forest surveyors visited it for appraisal. The Malayan Nature Society heard the rumors, approached the appropriate Ministry and argued the case for conservation, without much hope that anything could be done to roll back the tide of agricultural development or to prevent extraction of timber. The case was later published in the Malayan Nature Journal (vol. 24, 111-259). Contributions were made by many different professional biologists. It was a powerful case and favorably convinced the Ministry involved with the issue. Now, five or six years later, the area of the National Park appears white on land development maps. It is left out of soil and other surveys. The idea has been accepted that it is legitimate to have a small fraction of the lowland rain forest set aside as a conservation area. Many additional local scientists subsequently swelled the ranks of conservationists, and they now felt able to publish articles and to participate in meetings in which the thesis of total conversion to agriculture or production forestry is questioned. Thus, in half a decade, the pendulum has swung strongly in the direction of conservation, following the injection of a temperate and well-argued case, backed by much evidence, that conservation is in the national interest. I think we can all take heart from the example of Malaya, even when things look bleak. Let us hope for equal success elsewhere.

Question: (Anonymous) It has been implicit in many of the presentations here, that the adequate protection of many species of wild organisms involves legal restrictions upon formerly legal activities. In your opinion, is this kind of curtailment of freedom an inevitable consequence of current events?

Dr. R. Goodland: Before turning this over to the panel, I would like to comment on it myself. In Brazil, rather than in all countries of Latin America, there is a firmly established rule that half of all biological collections should remain within the country. This is a timely rule which has led to controversy, and I am surprised to hear that it does not exist in other countries.

Dr. G. Budowski: If the collection is from an area where endemism is great, I think there should be stringent restrictions. If a collection is from an area with little endemism, the collector should not only be allowed to take out or even keep the specimens but to do much more. A visiting foreign scientist has a series of special privileges and is extended preferential treatment within the host country and, consequently, can be useful to that country. As such, he should fulfill his obligation to collect, but he has more than that to do. He becomes part of an important element to promote conservation and he also has a duty to the local counterparts. For example, he should encourage local biologists. He should make an effort to publish locally. He should ask the local conservationists and biologists what they want him to say if he is granted an audience with the Minister of Agriculture. The question was raised, I suspect, because there is a certain antagonism against the well-staffed, well-funded, well-equipped, foreign 'expert' who comes into the country and does a job that local biologists could do if they were also well-equipped, well-financed, well-supported, and well-recognized. The foreigner, therefore, has a duty to heed his local colleagues. He cannot excuse himself and say his terms of reference are very restricted and that he has no time for such duties. I feel that if he behaves in the responsible way suggested above, some of these feelings would not have started at the beginning.

Prof. P. Richards: I agree with what Dr. Budowski has just said. This Brazilian legislation is both sensible and wise, but I don't think it has very much to do with the preservation of endangered species. The fact is that a large number of the collections, the fundamental collections of Brazilian plants and animals which anybody working with has to study, are in Europe or in North America. If this goes on being so, any Brazilian scientist wishing to study his own flora or fauna must journey to Europe or America. One of the objects of this legislation, as I understand it, is to improve that situation so that there will be adequate collections, at least to some extent, in the country where the species grow.

Dr. R. Goodland: The next question is from Dr. Budowski and may be paraphrased thus: Developed countries have been using the scientific resources, such as plants of the tropics, for centuries, sometimes with adverse results. How can people or institutions in industrialized countries assist in building awareness and offering means toward adequate management, particularly in avoiding the loss of a unique heritage for science, education, and inspiration?

Prof. P. Richards: Well, this seems to be the 64-dollar question. It covers almost the whole theme of this meeting. I don't think that there is any short answer.

Dr. R. Goodland: Is it the feeling of this Symposium that some form of help towards developing regions, or reparation toward the export of biological materials or possible loss caused by scientific expeditions, should be accorded to the people and nations taken advantage of in times past? I personally feel that some form of reparation is is order.

Dr. B. MacBryde: This reminds me of the U.S. Federal legislation and indicates some possibilities where U.S. personnel or U.S. funding are involved. Section 7 of the Act says that any activity with U.S. Federal authorization or funding must respond to the U.S. Endangered Species Act. In that way, Congress has deeply committed itself to protect endangered species anywhere in the world when the United States is in any way involved.

Dr. T. Whitmore: There is still perhaps the temptation amongst some scientists and some biologists in temperate countries of developed parts of the world, to regard the tropics as a vast and completely unknown scientific region. A point of fact, however, is that forest ecology in the tropics has now reached the state of great and considerable advance and the kind of observations and experiments which are now needed are most sophisticated. There aren't too many areas remaining where one can go out and expect to find very much. You have arrived, as it were, on the other side of the moon. You make your observations and retreat to your temperate fox nests and wrap them all up.

In the tropics today, biology is carried on very much in consultation and in conjunction with local scientists already working in residence there, who know a great deal more about the local scene.

Dr. R. Goodland: The next question is from Dr. Prance and reads: There has been much criticism of *Eucalyptus* and pine plantation in the tropics. Aren't *Eucalyptus* sometimes helpful in saving endangered habitats? They grow in poor soil; they have been planted in already-disturbed areas of second growth and they yield a crop. Does this not tend to relieve the pressure on other areas?

Dr. G. Budowski: This is such a favorite theme of mine that I've already published some papers on the subject. If you condemn *Eucalyptus,* for instance, in the state of São Paulo in Brazil, you condemn a multi-billion dollar industry; you condemn a well-known national hero and you condemn much of the land, which without *Eucalyptus* would have been exposed to serious erosion. After coffee left much of the *terra roxa* in deplorable shape, it was the introduction of eucalyptus which allowed many of the areas to continue existing while at the same time restoring a certain amount of fertility in the soil. It is also the eucalyptus in the central valley of the Andes which enables the people to avoid burning cattle excrement and to avoid digging up some of the polster cushion plants for use as firewood. Likewise, a case can be made for eucalyptus in many other Latin American countries and even more so in Africa.

We have to distinguish carefully between two ways of using eucalyptus. One is condemnable: when highly interesting, heterogeneous and valuable associations of plants are replaced by eucalyptus plantations for quick, simple, economic benefit. The laudatory, positive use is when eucalyptus, which are usually not demanding, are planted on lands which have been cultivated, abandoned, and eroded. Such plantations, which provide the people with some of the necessary materials for their day-to-day work, also relieve the pressure on the natural forests. The possibility of relieving this pressure is one of the prime justifications of some of the eucalyptus plantations. The question could be posed: why plant eucalyptus rather than some native species? This sounds better in theory than it does in actual practice since often a tree from the forest, which is planted in some kind of plantation, becomes also a kind of exotic in its new environment, and often it does not grow in that environment. One can, as has been suggested, make mixed plantations but so far this has proven to be rather unsatisfactory. It is possible but not easy. In summary, the problem should be considered as a whole and not only as creating what has been called "biological deserts" by eucalyptus. Planting eucalyptus produces some negative effects, but there are also some most positive effects as I have pointed out. The thing to do is to try to balance the two and to judge which is better in the long term.

Dr. T. Whitmore: May I suggest that eucalyptus is planted for two reasons? One is on a peasant or villager scale to provide firewood on a small basis. If, however, it is planted on a commercial scale on big plantations, then, surely, it is like any other agricultural crop, on a 4-to-7 year rotation, and will, like other agricultural crops, do best on the best soil. It will need fertilizing, as is certainly the case in Malaya, for example, where there is pressure on good land, including natural forested land, by people who want to get a bigger return of cellulose per year. It is the foresters' obsession with economics — to earn the maximal return in terms of their investment and in terms of tons per hectare per year — that prefers eucalyptus rather than local tree species.

Dr. G. Budowski: My answer derives from an examination of the alternatives. If eucalyptus is used for chipboards in, for instance, El Salvador (not Malaysia), where there is practically no forest left, what will people do for wood and firewood, paper and pulp? Today, this is indeed one of the main reasons for eucalyptus plantations. There are several possibilities. One is to extract these essentials from the last remnants of the natural forests and to cut everything, including mangoes, as has already been done. The second alternative is to import these essentials, while the third is neither to eat nor use wood. All three choices are extremely painful to the country. I don't mean that industrial plantations are the ideal solution, but in some cases they are the lesser of various evils, and I emphasize that it should not compete for good agricultural land. It should certainly not be a substitution for the mixed heterogeneous forest, even if we know very well that eucalyptus produces so much more timber per hectare per year. In some cases, it should even be subsidized for many reasons since in rather poor areas this can create work, and under proper management, it can restore soil fertility.

I insist that there is no reason to believe that eucalyptus sterilizes soils. This is a controversial subject to which we can return. There are studies demonstrating that in many places eucalyptus improves physical and other properties of the soil and as such, this should always be the main argument. Our national policies therefore should be directed towards relieving the pressure on some of the natural forests, and should provide additional income for the people and for the country itself, not only for some multi-national corporations. If eucalyptus is planted to produce an export cash crop on good soils for a foreign industrialized country, it would be a very adverse decision. However, in the highlands of Peru, it has become the basic material for firewood, for mine timber, and even for boundaries on poorly-marked lands of the Indian peasants in central Peru. Eucalyptus can give rise to lesser industries, such as extracting essential oils. Ask 100 Peruvian Indians if they would think of living without eucalyptus, and all would reply that it has become an essential and necessary part of their lives and it does not compete with good farm soil. We should consider eucalyptus from the point of view of the people themselves and their best interests. We have to identify the disadvantages, some of which have been already pointed out, which benefit industrial corporations, those using good land, substitution of forests by eucalyptus, and the use of vast amounts of fertilizers. I agree that these are all drawbacks but there are also advantages, and the whole should be the balanced.

Dr. A. Fernández: I am sure that there are certain species of *Eucalyptus* whose leaves contain toxic substances that inhibit the development of other plants.[1]

The indiscriminate introduction of *Eucalyptus* in Colombia can cause extinction of many native plants, principally the herbaceous flora.

This question points to the danger of introducing exotic plants in monoculture into any tropical community. In Colombia, we have the problem that our economy depends on just such a plant, coffee, which is an exotic from Africa. But if you ask any campesino from the mountain regions what he would do without coffee, he would have no idea how to make a living. I am not so happy with the extensive use of coffee because we ought to develop the use of native plants.

Dr. T. Whitmore: Commenting briefly on eucalyptus: I don't know whether eucalyptus sterilizes the soil or not. Now it is maintained that coffee is grown because there is no substitute for coffee. Well, it is a matter of fact that the genus *Eucalyptus* adds greater weight of timber per year per hectare than any other tree and in that respect it is unique. This is the main reason why people like to grow it. There is no other tree in the world that grows with such a yield as the genus *Eucalyptus.* Until another tree like eucalyptus is

[1]Dr. Fernández later wrote to the editors to inform us that Miss Elizabeth Dalal provided him with the following bibliographic evidence of the toxic effect of *Eucalyptus:*
Yardeni, D. and M. Evenari. 1952. The germination inhibiting, growth inhibiting, and phytocidal effect of certain leaves and leaf extracts. Phyton **2** (1):11-16.
Moral del, R. and C. H. Muller. 1969. Fog drip: a mechanism of toxin transport from *Eucalyptus globulus.* Bull. Torrey Bot. Club **96**: 467-475.

found, it is going to be better than any other tree. In this respect, it is similar to coffee. There is no other drink quite like coffee.

Dr. A. Fernández: We don't know another yet, no one has found one. One should not assert that there is no substitute for coffee.

We have been talking in this Symposium about plants we don't know, undescribed taxa, those without scientific names. Some of these are even extinct. How about some of these unknown plants? We must discover them and investigate them. We must try to find out the potential of our native plants; we must not wait for someone else to plant exotics! We must persuade the governments of Latin America to provide money for research.

Dr. G. Budowski: I agree that dependence upon coffee alone, like dependence upon sugarcane alone is, indeed, an extremely dangerous thing. I am involved at present in a program of diversification to reduce dependence on coffee and sugarcane in Central American countries, which are so dependent on these two plants. We are using tree crops such as macadamia and many innovations. In this case, your point is correct. But we must clarify our ideas when you say eucalyptus kills grass. I have indeed seen grass being killed by eucalyptus in certain specific instances but not always. There are 500 species or more of eucalyptus, some of which are demanding of water, while others are drought-tolerant. Those which are water-demanding and present a very superficial root system do, indeed, deprive the substrate under the eucalyptus of water and therefore no grass grows. They are not killing it, but rather are just competing excessively for water. But there are other species of eucalyptus with great amounts of grass in the plantation, depending on the species and the amount of rainfall. One mistake is common. A species is introduced with some success, but subsequently that species is planted throughout the region and sometimes in exactly the wrong ecological sites. This is what has happened in Colombia. There are some sites where not one of the 500 species of eucalyptus should have been planted. But they are planted there nonetheless, and that is when adverse effects arise. This is a case of using better techniques. Extension agents should not promote such mistakes. They should, rather, recommend the right crop for the right place at the right time and sometimes for a relatively short time only, until better techniques are devised. I am asking for a critical review of eucalyptus, not a total condemnation, not an emotional total defense, but each case should be studied on its own merits.

Dr. F. R. Fosberg: I am strongly against eucalyptus for a number of reasons. In the first place, I don't think it is particularly effective in controlling erosion, and I feel there are other methods that are much better. I don't object because it is eucalyptus. Sometimes it is planted in the wrong place and I suspect that there are chemical substances produced by it that inhibit other plants. I've seen too much circumstantial evidence of this, although I haven't any analytical data or anything like that. My most important objection to eucalyptus is totally different from anything mentioned here, or that I've heard from anyone else and it applied to pine as well. The trouble is that foresters get in the habit of planting eucalyptus. It doesn't make any difference where or what, and they're perfectly willing to cut down the forest remnants in Hawaii and substitute eucalyptus when they have examples all around them demonstrating that the eucalyptus they are planting probably is not going to be any good for timber. Even if it were, I don't think we can afford to lose the little bits of native forest remaining after 200 years of deforestation. In Madagascar, Forest Service personnel showed me the one remaining bit of mid-elevation forest in the nation: a few acres, a few remnants of the native species that we are talking about, and I'll be darned if they didn't have pine planted all over this! They seem to be bragging about it! I think the fault is not with eucalyptus but with the foresters.

Mr. G. Lucas: There are other cases closely similar to Dr. Fosberg's case. The topic could be generally applied to where a plant is out of control. For example, *Acacia cyclops* in South Africa is, in fact, running up hillsides and it *is* removing all other plant species. It is a total effect, but these are usually accidental introductions. This is also true for *Akea.*

But *Acacia cyclops* doesn't just exclude plants; it competes with people as well! If a mud wall is constructed at the wrong season and incorporates seeds, humidity permeating the walls of the house will germinate the seeds very nicely on the inside, in large quantities, enough in fact to pull the wall down. So I don't think it's only the plants that are being pushed out, but Man himself in some situations. Plants are often in the wrong place accidently, or as Dr. Fosberg said, sometimes through sheer ignorance.

Dr. R. Goodland: The next question is from Dr. John Semple of Waterloo, Canada, who asks: How large an area of tropical forest must be preserved in order to contain a representative species sample?

Prof. P. Richards: This is a very important question. It is one of the great difficulties at the present time about conservation of tropical rain forest that the answer to this question is unclear. Of course, it depends upon whether you are talking about conserving the total ecosystem or the actual equilibrium. If a forest contains a herd of elephants, you have to make some estimate of the amount of land necessary to maintain a herd of elephants. If you think that the elephants are not necessary and you just want to include all the species of plants, then a smaller area might suffice. But this is not the right approach. There is a lot of ecological data which suggests that for some forests the presence of the elephants is necessary for the maintenance of the forest because of the clearances they make. Further information on this subject is beginning to appear. There has been some work done on birds in particular by Dr. John Terborgh from Princeton University. His data makes it possible to guess as to what sort of area of forest is the smallest that can be maintained for the survival of the total flora and fauna. One of the interesting findings from the same work shows that the famous Barro Colorado Island is, in fact, not large enough to be self-maintaining. It has actually lost many species of birds since it was established in about 1913, not because the species are persecuted, but because there simply is not a large enough population to maintain itself. Similar occurrences are reported in England. One of the special features of Wicken Fen Nature Reserve used to be the population of the swallowtail butterfly. During World War II, this population totally disappeared, probably not because of any deliberate persecution but simply as a result of ecological changes. The population was just not large enough to maintain itself. This sort of consideration also applies to tropical forest, and like so many of these questions, we just don't know enough to answer them. Few areas of tropical rain forest have a complete census of even the tree species, and some of those where we do suggest that quite large areas are needed. By quite large, I mean at least tens of square kilometers if we're going to protect the total ecosystem.

Dr. T. Whitmore: Well, there are various comments that one could make. The first one is: the more the better. I'm sure Professor Richards is absolutely right in saying that most critical — and first to go — are animals, especially those which require a very large area. Talking as I did earlier, in specific terms of the Malayan National Park, elephants over the year walk around in a huge circuit, and if part of that circuit is an oil palm plantation, then they will eat the oil palm. Tigers each need several square miles to exist. The National Park cannot support very many tigers, and one can discuss how many tigers are needed to maintain the Malayan race of them and then what size of park would be big enough to accommodate them. Now, I think that the sort of guideline that one should work on in thinking of conservation areas in tropical rain forest is that a small number of large areas is better than a large number of small areas, largely because of the number of animals involved. I think in many cases plant species would survive a lot of destruction of habitat, but even plant species do need to be *in situ* in order to maintain their population structure and breeding systems. In southeast Asia, there are very many fruit trees which have scarcely been selected away from their wild condition. Rambutan *(Nephelium lappaceum)* is an example. You can go and pick them up in primary forest, but you can find them also in the market. A survey in Malay[2] showed that the Rambutan

[2]Whitmore, T. C., 1971. Wild trees and some trees of pharmacological potential in the rain forest of Ulu Velantan. Malay. Nat. Jour. 24: 222-4.

trees grow at the rate of approximately thirty per square mile. In addition, they are never very common so if one is interested in genetic conservation, and if fruit trees are among the things one obviously wants to conserve, what is the area needed to conserve a representative sample of the total gene pool of *Nephelium lappaceum?* It must be a large number of square miles, possibly about 300. The answer to the question, therefore, must be that full conservation requires a very big area indeed. I think the other aspect, having said "right, we're going to have a few big areas rather than a lot of small ones," is that one would look for areas which include within themselves a range of different sorts of forests. The National Park of Malaya includes riverine forest, riverine swamp, lowland forest, and also forest right up to the tops of the highest mountains, on several different sorts of rock, so that in that Park there is a great diversity of ecosystems all occurring together.

Dr. G. Budowski: The question is more academic rather than of practical value. For example, how were reserves created? Which were the successful ones? What happened ten years later when the question was reiterated? Has the area really been made sufficiently large? Almost unavoidably, in every one of these biological reserves and national parks, the answer has been they have *not* been sufficiently large. What is more important is that it has always been found that some areas which have been added, perhaps with no good reason concerning the protection of one particular species, have always proven to be extremely useful in the long run. What I'm trying to say in another way is that we have not progressed very far. Decision-makers will allocate funds and infrastructure to create parks if some 'magical' number is supplied. First, because scientists themselves find so many variables in the sites they would hardly reach an agreement, or they may delay an agreement which could be fatal. Second, because there are so many other variables and so many other reasons for creating the park, which may have a much greater value than this specific one, that I would say use the most pragmatic approach to get the largest area from the very beginning. Experience in many countries, including Latin America, has shown that the greatest rewards are in the largest area, including the protection of specific species. But we should not use this argument alone. I feel we have to combine this with many other arguments and make the very strongest case for getting the largest surface area possible.

Dr. T. Whitmore: A final example of an area which is a lowland tropical rain forest found to be too small is the famous 'garden jungle' in the Singapore Botanical Garden, the type of locality where at least ten plant species exist. Flying over Singapore, a 154-foot- Dipterocarp-tree can clearly be seen surrounded by a leafy suburb. The area is about 200-by-300 meters and in the last 20 years, it has died on its feet. The biggest trees are Dipterocarpaceae and these trees are mortal: some have died, some have been struck by lightning; there have been no seed parents within reach. Instead of being gradually replaced, as they would be in a natural forest, by small seedling Dipterocarps growing in the gap created, there has been an invasion of open-ground pioneers. The tract is just literally disappearing before one's eyes because it's too tiny to maintain itself; there are not enough seed parents from the one or two living Dipterocarps when 20 or 30 Dipterocarps die. There are insufficient Dipterocarps present as mother trees to replace the ones that are there.

Dr. R. Goodland: The next question from Larry Morse of Harvard University concerns the Endangered Species List: Can you ensure the confidentiality of information supplied to you on endangered species population, for example, at specific localities and population sizes? Is this kind of information immune from release under the U.S. Freedom of Information Act?

Dr. B. MacBryde: We are beginning to face this problem. Legally, when we define a critical habitat in the Federal Register, we need to define some sort of boundaries. Legal precedence from archeological sites provides that the whereabouts need not be revealed beyond the county where it is located. Through the Freedom of Information Act, the Bureau would not do anything more than open the file within the office, then only to people with official permission. This would ensure some control so that the information doesn't get into the wrong hands.

Mr. J. Malter: The title of this session — "What Can Be Done to Save Plant Species in the American Tropics" — has prompted me to give some thought to the question of what can the people of the United States, especially professionals and people associated with the universities do to help? The general attitude at this Conference reflects a position approach to the solution of the gravest problems facing human society. As concerned scientists in the Americas, we must be cognizant of the fact that perhaps the greatest problem in saving endangered habitats and their biotic components lies not in the piece-meal creation of isolated preserves, but in the solution of social, economic, and political turmoil that besets South and Central America, as well as other areas throughout the world.

One important step towards developing better human relations and understanding among various social strata and ethnically-diverse populations *must* begin in the industrial nations. As the prime destroyer of habitats in the underdeveloped nations, the western countries defoliated much of their own lands in the centuries leading up to the mid-1800's. Having expanded into the tropics beginning in the latter half of the 19th century, they are now faced with the "beginning of the end" of areas for further desecration, as a result of modern industrial and agricultural development.

It is to these nations, particularly the United States, that the burden of repair and restitution must fall. The following proposals are offered as useful and immediately applicable suggestions:

1) Required conservation methods courses for university students, particularly those studying biology. This will lead to a greater number of professionally-trained personnel, educated to *conserve* the earth.

2) Large-scale government and private cooperative funding, similar to the Works Progress Administration in the United States in the 1930's. This would stimulate unemployed sectors of the populace, who would be initially educated in conservation and "para-scientific" methods (similar to paramedics), to be sent as a core of personnel under the supervision of scientists to work in tropical and adjacent areas. They could help train the even more poorly-educated peoples in the underdeveloped nations.

3) More funding of graduate students from industrial and non-industrial countries by the industrial nations for studies concerned with biological problems in the tropical and underdeveloped nations.

Much needs to be done and quickly. In 1868, George Marsh detailed the downfall of earlier civilizations — Greece, Rome, Egypt, Mesopotamia — due to ignorance of their destructive methods in agriculture, silviculture, and waste removal. With great insight, just at the beginning of the Industrial Revolution, he warned society against repeating its previous disasters caused by a lack of understanding of the natural world.

Today, with a better understanding of the problems that confront society, though these problems may be on a larger scale, they are in essence only of greater magnitude rather than basic. It is hoped that we will not lead ourselves down the same path of no return.

When one looks at the beauty of the forest, the sea, the infinite grace of the starry night, one can only hope and work, with an open mind and heart, to rebuild a world full of much goodness, and perhaps to even become one in which all humankind may lead fulfilling lives.

A STRATEGY FOR SAVING WILD PLANTS: EXPERIENCE FROM CENTRAL AMERICA

Gerardo Budowski

Head, Forestry Department, Tropical Agricultural Research and Training Center (CATIE), Turrialba, Costa Rica

Destruction of the tropical rain forest — that botanical paradise — is, from the very long-range point of view, one of the greatest tragedies of the twentieth century. It is among the saddest heritages that our generation will be handing down to the world's future citizens for centuries to come.

This destruction is all around us, for all to see. Each Central American country has its pathetically long list. There is the intense pace of timber exploitation, with the round logs of mahogany and Spanish cedar being shipped to the United States. The opening of the new roads in Costa Rica, and the removal of the last forest of hilltops there in order to extend sugarcane production continue unabated. Century-old oaks *(Quercus copeyensis* C. H. Muller, and other *Quercus* spp.) from Costa Rica's magnificent cloud forests, have been widely cut to produce charcoal. There can only be one ultimate result of these destructive practices: the exploited areas immediately become ripe for further devastation through shifting cultivation.

In Panama, jungles continue to be cleared for short-term cropping along the new road close to Portobelo, or along the river banks in Darien. In Honduras, the practice of burning pine stands *(Pinus oocarpa* Schiede) results in drab and plant-poor landscapes. In El Salvador, we are confronted with the encroachment of mangrove swamps, so critical for native fisheries. The rationale for this encroachment is that the areas are needed for the sun-dried salt mines, or for cotton planting (with its concomitant heavy use of insecticides). Overgrazing of drier areas in Nicaragua results in heavy erosion and the elimination of practically all the herbaceous vegetation. And, of course, we all are aware of the luxurious residential construction programs that endanger the highly vulnerable lakeshore vegetation around beautiful Lake Atitlan in Guatemala's highlands. This list can go on and on.

But the greatest destruction of all is caused by the continually accelerating pace of rain forest clearing in order to create pastures for extensive animal husbandry in all the lowland wet forests in five of the six Central American countries. El Salvador is excluded from the list because there are practically no wet forests remaining there. All were removed many years ago. Clearing the primary wet forest for extensive beef production is particularly destructive since most pastures established in areas of high rainfall (more than 2,500 mm, or about 100 inches) can hardly maintain themselves for more than a few years because of the ever-increasing invasion of woody weed species, favoured by soil compaction. But apparently in Central America, most governments are convinced that there are sufficiently large tracts still available to continue this kind of "shifting" pasture (Budowski, 1966).

The official justification, of course, for such shortsighted policies, is an economic one. Pasture establishment, even at the cost of destroying rain forest and with about one animal for every two to five hectares, is still a most profitable business because of the high price of lean meat paid by the United States for its hamburger consumption.

Certainly not all forest clearance is wasteful, however. Permanent settlements have a good chance to be successfully established in the cleared areas, assuming, of course, that soils are particularly good (fresh volcanic, rich alluvial, etc.), though we must bear in mind that such soils cover relatively small surface areas. But this pattern of land use is at best deceptive since the first settlers in any hitherto unpopulated area usually have an excellent inherent ecological knowledge, and, of course, they choose the very best soils for the first opening of colonization. Their success then attracts successive waves of settlers who occupy what is left, which usually means the much poorer lands with bad soils or steep slopes. Here is where the real damage occurs.

This process is actually exacerbated by archaic laws still on the books, which in effect penalize landowners who keep the forest intact. Taxes are levied against these landowners while laws continue to favour those who clear the forests, irrespective of climate, slope, soil, or other environmental conditions and without any regard for the fragility of the land itself.

How can we prevent further and, perhaps, even more widespread destruction? What can we do to avoid this irrevocable loss of forests, of plants, of our cherished tools that give us our reputation, our very "raison d'être" which justifies our profession as botanists? If we are to be more than voices crying in the wilderness, if we want to save species for all time, we must take a more careful and closer look at the reality of the man/plant relationship. We must find practical and effective ways of influencing today's decision-makers. We must promote and support viable solutions. In short, we no longer can continue to live in an ivory tower. We must adapt or die while witnessing the disappearance of "our" plants. The time has come for us to change our strategy!

A NEW STRATEGY FOR PLANT CONSERVATION

Our success will be measured by our ability to find allies in the countries where plants are disappearing. This does not mean that we must become political activists or that we have to play the politician's game, or perform in their circus and surrender our scientific integrity. But it does mean that we must use our scientific knowledge and experience as potent weapons with which to influence these politicians, decision-makers, and educators, as well as professionals like agronomists, dam builders, fishermen, and foresters, among others. We must try to influence them all, to get them on our side. But in order to do this, we also must arrive at a new strategy, a new approach.

The following are a few suggestions towards achieving this goal. No pretense is made for an exhaustive treatment; rather the intention here is to stimulate thinking and to promote remedial action by providing some examples so that we can better assess the challenges which lie ahead.

LAND USE, THE KEY FACTOR

Examination of land use capabilities in Central American countries — by applying the criterion of the land's ability to sustain permanent utilization for field crops, grass for cattle, or trees for timber production, without destroying the soil capital — reveals the following:

a) Possibly no more than 10-15 percent of the land area in Central America now qualifies for crops;

b) An additional 10-20 percent, mostly in the dry-wet region which originally was deciduous forest, can possibly be used for cattle-raising;

c) No more than 20 percent qualifies for timber production.

Again, let me emphasize that in the case of forest exploitation, only *sustained yield* is implied. This must not be confused with high-grading, which currently is the case of 95 percent of the timber exploitation, as practiced in the mixed broadleaf forests. Pine forests admittedly are different.

Probably not more than 50 percent of the total land remains — let us call it "wildland" — with no apparent lucrative economic use. But we know better. We know that those lands are certainly useful for watershed protection, which of course is of paramount importance to cities, dams, flood and drought control, and irrigation schemes. We know that these "wildlands" can be utilized for scientific endeavor, recreation, health (air purification), education (living laboratories), maintenance of a natural equilibrium and, last but not least — these lands can be used for the preservation of an immense treasure of genetic material of plants and animals, some of which already are ominously threatened with extinction.

Let me reiterate the question: "Can we use those wildlands?" Some products and services, of course, can be converted into economic values but it is utterly impossible to place a price on some of the others even though they are very valuable to us.

369

Who manages, or otherwise influences, land use on those huge wildlands which cover the wet areas, the steep slopes, the swamps, and the rocky lands? Sadly, the people most responsible for this degradation include agronomists, cattle specialists, and foresters themselves. For example, agronomists speak of better machines to clear the land. They propose fertilizers, pesticides and "miracle" seeds, but they are largely unable — at least area-wise — to convert their theoretical promises into practical schemes. Cattle specialists promote aggressive new African grasses, better pasture management methods, and better breeds of cattle. But so far they have largely failed in the wet areas, which cover more than 50 percent of all of Central America. Foresters vaguely refer to improvement cuttings and regeneration schemes, but they also have, heretofore, been unsuccessful in improving the primary mixed stands. Silviculture, in fact, has fared much better in secondary forests or through plantations.

Yet these wildlands need to be managed, and even more important, the pressure on them can be relieved. We can encourage family planning and I am all for that, but we also can do other things. We can broaden our concept of carrying capacity, for example, by promoting leak-proof systems and by favouring promising alternatives. And above all, we can rally potential allies in other professions. Let me cite some examples:

1. *Forest plantations to relieve the pressure on wildlands*
Instead of verbally blasting the concept of pine and eucalyptus plantations, let us welcome them as long as they are not established at the expense of the natural forests. There are, indeed, plenty of degraded savannas or ruined croplands that still are available for tree planting — remember São Paulo's receding coffee slopes or overgrazed hillsides! Pines, with their mycorrhizae, can thrive where other plants do very poorly. Various species of *Eucalyptus* provide fuel and poles in a comparatively short time (2-6 years, according to the species and the site), and they produce 10-50 times more utilizable wood than do the primary forests. By using these arguments, we, as plant conservationists, can make important allies of foresters instead of having them work against us. We can offer them this conservation-oriented alternative which involves tree planting — trees established close to settlements that will *relieve the pressure* on wildlands (see Budowski, 1975a).

2. *Why should new land be opened when food production on much of the presently settled land could be considerably increased at a much lower cost?*
First, we must admit that it is politically expedient to promote new colonization schemes. It provides a heroic "pioneer" image. Moreover, financial support is likely to come from banks and international organizations, and one can surely count on the wholehearted support of the large landowners, who fear the obvious "land reform" on their huge properties. "Why divide my land for those hungry farmers," they claim, "when there is plenty of virgin territory that is only awaiting man's technology to be opened?" Some of the best soils in Central America are presently the most poorly used. Large tracts of alluvial and level lands are still too frequently managed for beef cattle for the benefit of a few people. Better farming methods on good soils will also *relieve the pressure* on what are clearly marginal lands (Budowski, 1971).

3. *Delineating the unique or exceptional areas to be managed for conservation.*
A few weeks ago, the Central American Development Bank loaned two million United States dollars for the development and protection of the Volcán Poás National Park in Costa Rica. This includes areas rich in endemics, and, generally speaking, one of the most outstanding cloud and "elfin" forests in the region.

In 1974, over one million people in Puerto Rico visited the Luquillo National Forest to have a glimpse of the "rain forest," a good part of which is of secondary growth although still interesting and esthetically impressive. This provided a tremendous boost to the local economy, and 1974 was a typical year. Obviously, managing for conservation is not only possible but profitable, and it also leaves the options open for future land use decisions.

Botanists can help zoologists, geologists, archeologists, hydrologists, limnologists, and

370

fellow biologists in carefully selecting outstanding areas that provide exceptional benefits. Managing for conservation can also be described as promoting the right kind of "development," be it recreational, aesthetic, scientific, educational, for better physical and mental health, for watershed and soil protection, or for gene pool preservation (Budowski, 1972). This was witnessed at the recent Central American meeting on the conservation of natural and cultural resources held in San José, Costa Rica in December 1974 (IUCN, 1975b).

Another promising approach is to promote a consensus as to what lands are clearly not manageable on a sustained yield basis for food or timber production. The very rainy, the very steep, the very dry, or the very swampy areas qualify in this category.

All the Central American countries (as well as many South American and Caribbean countries) have been mapped according to the Holdridge life zone system. An analysis of habitats in need of preservation has also been discussed for Central America (Budowski, 1965) and a series of principles for economic development has been designed for the tropics (Dasmann et al. 1973). These useful tools for better land use planning are already being employed by a few farsighted economists and planners. There are indications that the present widespread policy of laissez-faire or continuation in projecting the past into the future is quickly changing. It is, of course, high time.

4. *Guidelines for development*

Most Latin American conservationists are forced to admit, unlike those in the United States, that it is not only futile, but also sometimes outright dangerous, to fight "development schemes" such as roads, dams, fertilizers, pesticides, housing projects, and the like. Rather, it is more rewarding to redefine development, or at least to guide it along ecological principles, and this can be most effective. Recently-established ecological guidelines for development in the American humid tropics (IUCN, 1975a) are increasingly being used in Central America and elsewhere in the tropics. As a matter of fact, the short version of the guidelines produced in Spanish ("normas ecológicas"), have become standard teaching and reference materials in various universities and planning offices of Central America and elsewhere.

GETTING BOTANY TO BE RECOGNIZED AS A USEFUL TOOL

Various aspects of botany could play a much more productive role in serving the cause of plant conservation. Those aspects connected with ecological relationships, for example, are receiving far too little attention today in spite of our increased awareness of the problem. It is, therefore, essential that attention now be focused on the usefulness and applicability of current botanical knowledge. Many ideas could be developed by us, working jointly with scientists from other disciplines. Examples are: the use of lichens as indicators to map air pollution levels close to cities or factories; helping national park managers to interpret attractive nature trails based on understanding botanical features; and generally emphasizing any exceptional botanical aspect in a region.

Botanists can also help in working out, together with meteorologists, the importance of cloud forests in trapping horizontal fog and increasing rainfall. They could cooperate with ornithologists, mammalogists, herpetologists, entomologists and other scientists in ascertaining the unique plant-animal relationships. Another attractive field, of course, is ethnobotany, with its immense treasure of untapped knowledge. Botanists can help save mangroves from being logged or drained. This can be accomplished by teaming up with fishery experts in focusing attention on the importance of leaf fall as the basis for a network of food-chains that are vital to the fishing industry over large coastal areas. How many botanists in Central America and elsewhere have contributed in writing impact assessments on proposed large-scale habitat modification?

Clearly, if botanists want to contribute to saving plants, they must come out from their narrow specialization. There are many non-botanists who will welcome this input from those who deal with the basic plant material, the essentials of life in its many manifestations.

Even the small farmer's lot can be improved by more botanical involvement. Let me cite only one example: more than 50 species of trees and shrubs are being successfully used in Central America for live fences, yet practically nothing is known about the botanical variation and morphology of branch development from large cuttings, the growth habits, the root behaviour, the ability of some species to fix nitrogen or otherwise improve the soil, and so on.

Floras based on strict taxonomical criteria will, of course, always be indispensable tools for a comprehensive understanding of plants in a region. But popular books for the general public, with good illustrations, are also badly needed in Central America as well as elsewhere in Latin America, since they appeal to a much wider public. The few books on trees of Costa Rica and Panama have already had a considerable beneficial impact.

Much can also be gained in publishing notes and comments on famous trees with exceptional diameter and height, strange growth, sheer beauty due to striking flowers, cauliflory habits, association with ants, or even because of their historical values.

HOW MAY EXTERNAL SUPPORT BEST BE GEARED?

In Central America as elsewhere in the American tropics, changes concerning attitudes and ways of handling natural vegetation will come about only if local leadership is intent on supporting those changes. In light of the pressing needs for increased food production, the educational and housing problems, and the continual struggle against poverty, it is unreasonable to assume that Governments will take remedial action, unless it is carefully engineered, particularly from within the country.

Therefore, United States and European-based organizations wishing to promote changes must be very careful in supporting *the right* individuals or projects. They can easily go wrong and be accused of scientific imperialism (Budowski, 1975b), if they do not demonstrate some sensitivity towards local needs and local pride. There is no single golden rule to follow in this respect except that whatever time and effort is spent to identify the most likely future leaders in plant conservation, is time well worth spending. Some imaginative action to follow this up could be, for example, to support the establishment of national parks and reserves proposed by local conservationists, to promote lectureships in local universities on the importance of nature conservation. The Universidad Nacional de Heredia in Costa Rica, for example, with its dedicated group of young conservation-minded professors, would be an excellent investment. It makes a tremendous difference if money can be forthcoming to publish and support educational activities by pointing out such things as: the importance of gene reservoirs, designing posters of endangered plants, producing filmstrips or comic strips that impart a conservation message, and so on. Outside support for worthwhile plant conservation projects with funds and moral backing is a most rewarding activity.

Many Central American conservationist groups ignore the fact that there are concerned organizations in the richer countries which would be willing to back them up, both morally and financially, but they have no idea how or to whom they should write for support. They have their own pride too, and this must be understood. Their means are modest and their voice is not loud. They have discovered, unfortunately, that speaking out against the present onslaught on natural forests can be politically dangerous.

It will pay handsomely if we take the initiative to identify these groups, which so desperately need our support and backing. In years to come, it will be they and not the outsiders who will be responsible for the management of their nation's precious botanical heritage. Strengthening such groups by training and research activities should, therefore, become an indispensable ingredient in any strategy for saving wild plants.

In summary, the present destructive trends can only be stopped if an imaginative concerted approach replaces the presently ineffective and negative method of merely complaining and denouncing. Such change of strategy demands sensitivity and appreciation of the poor peasant's problems, influencing land use policies, and carrying out assistance programs by strengthening the position of local conservation leaders.

LITERATURE CITED

Budowski, G. 1965. The choice and classification of natural habitats in need of preservation in Central America. Turrialba **15**: 238-246.

_____. 1966. Las colonizaciones de regiones húmedas en América Latina y sus implicaciones forestales. Actas 6° Congreso Forestal Mundial, Madrid, Tomo 3. pp. 3143-3148.

_____. 1971. Necesidad de planeamiento paisajista al incorporar áreas naturales al desarrollo agropecuario en los países tropicales. Defensa de la Naturaleza (Venezuela) N° **2**: 23-29.

_____. 1972. Conservation as a tool for development in tropical countries Geoforum (Germany) **10**: 7-14.

_____. 1975a. Cooperation between forestry and conservation of nature. Second Congress of the International Union of Societies of Foresters. Proceedings. Helsinki, Finland. 19-24 August 1974. pp. 79-88.

_____. 1975b. Scientific imperialism. Unasylva (FAO, Rome) **27**: 24-30.

Dasmann, R., J. Milton and P. Freeman. 1973. Ecological principles for economic development. Published for IUCN and the Conservation Foundation by John Wiley and Sons, New York and Chichester. 252 pp.

IUCN. 1975a. Ecological guidelines for development in the American humid tropics. Compiled by Duncan Poore. IUCN Occasional Paper N° 11. Morges, 1975. 38 pp.

_____. 1975b. Recommendations of the Central American meeting on management of natural and cultural resources. San José, Costa Rica, 9-14 December 1974. Special Supplement of IUCN Bulletin 6(2). 1975. 4 pp.

RESOLUTIONS

The Following Resolutions Were Accepted
Unanimously By The Symposium Participants

1. This Conference urges all governments to ratify the convention on international trade in endangered species of wild fauna and flora as soon as possible.
2. This Conference, aware of the urgent need for scientifically verified lists of threatened species on a worldwide scale, calls for and pledges its full support for the International Union for Conservation of Nature and Natural Resources (IUCN) Threatened Plant Committee in the compiling of such lists, recognizing this as a vital first step in many conservation action programs.
3. This Conference calls for the widest publicity for its deliberations, particularly in decision-making circles where future action lies, to ensure the survival of the largest number of plant species, and to conserve this unique economic, aesthetic, scientific, and educational heritage for the present and future benefit of mankind.
4. This Conference resolves that all development projects should be obliged to inform and involve local, national (and if necessary, international) scientific institutions in some form of environmental assessment early in the planning process.
5. This Conference urges the government of the United States of America to consider the amendment of the International Cooperation Section of their Endangered Species Act, so that provisions therein will apply equally to plants, fish, and wildlife where such application is appropriate.
6. This Conference, cognizant of the fact that the survival of at least ten percent of the plant species and their habitats may be in jeopardy, and in view of the unequal consideration of plants and animals in the Endangered Species Act, recommends that botanists, through appropriate recognized professional societies, should offer their services to governmental bodies in order to amend the said Endangered Species Act in the interest of plant species preservation.
7. This Conference, recognizing that the increasing pressure of the human population imperils the remaining natural areas on earth, and causes the extinction of at least some plant and animal species each year, and aware that the flora in many places on earth is yet poorly known, urges each government to engage botanists and ecologists in all appropriate plans and programs for development. All conferees offer their services in this regard. Furthermore, in view of the limited number of botanists and ecologists and the concentration of the knowledge about the survival of plants among these few, and in view of the magnitude and severity of the problem of plant conserva-tion and the value of plants in ecosystems and to mankind, this Conference also urges, in the strongest possible terms, that all governments give full and adequate support to botanical conservation, education, and research.
8. This Conference encourages and supports a broad natural-history approach to the study of threatened and endangered plant species, particularly with respect to the understanding and conservation of such pollinators, fruit or seed dispersers, fungal symbionts and other organisms necessary to the continued existence of these plant species as viable breeding populations.
9. This Conference urges all appropriate governments to ratify and implement fully the Convention on Nature Protection and Wildlife Preservation in the Western Hemisphere.

SPECIAL RECOMMENDATION

The following recommendation was also passed unanimously

The participants of this symposium urge the Governor of the State of Bahia, Brazil, to set aside the Una region as a reserve to preserve one of the last remnants of the species-rich coastal forest of that region.

RESOLUCIONES

Las siguientes resoluciones fueron aceptadas unánimemente por los participantes del Simposio

1. Esta conferencia urge a todos los gobiernos a que ratifiquen el convenio internacional sobre el comercio en especies de la flora y la fauna silvestre en peligro de extinción a la mayor brevedad posible.

2. Esta conferencia, conciente de la urgente necesidad de obtener a nivel mundial listas científicamente verificadas de las especies en peligro de extinción, hace un llamado y se compromete a apoyar al "Comité sobre Plantas en Peligro de Extinción" de la Unión Internacional para la Conservación de la Naturaleza y los Recursos Naturales (IUCN) en la preparación de las citadas listas, reconociendo ésto como el primer paso vital en muchos programas de acción conservacionista.

3. Esta conferencia hace un llamado para una amplia diseminación de sus deliberaciones, particularmente en la toma de decisiones y acciones específicas para asegurar la supervivencia del mayor número de especies y conservar este acerbo único, económico, estético, científico y educativo para el presente y futuro beneficio de la humanidad.

4. Esta conferencia propone que todos los proyectos de desarrollo se vean obligados a informar y a trabajar en conjunto con las instituciones científicas a nivel local, nacional (oy si es necesario internacional) para evaluar el impacto de tales obras sobre el medio-ambiente y así asegurarse un proceso de desarrollo planificado.

5. Esta conferencia urge al gobierno de los Estados Unidos de Norte América a considerar la enmendación de la Sección Internacional de cooperación del Acta de las Especies en Peligro de modo que las medidas halladas en él se apliquen con igualdad a las plantas, peces y vida silvestre donde tales aplicaciones sean más necesarias.

6. Esta conferencia, conciente del hecho de que la supervivencia de por lo menos un diez por ciento de las especies de plantas y sus hábitats se hallan en peligro y en vista de la desigual consideración de las plantas y los animales en el Acta de las Especies en Peligro de Extinción, recomienda que los científicos botánicos ofrezcan sus servicios a los organismos de gobierno a través de las Sociedades Profesionales debidamente reconocidas con el propósito de enmendar el citado Acta para promover la preservación de las especies de plantas.

7. Esta conferencia, reconociendo que la presión cada vez mayor de la población hace peligrar las áreas naturales restantes de la tierra y causa la extinción por lo menos de algunas plantas y animales cada año, advierte que la flora en muchos lugares de la tierra es todavía pobremente conocida, por lo que considera urgente que cada gobierno comprometa a botánicos y ecólogos en todos los planes y programas de desarrollo. Todos los conferenciantes ofrecen sus servicios para este fin. Además, dado el número limitado de botánicos y ecólogos y la concentración de los conocimientos sobre la supervivencia de las plantas entre éstos pocos científicos y en vista de la magnitud y gravedad de éste problema de conservación de plantas y su valor en los ecosistemas y a la humanidad, ésta conferencia también urge a todos los gobiernos que proporcionen su apoyo decidido y adecuado para la conservación, educación e investigación de los recursos vegetales.

8. Esta conferencia estimula y apoya con sentido amplio la Historia Natural como medio de aproximación al estudio de las especies de plantas en peligro, particularmente con respecto a un conocimiento y conservación de los agentes polinizadores, de los diseminadores de frutos semillas, de los hongos simbiontes y de los otros organismos necesarios para la existencia continua de estas especies como poblaciones procreadoras.

9. Esta conferencia urge a todos los gobiernos a propiados a que ratifiquen e implementen en su totalidad el Tratado sobre la Protección de la Naturaleza y Preservación de la vida silvestre en el hemisferio occidental.

RECOMENDACIÓN ESPECIAL
La siguiente recomendación también fue aceptada unánimemente

Los participantes de este Simposio urgen al Gobernador del Estado de Bahia, Brasil, disponer que la región de UNA sea una reserva para preservar uno de los últimos relictos del bosque costanero de dicha región que es rica en especies.

LIST OF CONTRIBUTORS

Paulo de Tarso Alvim

Cacau Research Institute, CEPLAC
Itabuna, Bahia, Brazil

Dárdano de Andrade-Lima

Instituto de Pesquisas Agronomicas
Recife, Pernambuco, Brazil

George Argus

Botany Division, Museum of Natural Sciences
Ottawa, Ontario, Canada

Lyman Benson

Pomona College
Claremont, California

Gerardo Budowski

Centro Tropical de Investigación y Enseñanza
Turrialba, Costa Rica

Angel L. Cabrera

Instituto Darwinion
Buenos Aires, Argentina

Kenton L. Chambers

Oregon State University
Corvallis, Oregon

William Countryman

Aquatec, Inc.
South Burlington, Vermont

Theodore J. Crovello

Notre Dame University
Notre Dame, Indiana

William G. D'Arcy

Missouri Botanical Garden
St. Louis, Missouri

Thomas S. Elias

The Cary Arboretum
Millbrook, New York

Alvaro Fernández-Pérez

Instituto de Ciencias Naturales de la
Universidad Nacional
Bogotá, Colombia

Ramon H. Ferreyra

Universidad Nacional Mayor de San Marcos
Museo de Historia Natural
Lima, Peru

George W. Folkerts

Auburn University
Auburn, Alabama

Alwyn H. Gentry

Missouri Botanical Garden
St. Louis, Missouri

Arturo Gómez-Pompa

Instituto de Investigaciones sobre Recursos
Bióticos a.c. (INIREB)
Xalapa, Veracruz, Mexico

Robert J. Goodland

The Cary Arboretum
Millbrook, New York

James W. Hardin

North Carolina State University
Raleigh, North Carolina

Calvin J. Heusser	New York University Tuxedo, New York
Richard A. Howard	Harvard University Cambridge, Massachusetts
Howard S. Irwin	New York Botanical Garden Bronx, New York
Marshall C. Johnston	University of Texas Austin, Texas
Klaus Kubitzki	Institut für Allgemeine Botanik und Botanischer Garten Hamburg, Federal Republic of Germany
Charles R. Long	New York Botanical Garden Bronx, New York
Grenville Ll. Lucas	Royal Botanic Gardens Kew, Richmond, Surrey, Great Britain
Bruce MacBryde	U.S. Fish and Wildlife Service U.S. Department of the Interior Washington, D.C.
Antonio Dantas Machado	Conselho Nacional de Desenvolvimento Científico e Tecnológico Brasília, D.F., Brazil
Meryl A. Miasek	New York Botanical Garden Bronx, New York
John T. Mickel	New York Botanical Garden Bronx, New York
Robert H. Mohlenbrock	Department of Botany Southern Illinois University Carbondale, Illinois
Harold E. Moore, Jr.	Cornell University Ithaca, New York
Carlos Muñoz Pizarro	Universidad de Chile Santiago, Chile
João Murça Pires	Museu Goeldi Belém, Pará, Brazil
Ghillean T. Prance	New York Botanical Garden Bronx, New York
Pierfelice Ravenna	Universidad de Chile Santiago, Chile
James Reveal	University of Maryland College Park, Maryland

Elias R. de la Sota

La Plata University
La Plata, Argentina

Julian A. Steyermark

Instituto Botanico
Caracas, Venezuela

Andrew P. Vovides

Instituto de Investigaciones sobre recursos
  Bióticos a.c. (INIREB)
Xalapa, Veracruz, Mexico

Carl Withner

Brooklyn College
Brooklyn, New York

George M. Woodwell

Ecosystems Center, Marine Biological Laboratory
Woods Hole, Massachusetts

LIST OF PARTICIPANTS

Rupert C. Barneby

New York Botanical Garden
Bronx, New York

Harriette V. Bartoo

Western Michigan University
Kalamazoo, Michigan

Barbara L. Bentley

State University of New York
Stony Brook, New York

Cory W. Berish

Yale University
New Haven, Connecticut

Dorrell Biddle

Camden County College
Blackwood, New Jersey

Karen Blumer

Bellport, New York

Frank J. Bonacconso

University of Maryland, European Division
APO, New York

George Bookman

New York Botanical Garden
Bronx, New York

Lynn Boshkov

Montreal, Quebec, Canada

Joan E. Canfield

University of Washington
Seattle, Washington

Susan Carey

New York Botanical Garden
Bronx, New York

Steve Carpenter

New York Botanical Garden
Bronx, New York

C. Ronald Carroll

State University of New York
Stony Brook, New York

Chester Chambers

New York Botanical Garden
Bronx, New York

Nona Chiariello

National Wildlife Federation
Washington, D.C.

Dean Christianson

University of Bridgeport
Bridgeport, Connecticut

Roland C. Clement

National Audubon Society
New York, New York

Elmire L. Conklin

Warwick, New Jersey

Claire Cooney-Souetts

McGill University
Montreal, Quebec, Canada

Lidio Coradin

New York Botanical Garden
Bronx, New York

381

Frank J. Coyle	Greenwich, Connecticut
Helen Crispe	Brooklyn Botanic Garden Brooklyn, New York
Garrett E. Crow	University of New Hampshire Durham, New Hampshire
Thomas J. Delendick	New York Botanical Garden Bronx, New York
Kent Dumont	New York Botanical Garden Bronx, New York
David Fairbrothers	Rutgers University New Brunswick, New Jersey
Kathyrn Fiske	Burnt Hills, New York
Milan D. Fiske	Burnt Hills, New York
Emily Ford	Greenwich, Connecticut
F. R. Fosberg	Smithsonian Institution Washington, D.C.
Jane Frick	Lincoln, Massachusetts
Hellen Gelband	Cary Arboretum Millbrook, New York
John H. Haines	New York State Museum and Science Service Albany, New York
David Harrington	New York Botanical Garden Bronx, New York
Tom D. Hayes	Yale University, Forestry Department New Haven, Connecticut
L. R. Heckard	Jepson Herbarium, University of California Berkeley, California
Josephine Henry	Henry Foundation of Botanical Research Gladwyne, Pennsylvania
Annette Hervey	New York Botanical Garden Bronx, New York
Robert Hill	New York Botanical Garden Bronx, New York
Noel H. Holmgren	New York Botanical Garden Bronx, New York
Patricia Holmgren	New York Botanical Garden Bronx, New York

Mary Y. Hough — Rutgers University, New Brunswick, New Jersey

Joel Huang — New York Botanical Garden, Bronx, New York

Phoebe Hunter — New York Botanical Garden, Bronx, New York

Jacquelyn Kallunki — New York Botanical Garden, Bronx, New York

Helen Kennedy — Field Museum of Natural History, Chicago, Illinois

Joseph H. Kirkbride, Jr. — Smithsonian Institution, Washington, D.C.

Alfred C. Koelling — Illinois State Museum, Springfield, Illinois

Ronald F. Kujawski — Simon's Rock Early College, Great Barrington, Massachusetts

Bettie E. Lauchis — Olu Pua Botanical Garden, Kalaheo, Hawaii

Elizabeth Lawson — University of Texas at Austin, Austin, Texas

Grace E. Lotowyca — Planting Fields Arboretum, Oyster Bay, New York

Thomas Lovejoy — World Wildlife Fund, Washington, D.C.

James L. Luteyn — New York Botanical Garden, Bronx, New York

Gary Lyons — Huntington Botanical Gardens, Altadena, California

Shannon Lyons — Huntington Botanical Gardens, Altadena, California

Bruce McAlpin — New York Botanical Garden, Bronx, New York

Paul J. M. Maas — Institute for Systematic Botany, Utrecht, Netherlands

Antônio Dantas Machado — Conselho, Nacional de Desenvolvimento Científico e Tecnológico, Brasília, D.F., Brazil

Jeffrey Malter — University of Tennessee, Knoxville, Tennessee

Paul S. Mankiewicz

New York Botanical Garden
Bronx, New York

Frances Maroncelli

New York Botanical Garden
Bronx, New York

Fred Merrill

Woodstock, Vermont

Joan Moody

National Parks and Conservation Association
Washington, D.C.

Scott A. Mori

New York Botanical Garden
Bronx, New York

Larry E. Morse

Gray Herbarium
Cambridge, Massachusetts

David F. Murray

University of Alaska
Fairbanks, Alaska

Bruce Nelson

University of Maine
Orono, Maine

Lorin I. Nevling, Jr.

Field Museum of Natural History
Chicago, Illinois

Karl Niklas

New York Botanical Garden
Bronx, New York

R. Henry Norweb, Jr.

Holden Arboretum
Mentor, Ohio

Peter J. O'Connor

New York Botanical Garden
Bronx, New York

T. D. Pennington

NITBOT, Clarke's Farmhouse
Northmoor, Oxon, Great Britain

Richard W. Pippen

Western Michigan University
Kalamazoo, Michigan

Frank Pollach

New York Botanical Garden
Bronx, New York

John Popenoe

Fairchild Tropical Gardens
Miami, Florida

John Reed

New York Botanical Garden
Bronx, New York

William A. Rhoads

E.G. and G. (ERDA)
Goleta, California

Edward R. Ricciuti

Connecticut Zoological Society
Killingworth, Connecticut

Paul W. Richards

University College of North Wales
Bangor, Gwynedd, Great Britain

Clark Rogerson — New York Botanical Garden
Bronx, New York

Irwin Rosenblum — New York Botanical Garden
Bronx, New York

Joseph Salerno

Martha S. Salk — Oak Ridge National Laboratory
Oak Ridge, Tennessee

Jack E. Schmautz — U.S. Forest Service
Arlington, Virginia

Eileen K. Schofield — New York Botanical Garden
Bronx, New York

Anne O. Seaman — White Plains, New York

John C. Semple — University of Waterloo
Waterloo, Ontario, Canada

Kathleen Shea Semple — University of Guelph
Ontario, Canada

Samuel P. Shaw — U.S. Forest Service
Downingtown, Pennsylvania

Nancy Shopes — Curator of Instruction,
Brooklyn Botanic Garden
Brooklyn, New York

Alex Shoumatoff — Marsh Memorial Sanctuary of Wildlife
Preserves, Inc.
Mt. Kisco, New York

Gary Smith — New York Botanical Garden
Bronx, New York

Victor G. Soukup — Vic Soukup and Associates
Cincinnati, Ohio

William C. Steere — New York Botanical Garden
Bronx, New York

James Stevenson — The Cary Arboretum
Millbrook, New York

A. L. Stoffers — Institute for Systematic Botany
Utrecht, Netherlands

Edward E. Terrell — Agricultural Research Service
U.S. Department of Agriculture
Beltsville, Maryland

Linda Thacher — McGill University Herbarium
Ste. Anne de Bellevue
Quebec, Canada

385

William L. Theobald

Pacific Tropical Botanical Garden
Lawai, Kauai, Hawaii

Robert Tillman

The Cary Arboretum
Millbrook, New York

Stephen Tim

Brooklyn Botanic Garden
Brooklyn, New York

Timothy C. Whitmore

Commonwealth Forestry Institute
Oxford, Great Britain

Paulo Gunter Windisch

Harvard University
Cambridge, Massachusetts

Dennis W. Woodland

McGill University Herbarium
Ste. Anne de Bellevue
Quebec, Canada

Appendices

Appendix I

ENDANGERED PLANT SPECIES OF THE WORLD AND THEIR ENDANGERED HABITATS: A SELECTED BIBLIOGRAPHY

Compiled by

C. R. Long and M. A. Miasek

Library of the New York Botanical Garden, Bronx, New York 10458, U.S.A.

June 1976

Allen, Robert. 1975. Flower smuggler, drop that pistil. New York Times. 31 May.

Arbose, Jules. 1974. To the rescue of endangered plants. New York Times. 8 December.

Ayensu, E. S. 1975. Endangered and threatened orchids of the United States. Amer. Orc. Soc. Bull. **44**: 384-394.

Benedict, R. C. 1928. How shall we save rare plant species from extinction? Wild Flower **5**: 45-46.

Benson, Bruce. 1976. Endangered plants list draws fire. Honolulu Advertiser. 4 February.

Benson, L. 1975. Cacti-bizarre, beautiful, but in danger. Natl. Park and Conserv. Mag. **49**(7): 17-21.

Birkmane, Kornelija. 1974. Protected plants in Latvia [in Russian]. Zinatne, Riga. 58 pp.

Bosackova, Eva. 1972. Actual state and protection of the low moor vegetation in Zitnij ostrov [in Russian]. Slovensky ustav pamiatkovej starostlivosti a ochrany prirody, Bratislava. 82 pp.

Boudet, G. 1972. Desertification de l'Afrique tropicale seche. Adansonia, Ser. 2, **12**: 505-524.

Boughey, A. S. 1960. Man and the African environment. Proc. Trans. Rhod. Sci. Assoc. **48**: 8-18.

Boyko, H. 1951. On regeneration problems of the vegetation in arid zones. Colloq. Inst. Union Biol. Sci. Ser. B, **9**: 62-80.

Bureau of Sport Fisheries and Wildlife. 1972. Proposed endangered species conservation act of 1972. Draft environmental impact statement. Washington, D. C. 27 pp.

_____. 1973. Proposed endangered species conservation act of 1972. Final environmental impact statement. Washington, D. C. 309 pp.

Burt, DeVere E. 1973. The geography of extinct and endangered species in the United States. The Explorer **15**(3): 4-10.

Chiariello, Nona. 1976. Plant endangerment and ecological stability. Conserv. News **41**(13): 10-13.

Comptes rendus de la conference internationale sur la consérvation de la nature et de ses ressources à Madagascar, Tananarive, Malagasy Republic, 1970. 1972. IUCN, Morges [Switzerland]. 239 pp.

The conservation of wild creatures and wild plants act 1975. 1975. Habitat **11**(8): 1-3.

Convention on international trade in endangered species of wild fauna and flora, Washington, D. C. 1973. H.M.S.O., London. 43 pp.

Copyk, V. I. 1970. Scientific grounds for protection of rare species of Ukranian flora. Ukr. Bot. Zh. **27**: 693-704.

Costin, A. B. and R. H. Groves, eds. 1973. Nature conservation in the Pacific: proceedings of symposium A-10, XII Pacific Science Congress, 1971. IUCN and Aust. Natl. Univ. Pr., Canberra. 337 pp.

Cottam, W. P. 1929. Man as a biotic factor illustrated by recent floristic and physiographic changes at Mountain Meadows, Washington County, Utah. Ecology **10**: 361-363.

Cowan, I. McTaggart. 1973. Vanishing species: habitat change and reconciling conflict, pp. 321-333. *In* Hugh F. I. Elliott, ed., Conservation for development. IUCN 12th techn. meeting Banff, Alberta, Canada, 1972. IUCN, Morges, Switzerland.

Cox, Clinton. 1976. Our unknown vanishing breeds. The Sciences **16**(2): 21-24.

Cronquist, Arthur. 1971. Adapt or die. Bull. Jard. Bot. Nat. Belg. **41**(1): 135-144.

de Vogel, E. F. 1976. IX Tropical Orchids as an endangered plant group. Fl. Males.
Bull. **29**: 2602-2604.

Deneven, W. M. 1973. Development and the imminent demise of the Amazon rain forest.
Prof. Geogr. **25**: 130-135.

Dodson, Calaway H. 1968. Conservation of orchids, p. 170. *In* Proceedings of the
Latin American Conference on the conservation of renewable natural resources,
San Carlos de Bariloche, Argentina, 1968. IUCN, Morges, Switzerland.

Drury, William H. 1974. Rare species. Biol. Conserv. **6**: 162-169.

Du Mond, David M. 1973. A guide for the selection of rare, unique and endangered
plants. Castanea **38**: 387-395.

Duncan, Wilbur H. 1973. Endangered, rare and uncommon wildflowers found in the
southern national forests. U. S. Dept. Agric., For. Serv., South Reg. 22 pp.

Dyrness, C. T. et al. 1975. Research natural area needs in the Pacific Northwest:
a contribution to land-use planning. U. S. Dept. Agric., For. Serv. et al. Pacific Northwest Forest and Range Experiment Station, Portland, 231 pp.

Elias, Thomas S. 1976. Extinction is forever. Gard. Jour. **26**: 52-55.

_____. 1975. Vascular plants, pp. 88-93. *In* Proceedings of the symposium on
endangered and threatened species of North America, Washington, D. C., 1974.
Wild Canid Survival and Research Center, St. Louis, Missouri.

Fairbrothers, David E. and Mary Y. Hough. 1973. Rare or endangered vascular plants
of New Jersey. New Jersey State Museum, Trenton. 53 pp.

Farnworth, Edward G. and Frank B. Golley. 1974. Fragile ecosystems: evaluation of
research and application in the Neotropics. A report of The Institute of Ecology
(TIE), June 1973. Springer, New York. 258 pp.

Fisher, James. 1969. Wildlife in danger. Viking, New York. 368 pp.

Fleming, Robert L. 1969. Nepal fauna and flora: comments on present status. Bull.
IUCN **2**(13): 108.

Fosberg, F. R. 1975. The deflowering of Hawaii. Natl. Parks Conserv. Mag. **49**(10):
4-10.

_____. and Derral Herbst. 1975. Rare and endangered species of Hawaiian
vascular plants. Allertonia **1**(1):1-72.

Frankel, O. H., ed. 1973. Survey of crop genetic resources in their centres of diversity.
FAO, Rome. 164 pp.

_____. and E. Bennett. 1970. Genetic resources in plants — their exploration
and conservation. F. A. Davis Co., Philadelphia. 554 pp.

Gagne, William. 1975. Hawaii's tragic dismemberment. Defenders Wildl. News **50**:
461-470.

Ghiselin, Jon. 1973-1974. Wilderness and the survival of species. Declining populations
lose options in recessive genes. The Living Wilderness Winter Issue: 22-27.

Gillette, R. 1973. Endangered species moving toward a cease-fire. Science **179**: 1107-1109.

Goldstein, J. 1976. How gardeners can help save endangered plants. Org. Gard. Farm.
23(2):110-112.

Gómez Pompa, A., C. Vazquez-Yanes and S. Guevara. 1972. The tropical rain-forest:
a non-renewable resource. Science **177**: 762-765.

Goodland, Robert J. A. and H. S. Irwin. 1975. Amazon jungle: green hell to red desert?
Elsevier, New York. 155 pp.

Goodwin, H. A. and E. P. Dawson. 1971. Status of endangered species program. *In*
James B. Trefethen, ed., Transactions of the 36th North American wildlife and
natural resource conference. Symposium VIII. Wildlife Management Institute,
Washington, D. C. 534 pp.

Gosnell, M. 1976. Please don't pick the butterworts. Natl. Wild. **14**(3): 32-37.

Griggs, Robert F. 1940. The ecology of rare plants. Bull. Torrey Bot. Club **67**: 575-594.

Gwynne, P. and M. Gosnell. 1975. Fading flowers. Newsweek **86**(2): 72.

H., C. 1975. Slow going on the endangered species front. Science **189**: 623.

Harrison, Elizabeth A. 1975. Endangered species: a bibliography with abstracts.
NTIS, Springfield, Va. 34 pp.

Harwill, A. M., Jr. 1973. Some new and very local populations of rare species in
Virginia. Castanea **38**: 305-307.

Hedberg, Inga and Olav Hedberg, eds. 1968. Conservation of vegetation in Africa south of the Sahara. Almquist and Wiksell, Uppsala. 320 pp.

Hernández, M. Ospina. 1968. The disappearance of valuable native orchids in Latin America, pp. 168-169. *In* Proceedings of the Latin American conference on the conservation of renewable natural resources, San Carlos de Bariloche, Argentina, 1968. IUCN, Morges, Switzerland.

Heslop-Harrison, J. 1974. Genetic resource conservation: the end and the means. Jour. R. Soc. Arts. **123:** 157-169.

_____. 1975. Man and the endangered plant. Int. Yearb., London, pp. XII-XVI.

_____. 1973. The plant kingdom: an exhaustible resource? Trans. Bot. Soc. Edinb. **42:** 1-15.

Hirano, R. T. 1973. Preservation of the Hawaiian flora. Arbor. Bot. Gard. Bull. **7**(1): 10-11.

How to save a wildflower. 1975. Natl. Parks Conserv. Mag. **49**(10): 4-10.

Hunt, P. F. 1969. Conservation of orchids. Bull. IUCN **2**(10): 76.

Hutcherson, Kate. 1976. Endangered species: the law and the land. Jour. For. **74:** 31-34.

Iltis, H. H. 1974. Flowers and human ecology, pp. 289-317. *In* Cyril Selmes, ed., New movements in the study and technology of biology. Maurice Temple Smith, London.

_____. 1970. Man first? Man last? The paradox of human ecology. BioScience **29:** 820.

_____. 1968. The optimum human environment and its relation to modern agricultural preoccupations. The Biologist **50:** 114-125.

_____. 1972. Shepherds leading sheep to slaughter: the extinction of species and the destruction of ecosystems. Am. Biol. Teach. **34:** 201-205.

International Union for Conservation of Nature and Natural Resources. 1975. A preliminary draft for the list of threatened and endemic plants for the countries of Europe, compiled by the IUCN Threatened Plants Committee Secretariat at the Royal Botanic Garden, Kew. IUCN, Morges, Switzerland. (192 pp.)

Janzen, D. H. 1972. The uncertain future of the tropics. Nat. Hist. **81**(9): 80-94.

Jenkins, D. W. 1975. At last — a brighter outlook for endangered plants. Natl. Parks Conserv. Mag. **49**(1): 13-17.

_____. 1973. List of rare and endangered plants of the United States. Smithsonian Institution, Washington, D. C. 3 pp.

_____. and E. S. Ayensu. 1975. One-tenth of our plant species may not survive. Smithsonian **5**(10): 92-96

Jenkins, Robert E. 1975. Endangered plant species: a soluble ecological problem. Nat. Conserv. News **25**(4): 20-21.

_____. 1973. The preservation of ecosystems. Atlant. Nat. **28**(2): 44, 7.

Johnson, Anne. 1968. Rare plants and the community in South East Asia, pp. 340-343. *In* Lee M. Talbot, ed., Conservation in tropical South East Asia. IUCN, Morges, Switzerland.

Jones, Clyde H. 1943. Studies in Ohio floristics. II. Rare plants of Ohio. Castanea **8:** 81-108.

Jordan, C. F. 1971. A world pattern in plant energetics. Amer. Sci. **59:** 425-433.

Kilburn, Paul D. 1961. Endangered relic trees. Nat. Hist. **70**(10): 56-63.

King, F. W. 1974. International trade and endangered species. Zoological Society of London, London. 456 pp.

Kinkead, Eugene. 1976. The search for *Betula uber.* New Yorker 12 January: 58-69.

Krieger, M. H. 1973. What's wrong with plastic trees? Science **179:** 446-455.

Lamoureux, C. H. 1973. Conservation problems in Hawaii, pp. 315-319. *In* A. B. Costin and R. H. Graves, eds., Nature conservation in the Pacific. Proceedings of symposium A-10 XII Pacific Science Congress, 1971. IUCN, Morges, Switzerland and Aust. Natl. Univ. Pr., Canberra.

_____. 1976. Endangered species in Hawaii, effect on other resource management. A response. Newsl. Haw. Bot. Soc. **15**(1): 14-21.

Layne, E. N. 1973. Who will save the cacti? Audubon **75**(4): 4.

Lek, Chew Wei. 1968. Conservation of habitats, pp. 337-339. *In* Lee M. Talbot, ed., Conservation in tropical South East Asia. IUCN, Morges, Switzerland.

Linton, Ron M. 1970. Terracide. Little, Brown, Boston. 376 pp.

Little, E. L., Jr. 1975. Our rare and endangered trees. Amer. For. **81** (7): 16-21, 55-57.

_____. 1975. Rare and local conifers in the United States. U. S. Dept. Agric. Conserv. Res. Rep. no. 19, Washington, D. C.

[**Long, Charles R.,** comp.] 1976. Endangered plant species of the world, habitat alteration and destruction: a selected bibliography. [Prepared for the Symposium on threatened and endangered species of plants . . . May 11-13, 1976] . New York Botanical Garden, Bronx, New York. 10 pp.

Lovejoy, Thomas. 1976. We must decide which species will go forever. Smithsonian **7**(4): 52-59.

Lucas, Grenville, Ll. 1975. Problems and approaches in the conservation of threatened plants and plant genetic resources. IUCN 13th tech. meeting, Kenshasa, Zaire. (Mimeographed).

_____. **and S. M. Walkis.** 1975. Progress report on the production of a list of threatened and endemic plants for the countries of the Council of Europe. (Mimeographed).

Lyons, G. 1972. Conservation: a waste of time? Cact. Succ. Jour. (U.S.) **44**:173-177.

MacBryde, Bruce. 1976. Adopt a plant. Natl. Wildl. **14**(3):38.

McCrone, Harry G. 1976. America's cacti are endangered and threatened. Cact. Succ. Jour. (U.S.) **48**:119.

McIntosh, D. H. 1962. The effect of man on the forests of the highlands of eastern New Guinea, pp. 123-126. *In* Symposium on the impact of man on humid tropics vegetation, Goroka, Papua-New Guinea, 1960. [A. J. Arthur, Commonwealth Gov't. Printer, Canberra.]

Maheshwari, J. K. 1970. The need for conservation of flora and floral provinces in Southeast Asia. pp. 89-94. *In* C. W. Holloway, ed., Problems of threatened species. IUCN 11th tech. meeting, New Delhi, India, 1969, vol. II. Morges, Switzerland.

Marshall, Adrian G. 1973. Conservation in West Malaysia: the potential for international cooperation. Biol. Conserv. **5**(2): 133, 138.

Mathews, William H. 1971. Man's impact on terrestrial and oceanic ecosystems. MIT Press, Cambridge. 540 pp.

Melville, R. 1969. Endangered plants and conservation in the islands of the Indian Ocean, pp. 103-107. *In* C. W. Holloway, ed., Problems of threatened species. IUCN 11th tech. meeting, New Delhi, India, 1969. Morges, Switzerland.

_____. 1970. Red data book, vol. 5. Angiospermae — flowering plants. IUCN, Morges, Switzerland.

_____. 1973. Relict plants in the Australian flora and their conservation, pp. 83-90. *In* A. B. Costin and R. H. Graves, eds., Nature conservation in the Pacific. Proceedings of symposium A-10 XII Pacific Science Congress, 1971. IUCN, Morges, Switzerland and Aust. Natl. Univ. Pr., Canberra.

Mennema, J. 1975. Threatened and protected plants in the Netherlands. Naturopa **22**: 10-13.

Merwe, Phillip van der. 1975a. Impossible to save the Marsh Rose Protea? Veld Flora **61**(1): 4-5.

_____. 1975b. Liaison conservation directory for endangered and threatened species. U. S. Dept. Inter. Fish Wildl. Serv. [Washington, D. C.] .

Michigan endangered plants, pp. 28-30. 1976. *In* Michigan's endangered and threatened species program. Appendix E. Annotated lists. Dept. Nat. Resour., East Lansing, Michigan. 30 pp.

Miller, R. S. and D. B. Botkin. 1974. Endangered species models and predictions. Amer. Sci. **62**(2): 172-180.

Minnesota. Dept. of Natural Resources. 1975. The uncommon ones: animals and plants which merit special consideration and management. Dept. of Natural Resources. St. Paul, Minnesota. 32 pp.

Misra, R. 1970. Save the tropical ecosystems. Intecol. Bull. **2**: 29.

Muenscher, W. C. 1937. Why our native wild flowers need protection. Wild Flower **14**: 72-75.

Muñoz Pizarro, Carlos. 1973. Chile: Plantas en extinción. Ed. Universitaria, Santiago. 248 pp.

Myers, Norman. 1976. An expanded approach to the problem of disappearing species. Science **193**: 198-202.

New England Wild Flower Preservation Society. [n.d.] Helpful hints in conserving wild flowers. 4 pp.

Nicholls, Frank G. 1968. Regulation and co-ordination of collections of flora and fauna. pp. 347-348. *In* Lee M. Talbot, ed., Conservation in tropical South East Asia. IUCN, Morges, Switzerland.

Nicot, J. 1973. Les champignons dans la destruction de la nature: les mycologues et la protection de l'environment. Rev. Mycol. **37**(1-2): 96-99.

Pickoff, L. J. 1975. Our role in conservation. Cact. Succ. Jour. (U.S.) **47**(1): 20-22.

Plant conservation. 1976. Bull. IUCN n.s. **7**(1): 3.

Preston, D. J. 1975. Endangered plants. Am. For. **81**(4): 8-11, 46-47.

Preston, F. W. 1948. The commonness and rarity of species. Ecology **29**: 254-283.

Prina, Lee L. 1975. These plants, and hundreds more may soon be extinct. New York Times. 9 February: Sect. D, p. 39.

Quisumbing, E. 1967. Philippine species of plants facing extinction. Areneta Jour. Agric. **14**(3):135-162.

_____. 1962. The vanishing species of plants in the Philippines, pp. 344-349. *In* Symposium on the impact of man on humid tropics vegetation, Goroka, Papua-New Guinea, 1960. [A. J. Arthur, Commonwealth Gov't. Printer, Canberra.]

Qureshi, I. M. and O. N. Kaul. 1969. Some endangered plants and threatened habitats in Southeast Asia, pp. 115-126. *In* C. W. Holloway, ed., Problems of threatened species. IUCN 11th tech. meeting, New Delhi, India, 1969. Morges, Switzerland.

Rare and endangered plants of the United States to be listed. 1973. Bull. IUCN **4**(8): 34.

Rare Plant Study Center, The University of Texas. 1974. Rare and endangered plants native to Texas. 3rd ed. Austin. 12 pp.

The right to exist — a report on our endangered wildlife. [1974] U. S. Dept. Inter., Fish and Wildlife Service, Bureau of Sport Fisheries and Wildlife, Resource Publication 69. Washington, D. C. 12 pp.

Rowley, G. 1973. Save the succulents! A practical step to aid conservation. Cact. Succ. Jour. (U. S.) **45**(1):8-11.

Sahni, K. C. 1970. Protection of rare and endangered plants in the Indian flora, pp. 95-102. *In* C. W. Holloway, ed., Problems of threatened species. IUCN 11th tech. meeting, New Delhi, India, 1969, vol. II. Morges, Switzerland.

Salisbury, E. J. 1942. The reproductive capacity of plants: Studies in quantitative biology. G. Bill, London. 244 pp.

Santapau, H. 1970. Endangered plant species and their habitats, pp. 83-88. *In* C. W. Holloway, ed., Problems of threatened species. IUCN 11th tech. meeting, New Delhi, India, 1969, vol. II. Morges, Switzerland.

Schofield, Eileen K. 1973a. A unique and threatened flora. Gard. Jour. **23**: 68-73.

_____. 1973b. Galápagos flora: the threat of introduced plants. Biol. Conserv. **5**(1): 48-51.

Scoby, Donald R. 1971. Environmental ethics. Burgess, Minneapolis. 239 pp.

Sears, P. B. 1959. Deserts on the march. 3rd ed. rev. Univ. of Oklahoma Pr., Norman. 178 pp.

Sherbrooke, W. C. and P. Paylore. 1973. World desertification: cause and effect: a literature review and annotated bibliography. Univ. of Arizona, Off. of Arid Land Studies, Tucson. 168 pp.

Smith, E. et al. 1973. Fossil Creek Springs. Natural area rep. no. 11. Arizona Acad. of Sci., Tempe. Az. 22 pp.

Smith, R. F. 1972. The impact of the green revolution on plant protection in tropical and sub-tropical areas. Bull. Entomol. Soc. Amer. **18**: 7-14.

Smithsonian Institution. 1974. Report on endangered and threatened plant species of the United States. U. S. Gov. Print. Off., Washington, D. C. 200 pp.

Smitinand, Tem. 1968. Some rare and vanishing plants of Thailand, pp. 344-346. *In* Lee M. Talbot, ed., Conservation in tropical South East Asia. IUCN, Morges, Switzerland.

Societa botanica italiana, Florence. Gruppo di lavoro per la conservazioni della natura. 1971. Censimento dei biotopi di rilevante interesse vegetazionale meritenoli di conservazione in Italia. Savini-Mercuri, Camerino. 668 pp.

Specht, R. L. 1974. Conservation of major plant communities in Australia and Papua, New Guinea. CSIRO, East Melbourne. 667 pp.

Stanev, Stefan Todorov. 1975. Stars in the mountains are vanishing: stories about our rare plants. [in Russian]. Zemizdot, Sofia. 129 pp.

State of New York. Department of Environmental Conservation. 1974. Environmental conservation law 9-1503. [Albany]. 2 pp.

Stebbins, G. L. 1942. The genetic approach to problems of rare and endemic species. Madroño 6(6): 240-258.

_____. 1967. The lone island of plant life. Calif. Nat. Pl. Soc. Newsl. 3(5): 1-2.

_____. 1972. The scientific and aesthetic value of plants and animals in unusual places. Calif. Nat. Pl. Soc. Newsl. 7(4): 3.

_____. 1966. Why preserve? Calif. Nat. Pl. Soc. Newsl. 2(2): 1.

Stuckey, Ronald L. 1976. Additions to selected list of references on the preservation of natural areas and rare organisms. College of Biological Sciences, Ohio State Univ., Columbus. 3 pp. (Typed).

_____. 1975. Selected list of references on the preservation of natural areas and rare organisms. College of Biological Sciences, Ohio State Univ., Columbus. 6 pp. (Mimeographed).

Subramanyan, K. and C. P. Sreemadhavan. 1969. Endangered plant species and their habitats — a review of the Indian situation, pp. 108-114. *In* C. W. Holloway, ed., Problems of threatened species. IUCN 11th tech. meeting, 1969, New Delhi, India. Morges, Switzerland.

Szijj, J. 1972. Some suggested criteria for determining the international importance of wetlands in the western palearctic. *In* E. Carp, ed., International conference on the conservation of wetlands and waterfowl, Ramsar, Iran, 1971. Intl. Wildfowl Res. Bur., Slimsbridge (Glos.) England. 303 pp.

Takhtadzhyan, A. L., ed. 1975. Red book: native plant species to be protected in the USSR [in Russian]. Nauka, Leningrad. 204 pp.

Tamm, C. O. 1956. Further observations on the survival and flowering of some perennial herbs. I. Oikos 7: 273-292.

_____. 1948. Observations on reproduction and survival of some perennial herbs. Bot. Notiser 101: 305-321.

Teltsch, Kathleen. 1974. U. S., Soviet join ecology parley. New York Times. 15 September.

Thomas, William L. 1965. Man's role in changing the face of the earth. Univ. of Chicago Pr., Chicago. 1193 pp.

Thompson, P. 1975. Should botanic gardens save rare plants? New Sci. 68(979): 628-636.

Togawa, Tom K. 1976. Endangered species in Hawaii, effect on other resource management. Newsl. Haw. Bot. Soc. 15(1): 7-14.

Tucker, Gary E. 1974. Threatened native plants of Arkansas, pp. 39-65. *In* Charles T. Crow, Arkansas natural area plan, State of Arkansas. Arkansas Dept. of Planning. Little Rock, Arkansas.

Twisselmann, Ernest C. 1969. Status of the rare plants of Kern County. Calif. Nat. Pl. Soc. Newsl. 5(3): 1-7.

Ulychna, K. O. and L. Y. Partyka. 1972. Rare species of bryo flora in the Ukraine and necessity of their protection. Ukr. Bot. Zh. 29: 581-585.

UNESCO. 1964. Scientific problems of the humid tropical zone deltas and their implications. Proceedings of the Dacca symposium, 1964. [Dacca] 422 pp.

_____. 1970. Use and conservation of the biosphere. Proceedings of the intergovernmental conference of experts on the scientific basis for rational use and conservation of the resource of the biosphere, 1968. Paris. 272 pp.

U. S. acts to protect endangered plants by publishing a list. 1976. New York Times. 19 June: 23.

U. S. Dept. of State. 1973a. World wildlife conference: efforts to save endangered species. Dept. of State Pub. 8729. Washington, D. C. 30 pp.

_____. 1973b. World wildlife conference: efforts to save endangered species. (Suppl.) Dept. of State Pub. 8729. Washington, D. C. 4 pp.

U. S. Dept. of the Interior. Fish and Wildlife Service. 1976. Convention on international trade in endangered species of wildlife and flora. Federal Register **41**: 24367-24382.

_____. 1975a. Endangered and threatened wildlife and plants. Federal Register **40**: 44412-44429.

_____. 1975b. Family lists of candidate endangered and threatened plant species in the continental United States. Washington, D. C.

U. S. Dept. of the Interior. Office of Endangered Species and International Activities. 1973. Threatened wildlife of the United States. Bureau of Sport Fisheries and Wildlife. Resource Publication 114. Washington, D. C. 289 pp.

The use of ecological guidelines for development in the American humid tropics. Proceedings of international meeting held at Caracas, Venezuela, February 1974. 1975. 249 pp. IUCN, Morges, Switzerland.

Van der Pijl, L. 1969. Principles of dispersal in higher plants. Springer, New York. 154 pp.

Van Dersal, William R. 1972. Why living organisms should not be exterminated. Atlant. Nat. **27**(1): 7-10.

Vanishing plant life. 1972. Ecol. Today **2**(1): 8.

Voss, E. G. 1972. The state of things. Mich. Acad. **5**(1):1-7.

Webster, Bayard. 1975. Flowers will go on the danger list. New York Times. 25 May.

_____. 1974. U. S.-Soviet accord on plants signed. New York Times. 9 December.

Weinberg, J. H. 1975. Botanocrats and the fading flora. Sci. News **108**(6): 92-95.

Westing, Arthur H. 1971. Ecological effects of military defoliation in the forests of S. Vietnam. BioScience **21**(17): 893, 6.

Wildlife legislation — the final hurdle. 1975. Habitat **11**(7):1.

Wiley, Leonard. 1969. Rare wild flowers of North America. 2nd ed. Portland, Oregon. 501 pp.

Wooliams, K. R. 1975. The propagation of Hawaiian endangered species. Hawaii Bot. Soc. Newsl. **14**(4): 59-68.

Zimmerman, James H. and Hugh H. Iltis. 1961. Conservation of rare plants and animals. Wis. Acad. Rev. 1961 (winter): 7-11.

NATURE PROTECTION AND WILDLIFE PRESERVATION IN THE WESTERN HEMISPHERE

Convention Between the United States of America and Other American Republics and Annex

Convention between the United States of America and other American Republics respecting nature protection and wildlife preservation in the Western Hemisphere. Opened for signature at the Pan American Union at Washington, October 12, 1940; signed for the United States of America, October 12, 1940; ratification advised by the Senate of the United States of America, April 7, 1941; ratified by the President of the United States, April 15, 1941; ratification of the United States deposited with the Pan American Union at Washington, April 28, 1941; proclaimed by the President of the United States, April 30, 1942.

––––––––––

By the President of the United States of America

A PROCLAMATION

Whereas a convention on nature protection and wildlife preservation in the Western Hemisphere was opened for signature at the Pan American Union on October 12, 1940, and was on that day signed by the respective plenipotentiaries of the United States of America, Bolivia, Cuba, the Dominican Republic, Ecuador, El Salvador, Nicaragua, Peru and Venezuela and was subsequently signed on behalf of Costa Rica on October 24, 1940, Mexico on November 20, 1940, Uruguay on December 9, 1940, Brazil on December 27, 1940, Colombia on January 17, 1941, Chile on January 22, 1941, Guatemala on April 9, 1941, Haiti on April 29, 1941, and Argentina on May 19, 1941, the original of which convention, being in English,[***] languages, is word for word as follows:

PREAMBLE

The governments of the American Republics, wishing to protect and preserve in their natural habitat representatives of all species and genera of their native flora and fauna, including migratory birds, in sufficient numbers and over areas extensive enough to assure them from becoming extinct through any agency within man's control; and

Wishing to protect and preserve scenery of extraordinary beauty, unusual and striking geologic formations, regions and natural objects of aesthetic, historic or scientific value, and areas characterized by primitive conditions in those cases covered by this Convention; and

Wishing to conclude a convention on the protection of nature and the preservation of flora and fauna to effectuate the foregoing purposes, have agreed upon the following Articles:

ARTICLE I

Description of terms used in the wording of this Convention.

1. The expression NATIONAL PARKS shall denote:

Areas established for the protection and preservation of superlative scenery, flora and fauna of national significance which the general public may enjoy and from which it may benefit when placed under public control.

2. The expression NATIONAL RESERVE shall denote:

Regions established for conservation and utilization of natural resources under government control, on which protection of animal and plant life will be afforded in so far as this may be consistent with the primary purpose of such reserves.

3. The expression NATURE MONUMENTS shall denote:
Regions, objects, or living species of flora or fauna of aesthetic, historic or scientific interest to which strict protection is given. The purpose of nature monuments is the protection of a specific object, or a species of flora or fauna, by setting aside an area, an object, or a single species, as an inviolate nature monument, except for duly authorized scientific investigations or government inspection.

4. The expression STRICT WILDERNESS RESERVES shall denote:
A region under public control characterized by primitive conditions of flora, fauna, transportation and habitation wherein there is no provision for the passage of motorized transportation and all commercial developments are excluded.

5. The expression MIGRATORY BIRDS shall denote:
Birds of those species, all or some of whose individual members, may at any season cross any of the boundaries between the American countries. Some of the species of the following families are examples of birds characterized as migratory: Charadriidae, Scolopacidae, Caprimulgidae, Hirundinidae.

ARTICLE II

1. The Contracting Governments will explore at once the possibility of establishing in their territories national parks, national reserves, nature monuments, and strict wilderness reserves as defined in the preceding article. In all cases where such establishment is feasible, the creation thereof shall be begun as soon as possible after the effective date of the present Convention.

2. If in any country the establishment of national parks, national reserves, nature monuments, or strict wilderness reserves is found to be impractical at present, suitable areas, objects or living species of fauna or flora, as the case may be, shall be selected as early as possible to be transformed into national parks, national reserves, nature monuments or strict wilderness reserves as soon as, in the opinion of the authorities concerned, circumstances will permit.

3. The Contracting Governments shall notify the Pan American Union of the establishment of any national parks, national reserves, nature monuments, or strict wilderness reserves, and of the legislation, including the methods of administrative control, adopted in connection therewith.

ARTICLE III

The Contracting Governments agree that the boundaries of national parks shall not be altered, or any portion thereof be capable of alienation, except by the competent legislative authority. The resources of these reserves shall not be subject to exploitation for commercial profit.

The Contracting Governments agree to prohibit hunting, killing and capturing of members of the fauna and destruction or collection of representatives of the flora in national parks except by or under the direction or control of the park authorities, or for duly authorized scientific investigations.

The Contracting Governments further agree to provide facilities for public recreation and education in national parks consistent with the purposes of this Convention.

ARTICLE IV

The Contracting Governments agree to maintain the strict wilderness reserves inviolate, as far as practicable, except for duly authorized scientific investigations or government inspection, or such uses as are consistent with the purposes for which the area was established.

ARTICLE V

1. The Contracting Governments agree to adopt, or to propose such adoption to their respective appropriate law-making bodies, suitable laws and regulations for the protection

and preservation of flora and fauna within their national boundaries, but not included in the national parks, national reserves, nature monuments, or strict wilderness reserves referred to in Article II hereof. Such regulations shall contain proper provisions for the taking of specimens of flora and fauna for scientific study and investigation by properly accredited individuals and agencies.

2. The Contracting Governments agree to adopt, or to recommend that their respective legislatures adopt, laws which will assure the protection and preservation of the natural scenery, striking geological formations, and regions and natural objects of aesthetic interest or historic or scientific value.

ARTICLE VI

The Contracting Governments agree to cooperate among themselves in promoting the objectives of the present Convention. To this end they will lend proper assistance, consistent with national laws, to scientists of the American Republics engaged in research and field study; they may, when circumstances warrant, enter into agreements with one another or with scientific institutions of the Americas in order to increase the effectiveness of this collaboration; and they shall make available to all American Republics equally through publication or otherwise the scientific knowledge resulting from each cooperative effort.

ARTICLE VII

The Contracting Governments shall adopt appropriate measures for the protection of migratory birds of economic or aesthetic value to prevent the threatened extinction of any given species. Adequate measures shall be adopted which will permit in so far as the respective governments may see fit, a rational utilization of migratory birds for the purpose of sports as well as food, commerce, and industry, and for specific study and investigation.

ARTICLE VIII

The protection of the species mentioned in the Annex to the present Convention,* is declared to be of special urgency and importance. Species included therein shall be protected as completely as possible, and their hunting, killing, capturing, or taking, shall be allowed only with the permission of the appropriate government authorities in the country. Such permission shall be granted only under special circumstances, in order to further scientific purposes, or when essential for the administration of the area in which the animal or plant is found.

ARTICLE IX

Each Contracting Government shall take the necessary measures to control and regulate the importation, exportation and transit of protected fauna or flora or any part thereof by the following means:

1. The issuing of certificates authorizing the exportation or transit or protected species of flora or fauna, or parts thereof.
2. The prohibition of the importation of any species of fauna or flora or any part thereof protected by the country of origin unless accompanied by a certificate of lawful exportation as provided for in Paragraph 1 of this Article.

---

*The Annex comprises the lists of species transmitted by interested Governments to the Pan American Union, Washington, DC, depository for the Convention. These lists are printed in Treaty Series 981, pages 27-77. It is understood by this Government that such lists are to be considered as flexible rather than permanent in character and may from time to time be altered by the respective Governments by the addition or removal of such species from their several lists as changes and conditions may seem to warrant.

ARTICLE X

1. The terms of this convention shall in no way be interpreted as replacing international agreement previously entered into by one or more of the High Contracting Powers.

2. The Pan American Union shall notify the Contracting Parties of any information relevant to the purposes of the present Convention communicated to it by any national museums or by any organizations, national or international, established within their jurisdiction and interested in the purposes of the Convention.

ARTICLE XI

1. The original of the present Convention in Spanish, English, Portugese and French shall be deposited with the Pan American Union and opened for signature by the American Governments on October 12, 1940.

2. The present Convention shall remain open for signature by the American Governments. The instruments of ratification shall be deposited with the Pan American Union, which shall notify their receipt and the dates thereof, and the terms of any accompanying declarations or reservations, to all participating Governments.

3. The present Convention shall come into force three months after the deposit of not less than five ratifications with the Pan American Union.

4. Any ratification received after the date of the entry into force of the Convention, shall take effect three months after the date of its deposit with the Pan American Union.

ARTICLE XII

1. Any Contracting Government may at any time denounce the present Convention by a notification in writing addressed to the Pan American Union. Such denunciation shall take effect one year after the date of the receipt of the notification by the Pan American Union, provided, however, that no denunciation shall take effect until the expiration of five years from the date of the entry into force of this Convention.

2. If, as the result of simultaneous or successive denunciations, the number of Contracting Governments is reduced to less than three, the Convention shall cease to be in force from the date on which the last of such denunciations takes effect in accordance with the provisions of the preceding paragraph.

3. The Pan American Union shall notify all of the American Governments of any denunciations and the date on which they take effect.

4. Should the Convention cease to be in force under the provisions of Paragraph 2 of this article, the Pan American Union shall notify all of the American Governments, indicating the date on which this will become effective.

IN WITNESS WHEREOF, the undersigned Plenipotentiaries, having deposited their full powers found to be in due and proper form, sign this Convention at the Pan American Union, Washington, D. C., on behalf of their respective Governments and affix hereto their seals on the dates appearing opposite their signatures.

WHEREAS it is stipulated in section 3 of article XI of the said convention that the convention shall come into force three months after the deposit of not less than five ratifications with the Pan American Union; and in section 4 of the said article XI that any ratification received after the date of the entry into force of the convention shall take effect three months after the date of its deposit with the Pan American Union;

WHEREAS the said convention has been ratified on the parts of the Governments of the United States of America, Guatemala, Venezuela, El Salvador, Haiti, the Dominican Republic, and Mexico, and the respective instruments of ratification of the Governments of those countries were deposited with the Pan American Union on days as follows, by the United States of America on April 28, 1941, by Guatemala on August 14, 1941, by Venezuela on November 3, 1941, by El Salvador on December 2, 1941, by Haiti on January 31, 1942, by the Dominican Republic on March 3, 1942, and by Mexico on March 27, 1942; and

WHEREAS pursuant to the aforesaid provision of section 3 of article XI of the said convention, the convention will come into force on April 30, 1942, three months after January 31, 1942, the date of deposit of the ratification of Haiti;

NOW, THEREFORE, be it known that I, Franklin D. Roosevelt, President of the United States of America, have caused the said convention to be made public to the end that the same and every article and clause thereof may be observed and fulfilled with good faith by the United States of America and the citizens thereof on and after April, 30, 1942.

IN WITNESS WHEREOF, I have hereunto set my hand and caused the seal of the United States of America to be affixed.

Done at the City of Washington this thirtieth day of April in the year of our Lord one thousand nine hundred and forty-two, and of the Independence of the United States of America the one hundred and sixty-sixth.

FRANKLIN D. ROOSEVELT

Appendix 3

CONVENTION ON INTERNATIONAL TRADE
IN ENDANGERED SPECIES OF WILD FAUNA AND FLORA

*Prepared and adopted by the Plenipotentiary Conference to Conclude
an International Convention on Trade in Certain Species of Wildlife
held at Washington, D.C., from 12 February to 2 March 1973*

The Contracting States,

RECOGNIZING that wild fauna and flora in their many beautiful and varied forms are an irreplaceable part of the natural systems of the earth which must be protected for this and the generations to come;

CONSCIOUS of the ever-growing value of wild fauna and flora from aesthetic, scientific, cultural, recreational and economic points of view;

RECOGNIZING that peoples and States are and should be the best protectors of their own wild fauna and flora;

RECOGNIZING, in addition, that international cooperation is essential for the protection of certain species of wild fauna and flora against over-exploitation through international trade;

CONVINCED of the urgency of taking appropriate measures to this end;

HAVE AGREED as follows:

ARTICLE I
Definitions

For the purpose of the present Convention, unless the context otherwise requires:

(a) "Species" means any species, subspecies, or geographically separate population thereof;

(b) "Specimen" means:

(i) any animal or plant, whether alive or dead;

(ii) in the case of an animal: for species included in Appendices I and II, any readily recognizable part or derivative thereof; and for species included in Appendix III, any readily recognizable part or derivative thereof specified in Appendix III in relation to the species; and

(iii) in the case of a plant: for species included in Appendix I, any readily recognizable part or derivative thereof; and for species included in Appendices II and III, any readily recognizable part or derivative thereof specified in Appendices II and III in relation to the species;

(c) "Trade" means export, re-export, import and introduction from the sea;

(d) "Re-export" means export of any specimen that has previously been imported;

(e) "Introduction from the sea" means transportation into a State of specimens of any species which were taken in the marine environment not under the jurisdiction of any State;

(f) "Scientific Authority" means a national scientific authority designated in accordance with Article IX;

(g) "Management Authority" means a national management authority designated in accordance with Article IX;

(h) "Party" means a State for which the present Convention has entered into force.

ARTICLE II
Fundamental Principles

1. Appendix I shall include all species threatened with extinction which are or may be affected by trade. Trade in specimens of these species must be subject to particularly strict regulation in order not to endanger further their survival and must only be authorized in exceptional circumstances.

2. Appendix II shall include:

(a) all species which although not necessarily now threatened with extinction may become so unless trade in specimens of such species is subject to strict regulation in order to avoid utilization incompatible with their survival; and

(b) other species which must be subject to regulation in order that trade in specimens of certain species referred to in sub-paragraph (a) of this paragraph may be brought under effective control.

3. Appendix III shall include all species which any Party identifies as being subject to regulation within its jurisdiction for the purpose of preventing or restricting exploitation, and as needing the cooperation of other parties in the control of trade.

4. The Parties shall not allow trade in specimens of species included in Appendices I, II and III except in accordance with the provisions of the present Convention.

ARTICLE III
Regulation of Trade in Specimens
of Species included in Appendix I

1. All trade in specimens of species included in Appendix I shall be in accordance with the provisions of this Article.

2. The export of any specimen of a species included in Appendix I shall require the prior grant and presentation of an export permit. An export permit shall only be granted when the following conditions have been met:

(a) a Scientific Authority of the State of export has advised that such export will not be detrimental to the survival of that species;

(b) a Management Authority of the State of export is satisfied that the specimen was not obtained in contravention of the laws of that State for the protection of fauna and flora;

(c) a Management Authority of the State of export is satisfied that any living specimen will be so prepared and shipped as to minimize the risk of injury, damage to health or cruel treatment; and

(d) a Management Authority of the State of export is satisfied that an import permit has been granted for the specimen.

3. The import of any specimen of a species included in Appendix I shall require the prior grant and presentation of an import permit and either an export permit or a re-export certificate. An import permit shall only be granted when the following conditions have been met:

(a) a Scientific Authority of the State of import has advised that the import will be for purposes which are not detrimental to the survival of the species involved;

(b) a Scientific Authority of the State of import is satisfied that the proposed recipient of a living specimen is suitably equipped to house and care for it; and

(c) a Management Authority of the State of import is satisfied that the specimen is not to be used for primarily commercial purposes.

4. The re-export of any specimen of a species included in Appendix I shall require the prior grant and presentation of a re-export certificate. A re-export certificate shall only be granted when the following conditions have been met:

(a) a Management Authority of the State of re-export is satisfied that the specimen was imported into that State in accordance with the provisions of the present Convention;

(b) a Management Authority of the State of re-export is satisfied that any living specimen will be so prepared and shipped as to minimize the risk of injury, damage to health or cruel treatment; and

(c) a Management Authority of the State of re-export is satisfied that an import permit has been granted for any living specimen.

5. The introduction from the sea of any specimen of a species included in Appendix I shall require the prior grant of a certificate from a Management Authority of the State of introduction. A certificate shall only be granted when the following conditions have been met:

(a) a Scientific Authority of the State of introduction advises that the introduction will not be detrimental to the survival of the species involved;

(b) a Management Authority of the State of introduction is satisfied that the proposed recipient of a living specimen is suitably equipped to house and care for it; and

(c) a Management Authority of the State of introduction is satisfied that the specimen is not to be used for primarily commercial purposes.

Article IV
Regulation of Trade in Specimens of Species included in Appendix II

1. All trade in specimens of species included in Appendix II shall be in accordance with the provisions of this Article.

2. The export of any specimen of a species included in Appendix II shall require the prior grant and presentation of an export permit. An export permit shall only be granted when the following conditions have been met:

(a) a Scientific Authority of the State of export has advised that such export will not be detrimental to the survival of that species;

(b) a Management Authority of the State of export is satisfied that the specimen was not obtained in contravention of the laws of that State for the protection of fauna and flora; and

(c) a Management Authority of the State of export is satisfied that any living specimen will be so prepared and shipped as to minimize the risk of injury, damage to health or cruel treatment.

3. A Scientific Authority in each Party shall monitor both the export permits granted by that State for specimens of species included in Appendix II and the actual exports of such specimens. Whenever a Scientific Authority determines that the export of specimens of any such species should be limited in order to maintain that species throughout its range at a level consistent with its role in the ecosystems in which it occurs and well above the level at which that species might become eligible for inclusion in Appendix I, the Scientific Authority shall advise the appropriate Management Authority of suitable measures to be taken to limit the grant of export permits for specimens of that species.

4. The import of any specimen of a species included in Appendix II shall require the prior presentation of either an export permit or a re-export certificate.

5. The re-export of any specimen of a species included in Appendix II shall require the prior grant and presentation of a re-export certificate. A re-export certificate shall only be granted when the following conditions have been met:

(a) a Management Authority of the State of re-export is satisfied that the specimen was imported into that State in accordance with the provisions of the present Convention; and

(b) a Management Authority of the State of re-export is satisfied that any living specimen will be so prepared and shipped as to minimize the risk of injury, damage to health or cruel treatment.

6. The introduction from the sea of any specimen of a species included in Appendix II shall require the prior grant of a certificate from a Management Authority of the State of introduction. A certificate shall only be granted when the following conditions have been met:

(a) a Scientific Authority of the State of introduction advises that the introduction will not be detrimental to the survival of the species involved; and

(b) a Management Authority of the State of introduction is satisfied that any living specimen will be so handled as to minimize the risk of injury, damage to health or cruel treatment.

7. Certificates referred to in paragraph 6 of this Article may be granted on the advice of a Scientific Authority, in consultation with other national scientific authorities or, when appropriate, international scientific authorities, in respect of periods not exceeding one year for total numbers of specimens to be introduced in such periods.

Article V
Regulation of Trade in Specimens of Species included in Appendix III

1. All trade in specimens of species included in Appendix III shall be in accordance with the provisions of this Article.

2. The export of any specimen of a species included in Appendix III from any State which has included that species in Appendix III shall require the prior grant and presentation of an export permit. An export permit shall only be granted when the following conditions have been met:

(a) a Management Authority of the State of export is satisfied that the specimen was not obtained in contravention of the laws of that State for the protection of fauna and flora; and

(b) a Management Authority of the State of export is satisfied that any living specimen will be so prepared and shipped as to minimize the risk of injury, damage to health or cruel treatment.

3. The import of any specimen of a species included in Appendix III shall require, except in circumstances to which paragraph 4 of this Article applies, the prior presentation of a certificate of origin and, where the import is from a State which has included that species in Appendix III, an export permit.

4. In the case of re-export, a certificate granted by the Management Authority of the State of re-export that the specimen was processed in that State or is being re-exported shall be accepted by the State of import as evidence that the provisions of the present Convention have been complied with in respect of the specimen concerned.

Article VI
Permits and Certificates

1. Permits and certificates granted under the provisions of Articles III, IV, and V shall be in accord-

ance with the provisions of this Article.

2. An export permit shall contain the information specified in the model set forth in Appendix IV, and may only be used for export within a period of six months from the date on which it was granted.

3. Each permit or certificate shall contain the title of the present Convention, the name and any identifying stamp of the Management Authority granting it and a control number assigned by the Management Authority.

4. Any copies of a permit or certificate issued by a Management Authority shall be clearly marked as copies only and no such copy may be used in place of the original, except to the extent endorsed thereon.

5. A separate permit or certificate shall be required for each consignment of specimens.

6. A Management Authority of the State of import of any specimen shall cancel and retain the export permit or re-export certificate and any corresponding import permit presented in respect of the import of that specimen.

7. Where appropriate and feasible a Management Authority may affix a mark upon any specimen to assist in identifying the specimen. For these purposes "mark" means any indelible imprint, lead seal or other suitable means of identifying a specimen, designed in such a way as to render its imitation by unauthorized persons as difficult as possible.

ARTICLE VII
Exemptions and Other Special Provisions Relating to Trade

1. The provisions of Articles III, IV and V shall not apply to the transit or trans-shipment of specimens through or in the territory of a Party while the specimens remain in Customs control.

2. Where a Management Authority of the State of export or re-export is satisfied that a specimen was acquired before the provisions of the present Convention applied to that specimen, the provisions of Articles III, IV and V shall not apply to that specimen where the Management Authority issues a certificate to that effect.

3. The provisions of Articles III, IV and V shall not apply to specimens that are personal or household effects. This exemption shall not apply where:

(a) in the case of specimens of a species included in Appendix I, they were acquired by the owner outside his State of usual residence, and are being imported into that State; or

(b) in the case of specimens of species included in Appendix II:

(i) they were acquired by the owner outside his State of usual residence and in a State where removal from the wild occurred;

(ii) they are being imported into the owner's State of usual residence; and

(iii) the State where removal from the wild occurred requires the prior grant of export permits before any export of such specimens;

unless a Management Authority is satisfied that the specimens were acquired before the provisions of the present Convention applied to such specimens.

4. Specimens of an animal species included in Appendix I bred in captivity for commercial purposes, or of a plant species included in Appendix I artificially propagated for commercial purposes, shall be deemed to be specimens of species included in Appendix II.

5. Where a Management Authority of the State of export is satisfied that any specimen of an animal species was bred in captivity or any specimen of a plant species was artificially propagated, or is a part of such an animal or plant or was derived therefrom, a certificate by that Management Authority to that

effect shall be accepted in lieu of any of the permits or certificates required under the provisions of Articles III, IV or V.

6. The provisions of Articles III, IV and V shall not apply to the non-commercial loan, donation or exchange between scientists or scientific institutions registered by a Management Authority of their State, of herbarium specimens, other preserved, dried or embedded museum specimens, and live plant material which carry a label issued or approved by a Management Authority.

7. A Management Authority of any State may waive the requirements of Articles III, IV and V and allow the movement without permits or certificates of specimens which form part of a travelling zoo, circus, menagerie, plant exhibition or other travelling exhibition provided that:

(a) the exporter or importer registers full details of such specimens with that Management Authority;

(b) the specimens are in either of the categories specified in paragraphs 2 or 5 of this Article; and

(c) the Management Authority is satisfied that any living specimen will be so transported and cared for as to minimize the risk of injury, damage to health or cruel treatment.

ARTICLE VIII
Measures to be Taken by the Parties

1. The Parties shall take appropriate measures to enforce the provisions of the present Convention and to prohibit trade in specimens in violation thereof. These shall include measures:

(a) to penalize trade in, or possession of, such specimens, or both; and

(b) to provide for the confiscation or return to the State of export of such specimens.

2. In addition to the measures taken under paragraph 1 of this Article, a Party may, when it deems it necessary, provide for any method of internal reimbursement for expenses incurred as a result of the confiscation of a specimen traded in violation of the measures taken in the application of the provisions of the present Convention.

3. As far as possible, the Parties shall ensure that specimens shall pass through any formalities required for trade with a minimum of delay. To facilitate such passage, a Party may designate ports of exit and ports of entry at which specimens must be presented for clearance. The Parties shall ensure further that all living specimens, during any period of transit, holding or shipment, are properly cared for so as to minimize the risk of injury, damage to health or cruel treatment.

4. Where a living specimen is confiscated as a result of measures referred to in paragraph 1 of this Article:

(a) the specimen shall be entrusted to a Management Authority of the State of confiscation;

(b) the Management Authority shall, after consultation with the State of export, return the specimen to that State at the expense of that State, or to a rescue centre or such other place as the Management Authority deems appropriate and consistent with the purposes of the present Convention; and

(c) the Management Authority may obtain the advice of a Scientific Authority, or may, whenever it considers it desirable, consult the Secretariat in order to facilitate the decision under subparagraph (b) of this paragraph, including the choice of a rescue centre or other place.

5. A rescue centre as referred to in paragraph 4 of this Article means an institution designated by a Management Authority to look after the welfare of living specimens, particularly those that have been confiscated.

6. Each Party shall maintain records of trade in specimens of species included in Appendices I, II and III which shall cover:

(a) the names and addresses of exporters and importers; and

(b) the number and type of permits and certificates granted; the States with which such trade occurred; the numbers or quantities and types of specimens, names of species as included in Appendices I, II and III and, where applicable, the size and sex of the specimens in question.

7. Each Party shall prepare periodic reports on its implementation of the present Convention and shall transmit to the Secretariat:

(a) an annual report containing a summary of the information specified in sub-paragraph (b) of paragraph 6 of this Article; and

(b) a biennial report on legislative, regulatory and administrative measures taken to enforce the provisions of the present Convention.

8. The information referred to in paragraph 7 of this Article shall be available to the public where this is not inconsistent with the law of the Party concerned.

ARTICLE IX
Management and Scientific Authorities

1. Each Party shall designate for the purposes of the present Convention:

(a) one or more Management Authorities competent to grant permits or certificates on behalf of that Party; and

(b) one or more Scientific Authorities.

2. A State depositing an instrument of ratification, acceptance, approval or accession shall at that time inform the Depositary Government of the name and address of the Management Authority authorized to communicate with other Parties and with the Secretariat.

3. Any changes in the designations or authorizations under the provisions of this Article shall be communicated by the Party concerned to the Secretariat for transmission to all other Parties.

4. Any Management Authority referred to in paragraph 2 of this Article shall if so requested by the Secretariat or the Management Authority of another Party, communicate to it impression of stamps, seals or other devices used to authenticate permits or certificates.

ARTICLE X
Trade with States not Party to the Convention

Where export or re-export is to, or import is from, a State not a party to the present Convention, comparable documentation issued by the competent authorities in that State which substantially conforms with the requirements of the present Convention for permits and certificates may be accepted in lieu thereof by any Party.

ARTICLE XI
Conference of the Parties

1. The Secretariat shall call a meeting of the Conference of the Parties not later than two years after the entry into force of the present Convention.

2. Thereafter the Secretariat shall convene regular meetings at least once every two years, unless the Conference decides otherwise, and extraordinary meetings at any time on the written request of at least one-third of the Parties.

3. At meetings, whether regular or extraordinary, the Parties shall review the implementation of the present Convention and may:

(a) make such provision as may be necessary to enable the Secretariat to carry out its duties;

(b) consider and adopt amendments to Appendices I and II in accordance with Article XV;

(c) review the progress made towards the restoration and conservation of the species included in Appendices I, II and III;

(d) receive and consider any reports presented by the Secretariat or by any Party; and

(e) where appropriate, make recommendations for improving the effectiveness of the present Convention.

4. At each regular meeting, the Parties may determine the time and venue of the next regular meeting to be held in accordance with the provisions of paragraph 2 of this Article.

5. At any meeting, the Parties may determine and adopt rules of procedure for the meeting.

6. The United Nations, its Specialized Agencies and the International Atomic Energy Agency, as well as any State not a Party to the present Convention, may be represented at meetings of the Conference by observers, who shall have the right to participate but not to vote.

7. Any body or agency technically qualified in protection, conservation or management of wild fauna and flora, in the following categories, which has informed the Secretariat of its desire to be represented at meetings of the Conference by observers, shall be admitted unless at least one-third of the Parties present object:

(a) international agencies or bodies, either governmental or non-governmental, and national governmental agencies and bodies; and

(b) national non-governmental agencies or bodies which have been approved for this purpose by the State in which they are located. Once admitted, these observers shall have the right to participate but not to vote.

ARTICLE XII
The Secretariat

1. Upon entry into force of the present Convention, a Secretariat shall be provided by the Executive Director of the United Nations Environment Programme. To the extent and in the manner he considers appropriate, he may be assisted by suitable inter-governmental or non-governmental international or national agencies and bodies technically qualified in protection, conservation and management of wild fauna and flora.

2. The functions of the Secretariat shall be:

(a) to arrange for and service meetings of the Parties;

(b) to perform the functions entrusted to it under the provisions of Articles XV and XVI of the present Convention;

(c) to undertake scientific and technical studies in accordance with programmes authorized by the Conference of the Parties as will contribute to the implementation of the present Convention, including studies concerning standards for appropriate preparation and shipment of living specimens and the means of identifying specimens;

(d) to study the reports of Parties and to request from Parties such further information with respect thereto as it deems necessary to ensure implementation of the present Convention;

(e) to invite the attention of the Parties to any matter pertaining to the aims of the present Convention;

(f) to publish periodically and distribute to the Parties current editions of Appendices I, II and III together with any information which will facilitate identification of specimens of species included in those Appendices.

(g) to prepare annual reports to the Parties on its work and on the implementation of the present Convention and such other reports as meetings of the Parties may request;

(h) to make recommendations for the implementation of the aims and provisions of the present Convention, including the exchange of information of a scientific or technical nature;

(i) to perform any other function as may be entrusted to it by the Parties.

ARTICLE XIII
International Measures

1. When the Secretariat in the light of information received is satisfied that any species included in Appendices I or II is being affected adversely by trade in specimens of that species or that the provisions of the present Convention are not being effectively implemented, it shall communicate such information to the authorized Management Authority of the Party or Parties concerned.

2. When any Party receives a communication as indicated in paragraph 1 of this Article, it shall, as soon as possible, inform the Secretariat of any relevant facts insofar as its laws permit and, where appropriate, propose remedial action. Where the Party considers that an inquiry is desirable, such inquiry may be carried out by one or more persons expressly authorized by the Party.

3. The information provided by the Party or resulting from any inquiry as specified in paragraph 2 of this Article shall be reviewed by the next Conference of the Parties which may make whatever recommendations it deems appropriate.

ARTICLE XIV
Effect on Domestic Legislation and International Conventions

1. The provisions of the present Convention shall in no way affect the right of Parties to adopt:

(a) stricter domestic measures regarding the conditions for trade, taking possession or transport of specimens of species included in Appendices I, II and III, or the complete prohibition thereof; or

(b) domestic measures restricting or prohibiting trade, taking possession, or transport of species not included in Appendices I, II or III.

2. The provisions of the present Convention shall in no way affect the provisions of any domestic measures or the obligations of Parties deriving from any treaty, convention, or international agreement relating to other aspects of trade, taking, possession, or transport of specimens which is in force or subsequently may enter into force for any Party including any measure pertaining to the Customs, public health, veterinary or plant quarantine fields.

3. The provisions of the present Convention shall in no way affect the provisions of, or the obligations deriving from, any treaty, convention or international agreement concluded or which may be concluded between States creating a union or regional trade agreement establishing or maintaining a common external customs control and removing customs control between the parties thereto insofar as they relate to trade among the States members of that union or agreement.

4. A State party to the present Convention, which is also a party to any other treaty, convention or international agreement which is in force at the time of the coming into force of the present Convention and under the provisions of which protection is afforded to marine species included in Appendix II, shall be relieved of the obligations imposed on it under the provisions of the present Convention with respect to trade in specimens of species included in Appendix II that are taken by ships registered in that State and in accordance with the provisions of such other treaty, convention or international agreement.

5. Notwithstanding the provisions of Articles III, IV and V, any export of a specimen taken in accordance with paragraph 4 of this Article shall only require a certificate from a Management Authority of the State of introduction to the effect that the specimen was taken in accordance with the provisions of the other treaty, convention or international agreement in question.

6. Nothing in the present Convention shall prejudice the codification and development of the law of the sea by the United Nations Conference on the Law of the Sea convened pursuant to Resolution 2750 C (XXV) of the General Assembly of the United Nations nor the present or future claims and legal views of any State concerning the law of the sea and the nature and extent of coastal and flag State jurisdiction.

ARTICLE XV
Amendments to Appendices I and II

1. The following provisions shall apply in relation to amendments to Appendices I and II at meetings of the Conference of the Parties:

(a) Any Party may propose an amendment to Appendix I or II for consideration at the next meeting. The text of the proposed amendment shall be communicated to the Secretariat at least 150 days before the meeting. The Secretariat shall consult the other Parties and interested bodies on the amendment in accordance with the provisions of sub-paragraphs (b) and (c) of paragraph 2 of this Article and shall communicate the response to all Parties not later than 30 days before the meeting.

(b) Amendments shall be adopted by a two-thirds majority of Parties present and voting. For these purposes "Parties present and voting" means Parties present and casting an affirmative or negative vote. Parties abstaining from voting shall not be counted among the two-thirds required for adopting an amendment.

(c) Amendments adopted at a meeting shall enter into force 90 days after that meeting for all Parties except those which make a reservation in accordance with paragraph 3 of this Article.

2. The following provisions shall apply in relation to amendments to Appendices I and II between meetings of the Conference of the Parties:

(a) Any Party may propose an amendment to Appendix I or II for consideration between meetings by the postal procedures set forth in this paragraph.

(b) For marine species, the Secretariat shall, upon receiving the text of the proposed amendment, immediately communicate it to the Parties. It shall also consult inter-governmental bodies having a function in relation to those species especially with a view to obtaining scientific data these bodies may be able to provide and to ensuring coordination with any conservation measures enforced by such bodies. The Secretariat shall communicate the views expressed and data provided by these bodies and its own findings and recommendations to the Parties as soon as possible.

(c) For species other than marine species, the Secretariat shall, upon receiving the text of the proposed amendment, immediately communicate it to the Parties, and, as soon as possible thereafter, its own recommendations.

(d) Any Party may, within 60 days of the date on which the Secretariat communicated its recommendations to the Parties under sub-paragraphs (b) or (c) of this paragraph, transmit to the Secretariat any comments on the proposed amendment together with any relevant scientific data and information.

(e) The Secretariat shall communicate the replies received together with its own recommendations to the Parties as soon as possible.

(f) If no objection to the proposed amendment is received by the Secretariat within 30 days of the date the replies and recommendations were communicated under the provisions of sub-paragraph (e) of this paragraph, the amendment shall enter into force 90 days later for all Parties except those which make a reservation in accordance with paragraph 3 of this Article.

(g) If an objection by any Party is received by the Secretariat, the proposed amendment shall be submitted to a postal vote in accordance with the provisions of sub-paragraphs (h), (i) and (j) of this paragraph.

(h) The Secretariat shall notify the Parties that notification of objection has been received.

(i) Unless the Secretariat receives the votes for, against or in abstention from at least one-half of the Parties within 60 days of the date of notification under sub-paragraph (h) of this paragraph, the proposed amendment shall be referred to the next meeting of the Conference for further consideration.

(j) Provided that votes are received from one-half of the Parties, the amendment shall be adopted by a two-thirds majority of Parties casting an affirmative or negative vote.

(k) The Secretariat shall notify all Parties of the result of the vote.

(l) If the proposed amendment is adopted it shall enter into force 90 days after the date of the notification by the Secretariat of its acceptance for all Parties except those which make a reservation in accordance with paragraph 3 of this Article.

3. During the period of 90 days provided for by sub-paragraph (c) of paragraph 1 or sub-paragraph (l) of paragraph 2 of this Article any Party may by notification in writing to the Depositary Government make a reservation with respect to the amendment. Until such reservation is withdrawn the Party shall be treated as a State not a party to the present Convention with respect to trade in the species concerned.

ARTICLE XVI
Appendix III and Amendments thereto

1. Any party may at any time submit to the Secretariat a list of species which it identifies as being subject to regulation within its jurisdiction for the purpose mentioned in paragraph 3 of Article II. Appendix III shall include the names of the Parties submitting the species for inclusion therein, the scientific names of the species so submitted, and any parts or derivatives of the animals or plants concerned that are specified in relation to the species for the purposes of sub-paragraph (b) of Article I.

2. Each list submitted under the provisions of paragraph 1 of this Article shall be communicated to the Parties by the Secretariat as soon as possible after receiving it. The list shall take effect as part of Appendix III 90 days after the date of such communication. At any time after the communication of such list, any Party may by notification in writing to the Depositary Government enter a reservation with respect to any species or any parts or derivatives, and until such reservation is withdrawn, the State shall be treated as a State not a Party to the present Convention with respect to trade in the species or part or derivative concerned.

3. A Party which has submitted a species for inclusion in Appendix III may withdraw it at any time by notification to the Secretariat which shall communicate the withdrawal to all Parties. The withdrawal shall take effect 30 days after the date of such communication.

4. Any Party submitting a list under the provisions of paragraph 1 of this Article shall submit to the Secretariat a copy of all domestic laws and regulations applicable to the protection of such species, together with any interpretations which the Party may deem appropriate or the Secretariat may request. The Party shall, for as long as the species in question is included in Appendix III, submit any amendments of such laws and regulations or any new interpretations as they are adopted.

ARTICLE XVII
Amendment of the Convention

1. An extraordinary meeting of the Conference of the Parties shall be convened by the Secretariat on the written request of at least one-third of the Parties to consider and adopt amendments to the present Convention. Such amendments shall be adopted by a two-thirds majority of Parties present and voting. For these purposes "Parties present and voting" means Parties present and casting an affirmative or negative vote. Parties abstaining from voting shall not be counted among the two-thirds required for adopting an amendment.

2. The text of any proposed amendment shall be communicated by the Secretariat to all Parties at least 90 days before the meeting.

3. An amendment shall enter into force for the Parties which have accepted it 60 days after two-thirds of the Parties have deposited an instrument of acceptance of the amendment with the Depositary Government. Thereafter, the amendment shall enter into force for any other Party 60 days after that Party deposits its instrument of acceptance of the amendment.

ARTICLE XVIII
Resolution of Disputes

1. Any dispute which may arise between two or more Parties with respect to the interpretation or application of the provisions of the present Convention shall be subject to negotiation between the Parties involved in the dispute.

2. If the dispute cannot be resolved in accordance with paragraph 1 of this Article, the Parties may, by mutual consent, submit the dispute to arbitration, in particular that of the Permanent Court of Arbitration at The Hague, and the Parties submitting the dispute shall be bound by the arbitral decision.

ARTICLE XIX
Signature

The present Convention shall be open for signature at Washington until 30th April 1973 and thereafter at Berne until 31st December 1974.

ARTICLE XX
Ratification, Acceptance, Approval

The present Convention shall be subject to ratification, acceptance or approval. Instruments of ratification, acceptance or approval shall be deposited with the Government of the Swiss Confederation which shall be the Depositary Government.

ARTICLE XXI
Accession

The present Convention shall be open indefinitely for accession. Instruments of accession shall be deposited with the Depositary Government.

ARTICLE XXII
Entry into Force

1. The present Convention shall enter into force 90 days after the date of deposit of the tenth instrument of ratification, acceptance, approval or accession, with the Depositary Government.

2. For each State which ratifies, accepts or approves the present Convention or accedes thereto after the deposit of the tenth instrument of ratification, acceptance, approval or accession, the present Convention shall enter into force 90 days after the deposit by such State of its instrument of rati-

fication, acceptance, approval or accession.

ARTICLE XXIII
Reservations

1. The provisions of the present Convention shall not be subject to general reservations. Specific reservations may be entered in accordance with the provisions of this Article and Articles XV and XVI.

2. Any State may, on depositing its instrument of ratification, acceptance, approval or accession, enter a specific reservation with regard to:

(a) any species included in Appendix I, II or III; or

(b) any parts or derivatives specified in relation to a species included in Appendix III.

3. Until a Party withdraws its reservation entered under the provisions of this Article, it shall be treated as a State not a party to the present Convention with respect to trade in the particular species or parts or derivatives specified in such reservation.

ARTICLE XXIV
Denunciation

Any Party may denounce the present Convention by written notification to the Depositary Government at any time. The denunciation shall take effect twelve months after the Depositary Government has received the notification.

ARTICLE XXV
Depositary

1. The original of the present Convention, in the Chinese, English, French, Russian and Spanish languages, each version being equally authentic, shall be deposited with the Depositary Government, which shall transmit certified copies thereof to all States that have signed it or deposited instruments of accession to it.

2. The Depositary Government shall inform all signatory and acceding States and the Secretariat of signatures, deposit of instruments of ratification, acceptance, approval or accession, entry into force of the present Convention, amendments thereto, entry and withdrawal of reservations and notifications of denunciation.

3. As soon as the present Convention enters into force, a certified copy thereof shall be transmitted by the Depositary Government to the Secretariat of the United Nations for registration and publication in accordance with Article 102 of the Charter of the United Nations.

IN WITNESS WHEREOF the undersigned Plenipotentiaries, being duly authorized to that effect, have signed the present Convention.

DONE at Washington this third day of March, One Thousand Nine Hundred and Seventy-three.

Editor's note: The model export permit referred to in Article VI of the Convention as Appendix IV is not reproduced in this Supplement.

APPENDIX I

Interpretation:
1. Species included in this Appendix are referred to:
   (a) by the name of the species; or
   (b) as being all of the species included in a higher taxon or designated part thereof.
2. The abbreviation "spp." is used to denote all species of a higher taxon.
3. Other references to taxa higher than species are for the purposes of information or classification only.
4. An asterisk (*) placed against the name of a species or higher taxon indicates that one or more geographically separate populations, sub-species or species of that taxon are included in Appendix II and that these populations, sub-species or species are excluded from Appendix I.
5. The symbol (−) followed by a number placed against the name of a species or higher taxon indicates the exclusion from that species or taxon of designated geographically separate populations, sub-species or species as follows:
   − 101 *Lemur catta*
   − 102 Australian population
6. The symbol (+) followed by a number placed against the name of a species denotes that only a designated geographically separate population or sub-species of that species is included in this Appendix, as follows:
   + 201 Italian population only
7. The symbol (/) placed against the name of a species or higher taxon indicates that the species concerned are protected in accordance with the International Whaling Commission's schedule of 1972.

FAUNA

Mammalia

MARSUPIALIA

Macropodidae
　　Macropus parma
　　Onychogalea frenata
　　Onychogalea lunata
　　Lagorchestes hirsutus
　　Lagostrophus fasciatus
　　Caloprymnus campestris
　　Bettongia penicillata
　　Bettongia lesueur
　　Bettongia tropica
Phalangeridae
　　Wyulda squamicaudata
Burramyidae
　　Burramys parvus
Vombatidae
　　Lasiorhinus gillespiei
Peramelidae
　　Perameles bougainville
　　Chaeropus ecaudatus
　　Macrotis lagotis
　　Macrotis leucura
Dasyuridae
　　Planigale tenuirostris
　　Planigale subtilissima
　　Sminthopsis psammophila
　　Sminthopsis longicaudata
　　Antechinomys laniger
　　Myrmecobius fasciatus rufus
Thylacinidae
　　Thylacinus cynocephalus

PRIMATES

Lemuridae
Lemur spp. * −101
Lepilemur spp.
Hapalemur spp.
Allocebus spp.
Cheirogaleus spp.
Microcebus spp.
Phaner spp.
Indriidae
Indri spp.
Propithecus spp.
Avahi spp.
Daubentoniidae
Daubentonia madagascariensis
Callithricidae
Leontopithecus (Leontideus) spp.
Callimico goeldii
Cebidae
Saimiri oerstedii
Chiropotes albinasus
Cacajao spp.
Alouatta palliata (villosa)
Ateles geoffroyi frontatus
Ateles geoffroyi panamensis
Brachyteles arachnoides
Cercopithecidae
Cercocebus galeritus galeritus
Macaca silenus
Colobus badius rufomitratus
Colobus badius kirkii
Presbytis geei
Presbytis pileatus
Presbytis entellus
Nasalis larvatus
Simias concolor
Pygathrix nemaeus
Hylobatidae
Hylobates spp.
Symphalengus syndactylus
Pongidae
Pongo pygmaeus pygmaeus
Pongo pygmaeus abelii
Gorilla gorilla

EDENTATA

Dasypodidae
Priodontes giganteus (= maximus)

PHOLIDOTA

Manidae
Manis temmincki

LAGOMORPHA

Leporidae
Romerolagus diazi
Caprolagus hispidus

RODENTIA

Sciuridae
Cynomys mexicanus
Castoridae
Castor fiber birulaia
Castor canadensis mexicanus
Muridae
Zyzomys pedunculatus
Leporillus conditor
Pseudomys novaehollandiae
Pseudomys praeconis

Muridae (cont'd.)
Pseudomys shortridgei
Pseudomys fumeus
Pseudomys occidentalis
Pseudomys fieldi
Notomys aquilo
Xeromys myoides
Chinchillidae
Chinchilla brevicaudata boliviana

CETACEA

Platanistidae
Platanista gangetica
Eschrichtidae
Eschrichtius robustus (glaucus) ∤
Balaenopteridae
Balaenoptera musculus ∤
Megaptera novaeangliae ∤
Balaenidae
Balaena mysticetus ∤
Eubalaena spp. ∤

CARNIVORA

Canidae
Canis lupus monstrabilis
Vulpes velox hebes
Viverridae
Prionodon pardicolor
Ursidae
Ursus americanus emmonsii
Ursus arctos * + 201
Ursus arctos pruinosus
Ursus arctos nelsoni
Mustelidae
Mustela nigripes
Lutra longicaudis (platensis/annectens)
Lutra felina
Lutra provocax
Pteronura brasiliensis
Aonyx microdon
Enhydra lutris nereis
Hyaenidae
Hyaena brunnea
Felidae
Felis planiceps
Felis nigripes
Felis concolor coryi
Felis concolor costaricensis
Felis concolor cougar
Felis temmincki
Felis bengalensis bengalensis
Felis yagouaroundi cacomitli
Felis yagouaroundi fossata
Felis yagouaroundi panamensis
Felis yagouaroundi panamensis
Felis yagouaroundi tolteca
Felis pardalis mearnsi
Felis pardalis mitis
Felis wiedii nicaraguae
Felis wiedii salvinia
Felis tigrina oncilla
Fellis marmorata
Felis jacobita
Felis (Lynx) rufa escuinapae
Neofelis nebulosa
Panthera tigris *
Panthera pardus
Panthera uncia
Panthera onca
Acinonyx jubatus

PINNIPEDIA
Phocidae
Monachus spp.
Mirounga angustirostris

PROBOSCIDEA
Elephantidae
Elephas maximus

SIRENIA
Dugongidae
Dugong dugon* − 102
Trichechidae
Trichechus manatus
Trichechus inunguis

PERISSODACTYLA
Equidae
Equus przewalskii
Equus hemionus hemionus
Equus hemionus khur
Equus zebra zebra
Tapiridae
Tapirus pinchaque
Tapirus bairdii
Tapirus indicus
Rhinocerotidae
Rhinoceros unicornis
Rhinoceros sondaicus
Didermocerus sumatrensis
Ceratotherium simum cottoni

ARTIODACTYLA
Suidae
Sus salvanius
Babyrousa babyrussa
Camelidae
Vicugna vicugna
Camelus bactrianus
Cervidae
Moschus moschiferus moschiferus
Axis (Hyelaphus) porcinus annamiticus
Axis (Hyelaphus) calamianensis
Axis (Hyelaphus) kuhlii
Cervus duvauceli
Cervus eldi
Cervus elaphus hanglu
Hippocamelus bisulcus
Hippocamelus antisensis
Blastoceros dichotomus
Ozotoceros bezoarticus
Pudu pudu
Antilocapidae
Antilocapra americana sonoriensis
Antilocapra americana peninsularis
Bovidae
Bubalus (Anoa) mindorensis
Bubalus (Anoa) depressicornis
Bubalus (Anoa) quarlesi
Bos gaurus
Bos (grunniens) mutus
Novibos (Bos) sauveli
Bison bison athabascae
Kobus leche
Hippotragus niger variani
Oryx leucoryx
Damaliscus dorcas dorcas
Saiga tatarica mongolica

Bovidae (cont'd.)
Nemorhaedus goral
Capricornis sumatraensis
Rupicapra rupicapra ornata
Capra falconeri jerdoni
Capra falconeri megaceros
Capra falconeri chiltanensis
Ovis orientalis ophion
Ovis ammon hodgsoni
Ovis vignei

Aves

TINAMIFORMES
Tinamidae
Tinamus solitarius

PODICIPEDIFORMES
Podicipedidae
Podilymbus gigas

PROCELLARIFORMES
Diomedeidae
Diomedea albatrus

PELECANIFORMES
Sulidae
Sula abbotti
Fregatidae
Fregata andrewsi

CICONIFORMES
Ciconiidae
Ciconia ciconia boyciana
Threskiornithidae
Nipponia nippon

ANSERIFORMES
Anatidae
Anas aucklandica nesiotis
Anas oustaleti
Anas laysanensis
Anas diazi
Cairina scutulata
Rhodonessa caryophyllacea
Branta canadensis leucopareia
Branta sandvicensis

FALCONIFORMES
Cathartidae
Vultur gryphus
Gymnogyps californianus
Accipitridae
Pithecophaga jefferyi
Harpia harpyja
Haliaetus leucocephalus leucocephalus
Haliaetus heliaca adalberti
Haliacetus albicilla groenlandicus
Falconidae
Falco peregrinus anatum
Falco peregrinus tundrius
Falco peregrinus peregrinus
Falco peregrinus babylonicus

GALLIFORMES
Megapodiidae
Macrocephalon maleo

Galliformes (cont'd.)
Cracidae
  Crax blumenbachii
  Pipile pipile pipile
  Pipile jacutinga
  Mitu mitu mitu
  Oreophasis derbianus
Tetraonidae
  Tympanuchus cupido attwateri
Phasianidae
  Colinus virginianus ridgwayi
  Tragopan blythii
  Tragopan caboti
  Tragopan melanocephalus
  Lophophorus sclateri
  Lophophorus lhuysii
  Lophophorus impejanus
  Crossoptilon mantchuricum
  Crossoptilon crossopilon
  Lophura swinhoii
  Lophura imperialis
  Lophura edwardsii
  Syrmaticus ellioti
  Syrmaticus humiae
  Syrmaticus mikado
  Polyplectron emphanum
  Tetraogallus tibetanus
  Tetraogallus caspius
  Cyrtonyx montezumae merriami

GRUIFORMES

Gruidae
  Grus japonensis
  Grus leucogeranus
  Grus americana
  Grus canadensis pulla
  Grus canadensis nesiotes
  Grus nigricollis
  Grus vipio
  Grus monacha
Rallidae
  Tricholimnas sylvestris
Rhynochetidae
  Rhynochetos jubatus
Otididae
  Eupodotis bengalensis

CHARADRIIFORMES

Scolopacidae
  Numenius borealis
  Tringa guttifer
Laridae
  Larus relictus

COLUMBIFORMES

Columbidae
  Ducula mindorensis

PSITTACIFORMES

Psittacidae
  Strigops habroptilus
  Rhynchopsitta pachyrhyncha
  Amazona leucocephala
  Amazona vittata
  Amazona guildingii
  Amazona versicolor
  Amazona imperialis
  Amazona rhodocorytha
  Amazona petrei petrei
  Amazona vinacea

Psittacidae (cont'd.)
  Pyrrhura cruentata
  Anodorhynchus glaucus
  Anodorhynchus leari
  Cyanopsitta spixii
  Pionopsitta pileata
  Aratinga guaruba
  Psittacula krameri echo
  Psephotus pulcherrimus
  Psephotus chrysopterygius
  Neophema chrysogaster
  Neophema splendida
  Cyanoramphus novaezelandiae
  Cyanoramphus auriceps forbesi
  Geopsittacus occidentalis
  Psittacus erithacus princeps

APODIFORMES

Trochilidae
  Ramphodon dohrnii

TROGONIFORMES

Trogonidae
  Pharomachrus mocinno mocinno
  Pharomachrus mocinno costaricensis

STRIGIFORMES

Strigidae
  Otus gurneyi

CORACIIFORMES

Bucerotidae
  Rhinoplax vigil

PACIFORMES

Picidae
  Dryocopus javensis richardsii
  Campephilus imperialis

PASSERIFORMES

Cotingidae
  Cotinga maculata
  Xipholena atro-purpurea
Pittidae
  Pitta kochi
Atrichornithidae
  Atrichornis clamosa
Muscicapidae
  Picathartes gymnocephalus
  Picathartes oreas
  Psophodes nigrogularis
  Amytornis goyderi
  Dasyornis brachypterus longirostris
  Dasyornis broadbenti littoralis
Sturnidae
  Leucopsar rothschildi
Meliphagidae
  Meliphaga cassidix
Zosteropidae
  Zosterops albogularis
Fringillidae
  Spinus cucullatus

Amphibia

URODELA

Cryptobranchidae
  Andrias (= Megalobatrachus)
    davidianus japonicus

Urodela (cont'd.)

Andrias (= Megalobatrachus)
davidianus davidianus

SALIENTIA

Bufonidae
Bufo superciliaris
Bufo periglenes
Nectophrynoides spp.
Atelopodidae
Atelopus varius zeteki

Reptilia

CROCODYLIA

Alligatoridae
Alligator mississippiensis
Alligator sinensis
Malanosuchus niger
Caiman crocodilus apaporiensis
Caiman latirostris
Crocodylidae
Tomistoma schlegelii
Osteolaemus tetraspis tetraspis
Osteolaemus tetraspis osborni
Crocodylus cataphractus
Crocodylus siamensis
Crocodylus palustris palustris
Crocodylus palustris kimbula
Crocodylus novaeguineae mindorensis
Crocodylus intermedius
Crocodylus rhombifer
Crocodylus moreletii
Crocodylus niloticus
Gavialidae
Gavialis gangeticus

TESTUDINATA

Emydidae
Batagur baska
Geoclemmys (= Damonia) hamiltonii
Geoemyda (= Nicoria) tricarinata
Kachuga tecta tecta
Morenia ocellata
Terrapene coahuila
Testudinidae
Geochelone (= Testudo) elephantopus
Geochelone (= Testudo) geometrica
Geochelone (= Testudo) radiata
Geochelone (= Testudo) yniphora
Cheloniidae
Eretmochelys imbricata imbricata
Lepidochelys kempii
Trionychidae
Lissemys punctata punctata
Trionyx ater
Trionyx nigricans
Trionyx gangeticus
Trionyx hurum
Chelidae
Pseudemydura umbrina

LACERTILIA

Varanidae
Varanus komodoensis
Varanus flavescens
Varanus bengalensis
Varanus griseus

SERPENTES

Boidae
Epicrates inornatus inornatus
Epicrates subflavus
Python molurus molurus

RHYNCHOCEPHALIA

Sphenodontidae
Sphenodon punctatus

Pisces

ACIPENSERIFORMES

Acipenseridae
Acipenser brevirostrum
Acipenser oxyrhynchus

OSTEOGLOSSIFORMES

Osteoglossidae
Scleropages formosus

SALMONIFORMES

Salmonidae
Coregonus alpenae

CYPRINIFORMES

Catostomidae
Chasmistes cujus
Cyprinidae
Probarbus jullieni

SILURIFORMES

Schilbeidae
Pangasianodon gigas

PERCIFORMES

Percidae
Stizostedion vitreum glaucum

Mollusca

NAIADOIDA

Unionidae
Conradilla caelata
Dromus dromas
Epioblasma (= Dysnomia)
florentina curtisi
Epioblasma (= Dysnomia)
florentina florentina
Epioblasma (= Dysnomia) sampsoni
Epioblasma (= Dysnomia) sulcata
perobliqua
Epioblasma (= Dysnomia) torulosa
gubernaculum
Epioblasma (= Dysnomia) torulosa
torulosa
Epioblasma (= Dysnomia) turgidula
Epioblasma (= Dysnomia) walkeri
Fusconaia cuneolus
Fusconaia edgariana
Lampsilis higginsi
Lampsilis orbiculata orbiculata
Lampsilis satura
Lampsilis virescens
Plethobasis cicatricosus
Plethobasis cooperianus

Unionidae (cont'd.)
>Pleurobema plenum
>Potamilus (= Proptera) capax
>Quadrula intermedia
>Quadrula sparsa
>Toxolasma (= Carunculina) cylindrella
>Unio (Megalonaias/?/) nickliniana
>Unio (Lampsilis/?/) tampicoensis
>tecomatensis
>Villosa (= Micromya) trabalis

FLORA

ARACEAE
>Alocasia sanderiana
>Alocasia zebrina

CARYOCARACEAE
>Caryocar costaricense

CARYOPHYLLACEAE
>Gymnocarpos przewalskii
>Melandrium mongolicum
>Silene mongolica
>Stellaria pulvinata

CUPRESSACEAE
>Pilgerodendron uviferum

CYCADACEAE
>Encephalartos spp.
>Microcycas calocoma
>Stangeria eriopus

GENTIANACEAE
>Prepusa hookeriana

HUMIRIACEAE
>Vantanea barbourii

JUGLANDACEAE
>Engelhardtia pterocarpa

LEGUMINOSAE
>Ammopiptanthus mongolicum
>Cynometra hemitomophylla
>Platymiscium pleiostachyum

LILIACEAE
>Aloe albida
>Aloe pillansii

Liliaceae (cont'd.)
>Aloe polyphylla
>Aloe thorncroftii
>Aloe vossii

MELASTOMATACEAE
>Lavoisiera itambana

MELIACEAE
>Guarea longipetiola
>Tachigalia versicolor

MORACEAE
>Batocarpus costaricense

ORCHIDACEAE
>Cattleya jongheana
>Cattleya skinneri
>Cattleya trianae
>Didiciea cunninghamii
>Laelia lobata
>Lycaste virginalis var. alba
>Peristeria elata

PINACEAE
>Abies guatamalensis
>Abies nebrodensis

PODOCARPACEAE
>Podocarpus costalis
>Podocarpus parlatorei

PROTEACEAE
>Orothamnus zeyheri
>Protea odorata

RUBIACEAE
>Balmea stormae

SAXIFRAGACEAE (GROSSULARIACEAE)
>Ribes sardoum

TAXACEAE
>Fitzroya cupressoides

ULMACEAE
>Celtis aetnensis

WELWITSCHIACEAE
>Welwitschia bainesii

ZINGIBERACEAE
>Hedychium philippinense

APPENDIX II

Interpretation:

1. Species included in this Appendix are referred to:
>(a) by the name of the species; or
>(b) as being all of the species included in a higher taxon or designated part thereof.

2. The abbreviation "spp." is used to denote all the species of a higher taxon.

3. Other references to taxa higher than species are for the purposes of information or classification only.

4. An asterisk (*) placed against the name of a species or higher taxon indicates that one or more geographically separate populations, sub-species or species of that taxon are included in Appendix I and that these populations, sub-species or species are excluded from Appendix II.

5. The symbol (#) followed by a number placed against the name of a species or higher taxon designates parts or derivatives which are specified in relation thereto for the purposes of the present Convention as follows:
># 1 designates root
># 2 designates timber
># 3 designates trunks

6. The symbol (−) followed by a number placed against the name of species or higher taxon indicates the exclusion from that species or taxon of designated geographically separate populations, sub-species, species or groups of species as follows:
>− 101 Species which are not succulents

7. The symbol (+) followed by a number placed against the name of a species or higher taxon denotes that only designated geographically separate populations, sub-species or species of that species or taxon are included in this Appendix as follows:
>+ 201 All North American sub-species
>+ 202 New Zealand species
>+ 203 All species of the family in the Americas
>+ 204 Australian population.

FAUNA

Mammalia

MARSUPIALIA

Macropodidae
  Dendrolagus inustus
  Dendrolagus ursinus

INSECTIVORA

Erinaceidae
  Erinaceus frontalis

PRIMATES

Lemuridae
  Lemur catta*
Lorisidae
  Nycticebus coucang
  Loris tardigradus
Cebidae
  Cebus capucinus
Cercopithecidae
  Macaca sylvanus
  Colobus badius gordonorum
  Colobus verus
  Rhinopithecus roxellanae
  Presbytis johnii
Pongidae
  Pan paniscus
  Pan troglodytes

EDENTATA

Myrmecophagidae
  Myrmecophaga tridactyla
  Tamandua tetradactyla chapadensis
Bradypodidae
  Bradypus boliviensis

PHOLIDOTA

Manidae
  Manis crassicaudata
  Manis pentadactyla
  Manis javanica

LAGOMORPHA

Leporidae
  Nesolagus netscheri

RODENTIA

Heteromyidae
  Dipodomys phillipsii phillipsii
Sciuridae
  Ratufa spp.
  Lariscus hosei
Castoridae
  Castor canadensis frondator
  Castor canadensis repentinus
Cricetidae
  Ondatra zibethicus bernardi

CARNIVORA

Canidae
  Canis lupus pallipes
  Canis lupus irremotus
  Canis lupus crassodon
  Chrysocyon brachyurus
  Cuon alpinus

Ursidae
  Ursus (Thalarctos) maritimus
  Ursus arctos* + 201
  Helarctos malayanus
Procyonidae
  Ailurus fulgens
Mustelidae
  Martes americana atrata
Viveridae
  Prionodon linsang
  Cynogale bennetti
  Helogale derbianus
Felidae
  Felis yagouaroundi*
  Felis colocolo pajeros
  Felis colocolo crespoi
  Felis colocolo budini
  Felis concolor missoulensis
  Felis concolor mayensis
  Felis concolor azteca
  Felis serval
  Felis lynx isabellina
  Felis wiedii*
  Felis pardalis*
  Felis tigrina*
  Felis (= Caracal) caracal
  Panthera leo persica
  Panthera tigris altaica (= amurensis)

PINNIPEDIA

Otariidae
  Arctocephalus australis
  Arctocephalus galapagoensis
  Arctocephalus philippii
  Arctocephalus townsendi
Phocidae
  Mirounga australis
  Mirounga leonina

TUBULIDENTATA

Orycteropidae
  Orycteropus afer

SIRENIA

Dugongidae
  Dugong dugon* + 204
Trichechidae
  Trichechus senegalensis

PERISSODACTYLA

Equidae
  Equus hemionus*
Tapiridae
  Tapirus terrestris
Rhinocerotidae
  Diceros bicornis

ARTIODACTYLA

Hippopotamidae
  Choeropsis liberiensis
Cervidae
  Cervus elaphus bactrianus
  Pudu mephistophiles
Antilocapridae
  Antilocapra americana mexicana
Bovidae
  Cephalophus monticola
  Oryx (tao) dammah

Bovidae (cont'd.)
Addax nasomaculatus
Pantholops hodgsoni
Capra faconeri*
Ovis ammon*
Ovis canadensis

Aves

SPHENISCIFORMES

Spheniscidae
Spheniscus demersus

RHEIFORMES

Rheidae
Rhea americana albescens
Pterocnemia pennata pennata
Pterocnemia pennata garleppi

TINAMIFORMES

Tinamidae
Rhynchotus rufescens rufescens
Rhynchotus rufescens pallescens
Rhynchotus rufescens maculicollis

CICONIIFORMES

Ciconiidae
Ciconia nigra
Threskiornithidae
Geronticus calvus
Platalea leucorodia
Phoenicopteridae
Phoenicopterus ruber chilensis
Phoenicoparrus andinus
Phoenicoparrus jamesi

PELECANIFORMES

Pelecanidae
Pelecanus crispus

ANSERIFORMES

Anatidae
Anas aucklandica aucklandica
Anas aucklandica chlorotis
Anas bernieri
Dendrocygna arborea
Sarkidiornis melanotos
Anser albifrons gambelli
Cygnus bewickii jankowskii
Cygnus melancoryphus
Coscoroba coscoroba
Branta ruficollis

FALCONIFORMES

Accipitridae
Gypaetus barbatus meridionalis
Aquila chrysaetos
Falconidae
Spp.*

GALLIFORMES

Megapodiidae
Megapodius freycinet nicobariensis
Megapodius freycinet abbotti
Tetraonidae
Tympanuchus cupido pinnatus
Phasianidae
Francolinus ochropectus

Phasianidae (cont'd.)
Francolinus swierstrai
Catreus wallichii
Polyplectron malacense
Polyplectron germaini
Polyplectron bicalcaratum
Gallus sonneratii
Argusianus argus
Ithaginus cruentus
Cyrtonyx montezumae montezumae
Cyrtonyx montezumae mearnsi

GRUIFORMES

Gruidae
Balearica regulorum
Grus canadensis pratensis
Rallidae
Gallirallus australis hectori
Otididae
Chlamydotis undulata
Choriotis nigriceps
Otis tarda

CHARADRIIFORMES

Scolopacidae
Numenius tenuirostris
Numenius minutus
Laridae
Larus brunneicephalus

COLUMBIFORMES

Columbidae
Gallicolumba luzonica
Goura cristata
Goura scheepmakeri
Goura victoria
Caloenas nicobarica pelewensis

PSITTACIFORMES

Psittacidae
Coracopsis nigra barklyi
Prosopeia personata
Eunymphicus cornutus
Cyanoramphus unicolor
Cyanoramphus novaezelandiae
Cyanoramphus malherbi
Poicephalus robustus
Tanygnathus luzoniensis
Probosciger alteriimus

CUCULIFORMES

Musophagidae
Turaco corythaix
Gallirex porphyreolophus

STRIGIFORMES

Strigidae
Otus nudipes newtoni

CORACIIFORMES

Bucerotidae
Buceros rhinoceros rhinoceros
Buceros bicornis
Buceros hydrocorax hydrocorax
Aceros narcondami

PICIFORMES

Picidae
Picus squamatus flavirostris

PASSERIFORMES

Cotingidae
    Rupicola rupicola
    Rupicola peruviana
Pittidae
    Pitta brachyura nympha
Hirundinidae
    Pseudochelidon sirintarae
Paradisaeidae
  Spp.
Muscicapidae
    Muscicapa ruecki
Fringillidae
    Spinus yarellii

Amphibia

URODELA

Ambystomidae
    Ambystoma mexicanum
    Ambystoma dumerillii
    Ambystoma lermaensis

SALIENTIA

Bufonidae
    Bufo retiformis

Reptilia

CROCODYLIA

Alligatoridae
    Caiman crocodilus crocodilus
    Caiman crocodilus yacare
    Caiman crocodilus fuscus (chiapasius)
    Paleosuchus palpebrosus
    Paleosuchus trigonatus
Crocodylidae
    Crocodylus johnsoni
    Crocodylus novaeguineae novaeguineae
    Crocodylus porosus
    Crocodylus acutus

TESTUDINATA

Emydidae
    Clemmys muhlenbergi
    Testudinidae
    Chersine spp.
    Geochelone spp.*
    Gopherus spp.
    Homopus spp.
    Kinixys spp.
    Malacochersus spp.
    Pyxis spp.
    Testudo spp.*
Cheloniidae
    Caretta caretta
    Chelonia mydas
    Chelonia depressa
    Eretmochelys imbricata bissa
    Lepidochelys olivacea
Dermochelidae
    Dermochelys coriacea
Pelomedusidae
    Podocnemis spp.

LACERTILIA

Teiidae
    Cnemidophorus hyperythrus

Iguanidae
    Conolophus pallidus
    Cololophus subcristatus
    Amblyrhynchus cristatus
    Phrynosoma coronatum blainvillei
Helodermatidae
    Heloderma suspectum
    Heloderma horridum
Varanidae
    Varanus spp.*

SERPENTES

Boidae
    Epicrates cenchris cenchris
    Eunectes notaeus
    Constrictor constrictor
    Python spp.*
Colubridae
    Cyclagras gigas
    Pseudoboa cloelia
    Elachistodon westermanni
    Thamnophis elegans hammondi

Pisces

ACIPENSERIFORMES

Acipenseridae
    Acipenser fulvescens
    Acipenser sturio

OSTEOGLOSSIFORMES

Osteoglossidae
    Arapaima gigas

SALMONIFORMES

Salmonidae
    Stenodus leucichthys leucichthys
    Salmo chrysogaster

CYPRINIFORMES

Cyprinidae
    Plagopterus argentissimus
    Ptychocheilus lucius

ATHERINIFORMES

Cyprinodontidae
    Cynolebias constanciae
    Cynolebias marmoratus
    Cynolebias minimus
    Cynolebias opalescens
    Cynolebias splendens
Poeciliidae
    Xiphophorus couchianus

COELACANTHIFORMES

Coelacanthidae
    Latimeria chalumnae

CERATODIFORMES

Ceratodidae
    Neoceratodus forsteri

Mollusca

NAIADOIDA

Unionidae
    Cyprogenia aberti

Unionidae (cont'd.)
>	*Epioblasma (= Dysnomia) torulosa*
>		*rangiana*
>	*Fusconaia subrotunda*
>	*Lampsilis brevicula*
>	*Lexingtonia dolabelloides*
>	*Pleorobema clava*

STYLOMMATOPHORA

Camaenidae
>	*Papustyla (= Papuina) pulcherrima*

Paraphantidae
>	*Paraphanta* spp. + 202

PROSOBRANCHIA

Hydrobiidae
>	*Coahuilix hubbsi*
>	*Cochliopina milleri*
>	*Durangonella coahuilae*
>	*Mexipyrgus carranzae*
>	*Mexipyrgus churinceanus*
>	*Mexipyrgus escobedae*
>	*Mexipyrgus lugoi*
>	*Mexipyrgus mojarralis*
>	*Mexipyrgus multilineatus*
>	*Mexithauma quadripaludium*
>	*Nymphophilus minckleyi*
>	*Paludiscala caramba*

Insecta

LEPIDOPTERA

Papilionidae
>	*Parnassius apollo apollo*

FLORA

APOCYNACEAE
>	*Pachypodium* spp.

ARALIACEAE
>	*Panax quinquefolium* # 1

ARAUCARIACEAE
>	*Araucaria araucana* # 2

CACTACEAE
Cactaceae spp. + 203
>	*Rhipsalis* spp.

COMPOSITAE
>	*Saussurea lappa* # 1

CYATHEACEAE
>	*Cyathea (Hemitelia) capensis* #3
>	*Cyathea dredgei* # 3
>	*Cyathea mexicana* # 3
>	*Cyathea (Alsophila) salvinii* # 3

DIOSCOREACEAE
>	*Dioscorea deltoidea* # 1

EUPHORBIACEAE
>	*Euphorbia* spp. − 101

FAGACEAE
>	*Quercus copeyensis* # 2

LEGUMINOSAE
>	*Thermopsis mongolica*

LILIACEAE
>	*Aloe* spp.*

MELIACEAE
>	*Swietenia humilis* # 2

ORCHIDACEAE
>	Spp.*

PALMAE
>	*Arenga ipot*
>	*Phoenix hanceana* var. *philippinensis*
>	*Zalacca clemensiana*

PORTULACACEAE
>	*Anacampseros* spp.

PRIMULACEAE
>	*Cyclamen* spp.

SOLANACEAE
>	*Solanum sylvestris*

STERCULIACEAE
>	*Basiloxylon excelsum* # 2

VERBENACEAE
>	*Caryopteris mongolica*

ZYGOPHYLLACEAE
>	*Guaiacum sanctum* # 2

Appendix 4

Public Law 93-205
93rd Congress, S. 1983
December 28, 1973

An Act

87 STAT. 884

To provide for the conservation of endangered and threatened species of fish, wildlife, and plants, and for other purposes.

Be it enacted by the Senate and House of Representatives of the United States of America in Congress assembled, That this Act may be cited as the "Endangered Species Act of 1973".

Endangered Species Act of 1973.

TABLE OF CONTENTS

FINDINGS, PURPOSES, AND POLICY

SEC. 2. (a) FINDINGS.--The Congress finds and declares that—

(1) various species of fish, wildlife, and plants in the United States have been rendered extinct as a consequence of economic growth and development untempered by adequate concern and conservation;

(2) other species of fish, wildlife, and plants have been so depleted in numbers that they are in danger of or threatened with extinction;

(3) these species of fish, wildlife, and plants are of esthetic, ecological, educational, historical, recreational, and scientific value to the Nation and its people;

(4) the United States has pledged itself as a sovereign state in the international community to conserve to the extent practicable the various species of fish or wildlife and plants facing extinction, pursuant to—

(A) migratory bird treaties with Canada and Mexico;

(B) the Migratory and Endangered Bird Treaty with Japan;

(C) the Convention on Nature Protection and Wildlife Preservation in the Western Hemisphere; 56 Stat. 1354.

(D) the International Convention for the Northwest Atlantic Fisheries; 1 UST 477.

(E) the International Convention for the High Seas Fisheries of the North Pacific Ocean; 4 UST 380.

(F) the Convention on International Trade in Endangered Species of Wild Fauna and Flora; and

(G) other international agreements.

(5) encouraging the States and other interested parties, through Federal financial assistance and a system of incentives, to develop and maintain conservation programs which meet national and international standards is a key to meeting the

26-287 (253) O - 74

417

Nation s international commitments and to better safeguarding, for the benefit of all citizens, the Nation's heritage in fish and wildlife.

(b) PURPOSES.—The purposes of this Act are to provide a means whereby the ecosystems upon which endangered species and threatened species depend may be conserved, to provide a program for the conservation of such endangered species and threatened species, and to take such steps as may be appropriate to achieve the purposes of the treaties and conventions set forth in subsection (a) of this section.

(c) POLICY.—It is further declared to be the policy of Congress that all Federal departments and agencies shall seek to conserve endangered species and threatened species and shall utilize their authorities in furtherance of the purposes of this Act.

DEFINITIONS

SEC. 3. For the purposes of this Act—

(1) The term "commercial activity" means all activities of industry and trade, including, but not limited to, the buying or selling of commodities and activities conducted for the purpose of facilitating such buying and selling.

(2) The terms "conserve", "conserving", and "conservation" mean to use and the use of all methods and procedures which are necessary to bring any endangered species or threatened species to the point at which the measures provided pursuant to this Act are no longer necessary. Such methods and procedures include, but are not limited to, all activities associated with scientific resources management such as research, census, law enforcement, habitat acquisition and maintenance, propagation, live trapping, and transplantation, and, in the extraordinary case where population pressures within a given ecosystem cannot be otherwise relieved, may include regulated taking.

(3) The term "Convention" means the Convention on International Trade in Endangered Species of Wild Fauna and Flora, signed on March 3, 1973, and the appendices thereto.

(4) The term "endangered species" means any species which is in danger of extinction throughout all or a significant portion of its range other than a species of the Class Insecta determined by the Secretary to constitute a pest whose protection under the provisions of this Act would present an overwhelming and overriding risk to man.

(5) The term "fish or wildlife" means any member of the animal kingdom, including without limitation any mammal, fish, bird (including any migratory, nonmigratory, or endangered bird for which protection is also afforded by treaty or other international agreement), amphibian, reptile, mollusk, crustacean, arthropod or other invertebrate, and includes any part, product, egg, or offspring thereof, or the dead body or parts thereof.

(6) The term "foreign commerce" includes, among other things, any transaction—

    (A) between persons within one foreign country;

    (B) between persons in two or more foreign countries;

    (C) between a person within the United States and a person in a foreign country; or

    (D) between persons within the United States, where the fish and wildlife in question are moving in any country or countries outside the United States.

(7) The term "import" means to land on, bring into, or introduce into, or attempt to land on, bring into, or introduce into, any

place subject to the jurisdiction of the United States, whether or not such landing, bringing, or introduction constitutes an importation within the meaning of the customs laws of the United States.

(8) The term "person" means an individual, corporation, partnership, trust, association, or any other private entity, or any officer, employee, agent, department, or instrumentality of the Federal Government, of any State or political subdivision thereof, or of any foreign government.

(9) The term "plant" means any member of the plant kingdom, including seeds, roots and other parts thereof.

(10) The term "Secretary" means, except as otherwise herein provided, the Secretary of the Interior or the Secretary of Commerce as program responsibilities are vested pursuant to the provisions of Reorganization Plan Numbered 4 of 1970; except that 84 Stat. 2090. with respect to the enforcement of the provisions of this Act and 5 USC app. the Convention which pertain to the importation or exportation of terrestrial plants, the term means the Secretary of Agriculture.

(11) The term "species" includes any subspecies of fish or wildlife or plants and any other group of fish or wildlife of the same species or smaller taxa in common spatial arrangement that interbreed when mature.

(12) The term "State" means any of the several States, the District of Columbia, the Commonwealth of Puerto Rico, American Samoa, the Virgin Islands, Guam, and the Trust Territory of the Pacific Islands.

(13) The term "State agency" means the State agency, department, board, commission, or other governmental entity which is responsible for the management and conservation of fish or wildlife resources within a State.

(14) The term "take" means to harass, harm, pursue, hunt, shoot, wound, kill, trap, capture, or collect, or to attempt to engage in any such conduct.

(15) The term "threatened species" means any species which is likely to become an endangered species within the foreseeable future throughout all or a significant portion of its range.

(16) The term "United States", when used in a geographical context, includes all States.

DETERMINATION OF ENDANGERED SPECIES AND THREATENED SPECIES

SEC. 4. (a) GENERAL.—(1) The Secretary shall by regulation determine whether any species is an endangered species or a threatened species because of any of the following factors:

(1) the present or threatened destruction, modification, or curtailment of its habitat or range;

(2) overutilization for commercial, sporting, scientific, or educational purposes;

(3) disease or predation;

(4) the inadequacy of existing regulatory mechanisms; or

(5) other natural or manmade factors affecting its continued existence.

(2) With respect to any species over which program responsibilities have been vested in the Secretary of Commerce pursuant to Reorganization Plan Numbered 4 of 1970—

(A) in any case in which the Secretary of Commerce determines that such species should—

(i) be listed as an endangered species or a threatened species, or

(ii) be changed in status from a threatened species to an endangered species,

he shall so inform the Secretary of the Interior, who shall list such species in accordance with this section;

(B) in any case in which the Secretary of Commerce determines that such species should—

(i) be removed from any list published pursuant to subsection (c) of this section, or

(ii) be changed in status from an endangered species to a threatened species,

he shall recommend such action to the Secretary of the Interior, and the Secretary of the Interior, if he concurs in the recommendation, shall implement such action; and

(C) the Secretary of the Interior may not list or remove from any list any such species, and may not change the status of any such species which are listed, without a prior favorable determination made pursuant to this section by the Secretary of Commerce.

(b) BASIS FOR DETERMINATIONS.—(1) The Secretary shall make determinations required by subsection (a) of this section on the basis of the best scientific and commercial data available to him and after consultation, as appropriate, with the affected States, interested persons and organizations, other interested Federal agencies, and, in cooperation with the Secretary of State, with the country or countries in which the species concerned is normally found or whose citizens harvest such species on the high seas; except that in any case in which such determinations involve resident species of fish or wildlife, the Secretary of the Interior may not add such species to, or remove such species from, any list published pursuant to subsection (c) of this section, unless the Secretary has first—

Notice, publication in Federal Register.

(A) published notice in the Federal Register and notified the Governor of each State within which such species is then known to occur that such action is contemplated;

(B) allowed each such State 90 days after notification to submit its comments and recommendations, except to the extent that such period may be shortened by agreement between the Secretary and the Governor or Governors concerned; and

(C) published in the Federal Register a summary of all comments and recommendations received by him which relate to such proposed action.

(2) In determining whether or not any species is an endangered species or a threatened species, the Secretary shall take into consideration those efforts, if any, being made by any nation or any political subdivision of any nation to protect such species, whether by predator control, protection of habitat and food supply, or other conservation practices, within any area under the jurisdiction of any such nation or political subdivision, or on the high seas.

(3) Species which have been designated as requiring protection from unrestricted commerce by any foreign country, or pursuant to any international agreement, shall receive full consideration by the Secretary to determine whether each is an endangered species or a threatened species.

Publication in Federal Register.

(c) LISTS.—(1) The Secretary of the Interior shall publish in the Federal Register, and from time to time he may by regulation revise, a list of all species determined by him or the Secretary of Commerce to be endangered species and a list of all species determined by him or the Secretary of Commerce to be threatened species. Each list shall refer to the species contained therein by scientific and common name or names, if any, and shall specify with respect to each such species over what portion of its range it is endangered or threatened.

(2) The Secretary shall, upon the petition of an interested person under subsection 553(e) of title 5, United States Code, conduct a review of any listed or unlisted species proposed to be removed from or added to either of the lists published pursuant to paragraph (1) of this subsection, but only if he makes and publishes a finding that such person has presented substantial evidence which in his judgment warrants such a review.

Review.
80 Stat. 383.

(3) Any list in effect on the day before the date of the enactment of this Act of species of fish or wildlife determined by the Secretary of the Interior, pursuant to the Endangered Species Conservation Act of 1969, to be threatened with extinction shall be republished to conform to the classification for endangered species or threatened species, as the case may be, provided for in this Act, but until such republication, any such species so listed shall be deemed an endangered species within the meaning of this Act. The republication of any species pursuant to this paragraph shall not require public hearing or comment under section 553 of title 5, United States Code.

80 Stat. 926;
83 Stat. 275,
283.
16 USC 668aa
note.

(d) PROTECTIVE REGULATIONS.—Whenever any species is listed as a threatened species pursuant to subsection (c) of this section, the Secretary shall issue such regulations as he deems necessary and advisable to provide for the conservation of such species. The Secretary may by regulation prohibit with respect to any threatened species any act prohibited under section 9(a)(1), in the case of fish or wildlife, or section 9(a)(2), in the case of plants, with respect to endangered species; except that with respect to the taking of resident species of fish or wildlife, such regulations shall apply in any State which has entered into a cooperative agreement pursuant to section 6(a) of this Act only to the extent that such regulations have also been adopted by such State.

Post, p. 893.

(e) SIMILARITY OF APPEARANCE CASES.—The Secretary may, by regulation, and to the extent he deems advisable, treat any species as an endangered species or threatened species even though it is not listed pursuant to section 4 of this Act if he finds that—

(A) such species so closely resembles in appearance, at the point in question, a species which has been listed pursuant to such section that enforcement personnel would have substantial difficulty in attempting to differentiate between the listed and unlisted species;

(B) the effect of this substantial difficulty is an additional threat to an endangered or threatened species; and

(C) such treatment of an unlisted species will substantially facilitate the enforcement and further the policy of this Act.

(f) REGULATIONS.—(1) Except as provided in paragraphs (2) and (3) of this subsection and subsection (b) of this section, the provisions of section 553 of title 5, United States Code (relating to rulemaking procedures), shall apply to any regulation promulgated to carry out the purposes of this Act.

(2)(A) In the case of any regulation proposed by the Secretary to carry out the purposes of this Act—

(i) the Secretary shall publish general notice of the proposed regulation (including the complete text of the regulation) in the Federal Register not less than 60 days before the effective date of the regulation; and

Notice, publication in Federal Register.

(ii) if any person who feels that he may be adversely affected by the proposed regulation files (within 45 days after the date of publication of general notice) objections thereto and requests a public hearing thereon, the Secretary may grant such request, but shall, if he denies such request, publish his reasons therefor in the Federal Register.

Hearing request.

Publication in Federal Register.

80 Stat. 383.

(B) Neither subparagraph (A) of this paragraph nor section 553 of title 5, United States Code, shall apply in the case of any of the following regulations and any such regulation shall, at the discretion of the Secretary, take effect immediately upon publication of the regulation in the Federal Register:

(i) Any regulation appropriate to carry out the purposes of this Act which was originally promulgated to carry out the Endangered Species Conservation Act of 1969.

80 Stat. 926;
83 Stat. 275, 283.
16 USC 668aa note.

(ii) Any regulation (including any regulation implementing section 6(g)(2)(B)(ii) of this Act) issued by the Secretary in regard to any emergency posing a significant risk to the well-being of any species of fish or wildlife, but only if (I) at the time of publication of the regulation in the Federal Register the Secretary publishes therein detailed reasons why such regulation is necessary, and (II) in the case such regulation applies to resident species of fish and wildlife, the requirements of subsection (b) (A), (B), and (C) of this section have been complied with. Any regulation promulgated under the authority of this clause (ii) shall cease to have force and effect at the close of the 120-day period following the date of publication unless, during such 120-day period, the rulemaking procedures which would apply to such regulation without regard to this subparagraph are complied with.

Statement by the Secretary.

(3) The publication in the Federal Register of any proposed or final regulation which is necessary or appropriate to carry out the purposes of this Act shall include a statement by the Secretary of the facts on which such regulation is based and the relationship of such facts to such regulation.

LAND ACQUISITION

SEC. 5. (a) PROGRAM.—The Secretary of the Interior shall establish and implement a program to conserve (A) fish or wildlife which are listed as endangered species or threatened species pursuant to section 4 of this Act; or (B) plants which are concluded in Appendices to the Convention. To carry out such program, he—

70 Stat. 1119.
16 USC 742a note.
60 Stat. 1080;
72 Stat. 563.
16 USC 661 note.

(1) shall utilize the land acquisition and other authority under the Fish and Wildlife Act of 1956, as amended, the Fish and Wildlife Coordination Act, as amended, and the Migratory Bird Conservation Act, as appropriate; and

(2) is authorized to acquire by purchase, donation, or otherwise, lands, waters, or interest therein, and such authority shall be in addition to any other land acquisition authority vested in him.

45 Stat. 1222.
16 USC 4601-4 note.

(b) ACQUISITIONS.—Funds made available pursuant to the Land and Water Conservation Fund Act of 1965, as amended, may be used for the purpose of acquiring lands, waters, or interests therein under subsection (a) of this section.

COOPERATION WITH THE STATES

SEC. 6. (a) GENERAL.—In carrying out the program authorized by this Act, the Secretary shall cooperate to the maximum extent practicable with the States. Such cooperation shall include consultation with the States concerned before acquiring any land or water, or interest therein, for the purpose of conserving any endangered species or threatened species.

(b) MANAGEMENT AGREEMENTS.—The Secretary may enter into agreements with any State for the administration and management of any area established for the conservation of endangered species or

threatened species. Any revenues derived from the administration of
such areas under these agreements shall be subject to the provisions
of section 401 of the Act of June 15, 1935 (49 Stat. 383; 16 U.S.C.
715s).

78 Stat. 701.

(c) COOPERATIVE AGREEMENTS.—In furtherance of the purposes of
this Act, the Secretary is authorized to enter into a cooperative agreement in accordance with this section with any State which establishes
and maintains an adequate and active program for the conservation of
endangered species and threatened species. Within one hundred and
twenty days after the Secretary receives a certified copy of such a proposed State program, he shall make a determination whether such
program is in accordance with this Act. Unless he determines, pursuant
to this subsection, that the State program is not in accordance with this
Act, he shall enter into a cooperative agreement with the State for the
purpose of assisting in implementation of the State program. In order
for a State program to be deemed an adequate and active program for
the conservation of endangered species and threatened species, the Secretary must find, and annually thereafter reconfirm such finding, that
under the State program—

(1) authority resides in the State agency to conserve resident
species of fish or wildlife determined by the State agency or the
Secretary to be endangered or threatened;

(2) the State agency has established acceptable conservation
programs, consistent with the purposes and policies of this Act,
for all resident species of fish or wildlife in the State which are
deemed by the Secretary to be endangered or threatened, and has
furnished a copy of such plan and program together with all
pertinent details, information, and data requested to the
Secretary;

(3) the State agency is authorized to conduct investigations to
determine the status and requirements for survival of resident
species of fish and wildlife;

(4) the State agency is authorized to establish programs, including the acquisition of land or aquatic habitat or interests
therein, for the conservation of resident endangered species or
threatened species; and

(5) provision is made for public participation in designating
resident species of fish or wildlife as endangered or threatened.

(d) ALLOCATION OF FUNDS.—(1) The Secretary is authorized to
provide financial assistance to any State, through its respective State
agency, which has entered into a cooperative agreement pursuant to
subsection (c) of this section to assist in development of programs for
the conservation of endangered and threatened species. The Secretary
shall make an allocation of appropriated funds to such States based
on consideration of—

(A) the international commitments of the United States to
protect endangered species or threatened species;

(B) the readiness of a State to proceed with a conservation
program consistent with the objectives and purposes of this Act;

(C) the number of endangered species and threatened species
within a State;

(D) the potential for restoring endangered species and threatened species within a State; and

(E) the relative urgency to initiate a program to restore and
protect an endangered species or threatened species in terms of
survival of the species.

So much of any appropriated funds allocated for obligation to any
State for any fiscal year as remains unobligated at the close thereof is

authorized to be made available to that State until the close of the succeeding fiscal year. Any amount allocated to any State which is unobligated at the end of the period during which it is available for expenditure is authorized to be made available for expenditure by the Secretary in conducting programs under this section.

(2) Such cooperative agreements shall provide for (A) the actions to be taken by the Secretary and the States; (B) the benefits that are expected to be derived in connection with the conservation of endangered or threatened species; (C) the estimated cost of these actions; and (D) the share of such costs to be borne by the Federal Government and by the States; except that—

(i) the Federal share of such program costs shall not exceed 66⅔ per centum of the estimated program cost stated in the agreement; and

(ii) the Federal share may be increased to 75 per centum whenever two or more States having a common interest in one or more endangered or threatened species, the conservation of which may be enhanced by cooperation of such States, enter jointly into an agreement with the Secretary.

The Secretary may, in his discretion, and under such rules and regulations as he may prescribe, advance funds to the State for financing the United States pro rata share agreed upon in the cooperative agreement. For the purposes of this section, the non-Federal share may, in the discretion of the Secretary, be in the form of money or real property, the value of which will be determined by the Secretary, whose decision shall be final.

(e) REVIEW OF STATE PROGRAMS.—Any action taken by the Secretary under this section shall be subject to his periodic review at no greater than annual intervals.

(f) CONFLICTS BETWEEN FEDERAL AND STATE LAWS.—Any State law or regulation which applies with respect to the importation or exportation of, or interstate or foreign commerce in, endangered species or threatened species is void to the extent that it may effectively (1) permit what is prohibited by this Act or by any regulation which implements this Act, or (2) prohibit what is authorized pursuant to an exemption or permit provided for in this Act or in any regulation which implements this Act. This Act shall not otherwise be construed to void any State law or regulation which is intended to conserve migratory, resident, or introduced fish or wildlife, or to permit or prohibit sale of such fish or wildlife. Any State law or regulation respecting the taking of an endangered species or threatened species may be more restrictive than the exemptions or permits provided for in this Act or in any regulation which implements this Act but not less restrictive than the prohibitions so defined.

"Establishment period."

(g) TRANSITION.—(1) For purposes of this subsection, the term "establishment period" means, with respect to any State, the period beginning on the date of enactment of this Act and ending on whichever of the following dates first occurs: (A) the date of the close of the 120-day period following the adjournment of the first regular session of the legislature of such State which commences after such date of enactment, or (B) the date of the close of the 15-month period following such date of enactment.

(2) The prohibitions set forth in or authorized pursuant to sections 4(d) and 9(a)(1)(B) of this Act shall not apply with respect to the taking of any resident endangered species or threatened species (other than species listed in Appendix I to the Convention or otherwise specifically covered by any other treaty or Federal law) within any State—

(A) which is then a party to a cooperative agreement with the Secretary pursuant to section 6(c) of this Act (except to the extent that the taking of any such species is contrary to the law of such State) ; or

(B) except for any time within the establishment period when—

(i) the Secretary applies such prohibition to such species at the request of the State, or

(ii) the Secretary applies such prohibition after he finds, and publishes his finding, that an emergency exists posing a significant risk to the well-being of such species and that the prohibition must be applied to protect such species. The Secretary's finding and publication may be made without regard to the public hearing or comment provisions of section 553 of title 5, United States Code, or any other provision of this Act; but such prohibition shall expire 90 days after the date of its imposition unless the Secretary further extends such prohibition by publishing notice and a statement of justification of such extension.

80 Stat. 383.

(h) REGULATIONS.—The Secretary is authorized to promulgate such regulations as may be appropriate to carry out the provisions of this section relating to financial assistance to States.

(i) APPROPRIATIONS.—For the purposes of this section, there is authorized to be appropriated through the fiscal year ending June 30, 1977, not to exceed $10,000,000.

INTERAGENCY COOPERATION

SEC. 7. The Secretary shall review other programs administered by him and utilize such programs in furtherance of the purposes of this Act. All other Federal departments and agencies shall, in consultation with and with the assistance of the Secretary, utilize their authorities in furtherance of the purposes of this Act by carrying out programs for the conservation of endangered species and threatened species listed pursuant to section 4 of this Act and by taking such action necessary to insure that actions authorized, funded, or carried out by them do not jeopardize the continued existence of such endangered species and threatened species or result in the destruction or modification of habitat of such species which is determined by the Secretary, after consultation as appropriate with the affected States, to be critical.

INTERNATIONAL COOPERATION

SEC. 8. (a) FINANCIAL ASSISTANCE.—As a demonstration of the commitment of the United States to the worldwide protection of endangered species and threatened species, the President may, subject to the provisions of section 1415 of the Supplemental Appropriation Act, 1953 (31 U.S.C. 724), use foreign currencies accruing to the United States Government under the Agricultural Trade Development and Assistance Act of 1954 or any other law to provide to any foreign country (with its consent) assistance in the development and management of programs in that country which the Secretary determines to be necessary or useful for the conservation of any endangered species or threatened species listed by the Secretary pursuant to section 4 of this Act. The President shall provide assistance (which includes, but is not limited to, the acquisition, by lease or otherwise, of lands, waters, or interests therein) to foreign countries under this section under such terms and conditions as he deems appropriate. Whenever foreign currencies are available for the provision of assistance

66 Stat. 662.

68 Stat. 454.
7 USC 1691.

under this section, such currencies shall be used in preference to funds appropriated under the authority of section 15 of this Act.

(b) ENCOURAGEMENT OF FOREIGN PROGRAMS.—In order to carry out further the provisions of this Act, the Secretary, through the Secretary of State, shall encourage—

(1) foreign countries to provide for the conservation of fish or wildlife including endangered species and threatened species listed pursuant to section 4 of this Act;

(2) the entering into of bilateral or multilateral agreements with foreign countries to provide for such conservation; and

(3) foreign persons who directly or indirectly take fish or wildlife in foreign countries or on the high seas for importation into the United States for commercial or other purposes to develop and carry out with such assistance as he may provide, conservation practices designed to enhance such fish or wildlife and their habitat.

(c) PERSONNEL.—After consultation with the Secretary of State, the Secretary may—

(1) assign or otherwise make available any officer or employee of his department for the purpose of cooperating with foreign countries and international organizations in developing personnel resources and programs which promote the conservation of fish or wildlife; and

(2) conduct or provide financial assistance for the educational training of foreign personnel, in this country or abroad, in fish, wildlife, or plant management, research and law enforcement and to render professional assistance abroad in such matters.

(d) INVESTIGATIONS.—After consultation with the Secretary of State and the Secretary of the Treasury, as appropriate, the Secretary may conduct or cause to be conducted such law enforcement investigations and research abroad as he deems necessary to carry out the purposes of this Act.

(e) CONVENTION IMPLEMENTATION.—The President is authorized and directed to designate appropriate agencies to act as the Management Authority or Authorities and the Scientific Authority or Authorities pursuant to the Convention. The agencies so designated shall thereafter be authorized to do all things assigned to them under the Convention, including the issuance of permits and certificates. The agency designated by the President to communicate with other parties to the Convention and with the Secretariat shall also be empowered, where appropriate, in consultation with the State Department, to act on behalf of and represent the United States in all regards as required by the Convention. The President shall also designate those agencies which shall act on behalf of and represent the United States in all regards as required by the Convention on Nature Protection and Wildlife Preservation in the Western Hemisphere.

56 Stat. 1354.

PROHIBITED ACTS

SEC. 9. (a) GENERAL.—(1) Except as provided in sections 6(g)(2) and 10 of this Act, with respect to any endangered species of fish or wildlife listed pursuant to section 4 of this Act it is unlawful for any person subject to the jurisdiction of the United States to—

(A) import any such species into, or export any such species from the United States;

(B) take any such species within the United States or the territorial sea of the United States;

(C) take any such species upon the high seas;

(D) possess, sell, deliver, carry, transport, or ship, by any means whatsoever, any such species taken in violation of sub-paragraphs (B) and (C);

(E) deliver, receive, carry, transport, or ship in interstate or foreign commerce, by any means whatsoever and in the course of a commercial activity, any such species;

(F) sell or offer for sale in interstate or foreign commerce any such species; or

(G) violate any regulation pertaining to such species or to any threatened species of fish or wildlife listed pursuant to section 4 of this Act and promulgated by the Secretary pursuant to authority provided by this Act.

(2) Except as provided in sections 6(g)(2) and 10 of this Act, with respect to any endangered species of plants listed pursuant to section 4 of this Act, it is unlawful for any person subject to the jurisdiction of the United States to—

(A) import any such species into, or export any such species from, the United States;

(B) deliver, receive, carry, transport, or ship in interstate or foreign commerce, by any means whatsoever and in the course of a commercial activity, any such species;

(C) sell or offer for sale in interstate or foreign commerce any such species; or

(D) violate any regulation pertaining to such species or to any threatened species of plants listed pursuant to section 4 of this Act and promulgated by the Secretary pursuant to authority provided by this Act.

(b) SPECIES HELD IN CAPTIVITY OR CONTROLLED ENVIRONMENT.— The provisions of this section shall not apply to any fish or wildlife held in captivity or in a controlled environment on the effective date of this Act if the purposes of such holding are not contrary to the purposes of this Act; except that this subsection shall not apply in the case of any fish or wildlife held in the course of a commercial activity. With respect to any act prohibited by this section which occurs after a period of 180 days from the effective date of this Act, there shall be a rebuttable presumption that the fish or wildlife involved in such act was not held in captivity or in a controlled environment on such effective date.

(c) VIOLATION OF CONVENTION.—(1) It is unlawful for any person subject to the jurisdiction of the United States to engage in any trade in any specimens contrary to the provisions of the Convention, or to possess any specimens traded contrary to the provisions of the Convention, including the definitions of terms in article I thereof.

(2) Any importation into the United States of fish or wildlife shall, if—

(A) such fish or wildlife is not an endangered species listed pursuant to section 4 of this Act but is listed in Appendix II to the Convention,

(B) the taking and exportation of such fish or wildlife is not contrary to the provisions of the Convention and all other applicable requirements of the Convention have been satisfied,

(C) the applicable requirements of subsections (d), (e), and (f) of this section have been satisfied, and

(D) such importation is not made in the course of a commercial activity,

be presumed to be an importation not in violation of any provision of this Act or any regulation issued pursuant to this Act.

(d) IMPORTS AND EXPORTS.—(1) It is unlawful for any person to engage in business as an importer or exporter of fish or wildlife (other than shellfish and fishery products which (A) are not listed pursuant to section 4 of this Act as endangered species or threatened species, and (B) are imported for purposes of human or animal consumption or taken in waters under the jurisdiction of the United States or on the high seas for recreational purposes) or plants without first having obtained permission from the Secretary.

Recordkeeping

(2) Any person required to obtain permission under paragraph (1) of this subsection shall—

(A) keep such records as will fully and correctly disclose each importation or exportation of fish, wildlife, or plants made by him and the subsequent disposition made by him with respect to such fish, wildlife, or plants;

(B) at all reasonable times upon notice by a duly authorized representative of the Secretary, afford such representative access to his places of business, an opportunity to examine his inventory of imported fish, wildlife, or plants and the records required to be kept under subparagraph (A) of this paragraph, and to copy such records; and

(C) file such reports as the Secretary may require.

(3) The Secretary shall prescribe such regulations as are necessary and appropriate to carry out the purposes of this subsection.

(e) REPORTS.—It is unlawful for any person importing or exporting fish or wildlife (other than shellfish and fishery products which (1) are not listed pursuant to section 4 of this Act as endangered or threatened species, and (2) are imported for purposes of human or animal consumption or taken in waters under the jurisdiction of the United States or on the high seas for recreational purposes) or plants to fail to file any declaration or report as the Secretary deems necessary to facilitate enforcement of this Act or to meet the obligations of the Convention.

(f) DESIGNATION OF PORTS.—(1) It is unlawful for any person subject to the jurisdiction of the United States to import into or export from the United States any fish or wildlife (other than shellfish and fishery products which (A) are not listed pursuant to section 4 of this Act as endangered species or threatened species, and (B) are imported for purposes of human or animal consumption or taken in waters under the jurisdiction of the United States or on the high seas for recreational purposes) or plants, except at a port or ports designated by the Secretary of the Interior. For the purpose of facilitating enforcement of this Act and reducing the costs thereof, the Secretary of the Interior, with approval of the Secretary of the Treasury and after notice and opportunity for public hearing, may, by regulation, designate ports and change such designations. The Secretary of the Interior, under such terms and conditions as he may prescribe, may permit the importation or exportation at nondesignated ports in the interest of the health or safety of the fish or wildlife or plants, or for other reasons if, in his discretion, he deems it appropriate and consistent with the purpose of this subsection.

(2) Any port designated by the Secretary of the Interior under the authority of section 4(d) of the Act of December 5, 1969 (16 U.S.C. 666cc–4(d)), shall, if such designation is in effect on the day before the date of the enactment of this Act, be deemed to be a port designated by the Secretary under paragraph (1) of this subsection until such time as the Secretary otherwise provides.

83 Stat. 277.
16 USC 668cc–4.

(g) VIOLATIONS.—It is unlawful for any person subject to the jurisdiction of the United States to attempt to commit, solicit another to commit, or cause to be committed, any offense defined in this section.

Sec. 10. (a) Permits.—The Secretary may permit, under such terms
and conditions as he may prescribe, any act otherwise prohibited by
section 9 of this Act for scientific purposes or to enhance the propagation or survival of the affected species.

(b) Hardship Exemptions.—(1) If any person enters into a contract with respect to a species of fish or wildlife or plant before the
date of the publication in the Federal Register of notice of consideration of that species as an endangered species and the subsequent listing
of that species as an endangered species pursuant to section 4 of this
Act will cause undue economic hardship to such person under the contract, the Secretary, in order to minimize such hardship, may exempt
such person from the application of section 9(a) of this Act to the
extent the Secretary deems appropriate if such person applies to him
for such exemption and includes with such application such information as the Secretary may require to prove such hardship; except that
(A) no such exemption shall be for a duration of more than one year
from the date of publication in the Federal Register of notice of consideration of the species concerned, or shall apply to a quantity of fish
or wildlife or plants in excess of that specified by the Secretary; (B)
the one-year period for those species of fish or wildlife listed by the
Secretary as endangered prior to the effective date of this Act shall
expire in accordance with the terms of section 3 of the Act of December 5, 1969 (83 Stat. 275); and (C) no such exemption may be granted   16 USC 668cc-3.
for the importation or exportation of a specimen listed in Appendix
I of the Convention which is to be used in a commercial activity.

(2) As used in this subsection, the term "undue economic hardship"   "Undue economic
shall include, but not be limited to: hardship."

(A) substantial economic loss resulting from inability caused by
this Act to perform contracts with respect to species of fish and
wildlife entered into prior to the date of publication in the Federal Register of a notice of consideration of such species as an
endangered species;

(B) substantial economic loss to persons who, for the year prior
to the notice of consideration of such species as an endangered
species, derived a substantial portion of their income from the
lawful taking of any listed species, which taking would be made
unlawful under this Act; or

(C) curtailment of subsistence taking made unlawful under
this Act by persons (i) not reasonably able to secure other sources
of subsistence; and (ii) dependent to a substantial extent upon
hunting and fishing for subsistence; and (iii) who must engage
in such curtailed taking for subsistence purposes.

(3) The Secretary may make further requirements for a showing of
undue economic hardship as he deems fit. Exceptions granted under this
section may be limited by the Secretary in his discretion as to time,
area, or other factor of applicability.

(c) Notice and Review.—The Secretary shall publish notice in the   Publication in
Federal Register of each application for an exemption or permit which   Federal Register.
is made under this subsection. Each notice shall invite the submission
from interested parties, within thirty days after the date of the notice,
written data, views, or arguments with respect to the application.
Information received by the Secretary as a part of any application shall
be available to the public as a matter of public record at every stage
of the proceeding.

(d) Permit and Exemption Policy.—The Secretary may grant
exceptions under subsections (a) and (b) of this section only if he finds

and publishes his finding in the Federal Register that (1) such exceptions were applied for in good faith, (2) if granted and exercised will not operate to the disadvantage of such endangered species, and (3) will be consistent with the purposes and policy set forth in section 2 of this Act.

(e) ALASKA NATIVES.— (1) Except as provided in paragraph (4) of this subsection the provisions of this Act shall not apply with respect to the taking of any endangered species or threatened species, or the importation of any such species taken pursuant to this section, by—

(A) any Indian, Aleut, or Eskimo who is an Alaskan Native who resides in Alaska; or

(B) any non-native permanent resident of an Alaskan native village;

if such taking is primarily for subsistence purposes. Non-edible byproducts of species taken pursuant to this section may be sold in interstate commerce when made into authentic native articles of handicrafts and clothing; except that the provisions of this subsection shall not apply to any non-native resident of an Alaskan native village found by the Secretary to be not primarily dependent upon the taking of fish and wildlife for consumption or for the creation and sale of authentic native articles of handicrafts and clothing.

(2) Any taking under this subsection may not be accomplished in a wasteful manner.

Definitions.

(3) As used in this subsection—

(i) The term "subsistence" includes selling any edible portion of fish or wildlife in native villages and towns in Alaska for native consumption within native villages or towns; and

(ii) The term "authentic native articles of handicrafts and clothing" means items composed wholly or in some significant respect of natural materials, and which are produced, decorated, or fashioned in the exercise of traditional native handicrafts without the use of pantographs, multiple carvers, or other mass copying devices. Traditional native handicrafts include, but are not limited to, weaving, carving, stitching, sewing, lacking, beading, drawing, and painting.

(4) Notwithstanding the provisions of paragraph (1) of this subsection, whenever the Secretary determines that any species of fish or wildlife which is subject to taking under the provisions of this subsection is an endangered species or threatened species, and that such taking materially and negatively affects the threatened or endangered species, he may prescribe regulations upon the taking of such species by any such Indian, Aleut, Eskimo, or non-Native Alaskan resident of an Alaskan native village. Such regulations may be established with reference to species, geographical description of the area included, the season for taking, or any other factors related to the reason for establishing such regulations and consistent with the policy of this Act. Such regulations shall be prescribed after a notice and hearings in the affected judicial districts of Alaska and as otherwise

86 Stat. 1033.
16 USC 1373.

required by section 103 of the Marine Mammal Protection Act of 1972, and shall be removed as soon as the Secretary determines that the need for their impositions has disappeared.

PENALTIES AND ENFORCEMENT

SEC. 11. (a) CIVIL PENALTIES.— (1) Any person who knowingly violates, or who knowingly commits an act in the course of a commercial activity which violates, any provision of this Act, or any provision of any permit or certificate issued hereunder, or of any regulation issued in order to implement subsection (a)(1) (A), (B),

(C), (D), (E). or (F). (a)(2) (A), (B), or (C). (c), (d) (other than regulation relating to recordkeeping or filing of reports), (f) or (g) of section 9 of this Act, may be assessed a civil penalty by the Secretary of not more than $10,000 for each violation. Any person who knowingly violates, or who knowingly commits an act in the course of a commercial activity which violates, any provision of any other regulation issued under this Act may be assessed a civil penalty by the Secretary of not more than $5,000 for each such violation. Any person who otherwise violates any provision of this Act, or any regulation, permit, or certificate issued hereunder, may be assessed a civil penalty by the Secretary of not more than $1,000 for each such violation. No penalty may be assessed under this subsection unless **Notice;** such person is given notice and opportunity for a hearing with respect **hearing.** to such violation. Each violation shall be a separate offense. Any such civil penalty may be remitted or mitigated by the Secretary. Upon any failure to pay a penalty assessed under this subsection, the Secretary may request the Attorney General to institute a civil action in a district court of the United States for any district in which such person is found, resides, or transacts business to collect the penalty and such court shall have jurisdiction to hear and decide any such action. The court shall hear such action on the record made before the Secretary and shall sustain his action if it is supported by substantial evidence on the record considered as a whole.

(2) Hearings held during proceedings for the assessment of civil penalties authorized by paragraph (1) of this subsection shall be conducted in accordance with section 554 of title 5, United States Code. **80 Stat. 384.** The Secretary may issue subpenas for the attendance and testimony of witnesses and the production of relevant papers, books, and documents, and administer oaths. Witnesses summoned shall be paid the same fees and mileage that are paid to witnesses in the courts of the United States. In case of contumacy or refusal to obey a subpena served upon any person pursuant to this paragraph, the district court of the United States for any district in which such person is found or resides or transacts business, upon application by the United States and after notice to such person. shall have jurisdiction to issue an order requiring such person to appear and give testimony before the Secretary or to appear and produce documents before the Secretary, or both, and any failure to obey such order of the court may be punished by such court as a contempt thereof.

(b) CRIMINAL VIOLATIONS.—(1) Any person who willfully commits an act which violates any provision of this Act, of any permit or certificate issued hereunder, or of any regulation issued in order to implement subsection (a)(1) (A), (B). (C), (D), (E), or (F); (a) (2) (A), (B), or (C). (c), (d) (other than a regulation relating to recordkeeping, or filing of reports), (f), or (g) of section 9 of this Act shall, upon conviction, be fined not more than $20,000 or imprisoned for not more than one year, or both. Any person who willfully commits an act which violates any provision of any other regulation issued under this Act shall, upon conviction, be fined not more than $10,000 or imprisoned for not more than six months, or both.

(2) The head of any Federal agency which has issued a lease, license, permit, or other agreement authorizing the use of Federal lands, including grazing of domestic livestock, to any person who is convicted of a criminal violation of this Act or any regulation, permit, or certificate issued hereunder may immediately modify, suspend, or revoke each lease, license, permit, or other agreement. The Secretary shall also suspend for a period of up to one year, or cancel, any Federal hunting or fishing permits or stamps issued to any person who is convicted of a criminal violation of any provision of this Act or any

431

regulation, permit, or certificate issued hereunder. The United States shall not be liable for the payments of any compensation, reimbursement, or damages in connection with the modification, suspension, or revocation of any leases, licenses, permits, stamps, or other agreements pursuant to this section.

65 Stat. 725;
72 Stat. 348.

(c) DISTRICT COURT JURISDICTION.—The several district courts of the United States, including the courts enumerated in section 460 of title 28, United States Code, shall have jurisdiction over any actions arising under this Act. For the purpose of this Act, American Samoa shall be included within the judicial district of the District Court of the United States for the District of Hawaii.

(d) REWARDS.—Upon the recommendation of the Secretary, the Secretary of the Treasury is authorized to pay an amount equal to one-half of the civil penalty or fine paid, but not to exceed $2,500, to any person who furnishes information which leads to a finding of civil violation or a conviction of a criminal violation of any provision of this Act or any regulation or permit issued thereunder. Any officer or employee of the United States or of any State or local government who furnishes information or renders service in the performance of his official duties shall not be eligible for payment under this section.

(e) ENFORCEMENT.—(1) The provisions of this Act and any regulations or permits issued pursuant thereto shall be enforced by the Secretary, the Secretary of the Treasury, or the Secretary of the Department in which the Coast Guard is operating, or all such Secretaries. Each such Secretary may utilize by agreement, with or without reimbursement, the personnel, services, and facilities of any other Federal agency or any State agency for purposes of enforcing this Act.

Federal and
State agencies,
utilization.

Warrants.

(2) The judges of the district courts of the United States and the United States magistrates may, within their respective jurisdictions, upon proper oath or affirmation showing probable cause, issue such warrants or other process as may be required for enforcement of this Act and any regulation issued thereunder.

Package in-
spection.

(3) Any person authorized by the Secretary, the Secretary of the Treasury, or the Secretary of the Department in which the Coast Guard is operating, to enforce this Act may detain for inspection and inspect any package, crate, or other container, including its contents, and all accompanying documents, upon importation or exportation. Such person may execute and serve any arrest warrant, search warrant, or other warrant or civil or criminal process issued by any officer or court of competent jurisdiction for enforcement of this Act. Such person so authorized may search and seize, with or without a warrant, as authorized by law. Any fish, wildlife, property, or item so seized shall be held by any person authorized by the Secretary, the Secretary of the Treasury, or the Secretary of the Department in which the Coast Guard is operating pending disposition of civil or criminal proceedings, or the institution of an action in rem for forfeiture of such fish, wildlife, property, or item pursuant to paragraph (4) of this subsection; except that the Secretary may, in lieu of holding such fish, wildlife, property, or item, permit the owner or consignee to post a bond or other surety satisfactory to the Secretary.

Search and
seizure, au-
thority.

(4)(A) All fish or wildlife or plants taken, possessed, sold, purchased, offered for sale or purchase, transported, delivered, received, carried, shipped, exported, or imported contrary to the provisions of this Act, any regulation made pursuant thereto, or any permit or certificate issued hereunder shall be subject to forfeiture to the United States.

(B) All guns, traps, nets, and other equipment, vessels, vehicles, aircraft, and other means of transportation used to aid the taking, possessing, selling, purchasing, offering for sale or purchase, transporting, delivering, receiving, carrying, shipping, exporting, or importing of any fish or wildlife or plants in violation of this Act, any regulation made pursuant thereto, or any permit or certificate issued thereunder shall be subject to forfeiture to the United States upon conviction of a criminal violation pursuant to section 11(b)(1) of this Act.

(5) All provisions of law relating to the seizure, forfeiture, and condemnation of a vessel for violation of the customs laws, the disposition of such vessel or the proceeds from the sale thereof, and the remission or mitigation of such forfeiture, shall apply to the seizures and forfeitures incurred, or alleged to have been incurred, under the provisions of this Act, insofar as such provisions of law are applicable and not inconsistent with the provisions of this Act; except that all powers, rights, and duties conferred or imposed by the customs laws upon any officer or employee of the Treasury Department shall, for the purposes of this Act, be exercised or performed by the Secretary or by such persons as he may designate.

(f) REGULATIONS.—The Secretary, the Secretary of the Treasury, and the Secretary of the Department in which the Coast Guard is operating, are authorized to promulgate such regulations as may be appropriate to enforce this Act, and charge reasonable fees for expenses to the Government connected with permits or certificates authorized by this Act including processing applications and reasonable inspections, and with the transfer, board, handling, or storage of fish or wildlife or plants and evidentiary items seized and forfeited under this Act. All such fees collected pursuant to this subsection shall be deposited in the Treasury to the credit of the appropriation which is current and chargeable for the cost of furnishing the services. Appropriated funds may be expended pending reimbursement from parties in interest.

(g) CITIZEN SUITS.—(1) Except as provided in paragraph (2) of this subsection any person may commence a civil suit on his own behalf—

　　(A) to enjoin any person, including the United States and any other governmental instrumentality or agency (to the extent permitted by the eleventh amendment to the Constitution), who is alleged to be in violation of any provision of this Act or regulation issued under the authority thereof; or

　　(B) to compel the Secretary to apply, pursuant to section 6(g)(2)(B)(ii) of this Act, the prohibitions set forth in or authorized pursuant to section 4(d) or section 9(a)(1)(B) of this Act with respect to the taking of any resident endangered species or threatened species within any State.

The district courts shall have jurisdiction, without regard to the **Jurisdiction.** amount in controversy or the citizenship of the parties, to enforce any such provision or regulation, as the case may be. In any civil suit commenced under subparagraph (B) the district court shall compel the Secretary to apply the prohibition sought if the court finds that the allegation that an emergency exists is supported by substantial evidence.

(2)(A) No action may be commenced under subparagraph (1)(A) of this section—

　　(i) prior to sixty days after written notice of the violation has been given to the Secretary, and to any alleged violator of any such provision or regulation;

(ii) if the Secretary has commenced action to impose a penalty pursuant to subsection (a) of this section; or

(iii) if the United States has commenced and is diligently prosecuting a criminal action in a court of the United States or a State to redress a violation of any such provision or regulation.

(B) No action may be commenced under subparagraph (1)(B) of this section—

(i) prior to sixty days after written notice has been given to the Secretary setting forth the reasons why an emergency is thought to exist with respect to an endangered species or a threatened species in the State concerned; or

(ii) if the Secretary has commenced and is diligently prosecuting action under section 6(g)(2)(B)(ii) of this Act to determine whether any such emergency exists.

(3)(A) Any suit under this subsection may be brought in the judicial district in which the violation occurs.

Intervention.

(B) In any such suit under this subsection in which the United States is not a party, the Attorney General, at the request of the Secretary, may intervene on behalf of the United States as a matter of right.

Litigation costs.

(4) The court, in issuing any final order in any suit brought pursuant to paragraph (1) of this subsection, may award costs of litigation (including reasonable attorney and expert witness fees) to any party, whenever the court determines such award is appropriate.

Injunctive relief.

(5) The injunctive relief provided by this subsection shall not restrict any right which any person (or class of persons) may have under any statute or common law to seek enforcement of any standard or limitation or to seek any other relief (including relief against the Secretary or a State agency).

(h) COORDINATION WITH OTHER LAWS.—The Secretary of Agriculture and the Secretary shall provide for appropriate coordination of the administration of this Act with the administration of the animal quarantine laws (21 U.S.C. 101–105, 111–135b, and 612–614) and section 306 of the Tariff Act of 1930 (19 U.S.C. 1306). Nothing in this Act or any amendment made by this Act shall be construed as superseding or limiting in any manner the functions of the Secretary of Agriculture under any other law relating to prohibited or restricted importations or possession of animals and other articles and no proceeding or determination under this Act shall preclude any proceeding or determination under any Act administered by the Secretary of Agriculture. Nothing in this Act shall be construed as superseding or limiting in any manner the functions and responsibilities of the Secretary of the Treasury under the Tariff Act of 1930, including, without limitation, section 527 of that Act (19 U.S.C. 1527), relating to the importation of wildlife taken, killed, possessed, or exported to the United States in violation of the laws or regulations of a foreign country.

46 Stat. 689.
19 USC 1654.

46 Stat. 741.

ENDANGERED PLANTS

SEC. 12. The Secretary of the Smithsonian Institution, in conjunction with other affected agencies, is authorized and directed to review (1) species of plants which are now or may become endangered or threatened and (2) methods of adequately conserving such species, and to report to Congress, within one year after the date of the enactment of this Act, the results of such review including recommendations for new legislation or the amendment of existing legislation.

Report to Congress.

87 STAT. 901
87 STAT. 902

CONFORMING AMENDMENTS

SEC. 13. (a) Subsection 4(c) of the Act of October 15, 1966 (80 Stat. 928, 16 U.S.C. 668dd(c)), is further amended by revising the second sentence thereof to read as follows: "With the exception of endangered species and threatened species listed by the Secretary pursuant to section 4 of the Endangered Species Act of 1973 in States wherein a cooperative agreement does not exist pursuant to section 6(c) of that Act, nothing in this Act shall be construed to authorize the Secretary to control or regulate hunting or fishing of resident fish and wildlife on lands not within the system."

Ante, p. 886.

(b) Subsection 10(a) of the Migratory Bird Conservation Act (45 Stat. 1224, 16 U.S.C. 715i(a)) and subsection 401(a) of the Act of June 15, 1935 (49 Stat. 383, 16 U.S.C. 715s(a)), are each amended by striking out "threatened with extinction," and inserting in lieu thereof the following: "listed pursuant to section 4 of the Endangered Species Act of 1973 as endangered species or threatened species,".

80 Stat. 929.
80 Stat. 930.

(c) Section 7(a)(1) of the Land and Water Conservation Fund Act of 1965 (16 U.S.C. 460l—9(a)(1)) is amended by striking out:
"THREATENED SPECIES.—For any national area which may be authorized for the preservation of species of fish or wildlife that are threatened with extinction."
and inserting in lieu thereof the following:
"ENDANGERED SPECIES AND THREATENED SPECIES.—For lands, waters, or interests therein, the acquisition of which is authorized under section 5(a) of the Endangered Species Act of 1973, needed for the purpose of conserving endangered or threatened species of fish or wildlife or plants."

78 Stat. 897;
86 Stat. 459.

(d) The first sentence of section 2 of the Act of September 28, 1962, as amended (76 Stat. 652, 16 U.S.C. 460k-1), is amended to read as follows:
"The Secretary is authorized to acquire areas of land, or interests therein, which are suitable for—

Conservation areas.
86 Stat. 1063.

"(1) incidental fish and wildlife-oriented recreational development,
"(2) the protection of natural resources,
"(3) the conservation of endangered species or threatened species listed by the Secretary pursuant to section 4 of the Endangered Species Act of 1973, or
"(4) carrying out two or more of the purposes set forth in paragraphs (1) through (3) of this section, and are adjacent to, or within the said conservation areas, except that the acquisition of any land or interest therein pursuant to this section shall be accomplished only with such funds as may be appropriated therefor by the Congress or donated for such purposes, but such

property shall not be acquired with funds obtained from the sale of Federal migratory bird hunting stamps."

87 STAT. 902
87 STAT. 903
86 Stat. 1027.

(e) The Marine Mammal Protection Act of 1972 (16 U.S.C. 1361–1407) is amended—

16 USC 1362.
Ante, p. 884.

(1) by striking out "Endangered Species Conservation Act of 1969" in section 3(1)(B) thereof and inserting in lieu thereof the following: "Endangered Species Act of 1973";

16 USC 1402.

(2) by striking out "pursuant to the Endangered Species Conservation Act of 1969" in section 101(a)(3)(B) thereof and inserting in lieu thereof the following: "or threatened species pursuant to the Endangered Species Act of 1973";

(3) by striking out "endangered under the Endangered Species Conservation Act of 1969" in section 102(b)(3) thereof and inserting in lieu thereof the following: "an endangered species or threatened species pursuant to the Endangered Species Act of 1973"; and

(4) by striking out "of the Interior such revisions of the Endangered Species List, authorized by the Endangered Species Conservation Act of 1969," in section 202(a)(6) thereof and inserting in lieu thereof the following: "such revisions of the endangered species list and threatened species list published pursuant to section 4(c)(1) of the Endangered Species Act of 1973".

86 Stat. 973.
7 USC 136
note.

(f) Section 2(l) of the Federal Environmental Pesticide Control Act of 1972 (Public Law 92–516) is amended by striking out the words "by the Secretary of the Interior under Public Law 91–135" and inserting in lieu thereof the words "or threatened by the Secretary pursuant to the Endangered Species Act of 1973".

REPEALER

80 Stat. 926.
83 Stat. 275.

SEC. 14. The Endangered Species Conservation Act of 1969 (sections 1 through 3 of the Act of October 15, 1966, and sections 1 through 6 of the Act of December 5, 1969; 16 U.S.C. 668aa—668cc–6), is repealed.

AUTHORIZATION OF APPROPRIATIONS

SEC. 15. Except as authorized in section 6 of this Act, there are authorized to be appropriated—

(A) not to exceed $4,000,000 for fiscal year 1974, not to exceed $8,000,000 for fiscal year 1975 and not to exceed $10,000,000 for fiscal year 1976, to enable the Department of the Interior to carry out such functions and responsibilities as it may have been given under this Act; and

(B) not to exceed $2,000,000 for fiscal year 1974, $1,500,000 for fiscal year 1975 and not to exceed $2,000,000 for fiscal year 1976, to enable the Department of Commerce to carry out such functions and responsibilities as it may have been given under this Act.

EFFECTIVE DATE

SEC. 16. This Act shall take effect on the date of its enactment.

MARINE MAMMAL PROTECTION ACT OF 1972

SEC. 17. Except as otherwise provided in this Act, no provision of
this Act shall take precedence over any more restrictive conflicting
provision of the Marine Mammal Protection Act of 1972.

Approved December 28, 1973.

86 Stat. 1027.
16 USC 1361
note.

LEGISLATIVE HISTORY:

HOUSE REPORTS: No. 93-412 (Comm. on Merchant Marine and Fisheries)
          and No. 93-740 (Comm. of Conference).
SENATE REPORT No. 93-307 (Comm. on Commerce).
CONGRESSIONAL RECORD, Vol. 119 (1973):
    July 24, considered and passed Senate.
    Sept. 18, considered and passed House, amended, in lieu of
        H. R. 37.
    Dec. 19, Senate agreed to conference report.
    Dec. 20, House agreed to conference report.
WEEKLY COMPILATION OF PRESIDENTIAL DOCUMENT, Vol. 10, No. 1 (1974):
    Dec. 28, 1973, Presidential statement.

Notes

Notes

Notes

Notes